Water Hazards and Management

Water Hazards and Management

Editor: Sarah Luck

R CALLISTO REFERENCE

www.callistoreference.com

Callisto Reference,
118-35 Queens Blvd., Suite 400,
Forest Hills, NY 11375, USA

Visit us on the World Wide Web at:
www.callistoreference.com

ISBN: 978-1-63239-827-7 (Hardback)

The publisher's policy is to use permanent paper from mills that operate a sustainable forestry policy. Furthermore, the publisher ensures that the text paper and cover boards used have met acceptable environmental accreditation standards.

Trademark Notice: Registered trademark of products or corporate names are used only for explanation and identification without intent to infringe.

Printed in the United States of America.

Cataloging-in-publication Data

Water hazards and management / edited by Sarah Luck.
    p. cm.
Includes bibliographical references and index.
ISBN 978-1-63239-827-7
1. Water quality. 2. Water quality management. 3. Water--Pollution. I. Luck, Sarah.
TD370 .W38 2017
628.161--dc23

Table of Contents

Preface

The hazards related to water are called water hazards; they are a subset of natural hazards. Water resource management is the activity of planning, developing, distributing and managing the optimum use of water resources. This book is a valuable compilation of topics, ranging from the basic to the most complex advancements in the field of water management and water hazards. It aims to shed light on the different techniques that could be useful in water management especially in the urban setting where the demand for water and its resources are soaring high. This book aims to equip students and experts with the advanced topics and upcoming concepts in this area. It includes contributions of experts and scientists which will provide innovative insights into this area of study.

The information shared in this book is based on empirical researches made by veterans in this field of study. The elaborative information provided in this book will help the readers further their scope of knowledge leading to advancements in this field.

Finally, I would like to thank my fellow researchers who gave constructive feedback and my family members who supported me at every step of my research.

Editor

The Potential Connectivity of Waterhole Networks and the Effectiveness of a Protected Area under Various Drought Scenarios

Georgina O'Farrill[1]*, **Kim Gauthier Schampaert**[3], **Bronwyn Rayfield**[2], **Örjan Bodin**[4], **Sophie Calmé**[3,5], **Raja Sengupta**[6], **Andrew Gonzalez**[2]

1 Ecology and Evolutionary Biology Department, University of Toronto, Toronto, Ontario, Canada, 2 Biology Department, McGill University, Montreal, Quebec, Canada, 3 Département de géomatique (KGS), Département de biologie (SC), Université de Sherbrooke, Sherbrooke, Québec, Canada, 4 Stockholm Resilience Centre, Stockholm University, Stockholm, Sweden, 5 Departamento de conservación de la biodiversidad, El Colegio de la Frontera Sur, Chetumal, Quintana Roo, Mexico, 6 Geography Department, McGill University, Montreal, Quebec, Canada

Abstract

Landscape connectivity is considered a priority for ecosystem conservation because it may mitigate the synergistic effects of climate change and habitat loss. Climate change predictions suggest changes in precipitation regimes, which will affect the availability of water resources, with potential consequences for landscape connectivity. The Greater Calakmul Region of the Yucatan Peninsula (Mexico) has experienced a 16% decrease in precipitation over the last 50 years, which we hypothesise has affected water resource connectivity. We used a network model of connectivity, for three large endangered species (Baird's tapir, white-lipped peccary and jaguar), to assess the effect of drought on waterhole availability and connectivity in a forested landscape inside and adjacent to the Calakmul Biosphere Reserve. We used reported travel distances and home ranges for our species to establish movement distances in our model. Specifically, we compared the effects of 10 drought scenarios on the number of waterholes (nodes) and the subsequent changes in network structure and node importance. Our analysis revealed that drought dramatically influenced spatial structure and potential connectivity of the network. Our results show that waterhole connectivity and suitable habitat (area surrounding waterholes) is lost faster inside than outside the reserve for all three study species, an outcome that may drive them outside the reserve boundaries. These results emphasize the need to assess how the variability in the availability of seasonal water resource may affect the viability of animal populations under current climate change inside and outside protected areas.

Editor: Cédric Sueur, Institut Pluridisciplinaire Hubert Curien, France

Funding: This study was carried out with the support of the Consejo Nacional de Ciencia y Tecnologia-Mexico (CONACYT- doctoral fellowship) and Fonds Québécois de la Recherche sur la Nature et les Technologies (FQRNT- postdoctoral fellowship) to GO, the National Science and Engineering Research Council from Canada (NSERC) Discovery Grants program to AG and RS, the Canada Research Chair Program (to AG), the Global Environmental and Climate Change Center (GEC3), the Swedish Foundation for Strategic Environmental Research (MISTRA), the strategic research program Ekoklim at Stockholm University, and the NASA-LCLUC program(NAG5-6046 and NAG5-11134) to the Southern Yucatán Peninsular Region(SYPR) project. The Walter Hitschfeld Geographic Information Centre in the Geography Department at McGill University provided the authors with the Quick Bird images. The funders had no role in study design, data collection and analysis, decision to publish, or preparation of the manuscript.

Competing Interests: The authors have declared that no competing interests exist.

* E-mail: georgina.ofarrill@gmail.com

Introduction

The synergistic effects of land use change, habitat loss and climate change are expected to affect species persistence by altering the distribution and connectivity of resources and habitat. These effects will have significant consequences for biodiversity conservation and management [1,2]. Landscape connectivity allows species movement and dispersal, influencing the distribution of genes, resources and populations of many species [3,4]. Landscape connectivity analysis encompasses both the ease with which an animal can move from one resource patch to another (the animal perspective, [5]), and the location and quality of resources (the landscape perspective) that will determine the species' motivation to move [6,7].

Changes in temperature and precipitation regimes predicted by climate change models [8] are likely to influence resource availability through changes in their abundance (e.g. fruits, water).

Resource connectivity analyses are particularly important when resources found within habitat patches vary in space and over short (within years) and long (between years) time scales (e.g. [9]). Hence highly variable fluctuations in temporal resource availability make connectivity within a resource network dynamic and stochastic in space and time [10], which is expected to influence species movement patterns [11].

Resource connectivity studies should incorporate the temporal variability in the availability and connectivity of resources given the current predictions of climate change worldwide, the species' differential use and accessibility to resources, and the amount of suitable habitat remaining after landscape connectivity is lost [12–14]. This is particularly relevant when areas with low or declining connectivity may not be able to support viable populations of some species over long periods of time [15]. Given current observations on the long-term effects of climate change, longer-term fluctuations in precipitation than the ones presented currently may affect water networks resulting in multi-annual trends in network connectivity [10,11].

In the Yucatan Peninsula of Mexico, climate change is causing an increase in drought periods [16], which seems to be influencing the availability of resources and the movement patterns of animals (e.g. [17]). The Greater Calakmul Region of the Yucatan Peninsula of Mexico is a continuous forested area in a highly seasonal tropical climate where, over the last 50 years, annual precipitation decreased by 16% while drought frequency increased [16]. Yearly fluctuations in precipitation show a decreasing pattern despite reports of years with high precipitation (Comisión Nacional del Agua, unpublished data). According to regional climate models, this area will increasingly suffer from extreme droughts in future years [8]. This area is a karstic upland area, where freshwater is only available to wildlife and people in superficial waterholes and small seasonal streams. Therefore, water is a scarce and dynamic resource in the area. Many small waterholes dry up during the dry season, and if the reduction in precipitation continues, we hypothesize this will further influence the availability of waterholes. In addition, we hypothesize that if drought events affect surface water availability (i.e., waterhole presence), waterhole connectivity will decrease given that waterholes that remain in the landscape will be located beyond species maximum travel capabilities or home ranges.

We used a network (graph theoretical) analysis to test our hypotheses about the change in the spatial distribution of waterholes in this network [18]. We treated these waterholes and short seasonal streams as nodes, and we defined a link between any two if they fell within the range of our study species' movement distances [19,20]. We modelled the movement of Baird's tapir (*Tapirus bairdii* Gill, 1865), white-lipped peccary (*Tayassu pecari* Link, 1795) and jaguar (*Panthera onca* Linnaeus, 1758). We used these species because they are of significant conservation concern, they rely on waterholes, and data about their movement are available. We assessed the connectivity of waterholes for each species given the temporal and spatial changes of water availability within and between years inside and outside the Calakmul Biosphere Reserve considering actual observations and climate change predictions of severe droughts for the area. We used reported travel distances for our study species, rather than measuring the actual movement of individuals; our results therefore represent the potential connectivity network of waterholes. Although other studies have evaluated the importance of seasonal water resources for species survival [11,12,21], none have addressed the connectivity of water resource under scenarios of climate change and how resource availability interacts with species movement capacities to modify the functional connectivity of a landscape.

Materials and Methods

Study Area

The waterhole network is situated in an area of approximately 750 km^2 located in the north-eastern part of the Greater Calakmul Region to the south of the Yucatan Peninsula of Mexico (19°15' to 17°50'N and 90°20' to 89°00'W). Fieldwork activities inside and in the areas surrounding the Calakmul Biosphere Reserve were authorized by the Director of the Reserve (Fernando Durand Siller- Permit no. D-RBC-020-10-07). Fieldwork outside the reserve was carried out in the Ejido of Nuevo Becal (communal land) and the Community commissioner authorized field activities after the assembly of the community was notified and agreed on allowing our visit to their land. Even though this study considers

movement capacities of endangered species, data were collected from the literature and field activities did not involve any of the endangered species studied. The Greater Calakmul Region (Fig. 1) is part of the Selva Maya, the second largest area of tropical forests in the Americas. Approximately 13% of the forest cover is disturbed by human activities in the region [22]. Of our study area, 30–40% lies within the Calakmul Biosphere Reserve, while the eastern section represents a continuous forested landscape that corresponds to communal land and is mostly a managed forest reserve. Forest cover within and outside the Calakmul Biosphere Reserve does not show any large difference [22]; however, there are human settlements outside the borders of the reserve, which represent a threat to species, i.e. hunting and logging activities.

Between 1953 and 1998 precipitation decreased by almost 16% in the Calakmul Region; mean precipitation declined from 1,300 mm in 1950s to 790 mm in 1990s [23] following the same pattern until 2005 (1955–2005; Comisión Nacional del Agua, unpublished data; [23]). In addition, climate models predict a further decline in precipitation and warmer temperatures, and suggest that most severe droughts in Mexico would occur during el Niño years affecting 50% of the area covered by deciduous tropical forest [8]. Due to its geomorphologic conditions (karst topography), the region does not have any rivers and the majority of the superficial water is stored in small depressions on the landscape: waterholes (locally called "*aguadas*") and small seasonal streams [24]. No waterholes are formed by underground water in this area. Therefore, during the dry season these superficial water bodies represent the only available water source for many animal species.

Study Species

In the Greater Calakmul Region, species such as the Baird's tapir, white-lipped peccary and jaguar are endangered [25] [26] [27]. These species are highly associated with water bodies for water consumption, to regulate their body temperature (tapirs and peccaries [28]) or to find their prey (jaguars [27]). Given that water is only present in waterholes and short streams, species such as these depend on seasonal water bodies and the associated habitat surrounding them [27,29–31].

The number of water bodies throughout the landscape used by each species depends on their home ranges and daily movements. For individuals of Baird's tapir, reported yearly mean home ranges encompass approximately 1.3 km^2 (± 0.73 km^2, SD) with a maximum home range of 2.3 km^2 in Costa Rica [32]. These authors reported that even though home ranges did not vary between seasons, during the wet season individual tapirs shared 26% of their annual home range with other tapirs, whereas overlap was usually null in the dry season; these observations suggest that water availability influences tapir use of the landscape. For white lipped peccaries, Reyna-Hurtado [30] estimated herd annual home ranges from 43.9 to 97.5 km^2 in the tropical semi-dry forest of the Calakmul Biosphere Reserve (home range estimates based on VHF data and 95% fixed kernel). White-lipped peccaries visit waterholes disproportionately more often during the dry season than in the wet season [17]. During this study, when water became scarce and was only available at the larger waterhole in the area, the groups remained at this waterhole and foraged in a radius of 6 km (mean distance <600 m). In the Calakmul Region, jaguars show differential habitat use by season based on the availability of water bodies, which affects the density and location of prey species [27]. Data obtained from satellite collars showed that the activity area for two males was about 1000 km^2 [27].

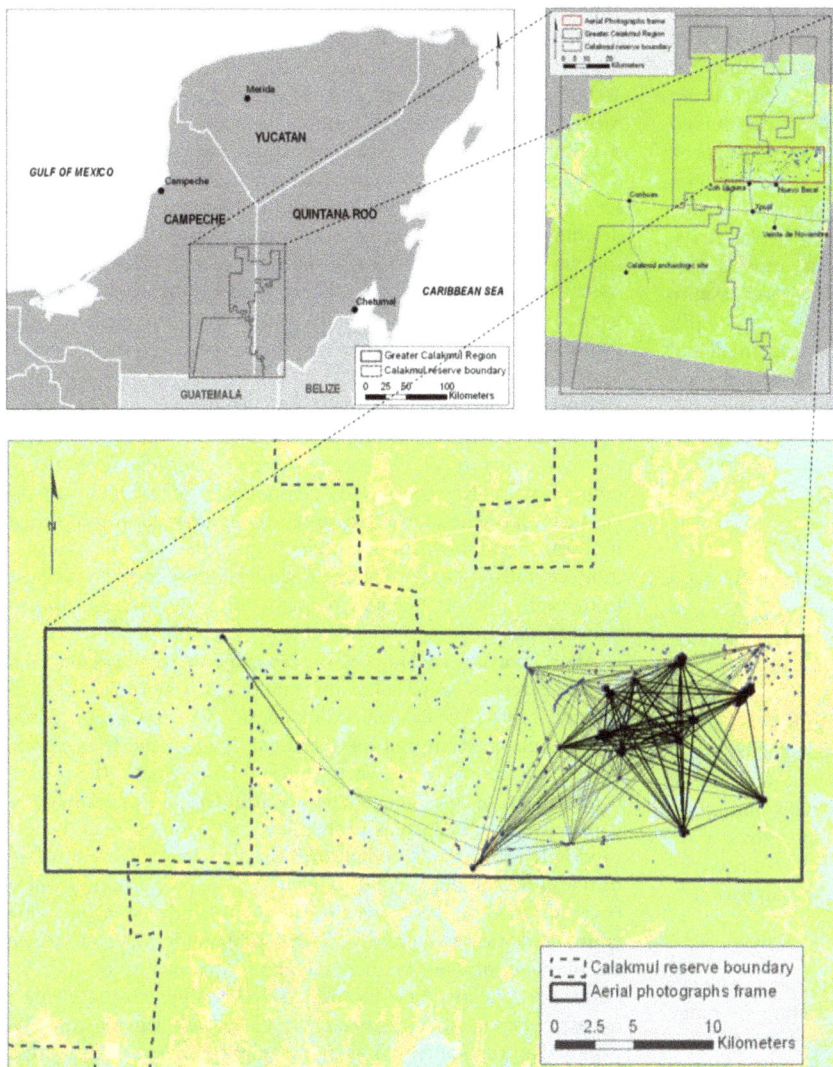

Figure 1. Greater Calakmul Region and study area (upper figures). This figure shows a representation of drought scenario D (grey links) and E (black links) for the 13 km travel distance. Links are lost inside and adjacent to the reserve and a narrow stepping-stone strip of waterholes remains, connecting the interior of the reserve with the smaller sub-network to the east, outside the reserve. The circle shows the waterhole that maintains the connectivity between the sub-networks.

In addition to home ranges, maximum travel capacities provide information about the ability of a species to reach distant water bodies if required. In Peru, Tobler [33] reported lowland tapir (*Tapirus terrestris*) individuals moving up to 13 km over a 24-hour period (GPS radio-collared data), with a mean movement distance in a 24-hr period of 5.2 km (range 3.6–6.7 km). This is the only formal study documenting tapir movements with GPS collars. When considering tapir movements, we used the travel distances reported for this species of tapir; however, we are confident these observations can be used for the Baird's tapir in our region as all tapir species depend on water for their survival [25]. Reyna-Hurtado *et al.* [17] found that white-lipped peccaries require visits to water bodies on an almost daily basis in our study region, performing search patterns at two spatial scales: they search one area intensively by moving no more than 3 km every day and occasionally perform long displacements (9 to >16 km) in a single direction that take them out of the previous searched area over the course of one to three days. For jaguars, the mean daily travelled

distance was 2.24 km with a maximal daily distance travelled of 10 km based on radio collar data [34].

In summary our model species have been reported to move distances between 3 and 13 km (minimum and maximum reported travel distances by tapirs), between 3 and 16 km (minimum and maximum travel distances by white-lipped peccaries) and between 2.24 and 10 km (minimum and maximum travel distances by jaguars). This range (3–16 km) provides the minimum and maximum potential movement capacities of these species. This suggests that these species have the ability to reach waterholes located further away than their daily home ranges.

Remotely Sensed Image Interpretation and Ground Truthing

For our analysis, we used remotely sensed imagery (orthophotographs from March 1998 and 2001) to obtain the most accurate locations of the waterholes in our study area (see File S1). During fieldwork, we observed that most of the waterholes smaller than 400 m^2 were dry at the beginning of the dry season, therefore we only digitized water bodies >400 m^2. We visited 15 waterholes observed in the orthophotos, confirming the presence of 80% of them in the field. All misidentified waterholes were small (< 700 m^2) and were subsequently excluded from the model. From 2006 to 2009, we repeatedly visited 15 waterholes and observed that large waterholes did not dry or were dry only during the peak of the dry season of very dry years. For example, in May 2006 a waterhole of 90,000 m^2 was observed to go dry for the first time since people re-colonized the area 40 years earlier (Nicolas Arias, comm. pers). These observations support climate models that suggest an increase in drought conditions in the area and capture the situation during a very dry year (2005) in the region [8]. Animal tracks (e.g., tapir, deer, peccaries) and dung near small and large waterholes are found at lower densities when waterholes are dry, suggesting that animals use waterholes mainly for water [35]. The intervening matrix surrounding waterholes was comprised of accessible forest for all species; no major physical barriers (e.g. roads, mountains, rivers) occur in the study area.

Waterhole Networks

Water resource connectivity was assessed for each of the three model species using species-specific waterhole networks. Various drought scenarios were simulated on these networks through node deletion sequences that reflected waterhole-drying patterns. Deletion sequences were based on observed waterhole drying during the dry season of 4 consecutive years (2006–2009). The implications of these drought scenarios on species-specific water resource connectivity were assessed by quantifying the structure of sub-networks, network-wide probability of connectivity, and access to waterhole-associated habitat. Our analysis thus provides an indication of potential future changes in waterhole connectivity based on current waterhole distribution and drying pattern.

Waterhole Network Delineation

The three species' networks consisted of the same set of nodes (waterholes), but each had a unique set of links reflecting the range of reported movement abilities of the three species, varying from 3 to 16 km (tapirs: 3–13 km; white-lipped peccaries: 3–16 km; and jaguars: 2.24–10 km). We identified links between waterhole centroids rather than waterhole edges due to high variability in waterhole edges with rainfall. Links that were longer than the maximum movement ability of each species were removed from their respective networks. Although we do not expect animals to necessarily travel their maximal distances in a day or constantly, these distances represent a reasonable estimate of the distance that these species might be able to travel to find water when it becomes limiting, e.g. to locate a new waterhole. Furthermore, in contrast to most resource networks where nodes represent patches of suitable habitat surrounded by an inhospitable matrix, in our study, nodes (waterholes) are separated by suitable habitat as the matrix landscape represents a continuous forested area. All waterhole networks were delineated using complete graph extraction in SELES (A Spatially Explicit Landscape Even Simulator; [36]).

Waterhole Network Deletion Sequences Based on Drought Scenarios

The impact of drought on the connectivity of the waterhole network was modelled by removing nodes in increasing order of waterhole area. We made this assumption based on field observations that showed the largest waterholes were the last to become dry at the peak of the dry season from 2006 to 2009. We created a base scenario (A) that included all waterholes larger than

700 m^2, and 10 drought scenarios (B–K) excluding waterholes smaller than or equal to (B) 1,000 m^2, (C) 2,500 m^2, (D) 5,625 m^2; (E) 10,000 m^2, (F) 15,625 m^2, (G) 22,500 m^2, (H) 30,600 m^2, (I) 40,000 m^2, (J) 50,600 m^2 and (K) 62,500 m^2. In the field, we measured the perimeter to evaluate the area of several waterholes that were monitored for water availability over the four years. The distribution of waterhole sizes allowed us to select our size class scenarios by providing maximum and minimum size limits. After deleting nodes in each scenario, only links shorter than the species-specific maximum movement distance were maintained within the species' waterhole network. These pruned waterhole networks were used in the calculation of connectivity for each drought scenario.

Waterhole Network Connectivity Analyses

Prior to modelling drought scenarios, we analysed the sub-network structure of the species-specific waterhole networks by enumerating the connected components (clusters). Network clusters represent connected areas of a network within which individuals can move among nodes (waterholes) via direct links or indirect paths [20,37]. Different clusters are effectively isolated from each other as no links or paths allow individuals to move between them.

After applying each drought scenario, we recorded the total area covered by waterholes, the number of waterholes, and the density of links. We calculated link density (L) as the proportion of actual links present out of the total number of links in the equivalent complete network (L≤1). To evaluate the overall connectivity of the network we calculated the probability of connectivity defined as the probability that two individuals located randomly within the landscape are found at waterholes that belong to the same component [38]. Given the low proportion of water-covered area with respect to the total study area, we only considered the numerator of the probability of connectivity calculations to allow for better comparisons of results [38]. We derived a link attribute for dispersal probability by applying a negative exponential dispersal kernel to link lengths (m). We assumed a probability of 0.05 that species could move further than their maximum movement distance (straight line) to parameterize the negative exponential dispersal kernel [20]. We used the patch removal technique [39] to assess the contribution of each node in maintaining network connectivity for each species and under each drought scenario, which allowed us to rank waterholes based on their node importance to the network. This was measured as delta probability of connectivity [38,40]. We calculated the area covered by the network or sub-networks using ArcGIS, and analysed network connectivity using the igraph package in R (R Development Core Team, 2005).

Waterhole-associated Habitat

Our model species are strongly associated with water bodies in the study area. Therefore, we assumed that their habitat must contain at least one waterhole and defined the suitable habitat as the area surrounding that waterhole. We assume that species are in general unwilling to travel long distances in search for resources on a daily basis, thus their movement patterns will likely occur mostly within a small area surrounding a waterhole as has been observed for the white-lipped peccary [17]. Therefore, to evaluate the species suitable habitat, we created 1 and 2 km radius buffers around the edge of each waterhole in each scenario using ArcGIS 9.3 Spatial Analyst Tool (ESRI 2009). By considering these radii, we assume species will move short distances during the day <3 km (minimum daily travel distance reported by all species) around a waterhole if all their requirements are fulfilled in that area. This analysis allowed us to estimate the loss of suitable habitat for the species and to complement the network connectivity analysis where we focus on the ability for the species to undertake long and rare movements. In addition, we quantified the percentage of total area covered by the suitable habitat in each scenario.

Results

The initial waterhole network (scenario A) included 187 waterholes that represented a total area covered by water of 3.14 km^2 and included 17,391 potential links (when no threshold distance was considered; Fig. 2). For all travel distances, link density abruptly decreased when all small waterholes (2,500 m^2) were deleted showing a threshold type response at an early stage of the model. The frequency of waterholes for each size class declines as a power function with many small waterholes (n = 115 of ≤ 2,500 m^2) and a few large waterholes (n = 15 of ≥40,000 m^2; Fig. S1). Our base scenario consisted of one connected network for all travel distances, except for the 3 km distance. This suggests that only when individuals move more than 3 km they will be able to reach the remaining waterholes in each scenario. As travel distance decreases, the number of clusters (2 or more linked waterholes) increases (Fig. 3).

Waterhole Network Structure

Drought dramatically influenced waterhole network structure and therefore the potential connectivity of waterholes and potential suitable habitat for our model species. An abrupt decrease in link density showing a threshold was observed when small waterholes (≤2,500 m^2) were removed, followed by stabilization in link density curve after scenario E, i.e., when medium size waterholes were removed (10000 m^2), with similar patterns for the different travel distances considered (Fig. 2). For example, for a

travel distance of 13 km (e.g., within tapir and peccaries daily travel capacities), the number of links in the base scenario (scenario A-that included all waterholes larger than 700 m^2) was 7,424 (42.7% of all potential links, i.e., without a threshold distance). For this same example, scenario C would correspond to a rapid decrease in link density with only 209 links left, which corresponds to 16.3% of the links present in the base scenario. In scenario F (when the link density curve stabilizes) we observed only 94 links left, i.e., 1.3% of all links found in the base scenario.

The removal of waterholes ≤2,500 m^2 (scenario C), which represents the most frequent waterhole size in our study area, resulted in a loss of 61.5% of the total number of waterholes included in the network (Fig. 2). Most of the waterholes of this size dry fast during the early dry season and are not a reliable source of water. The removal of 90.9% of the waterholes (scenario F) caused a loss of only 15% of the total area covered by water; however, this caused a drastic reduction of more than 95% of the links for all travel distances (Fig. 2, scenario F). For scenario C, network connectivity was maintained inside and outside the reserve when a minimum of 5 km travel distance was considered- a distance within all species movement range (Fig. 3); however, these waterholes were typically observed to dry every year in the field.

A travel distance of 5 km (within all species travel capacities) did not ensure connectivity for all drought scenarios; for example, the sub-network observed inside the reserve had a sparser density of waterholes linked compared to the sub-network outside the reserve, and its connectivity was maintained only when waterholes smaller than or equal to 10,000 m^2 persisted in the landscape (Fig. S2). In a conservative scenario of waterhole deletion (scenario E) and for a travel distance of 13 km (maximal reported travel distance by tapirs), the sub-network within the reserve collapsed, leaving only three waterholes within the boundary of the reserve linked to the sparse network outside the reserve (Fig. 3). In the

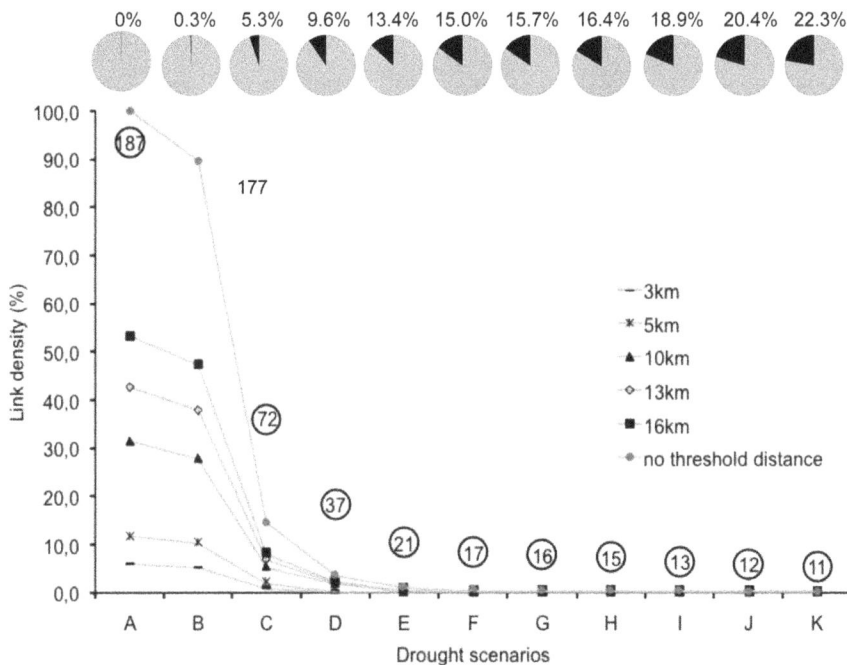

Figure 2. Number of waterholes (embedded in circle) and link density in each drought scenario by travel distances of species: tapirs (3–13 km), white-lipped peccaries (3–16 km) and jaguars (3–10 km). Drought scenarios correspond to waterhole removal based on waterhole size: A (all), B (≤1000 m^2), C (≤2500 m^2), D (≤5625 m^2), E (≤10000 m^2), F (≤15625 m^2), G (≤22500 m^2), H (≤30600 m^2), I (≤40000 m^2), J (≤50600 m^2), K (≤62500 m^2). Pie charts show the amount of waterhole area lost in each drought scenario. Lines between points are included only as reference to observe point-decreasing pattern.

Scenario C   Scenario E

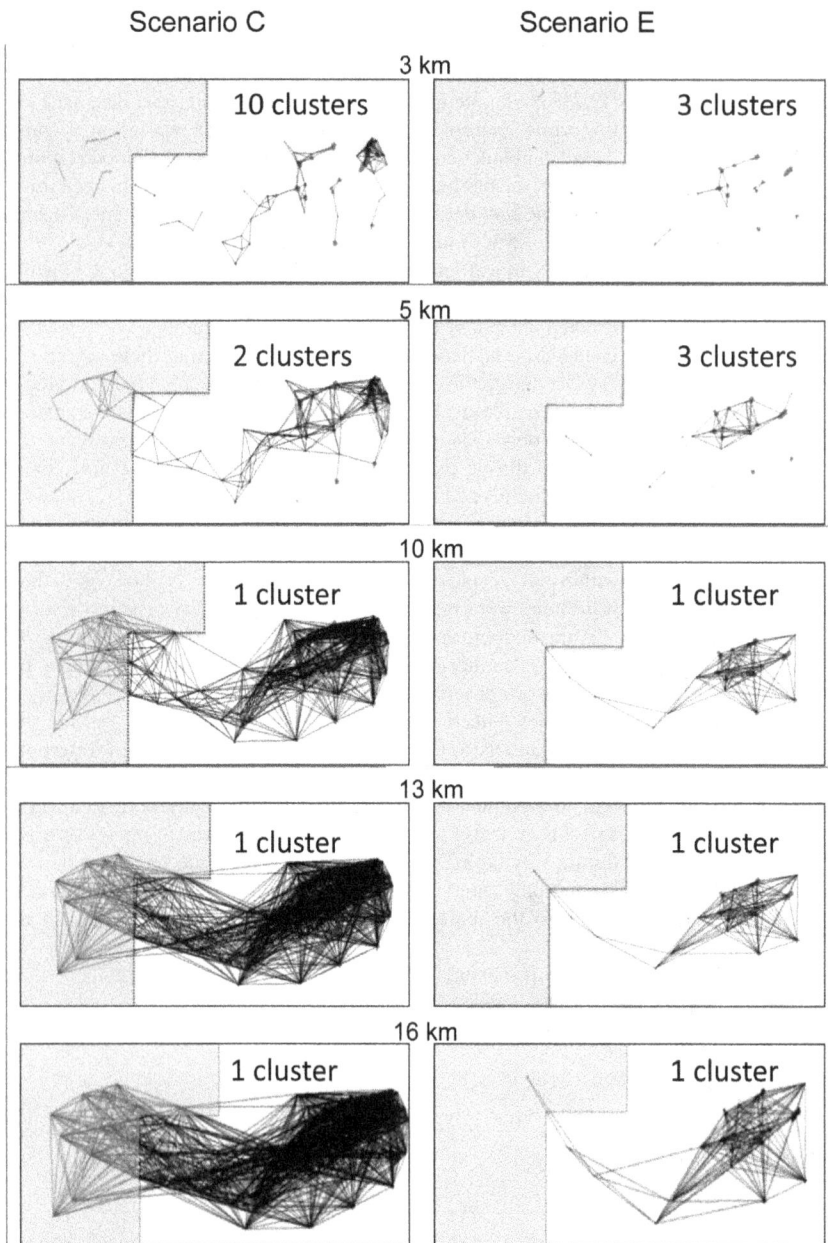

Figure 3. Representations of the network graphs showing the changes in network structure for Scenario C (removal of waterholes ≤2500 m²) and E (removal of waterholes ≤10000 m²) at 3, 5, 10, 13 and 16 km which represent a range of travel distances for our study species: tapirs (3–13 km), white-lipped peccaries (3–16 km), and jaguars (3–10 km). The two scenarios presented show abrupt changes in connectivity.

most severe drought scenario (K) waterholes became very sparsely connected outside the reserve and non-existent inside the reserve.

Probability of Connectivity and Node Importance

We observed a decrease in the probability of connectivity as waterholes were removed from the network for each travel distance. The probability of connectivity increased with increasing travel distance. The probability of connectivity for all scenarios was lower at the lower range of travelled distances of all three species (3 and 5 km; Fig. 4). However, we found that even though node importance (given by the probability of connectivity of each node) decreased with waterhole area (smaller waterholes were less

important than large waterholes), the location of waterholes further influenced their node importance; therefore, large waterholes located far from other waterholes had lower node importance than smaller waterholes located close to other waterholes. For example, a waterhole of 172,500 m² was ranked 9th with respect to its node importance even though it was larger than waterholes ranked 8th given its location (Fig. S3).

Suitable Habitat

The waterhole-associated area, or "suitable habitat" using a buffer radius of 1 km around waterholes, corresponded to 52% of the total area in the base scenario. The removal of waterholes ≤

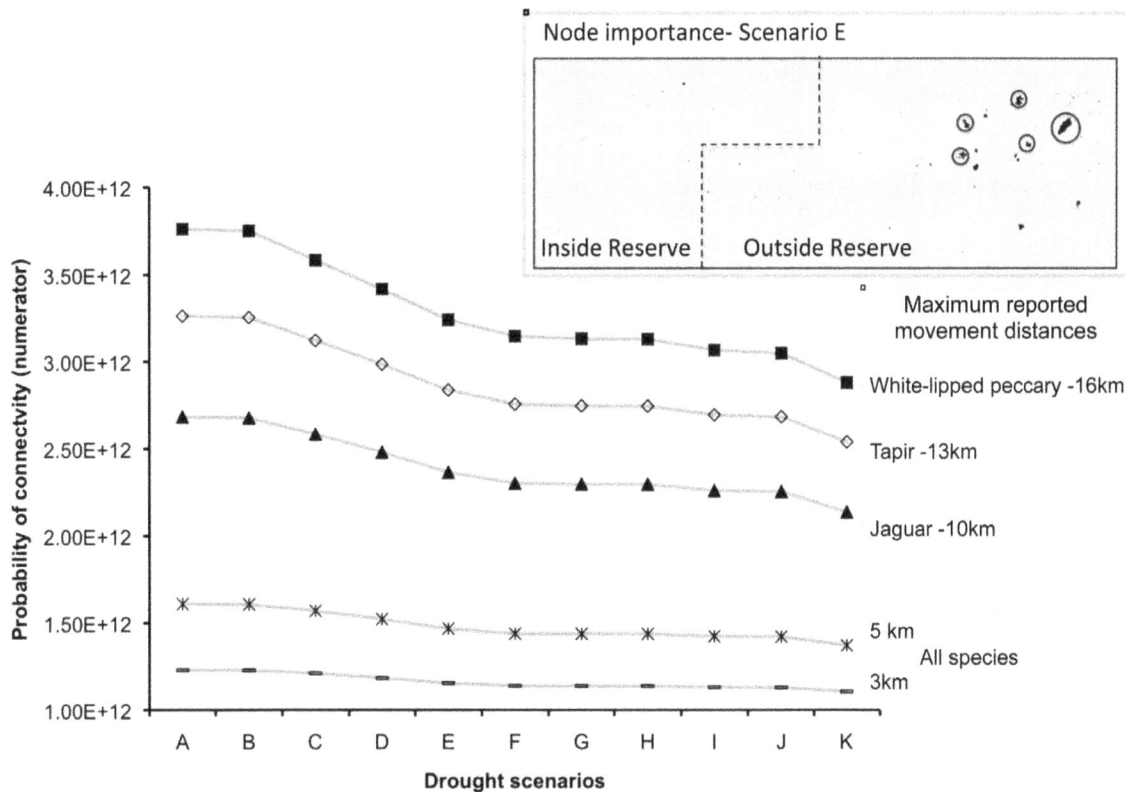

Figure 4. Changes in the probability of connectivity with different travel distances. Inset node importance map illustrates the five waterholes (encircled) with higher ranking based on their contribution to maintaining the overall probability of connectivity of the waterhole network in scenario E for all species (added probability of connectivity of each waterhole by all 5 distance thresholds used).

$10,000 \text{ m}^2$ (scenario E) caused a large reduction with only 21% of initial suitable habitat remaining (waterhole-associated area in the base scenario; Fig. S4). If we set our buffer radius to 2 km, the removal of waterholes $\leq 10,000 \text{ m}^2$ still caused a large reduction of suitable habitat with only 29% of the suitable area remaining. Of the total area covered by our network, only 9% was found within a radius of 1 km and 25% within a radius of 2 km from the existing waterholes in scenario F (simulating the peak of the dry season). Given a buffer radius of 1 km, we found that the number of connected patches of suitable habitat (components of buffered patches) decreased by half (from 13 to 6 connected patches) from scenario E to K. The same pattern was observed when considering buffers of 2 km; only three connected patches of suitable habitat remained in our last scenario, K (Fig. S4).

Discussion

Our results emphasize that: 1) waterhole number is seasonal and very sensitive to dry seasons, 2) changes in waterhole availability may not sustain a functionally connected network of waterholes for our endangered study species under present and future drought scenarios, and 3) network analysis can improve our understanding of reserve functioning and potential habitat connectivity in highly seasonal landscapes. Our analysis revealed that the potential connectivity of the waterhole network is very sensitive to drought for jaguars, white-lipped peccaries and Baird's tapirs. By using a range of reported travelled distances, we were able to model the effects of both daily movement patterns around waterholes (suitable habitat analysis) and long, rare movements (connectivity analysis). Our analyses highlight the potential negative effects for

three endangered species of observed trends of decreasing precipitation and future projections of changes in drought conditions in the area. The availability of water outside the protected area might result more attractive for water-dependant species, demonstrating the need for further species conservation programs in such a human dominated landscape.

Loss of Connectivity

Based on our models, maintaining waterholes smaller than $10,000 \text{ m}^2$ (scenarios A–D) is especially important for the potential connectivity of the landscape both inside and adjacent to the reserve. The distribution and abundance of these small waterholes ensure accessibility to water without large and costly movements. In addition, animals might use these small waterholes as stepping-stones en route to large waterholes, especially to move between sub-networks inside and outside the reserve (Fig. 3). Removing small waterholes caused a non-linear decrease in the connectivity of the network showing that link density abruptly decreases after a threshold of waterhole size removal. When only large waterholes remain in the landscape these are connected by few links (Fig. 3), showing a fragile network of waterholes available for species. Our results can be explained by percolation theory, which suggests that when random habitat loss occurs across more than 60% of the landscape, the largest habitat patch size decreases abruptly and no longer spans the landscape [41]. In our landscape, more than 60% of the waterholes are lost at an early stage of our deletion model (scenario C), which causes an abrupt decrease in link density and connectivity, with only a few large and clumped waterholes remaining.

The results from our network analyses do not imply that tapirs, white-lipped peccaries and jaguars move between waterholes on a daily basis but suggest that movement may be constrained when the species are forced to move to new waterholes due to seasonal or permanent waterhole drying (e.g., due to climate change) or to disturbance (e.g., logging). In addition, we do not expect species to walk in straight lines, but straight-line movements are general considerations relevant to all network connectivity analysis and represent the shortest distance our study species could cover to move between waterholes; therefore, our results showing changes in potential connectivity using maximal distances should be seen as representing the upper limit of potential connectivity. Additional information on species' water requirements would improve the assessment of the distances species move between waterholes; however this information is not available for these species. Our approach can be applied to test the potential functional resource connectivity of any other temporally and spatially dynamic resource [11,21,42,43].

The potential connectivity of waterholes in the area was severely reduced for our focal species, despite their body size and dispersal capabilities, given drought scenarios that represent the peak of the dry season in the driest years during the study period. Even though a small portion of the network remained connected (Fig. 3), this remnant network fell outside the reserve. Our study emphasizes the importance of considering the spatiotemporal dynamics of resources inside and around protected areas [13,44]. Even though our study only corresponds to a portion of the Greater Calakmul Region, our fieldwork suggests that these patterns may be common throughout the region.

Habitat Area vs. Quality?

Our results further emphasize that the spatio-temporal distribution of resources will likely determine the functional connectivity of the landscape. If resources become increasingly rare and then isolated, the chance that they fall outside the movement range of species will increase [45,46]. In this study, matrix habitat between waterholes is a homogenous-forested landscape, which allows free movement between waterholes. Our movement scenarios were not sensitive to estimates of the resistance to movement of the forested matrix, as is often the case in least-cost connectivity analyses in heterogeneous landscapes. Suitable habitat for the study species considers waterhole use; therefore habitat quality decreases as waterholes dry up causing species to perform longer and unusual movements. In our study, the changes in resource (waterhole) network connectivity in different drought scenarios show dynamic connectivity that will likely not be identified if only habitat area is considered in connectivity analyses; a result that suggests the importance of evaluating habitat quality in addition to habitat extent [11,47,48].

Tapirs, white-lipped peccaries, and jaguars possibly require more access to water when temperatures rise during the dry season, either to lower body temperature or, in the case of tapirs and white-lipped peccaries to compensate for the lower water content of their food items [28]. This shows that in very dry scenarios species are likely to be forced to move beyond their usual daily travel distances and in some cases even the maximum reported movement capacities of our model species might be insufficient for individuals to reach water resources. If species regularly perform short distance movements (no more than 2 km) suitable habitat may be seasonally limited in this continuous forested area. In addition, the number of connected patches of suitable area (overlapping buffer areas) decreases suggesting limited movement between suitable areas.

Even if our three model species are able to perform such long distance movements in less than three days [17,33,34], these movements are costly and performed rarely. Given the species' requirements, forested areas with only small waterholes, irrespective of the quality of the forest itself, are thus of decreasing quality as the dry season progresses. As observed in Figure 4, only a small percentage of suitable habitat remains during the most severe drought scenarios we studied. Such a situation was actually observed by Reyna-Hurtado et al. [17] in the Calakmul Biosphere Reserve. The four groups of white-lipped peccary they studied behaved like central-place foragers around, and foraged close to, the only remaining waterhole in their 240 km^2 study area. In addition, the loss of waterholes will not only affect jaguar habitat directly but it will indirectly influence prey availability near waterholes, which will have an overall effect on habitat quality. Therefore, habitat quality must be assessed in terms of resource availability for each species. More generally, although large reserves often contain critical areas of high quality habitat, this may not always be the case and analyses of the sort reported here can inform reserve design and management [49].

Reserve and Corridor Design and Management

The effect of drought was not uniform across the study region and was particularly apparent within the Calakmul Biosphere Reserve. Climate models predict an increase in temperature and a reduction in precipitation for the area [8]; therefore our study, based on patterns of waterhole drying observed in the field between 2006 and 2009, assumed a realistic drought scenario. Our study allowed us to show that the potential connectivity of the waterhole network is dependent on movement capabilities and is dynamic in space and time. Additionally, only one waterhole keeps the reserve connected to the network outside the reserve (Fig. 1), the loss of which would interrupt a potential resource corridor for individuals moving east to the waterhole network beyond the reserve boundary. The loss of waterholes of this size was observed in the field at the peak of the dry season during a very dry season (e.g., 2006).

Our study region is an important part of the Mesoamerican Biological Corridor and is of considerable conservation value because of its high diversity and area [50]. The Greater Calakmul Region contains large waterholes that ensure water availability throughout the dry season even in years of very low precipitation (e.g. 2005; personal observation). Our study area is represented by a homogenous-forested matrix, which allows free access to waterholes for most species. However, the network analysis revealed higher node importance of large waterholes and documents that the spatial distribution of these large waterholes was heterogeneous and aggregated beyond the boundary of the reserve (Fig. 3). Available habitat outside the reserve might represent a better habitat for species in terms of water resources. Habitat outside the protected area experiences threats caused by human activities such as hunting or habitat disruption by logging and agriculture which can hinder the survival of species. Waterholes located outside the reserve, and with a larger area, showed higher node importances in all scenarios, which emphasize their contribution to maintaining connectivity of the remaining network (Fig. 4). These large waterholes are all located outside the reserve are found close to each other, which suggests that this area might become a refuge for species if drought conditions continue as predicted.

Our results suggest that resource connectivity should be at the centre of reserve network design [51]. This will be especially important if trends of climate or land use change directly impact resource availability. In addition, human populations in the area also depend on water bodies. Critical waterholes important for humans and fauna in the region currently lack a sustainable management plan. The approach we used in this paper can be used to rapidly assess landscape viability and vulnerability for a range of species specifically considering variations across space and time. Our approach can also be used to initially prioritize habitat and resources for landscape management programs and to target further data collection and monitoring [47,49] in areas where critical resources are spatially and temporally dynamic. Our results emphasize the need to better understand the availability of water inside the reserve and the consequences for species survival in this protected area. Conservation actions are needed outside the reserve not only to ensure the survival of species in areas with low waterhole abundance, but also to identify areas of potential human-animal conflicts if animals move outside the reserve to find water (or other resources). A higher rate of hunting and crop-raid events might be expected outside the reserve; therefore, further studies on these topics are needed to inform conservation actions.

Given the rapid effects of climate change, which in some areas has now translated into altered precipitation regimes [21], we require new approaches to create dynamic reserve and corridor network designs that incorporate the temporal and spatial dynamics of resources [43,51]. This study considered the effect of climate change on species persistence by evaluating the effect of changes in precipitation in water resource availability and the connectivity of resources. We have provided science-based information that can inform future conservation programs in the area. These programs can be established for the protection of key water bodies inside and outside the reserve, for the conservation of areas where water bodies with higher node importance value were found, and to promote further studies on the movement capacities of these endangered species, their water requirements, and the potential consequences of a higher abundance of these species outside the protected area.

Future Considerations

Our analysis suggests that even though waterholes may remain connected during the wet season, resource connectivity is abruptly affected during the dry season. If current trends of precipitation continue, drought periods are expected to be longer and affect the waterhole network. In particular, our results suggest that the loss of a small number of water holes has a large effect on the network's structure and connectivity. These effects will increase in strength when dry years occur consecutively and may force species to move to unprotected areas beyond the reserve. Currently this effect is seasonal but it may reflect future scenarios with longer and more severe dry seasons. A permanent shift in conditions would have consequences for species persistence in and around the reserve. Climate and land use changes will dramatically alter the functional connectivity of this region, and hence the conservation capacity of the Calakmul Biosphere Reserve.

Our network approach allowed us to link field and GIS data to analyse the potential connectivity of a waterhole network for three large mammal species of conservation concern. Although this modelling approach has been applied to study the impacts of habitat fragmentation [4,19] few results are available for water resource networks [52] and their dynamics [53]. We recognize that detailed data and habitat use patterns are still missing from our model. However, our modelling approach is easily updated as movement data (e.g., from GPS collars) becomes available. Next steps will involve modelling demography to allow the identification of key features of the network that are critical for metapopulation persistence under climate change [11,20]. Furthermore, changing human land use in the area might influence species' movements in the future, so the inclusion of land use change (e.g. logging roads or human settlement) will also be important to evaluate change in landscape connectivity.

Supporting Information

Figure S1 Power relationship between waterhole size and frequency, $R^2 = 0.64$. Each dot corresponds to a drought scenario described in the text. Note the log scale of both axes.

Figure S2 Network graphs showing changes in network structure when considering a 5 km travel distance for scenarios A (all waterholes considered), C (waterholes ≤ 2500 m^2 removed), E (waterholes ≤ 10000 m^2 removed), and J (waterholes ≤ 50600 m^2 removed). The grey area corresponds to the reserve.

Figure S3 Area of patches with the ten highest node importance values. The symbols correspond to each of the 11 drought scenarios A to K.

Figure S4 Percentage of suitable habitat lost in each drought scenario. The upper figures show the buffer analysis graphs with waterholes remaining in drought scenario E and K and their 1 km and 2 km buffers.

Acknowledgments

We thank Marie-Josée Fortin, Patrick Leighton and Martin Lechowicz for their valuable comments on a previous version of this manuscript.

Author Contributions

Analyzed the data: GO KGS BR. Contributed reagents/materials/analysis tools: KGS OB RS AG. Wrote the paper: GO. Manuscript editing: BR SC RS AG.

References

1. Crooks KR, Sanjayan M (2006) Connectivity conservation: Cambridge University Press, Cambridge.

2. With KA, Crist TO (1995) Critical thresholds in species responses to landscape structure Ecology 76: 2446–2459.

3. Bodin O, Norberg J (2007) A network approach for analyzing spatially structured populations in fragmented landscape. Landscape Ecol 22: 31–44.

4. Kadoya T (2009) Assessing functional connectivity using empirical data. Popul Ecol 51: 5–15.

5. Hetherington DA, Miller DR, Macleod CD, Gorman ML (2008) A potential habitat network for the Eurasian lynx *Lynx lynx* in Scotland. Mamm Rev 38: 285–303.

6. Clobert J, Le Galliard JF, Cote J, Meylan S, Massot M (2009) Informed dispersal, heterogeneity in animal dispersal syndromes and the dynamics of spatially structured populations. Ecol Lett 12: 197–209.

7. Stevens VM, Baguette M (2008) Importance of habitat quality and landscape connectivity for the persistence of endangered natterjack toads. Conserv Biol 22: 1194–1204.

8. IPCC (2007) Climate Change 2007: Synthesis Report. Contribution of Working Groups I, II and III to the Fourth Assessment Report of the Intergovernmental Panel on Climate Change [Core Writing Team, Pachauri, R. K. and Reisinger, A. (eds.)]. IPCC, Geneva, Switzerland. 104.

9. Larson B, Sengupta R (2004) A spatial decision support system to identify species-specific critical habitats based on size and accessibility using US GAP data. Environ Model 19: 7–18.

10. Wright CK (2010) Spatiotemporal dynamics of prairie wetland networks: power-law scaling and implications for conservation planning. Ecology 91: 1924–1930.

11. Fortuna MA, Gomez-Rodriguez C, Bascompte J (2006) Spatial network structure and amphibian persistence in stochastic environments. Proc R Soc Lond, Ser B: Biol Sci 273: 1429–1434.

12. Chamaillé-Jammes S, Fritz H, Valeix M, Murindagomo F, Clobert J (2008) Resource variability, aggregation and direct density dependence in an open context: the local regulation of an African elephant population. J Anim Ecol 77.

13. Hansen AJ, DeFries R (2007) Ecological mechanisms linking protected areas to surrounding lands. Ecol Appl 17: 974–988.

14. Andren H (1994) Effects of habitat fragmentation on birds and mammals in landscapes with different proportions of suitable habitat: a review Oikos 71: 355–366.

15. Gonzalez A, Rayfield B, Lindo Z (2011) The disentangled bank: how habitat loss fragments and disassembles ecological networks. Am J Bot 98: 503–516.

16. Márdero S, Nickl E, Schmook B, Schneider L, Rogan J, et al. (2012) Sequías en el sur de la península de Yucatán: análisis de la variabilidad anual y estacional de la precipitación. Investigaciones Geográficas, Boletín del instituto de Geografía, UNAM, Mexico 78: 19–33.

17. Reyna-Hurtado R, Chapman CA, Calmé S, Pedersen EJ (2012) Searching in heterogeneous and limiting environments: foraging strategies of white-lipped peccaries (*Tayassu pecari*). J Mammal 93: 124–133.

18. Dale MRT, Fortin MJ (2010) From graphs to spatial graphs. Annu Rev Ecol, Evol Syst 41: 21–38.

19. Andersson E, Bodin O (2009) Practical tool for landscape planning? An empirical investigation of network based models of habitat fragmentation. Ecography 32: 123–132.

20. Urban D, Keitt T (2001) Landscape connectivity: A graph-theoretic perspective. Ecology 82: 1205–1218.

21. Valeix M, Fritz H, Chamaillé-Jammes S, Bourgarel M, Murindagomo F (2008) Fluctuations in abundance of large herbivore populations: insights into the influence of dry season rainfall and elephant numbers from long-term data. Anim Conserv 11: 391–400.

22. Vester HFM, Lawrence D, Eastman JR, Turner BL, Calmé S, et al. (2007) Land change in the southern Yucatan and Calakmul biosphere reserve: Effects on habitat and biodiversity. Ecol Appl 17: 989–1003.

23. Martínez E, Galindo-Leal C (2002) La vegetación de Calakmul, Campeche, México: clasificación, descripción y distribución. Boletín de la Sociedad Botánica de México 71: 7–32.

24. García-Gil G, Palacio JL, Ortiz MA (2002) Reconocimiento geomorfológico e hidrográfico de la Reserva de la Biosfera Calakmul, México. Investigación Geográficas Boletín del Instituto de Geografía, UNAM 48: 7–23.

25. Castellanos A, Foerster CR, Lizcano DJ, Naranjo E, Cruz-Aldan E, et al. (2008) *Tapirus bairdii*. IUCN 2009. IUCN Red List of Threatened Species: Version 2009.2.

26. Reyna-Hurtado R, Taber A, Altrichter M, Fragoso JMV, Keuroghlian A, et al. (2008) *Tayassu pecari*. *IUCN 2012*. IUCN Red List of Threatened Species. Version 2012.1.

27. Chávez Tovar JC (2010) Ecology and conservation of jaguar (*Panthera onca*) and puma (*Puma concolor*) in the Calakmul Region, and its implications for the conservation of the Yucatan peninsula: Ph.D Thesis. Departamento de Biologia Animal, Universidad de Granada.

28. Owen-Smith RN (1992) Megaherbivores: The influence of very large body size on ecology: Cambridge University Press.

29. Naranjo E, Bodmer RE (2002) Population ecology and conservation of Baird's tapir (*Tapirus bairdii*) in the Lacandon Forest, Mexico. Newsletter of the IUCN/SSC Tapir Specialist Group 11: 25–33.

30. Reyna-Hurtado R, Rojas-Flores E, Tanner GW (2009) Home range and habitat preferences of white-lipped peccaries (*Tayassu pecari*) in Calakmul, Campeche, Mexico. J Mammal 90: 1199–1209.

31. Caso A, Lopez-Gonzalez C, Payan E, Elzirik E, de Oliveira T, et al. (2008) *Panthera onca*. In: IUCN 2012. IUCN Red List of Threatened Species. Version 2012.1.

32. Foerster CR, Vaughan C (2002) Home range, habitat use, and activity of Baird's tapir in Costa Rica. Biotropica 34: 423–437.

33. Tobler MW (2008) The ecology of the lowland tapir in Madre de Dios, Peru: using new technologies to study large rainforest mammals: Ph. D. Thesis. Texas A&M University.

34. Colchero F, Conde DA, Manterola C, Chavez C, Rivera A, et al. (2011) Jaguars on the move: modeling movement to mitigate fragmentation from road expansion in the Mayan Forest. Anim Conserv 14: 158–166.

35. Perez-Cortez S, Enriquez PL, Sima-Panti D, Reyna-Hurtado R, Naranjo EJ (2012) Influence of water availability in the presence and abundance of *Tapirus bairdii* in the Calakmul forest, Campeche, Mexico. Revista Mexicana de Biodiversidad 83: 753–761.

36. Fall A, Fall J (2001) A domain-specific language for models of landscape dynamics. Ecol Model 141: 1–18.

37. Galpern P, Manseau M, Fall A (2011) Patch-based graphs of landscape connectivity: A guide to construction, analysis and application for conservation. Biol Conserv 144: 44–55.

38. Saura S, Pascual-Hortal L (2007) A new habitat availability index to integrate connectivity in landscape conservation planning: Comparison with existing indices and application to a case study. Landscape Urban Plann 83: 91–103.

39. Bodin O, Saura S (2010) Ranking individual habitat patches as connectivity providers: Integrating network analysis and patch removal experiments. Ecol Model 221: 2392–2405.

40. Pascual-Hortal L, Saura S (2006) Comparison and development of new graph-based landscape connectivity indices: towards the priorization of habitat patches and corridors for conservation. Landscape Ecol 21: 959–967.

41. Swift TL, Hannon SJ (2010) Critical thresholds associated with habitat loss: a review of the concepts, evidence, and applications. Biological Reviews 85: 35–53.

42. Gomez-Rodriguez C, Diaz-Paniagua C, Serrano L, Florencio M, Portheault A (2009) Mediterranean temporary ponds as amphibian breeding habitats: the importance of preserving pond networks. Aquat Ecol 43: 1179–1191.

43. Telleria JL, Ghaillani HEM, Fernandez-Palacios JM, Bartolome J, Montiano E (2008) Crocodiles *Crocodylus niloticus* as a focal species for conserving water resources in Mauritanian Sahara. Oryx 42: 292–295.

44. Pringle CM (2001) Hydrologic connectivity and the management of biological reserves: a global perspective. Ecol Appl 11: 981–998.

45. Loarie SR, Van Aarde RJ, Pimm SL (2009) Fences and artificial water affect African savannah elephant movement patterns. Biol Conserv 142: 3086–3098.

46. Sitters J, Heitkonig IMA, Holmgren M, Ojwang GSO (2009) Herded cattle and wild grazers partition water but share forage resources during dry years in East African savannas. Biol Conserv 142: 738–750.

47. Saura S, Rubio L (2010) A common currency for the different ways in which patches and links can contribute to habitat availability and connectivity in the landscape. Ecography 33: 523–537.

48. Metzger JP, Decamps H (1997) The structural connectivity threshold: An hypothesis in conservation biology at the landscape scale. Acta Oecologica-International Journal of Ecology 18: 1–12.

49. Calabrese JM, Fagan WF (2004) A comparison-shopper's guide to connectivity metrics. Front Ecol Environ 2: 529–536.

50. Neeti N, Rogan J, Christman Z, Eastman JR, Millones M, et al. (2012) Mapping seasonal trends in vegetation using AVHRR-NDVI time series in the Yucatan Peninsula, Mexico. Remote Sens Lett 3: 433–442.

51. Herbert ME, McIntyre PB, Doran PJ, Allan JD, Abell R (2010) Terrestrial reserve networks do not adequately represent aquatic ecosystems. Conserv Biol 24: 1002–1011.

52. Pereira M, Segurado P, Neves N (2011) Using spatial network structure in landscape management and planning: A case study with pond turtles. Landscape Urban Plann 100: 67–76.

53. Cote D, Kehler DG, Bourne C, Wiersma YF (2009) A new measure of longitudinal connectivity for stream networks. Landscape Ecol 24: 101–113.

2

Genetic Structure in the Seabuckthorn Carpenter Moth (*Holcocerus hippophaecolus*) in China: The Role of Outbreak Events, Geographical and Host Factors

Jing Tao[1,2], Min Chen[1,2], Shi-Xiang Zong[1,2], You-Qing Luo[1,2]*

1 Beijing Forestry University, Beijing, People's Republic of China, **2** Silviculture and Conservation of Ministry of Education, Beijing Forestry University, Beijing, China

Abstract

Understanding factors responsible for structuring genetic diversity is of fundamental importance in evolutionary biology. The seabuckthorn carpenter moth (*Holcocerus hippophaecolus* Hua) is a native species throughout the north of China and is considered the main threat to seabuckthorn, *Hippophae rhamnoides* L. We assessed the influence of outbreaks, environmental factors and host species in shaping the genetic variation and structure of *H. hippophaecolus* by using Amplified Fragment Length Polymorphism (AFLP) markers. We rejected the hypothesis that outbreak-associated genetic divergence exist, as evidenced by genetic clusters containing a combination of populations from historical outbreak areas, as well as non-outbreak areas. Although a small number of markers (4 of 933 loci) were identified as candidates under selection in response to population densities. *H. hippophaecolus* also did not follow an isolation-by-distance pattern. We rejected the hypothesis that outbreak and drought events were driving the genetic structure of *H. hippophaecolus*. Rather, the genetic structure appears to be influenced by various confounding bio-geographical factors. There were detectable genetic differences between *H. hippophaecolus* occupying different host trees from within the same geographic location. Host-associated genetic divergence should be confirmed by further investigation.

Editor: João Pinto, Instituto de Higiene e Medicina Tropical, Portugal

Funding: This research was supported by the National Nature Science Fund (project NO 30730075 and Fundamental Research Funds for the Central Universities (project NO YX2011-18). The funders had no role in study design, data collection and analysis, decision to publish, or preparation of the manuscript.

Competing Interests: The authors have declared that no competing interests exist.

* E-mail: youqingluo@126.com

Introduction

Pests with fluctuating population size are of major concern for forest security. Knowledge of a pest's population dynamics and associated influential factors is crucial for forest management. Habitat, weather, natural enemies and heritable traits are considered to play roles in insect population dynamics [1]. Despite many studies, the factors involved in the origin of insect outbreaks remain poorly understood. Multiple explanations have been proposed including: escape from natural enemies [2–5], favorable weather [6], changes in host quality and quantity [7–9], and genetic variation of pests [10–12].

The seabuckthorn carpenter moth, *Holcocerus hippophaecolus* Hua (Lepidoptera: Cossidae) is the main pest of seabuckthorn, *Hippophae rhamnoides* L. (Elaeagnaceae). It usually occurs on seabuckthorn, but can also occur on *Ulmus pumila* L. (Urticales: Ulmaceae) as well as a couple of species of Rosaceae [13]. The larvae seriously obstruct water transportation of seabuckthorn by boring into the trunk and roots. It has one generation every 3–4 years and larval stages occupy most of its life history. It is widely distributed throughout its hosts' range, with most damage being caused to trees more than 5 years old. The adult females have limited dispersal and lay their eggs in masses on nearby plants where the larvae feed gregariously. Berryman [14] has demonstrated that pests with low dispersal properties have short, intense, restricted outbreaks whereas those with high vagility have long, extended outbreaks. Consistent with the former pattern, the

seabuckthorn carpenter moth has limited dispersal ability and exhibits short but intense outbreaks that are geographically restricted [15]. Zhou reported that outbreaks of *H. hippophaecolus* can lead to more than 70% mortality of seabuckthorn in plantations in the Inner Mongolia Autonomous Region [16]. Limited mobility appears to play a role in the spatial restriction of the seabuckthorn carpenter moth. The outbreaks usually continue for one or two years before pest numbers decline [15,17].

Seabuckthorn is native to western and northern China, the northern Himalayas and northwestern Europe, through to central Asia and the Altai Mountains [18]. It is a native in 11 provinces (autonomous region, municipalities) in China, with less than 500 thousand hectares of natural forest in the 1950's [19]. Because of seabuckthorn's nitrogen-fixing symbionts, this plant serves to enrich and protect soils [18]. It has been promoted widely in western and northern China to prevent soil erosion and desertification. There are now 2,000,000 ha of seabuckthorn throughout 22 provinces in China, two-thirds of which are monoculture plantations. *H. hippophaecolus* was firstly reported as a pest of seabuckthorn in 1990 [20]. Today *H. hippophaecolus* is considered to be the main threat to seabuckthorn in China. It infests 133,000 ha of seabuckthorn and killed 67,000 ha during the 1990's. Most of the outbreak events occurred in Seabuckthorn monoculture plantations [16]. Prior to the spread of *H. rhamnoides* plantations in western and northern China, no outbreak events of

H. hippophaecolus had been recorded. *H. rhamnoides* was introduced as a novel plant to Jianping County (Liaoning Province) in the 1950's. Prior to that, *H. hippophaecolus* mainly fed on *U. pumila* [16,19]. Jianping has been cultivating seabuckthorn widely since the 1970's and used to have the largest area of seabuckthorn plantation. However, by 2001, the forests were heavily disturbed by the seabuckthorn carpenter moth.

Molecular markers are widely used for insect population genetic research and are a useful tool to study population structure. Correlation with ecological factors, to identify causes of the observed genetic structuring, is often possible using these techniques [21,22]. The Amplified Fragment Length Polymorphism (AFLP) technique generates a large number of fragments that are distributed throughout the genome, without requiring background knowledge of the genome [23]. In the present study, the AFLP technique was used to examine population genetic patterns of *H. hippophaecolus* to address the questions below.

Are genetic patterns a factor in causing outbreaks?

Natural selection, favoring certain genotypes at low densities and others at high densities, may contribute to the regulation of animal numbers [10]. Many studies have demonstrated that the genetic patterns may play a role in population dynamics. For example, changes of allelic frequencies of different loci have been shown to correlate with population fluctuations in *Microtrous ochgaster* [24]. Simchuk et al [25] suggested that esterase (*Est-4*) and protease (*Pts-4*) loci in *Tortrix viridana* L. were directly related to its population dynamics. Mormon crickets (*Anabrus simplex*) were found to consist of genetically distinct clusters that correspond with gregarious (outbreak) populations and solitary (non-outbreak) populations, respectively [11]. These genetic clades provided evidence that the differences of propensities to outbreak are likely due to genetic polymorphism. The seabuckthorn carpenter moth is a very destructive pest, with outbreaks reported in many areas within its range (outbreak areas), but it exists at low densities in other parts of its range (non-outbreak areas). *H. hippophaecolus* populations may therefore consist of genetically distinct clusters with different propensities for outbreak. Here we describe the genetic variation of *H. hippophaecolus* populations from 10 areas across its range with contrasting historical patterns of outbreak events [15–17,26,27] and assess whether the observed genetic patterns can be explained by outbreak history.

Do geographical distance, outbreaks and drought affect genetic population structure?

Identifying the factors responsible for structuring genetic diversity is important for a better understanding of the insect's evolutionary history. Such fundamental studies help to infer ecological characteristics that are crucial for establishing management strategies. Geographical distance, outbreaks, and drought were considered in this study. First, in the case of species with restricted dispersal abilities, we expect to observe a positive correlation between the genetic and geographical distance. Moreover, in the historical outbreak regions, *H. hippophaecolus* undergoes outbreak events that are followed by population decline when food plants become unavailable. Such population fluctuations may have large effects on genetic diversity within populations. Furthermore, non-genetic studies indicate recent outbreak events were related to a lack of rainfall [16]. Heavy periodic rainfalls could restrict *H. hippophaecolus* movement and genetic exchange within populations. Therefore, we might also expect lower genetic differentiation within populations during drought.

Does host-associated diversity exist in a common locality?

Herbivores and their host plants maintain an intimate relationship in feeding, oviposition, mate finding and predator avoidance. Distribution, availability, longevity and chemistry of host plants are major factors that affect the genetic differentiation of herbivorous populations [28]. Seabuckthorn was a novel host plant for insects following its introduction into Jianping. Herbivore populations can suffer from disruptive selection following shifts to novel host plants [29]. How did they adapt to the changes? Is there detectable genetic diversity among *H. hippophaecolus* populations feeding on different host trees in a common location? To answer these questions, we sampled larvae from sympatric populations of four host trees, sebuckthorn, *U. pumila*, *Prunus armeniaca* L, (Rosales: Rosaceae), *Pyrus pyrifolia* (Burman) Nakai (Rosales: Rosaceae), in Jianping.

Materials and Methods

Sample collection and DNA extraction

Individuals (n = 217) were collected from 10 locations across the carpenter moth range during the summer of 2008 (Table 1) by directly sampling under the bark of infested trees and byusing light and pheromone traps. Sampling locations represented two contrasting patterns of historical outbreak events, based on a literature survey and unpublished data (J. Zong, personal communication) (Figure 1). Populations from some areas have experienced outbreaks while in others population densities have been consistently low. In Jianping, a further 24 insects were collected from different hosts (*U. pumila* (JPY, n = 7), *Prunus armeniaca* (JPX, n = 8), *Pyrus pyrifolia* (JPL, n = 9)). Individuals were transported alive to the laboratory, and then kept at −80°C. Prior to DNA extraction, insects were washed in 80% ethanol. Total genomic DNA was isolated using the SDS-method of Zhang and Hewitt [30]. After extraction, DNA was dissolved in TE buffer and stored at −20°C until further use.

AFLP protocol

Amplified fragment length polymorphism (AFLP) analysis was used to assess genetic diversity among sampled populations of *H. hippophaecolus*. The AFLP procedure followed Vos et al. [23] with minor modifications. Genomic DNA was digested with *Eco*RI and *Mse*I restriction enzymes (New England Biolabs) and double stranded adapters were ligated to the sticky ends of the fragments. After 4 h incubation at 37°C, each sample was diluted 1:9 with H_2O and a two-step amplification strategy was used. Pre-selective amplification was performed for 3 min at 94°C, then 30 cycles of 30 s at 94°C, 30 s at 56°C and 1 min 72°C. A 20-µl Pre-selective amplification PCR mixture consisted of 30 mM $MgCl_2$, 4.5 mM dNTP, 0.6 U Taq DNA polymerase, 30 ng *Eco*RI-C and *Mse*I-A primer. In the selective amplification, we used the following nine primer combinations selected from 100 tested combinations [31]: *Eco*RI-AAC/*Mse*I-CAA, *Eco*RI-AAC/*Mse*I-CAC, *Eco*RI-AAC/*Mse*I-CCT, *Eco*RI-AAC/*Mse*I-CTT, *Eco*RI-AAG/*Mse*I-CCA, *Eco*RI-AAG/*Mse*I-CTG, *Eco*RI-CA/*Mse*I-CAA, *Eco*RI-CA/*Mse*I-CAC, *Eco*RI-CA/*Mse*I-CCT. The *Eco*RI primers were labeled with IRD-700. Selective amplification was performed with the following touchdown thermal profile: 3 min at 94°C; 12 touchdown cycles at 94°C for 30 s, 65°C for 30 s (decreasing the temperature by 0.7°C per cycle), and 72°C for 60 s; 30 cycles at 94°C for 30 s, 56°C for 30 s, 72°C for 1 min; 5 min at 72°C. The 10 µl PCR mixture contained 15 $MgCl_2$, 1.5 ng *Mse*I and *Eco*RI primer, 2 mM dNTP), 2 µl diluted (1:9) Pre-amplified

Table 1. Geographical location, average annual rainfall, and hosts for populations of *H. hippophaecolus*.

No.	Location	Coordinates	Average annual rainfall/mm	Population indentifier	Host plant	Sample Size
1	Liaoning, Jianping	119.71E/41.84N	478.37	JPS	*H.rhamnoides*	26
				JPY	*U. pumil*	7
				JPX	*P. armeniaca*	8
				JPL	*P. pyrifolia*	9
2	Inner mongolia, Linxi	118.23E/43.61N	378.98	LX	*H.rhamnoides*	16
3	Hebei, Fengning	116.61E/41.21N	535.16	FN	*H.rhamnoides*	25
4	Shanxi, Youyu	112.39E/39.96N	377.24	YY	*H.rhamnoides*	26
5	Inner mongolia, Dongsheng	111.25E/39.87N	410.52	DS	*H.rhamnoides*	15
6	Shanxi, Wuzhai	111.87E/38.88N	431.54	WZ	*H.rhamnoides*	18
7	Shanxi, Yulin	109.76E/38.28N	397.43	YL	*H.rhamnoides*	19
8	Shanxi, Wuqi	108.26E/36.90N	534.83	WQ	*H.rhamnoides*	29
9	Nixia, Yanchi	107.48E/37.89N	290.24	YC	*H.rhamnoides*	22
10	Nixia, Pengyang	106.50E/35.82N	498.67	PY	*H.rhamnoides*	21

DNA. All PCRs were conducted on a GeneAmp PCR System 9700 (USA Applied Biosystems).

Amplification products were separated on 6% polyacrylamide gels for 2.5 h on a LI-COR 4300 DNA Analyzer (LI-COR Biosciences, USA), using LI-COR 50–700 bp (Labeled with IRD-700) as a size standard. Fragments from 100–700 bp in size were scored as present (1) or absent (0) using SAGA MX (LI-COR Biosciences, USA), and exported for data analysis.

A blank control was carried out along with each set of DNA extractions and PCR amplifications to monitor any possible cross contamination. Poor-quality DNA samples that did not amplify were excluded from further analysis.

Data analysis

Genetic variation and structure of *H. hippophaecolus* populations. The diversity of geographic populations was assessed by estimating the percentage of polymorphic loci (%P) and Nei's heterozygosity. Percentage of polymorphic loci estimates were based on 99% criteria and heterozygosity estimates were made using the software TFPGA [32].

The genetic structure was examined by an analysis of molecular variance (AMOVA) performed by the software ARLEQUIN 3.1 [33]. This method was used to partition the genotypic variance among and within populations. Two separate analyses were performed to test the hypotheses of genetic structure attributable to variation: among individuals across the different localities feeding on *H. rhamnoides* and among individuals across different host plants in Jianping. An additional analysis of individuals feeding on *H. rhamnoides* compared to the group combining three other host plants in Jianping was also performed. Genetic differentiation coefficients between populations (both geographic and host-associated) were calculated as F_{ST}, with 95% confidence

intervals (CI) obtained by bootstrapping 1000 replicates over loci. The TFPGA software was also applied to calculate Nei's genetic distance (D) [34]. Neighbor-joining (NJ) trees were constructed based on D using MEGA4 [35].

Identification of candidate outbreak loci under selection. Outlier loci were identified using the Dfdist approach [36,37] in Mcheza program [38] (available at http://popgen.eu/soft/mcheza/). Allele frequencies are estimated in Dfdist based on Zhivotovsky's [39] Bayesian approach.

Because of our particular interest in outbreak-associated divergence, the Dfdist was run for two groups of populations (outbreaking population vs non-outbreaking population). A total of 50000 realizations were performed and maximum allowable allele frequency was 0.99. We chose the 0.995 confidence interval and set the significance level at 99%. The Benjamini and Hochberg false discovery rate (FDR) correction method was used to correct for the occurrence of false positives in loci identified as under selection [40]. Loci with significant *P*-values at FDR threshold of 50% were identified using the Benjamini and Hochberg method.

Testing outbreaks and environmental factors driving genetic structure. The following analysis tested outbreaks and environmental factors that potentially influenced genetic population structure. The effect of geographical distance was assessed using linear map distances between *H. hippophaecolus* populations. Secondly, outbreak patterns were scored with 1 indicating populations from areas where outbreaks had occurred and 0 representing populations in non-outbreaking areas. Finally, an index for the "degree of drought," represented by the average annual rainfall collected over 50 years was obtained (1955–2007, China meteorological data sharing service system http://cdc.cma.gov.cn/). Mantel tests were conducted with the software TFPGA to test the correlation between Euclidean distances for all the factors and genetic distances.

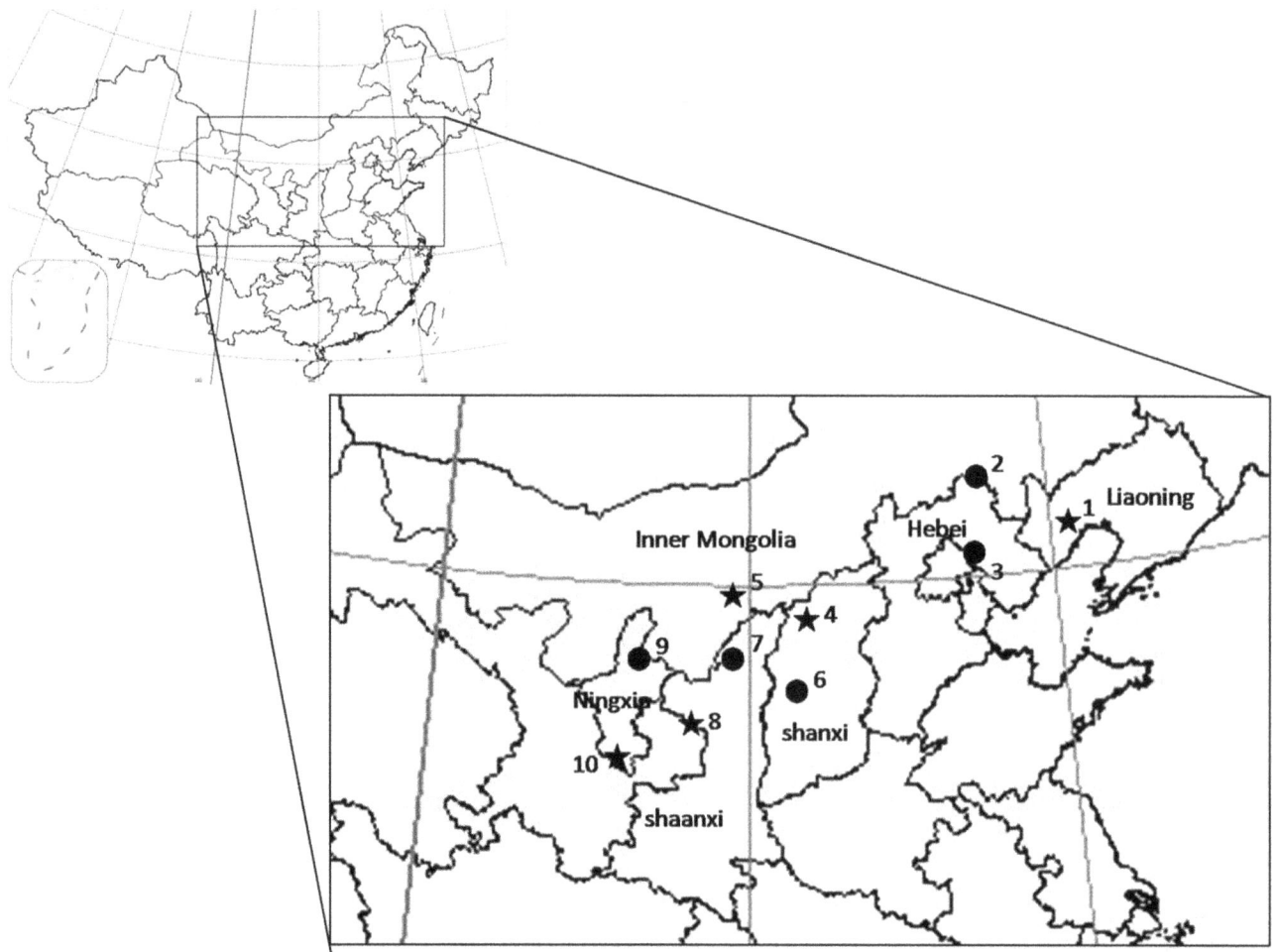

Figure 1. Locations of 10 *H. hippophaecolus* sites with known outbreak (star) or non-outbreak (circle) patterns. Outbreak patterns based on Luo et al. [16], population 1; Wang [26], poulation 4; Zhou [15], population 5; You [27], population 7; Zong et al. [17] population 10; Zong, personal communication, population 2,3,6,7,9. Identities and locations of populations see Table 1.

The general linear models (GLM) method was also used to test the effect of outbreak and drought on the genetic differentiation between populations. In this analysis the factor "drought" was defined as locations with less than 400 mm average annual rainfall. Values of 1 were used for drought locations (YY, YL, YC, LX) and 0 for other locations (PY, JP, WQ, DS, WZ, FN). The outbreak factor was standardized, as previously, for an outbreak area of 1 and a non-outbreak area of 0. We performed a GLM analysis of the heterozygosity with outbreak and drought as fixed factors. A *P*-value of <0.05 was used to indicate statistical significance. GLM was implemented using SPSS 16.0.

Results

Genetic variation and structure of *H. hippophaecolus* populations

The nine primer combinations produced a total of 933 bands. The global Θ among the 10 sites was 0.2106 (95% CI 0.1981–0.2230). Nei's heterozygosity for each geographical population was moderate and ranged from 0.1505~0.2042 (Table 2).

AMOVA conducted on AFLP markers confirmed the presence of moderate genetic differentiation showing that 22.54% of total variability was due to the variation among geographic populations

Table 2. Percentage of polymorphic loci (%P) and Nei's heteozygosity of *H. hippophaecolus* populations.

	Population indentifier									
	JP	LX	FN	YY	DS	WZ	YL	WQ	YC	PY
%Polymorphic loci (p)	72.56	51.23	77.60	58.41	74.49	68.27	70.85	66.34	56.38	61.74
Heterozygosity (H)	0.1854	0.1529	0.1702	0.1495	0.2042	0.1505	0.1872	0.1749	0.1604	0.1679

Table 3. Analysis of molecular variance (AMOVA) of *H. hippophaecolus* populations.

Source of variation	d.f.	SS	Percentage of variation (%)	P
Geographical grouping				
Among localities	9	5831.247	22.54	<0.0001
Individuals within localities	206	18395.396	77.46	<0.0001
Host-plant grouping				
Among host plants in Jianping	3	1537.343	31.73	<0.0001
Individuals within host plants in Jianping	30	3895.897	68.27	<0.0001
Among two host groups in Jianping	1	1249.317	34.82	<0.0001
Individuals within groups in Jianping	48	4183.923	65.18	<0.0001

($F_{ST} = 0.2254$, $P<0.0001$) (Table 3). The pair-wise comparisons between populations were characterized by values of F_{ST} ranging from 0.0424–0.3663 (Table 3). Most of the populations showed highly significant differences ($P<0.0001$) with the exception of the YY and LX populations ($P=0.0182$). This result indicates that most of the 10 sampled populations represent differentiated populations.

The Neighbor Joining phenogram (Figure 2) indicates that the clusters comprised populations with a mixture of outbreak patterns. For instance, populations from Dongsheng and Youyu were in two distinct NJ genetic clusters, although they have the same intensity of outbreak events.

Examination of the AFLP data using Dfdist in Mcheza sought to determine whether there was evidence of any highly differentiated loci. F_{ST} is plotted against heterozygosity in Figure 3. The outbreak and non-outbreak population comparison performed with Dfist resulted in four markers out of 993 (loci 93, 188, 223, 390) showing more differentiation than expected at the 99.5% confidence level. All these loci were detected as potential positive outliers at the 50% FDR threshold (Figure 3).

Testing outbreaks and drought as factors driving *H. hippophaecolus* genetic structure

The Mantel test based on the 10 localities gave an *r* value of 0.0554 ($P=0.3460$, for 10000 randomizations), indicating no correlation between geographic and genetic differences. The Nei's genetic distances between populations were not significantly correlated to outbreak differences in the Mantel test ($r=0.2516$, $P=0.0740$). The interaction between Euclidean distances for average annual rainfall and genetic distances was also not significant (Mantel test $r=0.1271$, $P=0.2070$). GLM analysis showed that the factors of outbreak and drought, and their interaction, did not have a significant effect on heterozygosity ($F_{1,10}=0.053$, $P=0.826$, $F_{1,10}=1.329$, $P=0.293$ and $F_{1,10}=2.904$, $P=0.139$ respectively).

Host-associated diversity

The host plant was found to have a larger effect on the genetic structure among populations than geographic location. The global Θ value among different hosts was 0.2785 (95% CI 0.2548–0.3024), higher than the value among 10 sites (0.2106). AMOVA

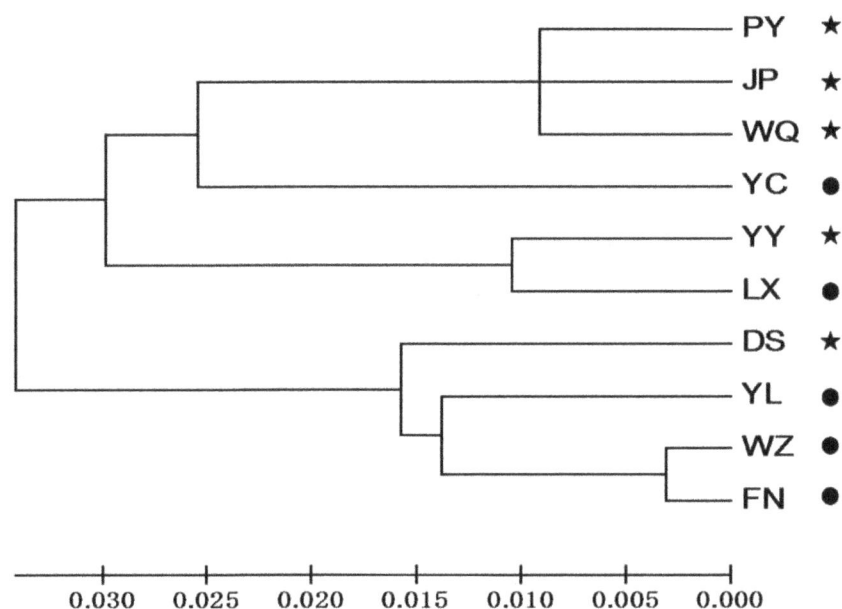

Figure 2. Neighbor-joining phenogram of Nei's genetic distances between *H. hippophaecolus* populations at different collection sites. Site names and outbreak patterns correspond to those in figure 1. Populations are depicted as outbreaking (star) or non-outbreaking (circle).

Fst/He

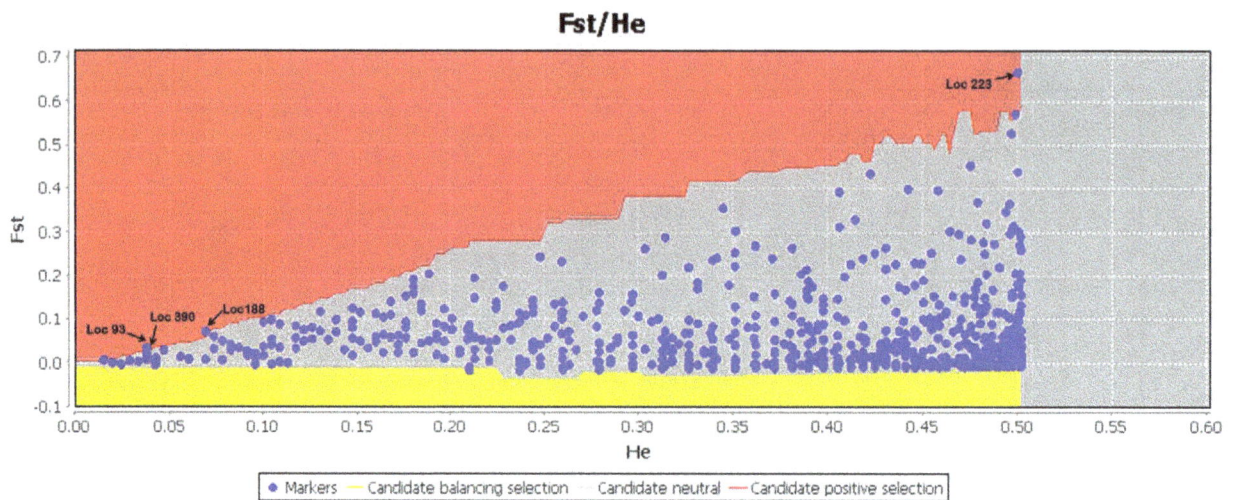

Figure 3. Result of Dfdist analyses in Mcheza. F_{ST} against heterozygosity plot is provided. Loci were identified as significant, corresponding to a 99% critical frequency and 99.5% confidence interval. Red and yellow lines represent 0.995th and 0.005th quantiles of the conditional distribution obtained from Dfdist simulations respectively. Each blue dot indicates AFLP marker; dots in gray areas represent neutral loci falling below the 0.995 quantile's broad line, whereas outlier loci in red areas are pointed out by arrows and accompanied by the locus number.

with ARLEQUIN found greater variation among populations in host-plant groupings (31.73%) than populations in geographical groupings (22.54%) (Table 3). Pairwise F_{ST} statistics between JPS and each other location population ranged from 0.0856 to 0.2978 (Table 4), while the genetic divergences were all highly significant 0.3510~0.3773 in the host-associated analysis (Table 5).

In Jianping, individuals feeding on *H. rhamnoides* had a great separation from individuals feeding on other host plants. When combined individuals feeding on *U. pumila*, *P. armeniaca* and *P. pyrifolia* as a group, compared to individuals feeding on *H. rhamnoide*, the variation among two groups rose up to 34.82% by AMOVA with ARLEQUIN. Pairwise comparisons of F_{ST} values between all host plant combinations further supported the pattern of genetic structure. F_{ST} values were much greater in comparisons between the *H. rhamnoides* feeders (0.3510–0.3773) and each other host-plant feeders (0.0527–0.1180) (Table 5). *H. rhamnoides* feeders showed strongly significant differences (*P*<0.0001) with the moth on other host plants (Table 5).

Discussion

Genetic patterns associated with outbreak events of *H. hippophaecolus*

Genetic clustering did not support distinct outbreak-associated genetic clades in *H. hippophaecolus*. NJ genetic population clusters contained a combination of populations from historical outbreak areas as well as non-outbreak areas (Figure 2). The outbreak effect may have been difficult to detect among different geographical populations due to various confounding biogeographical factors that also shape genetic structure in *H. hippophaecolus*. In addition, one cannot exclude the possibility that the outbreak and non-outbreak patterns are associated with a single genotype, but depend on the expression of different phenotypes in different environments.

Indeed, our results support the notion that outbreak events were likely to be endemic population changes from latent to epidemic rather than being due to insects with an outbreak-associated

Table 4. Nei's genetic distance and F_{ST} value between all geographic combinations.

	PY	JP	YY	WQ	DS	YL	WZ	YC	LX	FN
PY	—	0.0902	0.2409	0.0683	0.1589	0.2179	0.2929	0.2779	0.2601	0.2298
JP	0.0191	—	0.2682	0.0856	0.1969	0.2204	0.2978	0.2754	0.2737	0.2467
YY	0.0465	0.0587	—	0.2654	0.1115	0.2938	0.3663	0.3477	0.1047	0.2715
WQ	0.0170	0.0196	0.0583	—	0.1813	0.2048	0.2914	0.2404	0.2856.	0.2454
DS	0.0401	0.0442	0.0423	0.0468	—	0.1550	0.1828	0.2511	0.1089	0.0986
YL	0.0530	0.0559	0.0888	0.0527	0.0362	—	0.1129	0.1642	0.2520	0.1298
WZ	0.0816	0.0780	0.1247	0.0869	0.0375	0.0254	—	0.2998	0.3131	0.0424
YC	0.0538	0.0541	0.0774	0.0447	0.0617	0.0437	0.0871	—	0.3387	0.2819*
LX	0.0456	0.0578	0.0209	0.0606	0.0343	0.0678	0.0944	0.0731	—	0.2126
FN	0.0669	0.0652	0.0969	0.0740	0.0207	0.0298	0.0062	0.0827	0.0705	—

Nei's genetic distances are below the diagonal. F_{ST} value and their significance level are above the diagonal. Significance level of associated F_{ST} value are shown as: *0.01<P<0.05, unmarked mean P<0.0001.

Table 5. Pairwise comparisons of genetic divergence estimates (F_{ST}) between all host plants combinations.

	JPY	JPL	JPX	JPS
JPY	0.00000			
JPL	0.07590*	0.00000		
JPX	0.05271*	0.11808**	0.00000	
JPS	0.36080***	0.37730***	0.35109***	0.00000

Significance level of associated F_{ST} value are shown as:
*0 01<P<0.05;
**0.001<P<0.01;
***P<0.0001.

genotype spreading to outbreak areas. This conclusion is also consistent with the poor dispersal ability of *H. hippophaecolus*, which has been observed in non-genetic studies. Zong et al. [17] indicated female moths usually choose a nearby tree for mating after emergence, and that the body weight of fertilized female moths is too heavy for long-distance migration. Males are not attracted to sex pheromone traps located too far away from the infested forests (<100 m) [13]. Furthermore, adults live for only several days, which limits the degree of dispersal. Young seedlings (1–2 years) and seeds of seabuckthorn are often used for its introduction. However, *H. hippophaecolus* only harms seabuckthorn plants that are more than 5 years old. Therefore, *H. hippophaecolus* would not be dispersed long-distances by artificial movement of host plants.

We rejected the hypothesis of genetic difference associated with outbreaks in the seabuckthorn carpenter moth. However, habitat, weather, natural enemies are also considered as main factors affecting insect population dynamics. An outbreak occurs when the physiological state of the plant permits a herbivore phenotype with a high reproductive capacity to become dominant. Agricultural and forest monocultures consisting of extensive plantings of hosts with narrow genetic variability are havens for pest outbreaks [1]. Seabuckthorn monoculture plantations are optimal sites for survival of *H. hippophaecolus*, especially for those that were introduced as an exotic species growing under unfavorable conditions [41]. Weather and climatic conditions significantly affect population fluctuation. Unusual weather is known to have strong effects on the dynamics of insect populations [42,43]. Several authors associated the initiation of outbreaks in Jianping and Dongsheng with consecutive dry years before outbreak [15,19,]. Seabuckthorn plantations have plenty of nutrients, relatively few natural enemies and are highly vulnerable to drought or human disturbance, which may explain why outbreak events happen there.

The outlier analysis revealed that although differentiation for the majority of markers did not significantly deviate from neutral expectations, a small number of markers (n = 4) were identified as outlier loci. The false discovery rate test supported the conclusion that all identified outlier loci are under selection. These results do not allow us to reject the hypothesis that specific genome regions or genes are associated with outbreak events. Having indentified outlier loci in *H. hippophaecolus*, it will be necessary to try to find candidate genes that could correspond to these AFLP markers. Then we can characterize these genes with functional genomics analyses.

Factors influencing the population structure of *H. hippophaecolus*

We found evidence of limited gene flow among samples collected from 10 locations. The poor correlation between genetic

and geographical distances is an unexpected result given the previous assumption that populations are isolated-by-distance.

These results suggest that multiple factors other than simple geographic distance are influencing the genetic composition of populations. The Mantel analysis failed to support the idea that outbreak and drought are pertinent factors underlying genetic structure in *H. hippophaecolus*. Compared to other observations in insects, Nei's heterozygosity values for each population are moderate [44,45,46]. We found similar genetic diversity within outbreak and non-outbreak populations of the seabuckthorn carpenter moth. Similarly, GLM analysis showed that outbreaks had no significant effects on heterozygosity. Genetic diversity within outbreak populations was not strongly affected by increases in population size during outbreak periods. This result could be consistent with the theoretical prediction that long-term fluctuating populations correspond to the harmonic mean size over time, and should thus be closer in size to that during the remission period than during an outbreak [47,48]. We rejected the hypothesis that drought populations have lower genetic variation. GLM analysis showed that drought had no significant effects on heterozygosity.

Comparative studies of population structure in phytophagous insects show that genetic structure is mostly determined by the ability of the species to disperse [49]. The probability of successful dispersal is largely determined by habitat availability [50,51]. Louy et al. [52] have shown in experiments with three skipper species that dispersal ability and habitat availability determine the genetic structure of species. Whether a habitat is available for phytophagous insects strongly depends on the existence of host trees. Several forest insects have pronounced geographical structure that follows the distribution of their host tree species [46,53].

Louy et al. [52] suggest that limited habitat availability in combination with low dispersal capacity result in independent genetic structure with relative high genetic differentiations and low gene flow among populations. *H. hippophaecolus* is a specialist, which only feeds on a few plants (*U. pumila, P.armeniaca, P.pyrifolia*). Moreover, most of the principal host in China, *H. rhamnoides*, is in single species plantations which are suffering from human interference. Could this independent population structure of *H. hippophaecolus* be explained by low dispersal capacity and habitat fragmentation? Combining habitat landscape and population genetic analysis might answer this question in the future.

The role of the plant

Host races are genetically differentiated sympatric populations of parasites that use different hosts and between which there is limited gene flow [54]. Our analyses uncovered very high F_{ST} values (0.3510–0.3773) between JPS and other non-seabuckthorn populations. It is indicated that *H. rhamnoides* constitutes a barrier to gene flow between *H. hippophaecolus* populations from other host plants in Jianping. *H. hippophaecolus* feeding on *H. rhamnoides* in Jianping are more genetically differentiated than those from other hosts in sympatric rather than other geographically distant populations of seabuckthorn. Host races might therefore exist in seabuckthorn and other host plant used by *H. hippophaecolus*. Factors favoring host race formation include correlations between host choice and mate choice. Although host fidelity and assortative mating has not been fully explored in *H. hippophaecolus*, tests using both artificial and natural methods suggest female host preferences may exist. Adult emergence from the seabuckthorn roots confirmed oviposition preference on *H. rhamnoides*, rather than on *U. pumila* and *P. armeniaca* [55].

Seabuckthorn was an endemic perennial, sporadically growing in Inner Mongolia, Shanxi and areas of Liaoning province before it was widely promoted. The timing of host shifting of *H. hippophaecolus* in Jianping is likely due to the introduction of *H. rhamnoides*. However, how did host shifting occur in *H. hippophaecolus* in Jianping? When did host-associated genetic divergence initially occur in *H. hippophaecolus*? Data from many host utilization systems gave rise to a possible scenario that host shifts occur as a result of host plant's increased abundance and availability as a potential resource following human-mediated plant community changes [56,57]. If this is the case, our data suggests a local host shift and genetic differentiation of *H. hippophaecolus* following the introduction of seabuckthorn in Jianping. Though a rapid range expansion of *H. hippophaecolus* following human-mediated changes is possible, it does seem unlikely given the wide extent of genetic divergence observed during such a brief time. This scenario was also rejected by Sword et al in the *Hesperotettix viridis* host utilization system [58]. Another possibility is a genetic divergence of moth between *H. rhamnoides* and other hosts, prior to the host shift. Feder et al. [59] found genetic divergence between apple and hawthorne host races of *Rhagoletis pomonella* L. pre-dating the introduction of the apple to North America. Given the long life history of *H. hippophaecolus* and brief planting history of *H. rhamnoides* in Jianping, we suppose the latter scenario is the case. Seabuckthorn is native to parts of western and northern China although records for the historical host plant use by *H. hippophaecolus* are lacking. Our results indicate that an *H. hippophaecolus* lineage might have adapted to utilize *H. rhamnoides* in China prior its spread. The possibilities of an ancestral host shift and stable host-associated genetic divergence in seabuckthorn carpenter moth are suggested.

We found no fixed diagnostic differences in AFLP data between the different host-associated forms. Host-associated genetic divergence should also be further demonstrated by sampling additional populations feeding on different host plants in more locations. In future studies, more different genetic markers are recommended in this system. They should include co-dominant markers such as microsatellites (not currently available for this species) and incorporation of variable regions of the mitochondrial genome. Microsatellites are highly polymorphic, locus-specific and can show co-dominant inheritance. They may recover higher levels of variability than other markers, particularly if following a population bottleneck associated with host shift. Mitochondrial sequences can be analyzed to determine patterns of evolutionary relationships between different haplotypes. This may provide information on the historical evolution of host-associated forms in the seabuckthorn moth.

Acknowledgments

We are grateful to Jianwei Wang, Rong Wang, Zhizheng Wang for sample collection. We would also to thank Mark P. Miller who helped with data analysis using TFPGA software. We thank Michael Klein, Katie Robinson, Tamara Pulpitel, Julie-Anne Popple for English editing of manuscript.

Author Contributions

Conceived and designed the experiments: JT MC S-XZ Y-QL. Performed the experiments: JT MC. Analyzed the data: JT MC S-XZ Y-QL. Contributed reagents/materials/analysis tools: JT MC S-XZ Y-QL. Wrote the paper: JT. Collected samples: JT S-XZ.

References

1. Wallner WE (1987) Factors affecting insect population dynamics: differences between outbreak and non-outbreak species. Annual Review of Entomology 32: 317–340.
2. Price PW (1987) The role of natural enemies in insect populations, In: Barbosa P, Schultz J, eds. Insect outbreaks. New York: Academic Press. pp 208–312.
3. Myers JH (1988) Can a general hypothesis explain population cycles of forest Lepidoptera. San Diego: Academic Press Inc. pp 179–242.
4. Walsh PJ (1990) Site factors, predators and pine beauty moth mortality. In: Watt AD, Leather SR, Hunter MD, Kidd NAC, eds. Population dynamics of forest insects. Andover: Intercept. pp 242–252.
5. John LM, Susan H, Mary G (2001) Origin of an insect outbreak: escape in space or time from natural enemies? Oecologia 126: 595–602.
6. Martinat PJ (1987) The role of climatic variation and weather in forest insect outbreaks, In: Barbosa P, Schultz J, eds. Insect outbreaks. New York: Acedemic Press. pp 241–268.
7. White TCR (1984) The abundance of invertebrate herbivores in relation to the availability of nitrogen in stressed food plants. Oecologia 63: 90–105.
8. Mattson WJ, Haack RA (1987) The role of drought in outbreaks of plant-eating insects. Bio Science 37: 110–118.
9. Rossiter M (1992) The impact of resource variation on population quality in herbivorous insects: a critical aspect of population dynamics. In: Hunter MD, Ohguhi T, Price PW, eds. Effects of resource distribution on animal-plant interactions. New York: Academic Press. pp 13–42.
10. Chitty D (1971) The natrual selection of self-regulatory behavior in animal populations. In McLaren IA, ed. Natrual Regulation of Animal Populations. New York: Atherton Press.
11. Bailey NW, Gwynne DT, Ritchie MG (2005) Are solitary and gregarious Mormon crickets (*Anabrus simplex*, Orthoptera, Tettigoniidae) genetically distinct? Heredity 95: 166–173.
12. Chapuis MP, Estoup A, Auge SA, Foucart A, Lecoq M, et al. (2008) Genetic variation for parental effects on the propensity to gregarise in *Locusta migratoria*. BMC Evolutionary Biology 8: 37.
13. Zong SX, Luo YQ, Lu CK, Xu ZC, Zhang LS (2006) Prelininary Study on Biological Characteristic of *Holcocerus hippophaecolus*. Scientia Silvae Sinicae 41: 79–84.
14. Berryman AA (1978) Towards a theory of insect epidemiology. Research in Population Ecology 19: 181–196.
15. Luo YQ, Lu CK, Xu ZC (2003) Control strategies on a new serious forest pest insect seabuckthorn carpenterworm, *Holcocerus hippophaecolus*. Forest Pest Disease 5: 25–28.
16. Zhou ZY (2002) The causes of death and control strategies to deal with *Holcocerus hippophaecolus* in the east of Erdos City, Inner Mongolia Autonomous Region. Hippophae 15: 7–11.
17. Zong SX, Luo YQ, Xu ZC, Yao GL, Wang YY (2006) Harm characteristics and population dynamics of two important borers in *Hippophae rhamnoidea*. Forest Pest and Disease 25: 7–10.
18. Stewart PE, Pearson MC (1967) Nodulation and nitrogen-fixation by *Hippophae rhamnoides* L. in the field. Plant and Soil 26: 348–360.
19. Luo YQ, Lu CK, Xu ZC (2003) Control strategies on a new serious forest pest insect seabuckthorn carpenterworm, *Holcocerus hippophaecolus*. Forest Pest Disease 5: 25–28.
20. Hua BZ, Zhou Y, Fang DQ (1990) Chinese Cossidae. Beijing: Tianze Press. pp 56–57.
21. Behura SK (2006) Molecular marker systems in insects: current trends and future avenues. Molecular Ecology 15: 3087–3113.
22. Foll M, Gaggiotti O (2006) Identifying the Environmental Factors That Determine the Genetic Structure of Populations. Genetics 174: 857–891.
23. Vos P, Hogers R, Bleeker M, Reijans M, Lee Tvd, et al. (1995) AFLP: a new technique for DNA fingerprinting. Nucleic Acids Research 23: 4407–4414.
24. Gaines MS, McClenaghan LR, Rose RK (1978) Temporal patterns of allozymic variation in fluctuating populations of *Microtus ochrogaster*. Evolution 32: 723–739.
25. Simchuk APV, Ivashov A, Companiytsev VA (1999) Genetic patterns as possible factors causing population cycles in oak leafroller moth, *Tortrix Viridana* L. Forest Ecology and Management 113: 35–49.
26. Wang DS (2007) Research on the regularity for the change of the population of *Holcocerus hippophaecolus* in Youyu County, Shuozhou City. Sci-Tech Information Development & Economy 17: 170.
27. You Y (2000) The present situation of forest diseases and insect pests on Yulin and the control measures. Shaanxi Forest Science and Technology 4: 1–3.
28. Mopper S (1996) Adaptative genetic structure in phytophagous insect populations. Tree 11: 235–238.
29. Funk DJ, Filchak KE, Feder JL (2002) Herbivorous insects: model systems for the comparative study of speciation ecology. Genetica 116: 251–267.
30. Zhang DX, Hewitt GM (1998) Isolation of animal cellular total DNA. In: Karp A, Ssac PG, Ingram DS, eds. Molecular Tools for Screening Biodiversity: Plant and Animals. London: Chapman & Hall. pp 5–9.
31. Chen M, Tao J, Ma CD, Yin WL, Zhu YY, et al. (2008) Selection of primers and establishment of AFLP analysis system on *Holcocerus hippopaecolu*. Journal of Beijing Forestry University 30: 116–120.
32. Miller MP (1997) Tools for population genetic analysis (TFPGA) 1.3: A Windows program for the analysis of allozyme and molecular population genetic data. Department of Biological Sciences, Northern Arizona University: AZ, USA.
33. Schneider S, Roessli D, Excoffier L (2000) *Arlequin, Version 2.0: a Software for Population Genetic Data Analysis*. Genetics and Biometry Laboratory. University of Geneva: Switzerland.

34. Nei M (1978) Estimation of heterozygosity and genetic distance from a small number of individuals. Genetics 89: 583–590.

35. Tamura K, Dudley J, Nei M, Kumar S (2007) MEGA4: Molecular Evolutionary Genetics Analysis (MEGA) software version 4.0. Molecular Biology and Evolution 24: 1596–1599.

36. Beaumont MA, Nichols RA (1996) Evaluating Loci for Use in the Genetic Analysis of Population Structure. Proceedings: Biological Sciences 263: 1619–1626.

37. Beaumont MA, Balding DJ (2004) Identifying adaptive genetic divergence among populations from genome scans. Molecular Ecology 13: 969–980.

38. Antao T, Beaumout MA (2011) Mcheza: A workbench to detect selection using dominant markers. Bioinformatics 27: 1117–1118.

39. Zhivotovsky LA (1999) Estimating population structure in diploids with multilocus dominant DNA markers. Molecular Ecology 8: 907–913.

40. Benjamini Y, Hochberg Y (1995) Controlling the False Discovery Rate: A Practical and Powerful Approach to Multiple Testing. Journal of the Royal Statistical Society Series B (Methodological) 57: 289–300.

41. Zhou ZY, Yin WL, Liang HJ, Yu JM, Zhang Q (2007) Mechanism and the stand conditions of *Hippophae rhamnoides* resistance to *Holcocerus hippophaecolus*. Journal of Beijing Forestry University 29: 50–56.

42. Crister S (1991) Unusual weather and insect population dynamics: *Lygaeus equestris* during an extinction and recovery period. Oikos 60: 343–350.

43. Kennedy GG, Storer NP (2000) Life systems of polyphagous arthropod pests in temporally unstable cropping systems. Annual Review of Entomology 45: 467–493.

44. Salvato P, Battisti A, Concato S, Masutti L, Patarnello T, et al. (2002) Genetic differentiation in the winter pine processionary moth (*Thaumetopoea pityocampa - wilkinsoni* complex), inferred by AFLP and mitochondrial DNA markers. Molecular Ecology 11: 2435–2444.

45. Alessandro G, Sanna B, Leena L, Anne L, Johanna M (2005) The voyage of an invasive species across continents: genetic diversity of North American and European Colorado potato beetle populations. In Molecular ecology. pp 4207–4219.

46. Mock KE, Bentz BJ, O'neill EM, Chong JP, Orwin J, et al. (2007) Landscape-scale genetic variation in a forest outbreak species, the mountain pine beetle (*Dendroctonus ponderosae*). Molecular Ecology 16: 553–568.

47. Motro U, Thomson G (1982) On heterozygosity and the effective size of populations subject to size changes. Evolution 36: 1059–1066.

48. Chapuis MP, Loiseau A, Michalakis Y, Lecoq M, Franc A, et al. (2009) Outbreaks, gene flow and effective population size in the migratory locust, *Locusta migratoria*: a regional-scale comparative survey. Molecular Ecology 18: 792–800.

49. Peterson MA, Denno RF (1998) Life-history strategies and the genetic structure of phytophagous insect populations, In: Mopper S, Strauss SY, eds. Genetic Structure and Local Adaptation in Natural Insect Population New York, Chapman & Hall. pp 236–322.

50. Arnaud JF (2003) Metapopulation genetic structure and migration pathways in the land snail *Helix aspersa*: influence of landscape heterogeneity. Landscape Ecology 18: 333–346.

51. Sander AC, Purtauf T, Wolters V, Dauber J (2006) Landscape genetics of the widespread ground-beetle *Carabus auratus* in an agricultural region. Basic and Applied Ecology 7: 555–564.

52. Louy D, Habel J, Schmitt T, Assmann T, Meyer M, et al. (2007) Strongly diverging population genetic patterns of three skipper species: the role of habitat fragmentation and dispersal ability. Conservation Genetics 8: 671–681.

53. Cognato AI, Harlin AD, Fisher ML (2003) Genetic structure among pinyon pine beetle populations (Scolytinae: *Ips confusus*). Environmental Entomology 32: 1262–1270.

54. Drès M, Mallet J (2002) Host races in plant-feeding insects and their importance in sympatric speciation. Philosophical Transactions of the Royal Society of London. Series B: Biological Sciences 357: 471–492.

55. Wang ZZ, Wen JB, Yao GL, Zong SX, Luo YQ (2010) Oviposition preference of *Holcocerus hippophaecolus* to different tree species. Journal of Beijing Forestry University 32: 130–135.

56. Janike J (1990) Host specialization in phytophagous insects. Annual Review of Ecology and Systematics 21: 243–273.

57. Bernays EA, Chapman RF (1994) Host-plant selection by phytophagous Insects. New York: Chapman & Hall.

58. Sword GA, Joern A, Senior LB (2005) Host plant-associated genetic differentiation in the snakeweed grasshopper, *Hesperotettix viridis* (Orthoptera: Acrididae). Molecular Ecology 14: 2197–2205.

59. Feder JL, Berlocher SH, Roethele JB, Dambroski H, Smith JJ, et al. (2003) Allopatric genetic origins for sympatric host-plant shifts and race formation in *Rhagoletis*. Proceedings of the National Academy of Sciences of the United States of America 100: 10314–10319.

3

Predicting the Future Impact of Droughts on Ungulate Populations in Arid and Semi-Arid Environments

Clare Duncan[1], Aliénor L. M. Chauvenet[1,2], Louise M. McRae[1], Nathalie Pettorelli[1]*

1 Institute of Zoology, Zoological Society of London, London, United Kingdom, 2 Division of Biology, Imperial College London, Ascot, United Kingdom

Abstract

Droughts can have a severe impact on the dynamics of animal populations, particularly in semi-arid and arid environments where herbivore populations are strongly limited by resource availability. Increased drought intensity under projected climate change scenarios can be expected to reduce the viability of such populations, yet this impact has seldom been quantified. In this study, we aim to fill this gap and assess how the predicted worsening of droughts over the 21st century is likely to impact the population dynamics of twelve ungulate species occurring in arid and semi-arid habitats. Our results provide support to the hypotheses that more sedentary, grazing and mixed feeding species will be put at high risk from future increases in drought intensity, suggesting that management intervention under these conditions should be targeted towards species possessing these traits. Predictive population models for all sedentary, grazing or mixed feeding species in our study show that their probability of extinction dramatically increases under future emissions scenarios, and that this extinction risk is greater for smaller populations than larger ones. Our study highlights the importance of quantifying the current and future impacts of increasing extreme natural events on populations and species in order to improve our ability to mitigate predicted biodiversity loss under climate change.

Editor: Frank Seebacher, University of Sydney, Australia

Funding: The authors have no support or funding to report.

Competing Interests: The authors have declared that no competing interests exist.

* E-mail: Nathalie.Pettorelli@ioz.ac.uk

Introduction

In light of the current global extinction crisis, understanding how and where drivers of population decline will take effect has never been more important [1]. Climate change is expected to be a major driver of species extinctions in the 21st century [2,3]. Average changes in greenhouse gas concentrations are expected to produce directional changes in climatic conditions, and increase the level of inter-annual variability in these conditions [4]. Droughts are a significant component of such climatic variability, and can have a devastating impact on animal populations [5–8]. Through processes such as recurrent reductions in population numbers and the consequent genetic effects caused by demographic bottlenecks [5], droughts have the potential to lead populations, and entire species, to extinction.

In a recent publication, the Intergovernmental Panel on Climate Change (IPCC) reported a likely increase in droughts over the 21st century in various regions of the world, including southern Europe and the Mediterranean, central Europe, central North America, Central America and Mexico, northeast Brazil, and southern Africa [9]. This predicted increase is a potential cause for conservation concern: although the impact of droughts on population size fluctuations has been assessed by many [10–13], the potential impact of expected changes in drought conditions on wildlife populations has almost never been quantified (but see [14]). Moreover, to date no comparative study has been conducted to assess the impact of future changes in drought conditions for species exhibiting contrasting life histories. The risk of extinction of species in response to various threats is

partially shaped by their intrinsic biological characteristics, e.g., body mass, feeding strategy, reproductive strategy, territoriality and home range size [15,16]. Certain life history traits can be expected to make species more susceptible to increased droughts, such as strong dependence on permanent water-sources [17,18]; obligate grazing or mixed feeding (due to whole or partial dependence on drought-intolerant food species [19,20]); sedentary behaviour (due to being unable to escape the effects of drought conditions on resource availability [20–21]). However, how possessing these traits will shape the future susceptibility of populations to changes in climatic conditions is currently unknown. Such information is yet deemed necessary to improve our ability to mitigate predicted biodiversity loss under climate change.

As highlighted by the IPCC 2012 report [9], the need to quantitatively assess how predicted changes in drought conditions are likely to impact wildlife is particularly great for populations inhabiting semi-arid and arid regions. Primary productivity in these environments is already heavily limited by precipitation [22]. As a result, even a slight increase in drought duration, intensity or frequency in these regions has the potential to severely impact resource availability and thus herbivore abundance [12,23–25]. Increased drought conditions may also indirectly impact herbivore populations in these habitats as reduced forage and water availability can lead to increased vulnerability to predation, due to e.g., increased densities at sparse water points [26–27]. This study thus proposes to quantify the impact of future changes in drought conditions on future growth rate of terrestrial ungulate species with contrasted life histories inhabiting arid and semi-arid environments. Based on the knowledge available on ungulate susceptibility to drought [18–23], we hypothesised that population growth rates of species that are grazers or mixed feeders (H1), and that are relatively sedentary (H2) should show a significant negative relationship with drought intensity, and be more negatively impacted by drought intensity than population growth rates of other species groups.

Materials and Methods

Drought Data

An extreme natural event, such as a drought, can be defined as an event that is rare within its statistical reference distribution. It is normally as rare or rarer than the 10th or 90th percentile [4,28], and can be quantified in relation to a specific (possibly impact-related) threshold [9]. However, it is difficult to quantitatively define a drought as there are several different forms (e.g., meteorological drought, agricultural drought and hydrological drought; see [29]). As a result, several different drought indices have been developed [30,31]. One index developed to quantify meteorological drought is the Palmer Drought Severity Index (PDSI; [32]). The PDSI is a common meteorological drought index [33], which has been used to quantify both historical and projected long-term trends in global aridity (see [34]). The PDSI is a standardized index, incorporating both previous and current moisture supply (precipitation) and demand (potential evapotranspiration, PE), which ranges from -10 (dry) to $+10$ (wet).

Palmer's original PDSI was calibrated using fixed coefficients from limited data from the central United States, and to improve spatial comparability several attempts have been made to recalibrate the PDSI index since (see [35]). Self-calibrating PDSI (sc_PDSI) [36] has been found to be more spatially comparable than Palmer's original PDSI, while calculation of PE using the more sophisticated Penman-Monteith equation (PDSI_pm) as opposed to the original Thornwaite equation, has also improved the efficiency of the PDSI [35]. We used global PDSI data (hereafter sc_PDSI_pm) for the years 1970 to 2005. It is calculated using observed or modelled monthly air surface temperature and precipitation, calibrated with Penman-Monteith PE based on historical data and gridded to a 2.5°x2.5° grid ([35,36];http://www.cgd.ucar.edu/cas/catalog/climind/pdsi.html).

Study Species and Population Data

Only ungulate species occurring in arid and semi-arid areas where droughts are predicted to become more common and more intense over the 21st century were considered for this analysis. To identify relevant species, we first compiled ungulate species distribution data (genera Artiodactyla, Perissodactyla, Proboscidea) from the International Union for the Conservation of Nature (IUCN) mammal species dataset (Geographic Information Systems data available at http://www.iucnredlist.org/technical-documents/spatial-data), and then overlaid these species distribution maps with a map of global arid areas of the Köppen-Geiger climate classification ([37]; Geographic Information Systems data available at http://koeppen-geiger.vu-wien.ac.at/shifts.htm).

Time-series abundance data for 1970–2005 for the ungulate populations found in arid and semi-arid areas were collated from the WWF/ZSL Living Planet Index (LPI) database [38]. The database is comprised of yearly abundance data (either population size estimates, population density, relative abundance, biomass, index data, proxy data, samples or measures per unit effort) for vertebrate populations, collated from data in published scientific literature, unpublished reports and online databases. Data is only included in the database if the method of collection or estimation, the geographic location of the population, and the units of measurement are known, and if the data source is referenced and traceable.

As our study focused on the impact of drought on population growth rate, abundance records were only included in the analysis when population estimates were for two consecutive years; i.e. if two abundance records had a one or more years gap between them they were not included. Only species for which the sample size of growth rates over all populations was greater than 20 observations were included in the analysis. Moreover, populations for which the initial starting abundance record was smaller than the average herd size range for that given species were removed in order to ensure that we did not include unstable populations in our analysis. In addition, individual populations for which heavy management practices (e.g. African elephant (*Loxodonta africana*), African buffalo (*Syncerus caffer*), blue wildebeest (*Connochaetes taurinus*) and impala (*Aepyceros melampus*) in Kruger National Park, South Africa), or poaching activities (e.g. African elephant in Ruaha National Park, Tanzania and white rhinoceros (*Ceratotherium simum*) in Garamba National Park, Democratic Republic of Congo) were known to occur over the years of the abundance record (i.e. where references in the literature confirm the presence of such processes) were also excluded in order to eliminate potentially biased abundance data. Then, populations for which estimates were collated from a number of different sources and using a number of different estimation methods were also omitted due to the resulting high level of sampling bias. Finally, populations that contained growth rates higher than that which is physiologically possible for the given species (e.g., higher than possible if all members of the population in a given year were female, each gave birth to the maximum number of offspring possible for that species in one year, and there was no mortality) were not used (n = 4). Such high growth rates are very likely attributable to population increases caused by alternative processes such as immigration, the intentional introduction of individuals into national parks, or incorrect estimates of population size.

The resulting dataset comprised time-series abundance data from 71 populations of 12 ungulate species (Table S1). Most ungulate species currently covered by the LPI database occur in eastern and southern Africa, due to increased sampling effort in these areas. As a result, most of our study species and populations also occur in these regions. The geographic coordinates for each population were taken from the LPI database [38], and monthly sc_PDSI_pm values for each location were collated from the 2.5°x2.5° grid pixel in which the population fell and the years in which they were surveyed.

Calculating Drought Indices

A drought can be described by three axes: duration, frequency, and intensity [39]. Drought duration refers to the timescale of the drought occurrence, e.g., the length of the drought episode. Frequency refers to the average interval (or distance) between drought events at a given location, which can vary between two years in extreme arid regions and 100 years in extremely wet regions [39]. Intensity refers to the extent of the precipitation deficit, and is usually calculated in relation to the duration as the cumulative moisture deficiency across the drought duration [39,40].

For this study, a year is classified as a "drought year" if the sc_PDSI_pm value of at least one month within that year is below the value of the 10th percentile of its statistical reference distribution at a particular location (e.g., threshold θ). Drought intensity then refers to the number of consecutive months within a given "drought year" in which $q < \theta$ (where q is the sc_PDSI_pm value for a given month). In addition, drought frequency (or recurrence interval) is defined as the average distance (in years) between "drought years" across all study locations for each species between 1970 and 2005.

In order to test the impact of changes in drought conditions on the population growth rate of ungulate species in arid and semi-arid areas, we developed four potential predictor variables of drought: the total number of months of the preceding year (T), and the preceding two years (T_{t2}), in which $q<\theta$; the maximum number of consecutive months of the preceding year (C), and the preceding two years (C_{t2}), in which $q<\theta$. The variables C and C_{t2} were developed based on the definition of a drought event provided by Sheffield and Wood [40]. The variables T and T_{t2} were developed in order to create indices of drought that incorporated potential small breaks in drought occurrence, which could not be incorporated under the variables C and C_{t2}. The correlation between our chosen drought index and both annual average PDSI and annual modal PDSI across all study populations was tested using the Spearman's rank correlation. All analyses were carried out in R v. 2.14.2 [41].

Modelling the Future Impact of Drought on Ungulate Populations

Establishing a link between drought indices and growth rates. For all ungulate species considered, we calculated observed population growth rates (r_t) for a given year t between 1970 and 2005 as the change in abundance over time, normalised by the natural logarithm:

$$r_t = \ln\left(\frac{N_{t+1}}{N_t}\right)$$

where N_t is the population size at time t, and N_{t+1} is the population size at time $t+1$.

We then carried out single predictor regressions of observed population growth rate (r_t) against C, T, C_{t2} and T_{t2} using linear mixed effect models with population location and species as random effects in order to identify the best predictor of r_t for all the populations and species considered. This helped us identify the best drought indicator variable of the four.

In order to test our two hypotheses, we then created four subset species groups based on descriptions of individual species diet and movement behaviour in the literature [42]: (1) non-sedentary species (e.g., species that are described as being nomadic, migratory, displaying seasonal movements, or extremely wide-ranging) that wholly or partially depend on drought-intolerant food species (e.g., species that are described as pure grazers or 'mostly' grazers, and mixed feeders) (hereafter MG), (2) non-sedentary species that do not depend on drought-intolerant food species (e.g., species that are described as pure browsers or 'mostly' browsers, and omnivores) (hereafter MB), (3) sedentary species that wholly or partially depend on drought-intolerant food species (hereafter SG), and (4) sedentary species that do not depend on drought-intolerant food species (hereafter SB). We then carried out single predictor regressions of population growth rate (r_t) for each of these groups against the best drought indicator variable; we used linear mixed models with population location and species as random effects.

Model description. For each group of species (MG, MB, SG, and SB) for which our best performing variable of drought intensity was found to be a significant predictor of r_t we developed species-specific stochastic population models to predict the impact of future drought conditions on their viability. Because each species was divided into several populations, for which records of abundance were different, we built the model to project each

population separately, e.g., the initial population size N_0 was different per population. Our model took the form:

$$N_{t+1} = \lambda_t N_t$$

where

$$\lambda_t = \ln(R_t)$$

and

$$R_t = a_t + b_t D_t$$

R_t is the modelled population growth rate at time t. b_t is a coefficient describing the impact of drought conditions (D_t) on the modelled growth rate for each group (MG, MB, SG, and SB): it was estimated using the outputs of the linear mixed effects models for each group of ungulate species. a_t is the average growth rate in the absence of drought: this coefficient was estimated using the observed average growth rate (r_{av}) for each individual species. D_t describes the drought conditions of a given year t and reflects the structure of the best drought indicator variable found above.

While modelling D_t we aimed to reproduce observed drought patterns in our data. To do so, we first determined the average lengths of drought and non-drought episodes per years (i.e., the average number of months in droughts or not in drought per years) across all populations of the species for which we built a stochastic population model. This was to be able to simulate the length (in months) of a drought episode if a given modelled year was in drought. To determine if a modelled year experienced drought or not, we first generated an initial value D_t at time t_0 by comparing a random number sampled from a uniform distribution between 0 and 1, to an 'initial threshold' (Table S2). This initial threshold number was the probability that a given year t would be in drought based on observed drought patterns for each species between 1970 and 2005. If the random number was greater than the threshold number, year t was considered to be not in drought $(D_t=0)$. If the random number was lower than the threshold number, year t was considered to be in drought, and D_t was assigned a value between 1 and 12 to represent a number of months in drought. The value of D_t when in drought was randomly sampled from the observed distribution of our best performing drought indicator variable for the given group of species.

Then, at each time-step (i.e., each simulated year) we assigned a drought or non-drought status to the year based on new 'distance thresholds', representative of observed drought event length (in years) and observed non-drought event length (in years). D_{t+1} was computed by comparing a random number (again from a uniform distribution between 0 and 1) to these 'distance thresholds'; if in the preceding year, at time t, $D_t > 0$, then D_{t+1} would be computed by comparing the random number generated to the 'drought distance threshold' (Figure 1). The drought distance threshold for each group of species is the observed probability for each group that if year t was in drought, then year $t+1$ would also be in drought. However, if at time t, $D_t=0$, then D_{t+1} would be determined by comparing the random number generated to the 'non-drought distance threshold' (Figure 1). Similarly, the non-drought distance threshold is the observed probability for each group of species that if year t was not in drought $(D_t=0)$, then year

t+1 would also not be in drought. If $D_t = 0$ and the random number generated was greater than the non-drought distance threshold, then $D_{t+1} = 0$, or if D_t was >0 and the random number generated was greater than the drought distance threshold then the value of D_{t+1} was again generated from a random sample based on the probability distribution of our best performing drought variable, described above.

To test how well how the stochastic population models performed, we calculated the r-squared values between the observed abundances between 1970 and 2005 and the model-predicted abundances for each population of each species. Moreover, we assessed the ability of the models to reproduce observed patterns in the average length of drought and non-drought events (in years) by performing Wilcoxon signed-rank tests for the 95% confidence intervals of the median for both the observed and model predicted drought and non-drought event lengths (in years).

Simulations. For all groups of species for which a significant relationship was found between growth rates and the best performing drought indicator variable, we used emissions scenario predictions for the region in which each species occurs to train our species-specific model simulations. Three scenarios of future drought occurrence were considered: the first scenario assumed the continuation of current conditions (hereafter 20C). The two other scenarios were based on Sheffield and Wood's projections [40] for the southern African region. Under the second scenario, a doubling of short-term (4–6 months) droughts detectable by 2025 was considered (hereafter B1); under the third scenario, a tripling of short-term droughts detectable by 2040 was considered (hereafter A2) [40]. Changes in occurrence of medium-term (7–11 months) droughts were not investigated by Sheffield and Wood [40]. As we were modelling drought occurrence on a yearly basis (e.g., 1–12 months length), we made the assumption that medium-term droughts would exhibit similar increases as short-term droughts under scenarios B1 and A2, which make our results a cautious underestimate of future drought occurrence.

For each population of each species, we ran a model for which the initial abundance corresponded to that population's first abundance record. Because the model was stochastic, it was run for 5000 simulations in order to generate a large number of

possible population trajectories from which to draw a mean observation. The total number of time-steps for each population was (1) 2005-t_0 (as each population had a different starting year t_0), in order to reproduce observed abundance, and (2) 2099-t_0, in order to model future abundance up to 2099 under different climate change scenarios. A given population was considered extinct when its size was ≤ 5 individuals.

Results

Three (C, T and T_{t2}) of the four derived indices of drought conditions were found to show a significant negative relationship with observed growth rates (r_t) across all study species (C: slope $= -0.01$, $p<0.01$; T: slope $= -0.01$, $p<0.01$; T_{t2}: slope $= -0.01$, $p = 0.03$), while the relationship between the maximum number of consecutive months of drought ($q<\theta$) over the preceding year (C_{t2}) and observed growth rates (r_t) was not significant (slope $= -0.01$, $p = 0.11$). All four drought indices were highly correlated with each other, with C_{t2} showing the lowest correlations with all other variables (Table S3), potentially explaining the non-significance of its relationship with r_t. The two measures of drought intensity over the preceding year (C and T) displayed the most significant relationships with observed growth rates (r_t). Thus, based on previous definitions of drought occurrence in the literature [40], we elected to use C only in further analyses. There was a high degree of correlation between C and the annual average PDSI (rho $= -0.60$, $p<0.001$), and C and the annual modal PDSI (rho $= -0.59$, p<0.001) across all study populations.

As expected (H1 & H2), species with different life histories did not exhibit the same level of susceptibility to drought conditions: when modelling observed growth rates as a function of C for each species group (SB, SG, MB, MG), the only group of species for which C showed a significant negative relationship with growth rates was the group of sedentary species that either wholly or partially depend on drought-intolerant food species (SG group; slope $= -0.04$, $p = 0.001$) (Figure 2; Figure S1). Species that fell within this group were buffalo (*Syncerus caffer*), impala (*Aepyceros melampus*), hartebeest (*Alcelaphus buselaphus*) and waterbuck (*Kobus ellipsiprymnus*), and the population dynamics impacts of future

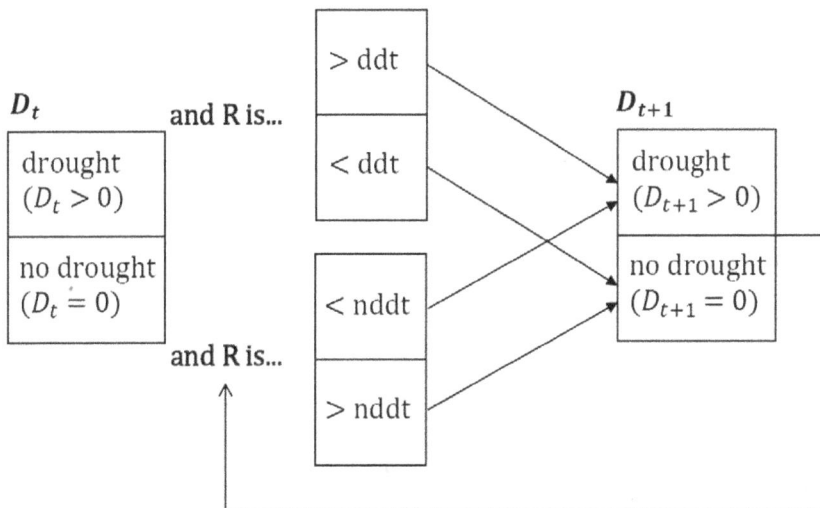

Figure 1. Model generation of D_{t+1} at each time step, where R is a random number sampled from a uniform distribution, and 'ddt' and 'nddt' correspond to the 'drought distance threshold' and the 'non-drought distance threshold' respectively.

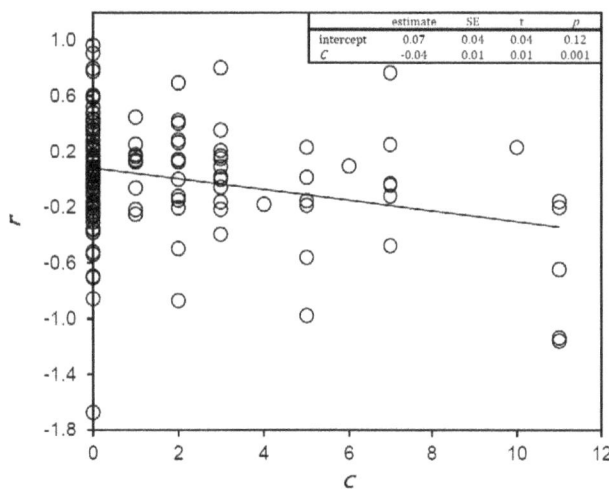

	estimate	SE	t	p
intercept	0.07	0.04	0.04	0.12
C	-0.04	0.01	0.01	0.001

Figure 2. Number of consecutive months of the preceding year in which $q<\theta$ (C) as a predictor of growth rates (r) for all sedentary, grazing or mixed feeding species (4 species, n = 148).

increases in drought conditions were therefore only investigated for these four species.

The stochastic population models showed a certain level of heterogeneity in their ability to mimic observed abundances across individual species and across individual populations. For example, our model explained 70% of the variance in abundance of buffalo in the Serengeti-Mara ecosystem and in Addo Elephant Park, but only 20% in the Narok District; it explained 36% of variance in abundance of waterbuck in Malilangwe Wildlife Reserve but only 2% in Kruger National Park (Table 1). However, observed average drought (observed: pseudomedian = 2.0, lower 95% Confidence Interval = 1.5, upper 95% CI = 2.0; predicted: pseudomedian = 2.0, lower 95% CI = 1.5, upper 95% CI = 2.0) and non-drought episode lengths (observed: pseudomedian = 3.0, lower interval = 2.5, upper 95% CI = 4.0; predicted: pseudomedian = 3.0, lower 95% CI = 2.5, upper 95% CI = 3.5) were well replicated.

Population projections for the four species in group SG showed that average population growth rates (λ) decreased, while extinction probabilities (E) and the variation in average growth rates steadily increased, when successively considering scenario 20C, B1 and A2 (Table 2; Table S4). There was however little variation in projected average growth rates and extinction risks between scenarios B1 and A2, likely due to the late onset of increases in drought intensity. Waterbuck populations had extremely low negative average growth rates (λ) even under

continued current conditions (scenario 20C), with the average chance of going extinct at the end of this century being 100%. Contrastingly, buffalo and impala displayed positive projected average growth rates, with both species showing a negligible drought-related average risk of extinction under all scenarios (Table 2). Hartebeest populations showed negative projected average growth rates under all scenarios, with a relatively high risk of extinction (51.1%) under continued current conditions which increased to 66.4% and 69.1% under scenarios B1 and A2 respectively (Table 2). In addition, our results also showed that smaller populations of all species will be put at higher risk of extinction from increasing future drought occurrence than larger populations (Table S4 and Table S5).

Discussion

Investigations into the potential future impacts of climate change on biodiversity mostly consider directional changes in environmental conditions (see e.g. [43,44]), and studies into the effects of climate extremes are few at present. Here we provide research to help close this gap, and present a model framework to quantitatively assess the impact of potential future increases in such highly variable and devastating events. Our study shows that future climate change will negatively impact certain ungulate species in arid and semi-arid environments, dramatically increasing extinction risk from drought occurrence for some of them. Our results also provide support to the hypotheses that the species most at risk from increasing future drought intensity are those that are relatively sedentary, and that are wholly or partially dependent on drought-intolerant food species (e.g., grazers and mixed feeders).

Our findings that sedentary ungulate species (as opposed to nomadic or migratory species), which are dependent on drought-intolerant food species (as opposed to browsers or omnivorous species) are more at risk from current and future drought conditions are in line with frequent reports in the literature of species exhibiting these life history traits suffering high mortality during individual drought events [13,20–22]. Our results suggest that at present the frequency and intensity of drought occurrence is sufficiently low that it is not inflicting a significantly negative impact on populations that are currently able to escape the effects of resource depletion in dry conditions. However, as drought intensity increases under future climate change its impact on such species may become significant.

The results of the predictive stochastic population models show that hartebeest and waterbuck will be put at extremely high risk from future increases in drought intensity under climate change over this century. Conversely, the chance of buffalo and impala populations going extinct will remain low. Those patterns are likely to reflect reported general population trends for these species

Table 1. R-squared values (and respective standard deviations) between observed and modelled abundance for all populations of sedentary grazer species.

Species	Addo	Karoo	Kruger	Lewa	Malilangwe	Mountain Zebra	Narok District	Serengeti-Mara	Umfolozi
buffalo	0.70 (0.29)	–	–	0.38 (0.16)	–	0.49 (0.30)	0.20 (0.22)	0.70 (0.15)	–
hartebeest	0.49 (0.27)	0.56 (0.29)	–	0.16 (0.17)	0.46 (0.25)	0.25 (0.17)	0.40 (0.23)	–	–
impala	–	–	–	0.31 (0.23)	0.05 (0.08)	–	0.37 (0.30)	–	0.37 (0.26)
waterbuck	–	–	0.02 (0.02)	0.03 (0.02)	0.37 (0.12)	–	0.34 (0.16)	–	0.19 (0.05)

Table 2. Modelled average growth rate (λ) and mean extinction probability (E; in %) to 2099 across all populations of each sedentary grazer species under model scenarios 20C, B1 and A2. SD stands for standard deviation.

scenario	Buffalo λ (SD)	E	Hartebeest λ (SD)	E	Impala λ (SD)	E	Waterbuck λ (SD)	E
20C	1.05 (0.08)	0.00	0.98 (0.08)	51.1	1.00 (0.08)	0.0004	0.92 (0.07)	1.00
B1	1.04 (0.10)	0.0001	0.97 (0.09)	66.1	0.99 (0.09)	0.01	0.92 (0.08)	1.00
A2	1.04 (0.10)	0.00	0.97 (0.09)	69.1	0.99 (0.09)	0.01	0.92 (0.07)	1.00

throughout their global ranges, as well as trends occurring at the level of the study populations. In fact, population trends for hartebeest, buffalo and waterbuck across their global ranges are reported by the IUCN to be generally decreasing at present [45–47]. In particular, according to our simulations, waterbuck have an extremely high probability of extinction and low average population growth rates even under continued present conditions. While reported decline in the species is often largely due to poaching [46], the species is also one of the most water-dependent of African ungulate species and has a high-protein dietary requirement [48]. As such it is extremely susceptible to drought conditions, explaining our simulations' results. The positive average growth rates and low risk of extinction displayed by buffalo under all scenarios, however, does not necessarily show an immunity of this species to drought conditions. Instead, the impact of drought on buffalo population dynamics could be dampened, or even counteracted, by alternative processes. For example, all study populations occur in large national parks where populations are doing well, with an increasing population trend in Addo National Park, South Africa, Mountain Zebra National Park, South Africa and Lewa Nature Conservancy, Kenya since 1991, 2002 and 1990 respectively [49,50]. In addition, the only years in which abundance data are available for two consecutive years for the population in the Serengeti-Mara Ecosystem, Tanzania occur over 1970–1977 when the population was undergoing a rapid come-back following the eradication of the rinderpest disease in the region [51]. As a result, our population projections for buffalo under the three scenarios over the 21st century may be considerably modest.

The results presented here do have some limitations. First, sample size was small for some of our study species due to populations being omitted for reasons listed above. Second, in the absence of data regarding potential alternative processes that may be acting on the study populations, the main assumption of this work is that drought intensity is the sole process acting on growth rates. However, management practices could also be acting on the population dynamics of studied populations, as many of them occur within protected areas and national parks and management actions are often listed as 'unknown' within the LPI database. For example, water- or food-provisioning [52–54], culling [21] or fencing [55–57] may serve to either increase or decrease the resilience of ungulate populations to drought occurrence. Similarly, processes such as poaching (particularly in the case of the black rhinoceros; [58]), predation [54,59], dispersal [13,60], disease outbreaks [61,62], or the impacts of other extreme natural events (e.g., wild fire or flooding; [5]), which also have an effect on some ungulate populations, cannot be accounted for in our data. The influence of such processes could be buffering the impact of the effect of drought on populations of our study species. Third, estimation methods of yearly abundance within our dataset also differed between individual populations of our study species, with

some methods resulting in coarse resolution data (e.g., scaling-up from walking transects), which may similarly have buffered the effect of drought occurrence on these populations. Individual population size estimates may also suffer some degree of inaccuracy due to the difficult nature of obtaining counts for game species [63–66]. Finally, the population models built to project species' abundance under future climate change scenarios remain relatively simple. For example, they ignore processes such as density-dependence, predation or the difference in survival and reproduction rates of individuals of different sexes and ages, which are known to impact population dynamics [67–69]. While our results provide a good first model of variation in future population abundance in response to drought, we acknowledge that additional more complex studies will be required in order to gain a complete understanding of the potential future impact of increases in drought intensity under climate change on the persistence of ungulate populations.

In the face of our results, should highly sedentary, grazing and mixed feeding ungulate species be targeted by park authorities in the future for management in drought years? Some such species are clearly highly susceptible, but potential management strategies for their benefit may in reality come at a cost to them and to other ungulate species in arid and semi-arid environments. Studies have found that provisioning with artificial water-holes can enable less sedentary species to expand their ranges within national parks and can promote increases in these populations, in turn resulting in heightened ungulate densities at such water points, and negatively impacting rarer sedentary species such as the waterbuck, roan antelope (*Hippotragus equinus*) and tsessebe (*Damaliscus lunatus*) through resource exhaustion and increased predation [53,54,70]. This effect has also been found to extend to the predators of ungulate species, with lions (*Panthera leo*) in the Kruger National Park, South Africa, benefitting from higher prey densities around artificial water-points and causing competitive exclusion of the less common brown hyena (*Hyaena brunnea*) [71]. Indeed, the cause for caution in determining effective spacing of artificial water-points for the maintenance of ungulate abundance and diversity, and ecosystem heterogeneity both within and outside protected areas has been highlighted extensively in the literature [53,72,73]. Such provisioning can also disrupt movement patterns of migratory ungulate species, and result in heightened inter-specific competition and die-off populations of these in dry years [20]. This raises the issue of whether the risk of increasing drought severity and frequency over the 21st century [9] may potentially further exacerbate the effect of artificial water-holes on both the competitive exclusion of rarer, more sedentary ungulate species and the dry season survival of migratory species. Climate change is predicted to alter the timing of migrations and the migration routes of terrestrial mammals, through altering the distribution of forage and surface-water availability [74–76]. Hence under future increases in drought intensity and frequency, both wet and dry season ranges of migratory ungulate species may become less able to support these populations and may be altered, which could have serious implications for the future conservation of such species. Fencing around national parks can have a severe impact on the survival of the migratory populations within, disrupting migratory pathways and heightening the impact of drought through disabling such populations to access their dry season ranges [55–57]. Under increasing future drought intensity, national parks should focus on the effective spacing of artificial water-points and on enabling greater connectivity for migratory ungulates. Such management strategies would assist in limiting the negative impacts of water-provisioning on both sedentary and migratory populations.

Altogether, our work illustrates that climate change and increased drought conditions could lead to the extinction of certain populations over the 21st century. Our findings provide further evidence that increasing future drought conditions will pose a greater risk to ungulate species that are highly sedentary, and that are wholly or partially dependent on drought-intolerant food species. Although none of our study species are threatened species and have been classified by the International Union for the Conservation of Nature (IUCN) as 'Least Concern' [45–47,77], our findings may have implications for some other highly threatened ungulate species and subspecies in areas where drought intensity is predicted to increase over the 21st century, such as Mountain zebra (*Equus zebra*), European bison (*Bison bonasus*), Cuvier's gazelle (*Gazella cuvieri*) and Tora hartebeest (*Alcelaphus buselaphus tora*) [78–81]. In addition, these drought-intolerant life history traits are also likely to cause enhanced susceptibility to increasing future drought intensity of certain species in other taxonomic groups. Our study clearly stresses the importance of long-term monitoring in order to provide a basis on which to explore the impacts of extreme natural events on animal populations under future climate change. Future studies should be conducted in order to determine the susceptibility of species in differing environments and taxonomic groups, particularly threatened species and small or isolated populations, to these increased climate extremes, in order to develop appropriate management strategies.

Supporting Information

Figure S1   *C* as a predictor of growth rates *r* for all (a) sedentary, browsing species and (b) migratory or nomadic, grazing or mixed-feeding species. *C* is the maximum number of consecutive months of the preceding year in which $q<\theta$. For sedentary browsing species, there were six species, n = 373, slope = −0.01, p = 0.20. For migratory or nomadic, grazing or mixed feeding species, there were 2 species, n = 165, slope = 0.001, p = 0.88.

References

1. Ceballos G, Ehrlich PR (2002) Mammal population losses and the extinction crisis. Science 296: 904–907.
2. Millennium Ecosystem Assessment (2005) Ecosystems and human well-being: biodiversity synthesis. Washington, D.C., USA: World Resources Institute. 86 p.
3. Foden W, Mace GM, Vié J-C, Angulo A, Butchart S, et al. (2008) Species susceptibility to climate change impacts. In: Vié J-C, Hilton-Taylor C, Stuart SN, editors. The 2008 review of the IUCN Red List of Threatened Species. Gland, Switzerland: IUCN/SSC. 77–88.
4. Intergovernmental Panel on Climate Change (2007) Climate change 2007: The physical science basis. Cambridge, United Kingdom and New York, USA: Cambridge University Press. 996 p.
5. Young TP (1994) Natural die-offs of large mammals: implications for conservation. Conserv Biol 8: 410–418.

Table S1   Ungulate species and populations used in this study (WWF/ZSL 2012).

Table S2   Probabilities under current conditions of a given year being in drought, and of different lengths of drought occurring if that's the case, for sedentary grazing or mixed feeding species. (a) Observed probabilities under current conditions of a given year being in drought (initial threshold; $C>0$) and not in drought ($C=0$) and that if one year is a drought year that the following year will also be a drought year (the drought distance threshold; ddt) or if one year is a non-drought year that the following year will also be a non-drought year (the non-drought distance threshold; nddt), across all populations of all sedentary grazing or mixed feeding species. (b) Observed probabilities under current conditions that if a given year is a drought year, C will be equal to each value between 1 and 12, across all populations of all sedentary grazing or mixed feeding species.

Table S3   Spearman's rank correlation analysis between the four predictor variables of drought intensity (C, T, C_{t2} and T_{t2} all $p<0.001$).

Table S4   Modelled average growth rate (λ) and extinction probability (E) of all populations of sedentary grazer species to 2099 under scenarios 20C, B1 and A2. SD stands for standard deviation.

Table S5   Relative starting abundance of populations of sedentary, grazing species (SG; relative abundance here is initial population abundance at year *t* in comparison to the initial abundance at year *t* for all populations of that species). Due to confidential data sources, raw abundance values cannot be published.

Acknowledgments

We are grateful to Stefanie Deinet for her help collating the LPI ungulate species data. We would like to thank the two anonymous reviewers for their help in improving the manuscript.

Author Contributions

Conceived and designed the experiments: NP ALMC. Performed the experiments: CD ALMC NP. Analyzed the data: CD ALMC NP. Contributed reagents/materials/analysis tools: LMM. Wrote the paper: CD NP ALMC.

6. Saltz D (2002) The dynamics of equid populations. In: Moehlmann P, IUCN/SSC Equid Specialist Group, editors. Zebra, asses, and horses: an action plan for the conservation of wild equids. Gland, Switzerland: IUCN/SSC. 110 p.
7. Epps CW, McCullough DR, Wehausen JR, Bleich VC, Rechel JL (2004) Effects of climate change on population persistence of desert-dwelling mountain sheep in California. Conserv Biol 18: 102–113.
8. Foley C, Pettorelli N, Foley L (2008) Severe drought and calf survival in elephants. Biol Lett 4: 541–4.
9. Intergovernmental Panel on Climate Change (2012) Managing the risks of extreme events and disasters to advance climate change adaptation. A special report of Working Groups I and II of the Intergovernmental Panel on Climate Change. Cambridge, United Kingdom and New York, USA: Cambridge University Press. 582 p.

10. Armbruster P, Lande R (1993) A population viability analysis of African elephant (*Loxodonta africana*): how big should reserves be? Conserv Biol 7: 602–610.

11. Owen-Smith N, Mason DR, Ogutu JO (2005) Correlates of survival rates for 10 African ungulate populations: density, rainfall and predation. J Anim Ecol 74: 774–788.

12. Ogutu JO, Piepho HP, Dublin HT, Bhola N, Reid RS (2008) Rainfall influences on ungulate population abundance in the Mara-Serengeti ecosystem. J Anim Ecol 77: 814–829.

13. Augustine DJ (2010) Response of native ungulates to drought in semi-arid Kenyan rangeland. Afr J Ecol 48: 1009–1020.

14. Saltz D, Rubenstein DI, White GC (2006) The impact of increased environmental stochasticity due to climate change on the dynamics of Asiatic wild ass. Conserv Biol 20: 1402–1409.

15. Isaac NJB, Cowlishaw G (2004) How species respond to multiple extinction threats. Philos Trans R Soc Lond B Biol Sci 271: 1135–1141.

16. Collen B, McRae LM, Deinet S, De Palma A, Carranza T, et al. (2011) Predicting how populations decline to extinction. Philos Trans R Soc Lond B Biol Sci 366: 2577–86.

17. Dunham KM (1994) The effect of drought on the large mammal populations of the Zambezi riverine woodlands. J Zool 234: 489–526.

18. Kay RNB (1997) Responses of African livestock and wild herbivores to drought. J Arid Environ 37: 683–694.

19. Hillman JC, Hillman AKK (1977) Mortality of wildlife in Nairobi National Park, during the drought of 1973–1974. Afr J Ecol 15: 1–18.

20. Knight MH (1995) Drought-related mortality of wildlife in the southern Kalahari and the role of man. Afr J Ecol 33: 377–394.

21. Walker BH, Emslie RH, Owen-Smith RN, Scholes RJ (1987) To cull or not to cull: Lessons from a southern African drought. J Appl Ecol 24: 381–401.

22. Rutherford MC (1980) Annual plant production-precipitation relations in arid and semi-arid regions. S Afr J Sci 76: 53–56.

23. Ogutu JO, Owen-Smith N (2003) ENSO, rainfall and temperature influences on extreme population declines among African savanna ungulates. Ecol Letts 6: 412–419.

24. Illius AW, O'Connor TG (2000) Resource heterogeneity and ungulate population dynamics. Oikos 89: 283–294.

25. Georgiadis N, Hack M, Turpin K (2003) The influence of rainfall on zebra population dynamics: implications for management. J Appl Ecol 40: 125–136.

26. Loveridge AJ, Hunt JE, Murindagomo F, Macdonald DW (2006) Influence of drought on predation of elephant (*Loxodonta africana*) calves by lions (*Panthera leo*) in an African wooded savannah. J Zool 270: 523–530.

27. Owen-Smith N, Mills MGL (2006) Manifold interactive influences on the population dynamics of a multispecies ungulate assemblage. Ecol Monogr 76: 73–92.

28. Intergovernmental Panel on Climate Change (2001) Climate change 2001: The scientific basis. Cambridge, United Kingdom and New York, USA: Cambridge University Press. 881 p.

29. AMS (American Meteorological Society) (1997) Meteorological drought-policy statement. B Am Meteorol Soc 78: 847–849.

30. Heim RR (2000) Drought indices: a review. In: Wilhite DA, editor. Drought: a global assessment. Oxford, United Kingdom: Taylor & Francis. 159–167.

31. Keyantash J, Dracup JA (2002) The quantification of drought: an evaluation of drought indices. B Am Meteorol Soc 83: 1167–1180.

32. Palmer WC (1965) Meteorological drought. Washington, D.C., USA: Weather Bureau Research Paper No. 45.

33. Heim RR (2002) A review of twentieth-century drought indices used in the United States. B Am Meteorol Soc 83: 1149–1165.

34. Dai A (2011) Drought under global warming: a review. Wiley Interdiscip Rev Clim Change 2: 45–65.

35. Dai A (2011) Characteristics and trends in various forms of the Palmer Drought Severity Index during 1900–2008. J Geophys Res 116: 1–26.

36. Wells N, Goddard S, Hayes MJ (2004) A self-calibrating Palmer Drought Severity Index. J Clim 17: 2335–2351.

37. Rubel F, Kottek M (2010) Observed and projected climate shifts 1901–2100 depicted by world maps of the Köppen-Geiger climate classification. Meteorol Z 19: 135–141.

38. WWF/ZSL (2012) The Living Planet Database. Available: www.livingplanetindex.org. Accessed 2012 Jul 26.

39. Ponce VM, Pandey RP, Ercan S (2000) Characterization of drought across climatic spectrum. J Hydrol Eng 5: 222–224.

40. Sheffield J, Wood EF (2008) Projected changes in drought occurrence under future global warming from multi-model, multi-scenario, IPCC AR4 simulations. Clim Dynam 31: 79–105.

41. Team RDC (2012) R: a language and environment for statistical computing, Vienna, Austria. Available: www.R-project.org. Accessed 2012 Feb 29.

42. Wilson DE, Mittermeier RA (2011) Handbook of the mammals of the world - volume 2: hoofed mammals. Barcelona, Spain: Lynx Edicions. 885 p.

43. Erasmus BFN, van Jaarsveld A, Chown SL, Kshatriya M, Wessels KJ (2002) Vulnerability of South African animal taxa to climate change. Glob Change Biol 8: 679–693.

44. Thuiller W, Broennimann O, Hughes G, Alkemade JRM, Midgley GF, et al. (2006) Vulnerability of African mammals to anthropogenic climate change under conservative land transformation assumptions. Glob Change Biol 12: 424–440.

45. IUCN/SSC Antelope Specialist Group (2008a) *Alcelaphus buselaphus*. In: IUCN, editors. IUCN Red List of Threatened Species. Version 2011.2. Available: www.iucnredlist.org. Accessed 2012 May 21.

46. IUCN/SSC Antelope Specialist Group (2008b) *Kobus ellipsiprymnus*. In: IUCN, editors. IUCN Red List of Threatened Species. Version 2011.2. Available: www.iucnredlist.org. Accessed 2012 May 21.

47. IUCN/SSC Antelope Specialist Group (2008c) *Syncerus caffer*. In: IUCN, editors. IUCN Red List of Threatened Species. Version 2011.2. Available: www.iucnredlist.org. Accessed 2012 May 21.

48. Spinage CA (1982) A territorial antelope: the Uganda waterbuck. London, UK: Academic Press. 334 p.

49. Chege G, Kisio E (2008) Lewa Wildlife Conservancy Research and Monitoring Annual Report. Nairobi, Kenya: Lewa Wildlife Conservancy. 27 p. Available: www.lewa.org. Accessed 2012 Oct 10.

50. South African National Parks (2008) South African National Parks Data Repository, SANParks. Available: www.sanparks.org. Accessed 2012 Oct 10.

51. Sinclair ARE, Mduma SAR, Hopcraft JGC, Fryxell JM, Hilborn R, et al. (2007) Long-term ecosystem dynamics in the Serengeti: lessons for conservation. Conserv Biol 21: 580–590.

52. van Dierendonck MC, Wallis de Vries MF (1996) Ungulate reintroductions: experiences with the Takhi or Przewalski horse (*Equus ferus przewalskii*) in Mongolia. Conserv Biol 10: 728–740.

53. Owen-Smith N (1996) Ecological guidelines for waterpoints in extensive protected areas. S Afr J Wildl Res 26: 107–112.

54. Harrington R, Owen-Smith N, Viljoen PC, Biggs HC, Mason DR, et al. (1999) Establishing the causes of the roan antelope decline in the Kruger National Park, South Africa. Biol Conserv 90: 69–78.

55. Ben-Shahar R (1993) Does fencing reduce the carrying-capacity for populations of large herbivores? J Trop Ecol 9: 249–253.

56. Boone RB, Hobbs NT (2004) Lines around fragments: effects of fencing on large herbivores. Afr J Range Forage Sci 21: 147–158.

57. Mbaiwa BYJE, Mbaiwa OI (2006) The effects of veterinary fences on wildlife populations in Okavango Delta, Botswana. Methods 12: 17–24.

58. Emslie R (2011) *Diceros bicornis*. In: IUCN, editors. IUCN Red List of Threatened Species. Version 2012.1. Available: www.iucnredlist.org. Accessed 2012 Jul 2.

59. Gasaway WC, Gasaway KT, Berry HH (2006) Persistent low densities of plains ungulates in Etosha National Park, Namibia: testing the food-regulating hypothesis. Can J Zool 74: 1556–1572.

60. Verlinden A (1998) Seasonal movement patterns of some ungulates in the Kalahari ecosystem of Botswana between 1990 and 1995. Afr J Ecol 36: 117–128.

61. Joly DO, Messier F (2004) Testing hypotheses of bison population decline (1970–1999) in Wood Buffalo National Park: synergism between exotic disease and predation. Can J Zool 82: 1165–1176.

62. Jolles AE, Cooper DV, Levin SA (2005) Hidden effects of chronic tuberculosis in African buffalo. Ecology 86: 2358–2364.

63. Bouché P, Lejeune P, Vermeulen C (2012) How to count elephants in West African savannahs? Synthesis and comparison of main game count methods. Biotechnol Agron Soc 16, 77–91.

64. Brockett BH (2002) Accuracy, bias and precision of helicopter-based counts of black rhinoceros in Pilanesberg National Park, South Africa. S Afr J Wildl Res 32, 121–134.

65. Jachmann H (2001) Estimating abundance of African wildlife: an aid to adaptive management. Boston, USA: Kluewer Academic Publishers. 285 p.

66. Magin CD (1989) Variability in total ground counts on a Kenyan game ranch. Afr J Ecol 27, 297–303.

67. Coulson T, Catchpole EA, Albon SD, Morgan BJT, Pemberton JM, et al. (2001) Age, sex, density, winter weather, and population crashes in Soay sheep. Science 292: 1528–1531.

68. Leirs H, Stenseth NC, Nichols JD, Hines JE, Verhagen R, et al. (1997) Stochastic seasonality and nonlinear density-dependent factors regulate population size in an African rodent. Nature 389: 178–180.

69. Clutton-Brock TH (1988) Reproductive success- studies of individual variation in contrasting breeding systems. Chicago: University of Chicago Press. 548 p.

70. Ogutu JO, Owen-Smith N, Piepho HP, Kuloba B, Edebe J (2012) Dynamics of ungulates in relation to climatic and land use changes in an insularized African savanna ecosystem. Biodivers Conserv 21: 1033–1053.

71. Gaylard A, Owen-Smith N, Redfern J (2003) Surface water availability: implications for heterogeneity and ecosystem processes. In: Du Toit JT, Biggs H, Rogers KH, editors. The Kruger experience: ecology and management of savanna heterogeneity. Washington, D.C., USA: Island Press. 171–188.

72. Du Toit JT, Cumming DHM (1999) Functional significance of ungulate diversity in African savannas and the ecological implications of the spread of pastoralism. Biodivers Conserv 8: 1643–1661.

73. de Leeuw J, Waweru MN, Okello OO, Maloba M, Nguru P, et al. (2000) Distribution and diversity of wildlife in northern Kenya in relation to livestock and permanent water points. Biol Conserv 100: 297–306.

74. UNEP/CMS (2006) Migratory species and climate change: impacts of a changing environment on wild animals. Bonn, Germany: UNEP/CMS Secretariat. 63 p.

75. Berger J (2004) The last mile: how to sustain long-distance migration in mammals. Conserv Biol 18: 320–331.

76. Robinson R, Crick HQP, Learmonth JA, Maclead IMD, Thomas CD, et al. (2009) Travelling through a warming world: climate change and migratory species. Endanger Species Res 7: 87–99.

77. IUCN/SSC Antelope Specialist Group (2008d) *Aepyceros melampus*. In: IUCN, editors. IUCN Red List of Threatened Species. Version 2011.2. Available: www.iucnredlist.org. Accessed 2012 May 21.

78. IUCN/SSC Antelope Specialist Group (2008e) *Alcelaphus buselaphus ssp. tora*. In: IUCN, editors. IUCN Red List of Threatened Species. Version 2012.1. Available: www.iucnredlist.org. Accessed 2012 Jul 2.

79. Mallon DP, Cuzin F (2008) *Gavella cuvieri*. In: IUCN, editors. IUCN Red List of Threatened Species. Version 2012.1. Available: www.iucnredlist.org. Accessed 2012 Jul 18.

80. Novellie P (2008) *Equus zebra*. In: IUCN, editors. IUCN Red List of Threatened Species. Version 2012.1. Available: www.iucnredlist.org. Accessed 2012 Jul 2.

81. Olech W (2008) *Bison bonasus*. In: IUCN, editors. IUCN Red List of Threatened Species. Version 2012.1. Available: www.iucnredlist.org. Accessed 2012 Jul 18.

4

Predicted Disappearance of *Cephalantheropsis obcordata* in Luofu Mountain Due to Changes in Rainfall Patterns

Xin-Ju Xiao[1,2,9], Ke-Wei Liu[1,3,9], Yu-Yun Zheng[1], Yu-Ting Zhang[1], Wen-Chieh Tsai[4], Yu-Yun Hsiao[5], Guo-Qiang Zhang[1], Li-Jun Chen[1]*, Zhong-Jian Liu[1,3,6]*

1 Shenzhen Key Laboratory for Orchid Conservation and Utilization, The National Orchid Conservation Center of China/The Orchid Conservation & Research Center of Shenzhen, Shenzhen, China, 2 Continuing Education College of Beijing Forestry University, Beijing, China, 3 The Center for Biotechnology and BioMedicine, Graduate School at Shenzhen, Tsinghua University, Shenzhen, China, 4 Institute of Tropical Plant Sciences, and Orchid Research Center, National Cheng Kung University, Taiwan, China, 5 Department of Life Sciences, National Cheng Kung University, Taiwan, China, 6 College of Forestry, South China Agricultural University, Guangzhou, China

Abstract

Background: In the past century, the global average temperature has increased by approximately 0.74°C and extreme weather events have become prevalent. Recent studies have shown that species have shifted from high-elevation areas to low ones because the rise in temperature has increased rainfall. These outcomes challenge the existing hypothesis about the responses of species to climate change.

Methodology/Principal Findings: With the use of data on the biological characteristics and reproductive behavior of *Cephalantheropsis obcordata* in Luofu Mountain, Guangdong, China, trends in the population size of the species were predicted based on several factors. The response of *C. obcordata* to climate change was verified by integrating it with analytical findings on meteorological data and an artificially simulated environment of water change. The results showed that *C. obcordata* can grow only in waterlogged streams. The species can produce fruit with many seeds by insect pollination; however, very few seeds can burgeon to become seedlings, with most of those seedlings not maturing into the sexually reproductive phase, and grass plants will die after reproduction. The current population's age pyramid is kettle-shaped; it has a Deevey type I survival curve; and its net reproductive rate, intrinsic rate of increase, as well as finite rate of increase are all very low. The population used in the artificial simulation perished due to seasonal drought.

Conclusions: The change in rainfall patterns caused by climate warming has altered the water environment of *C. obcordata* in Luofu Mountain, thereby restricting seed burgeoning as well as seedling growth and shortening the life span of the plant. The growth rate of the *C. obcordata* population is in descending order, and models of population trend predict that the population in Luofu Mountain will disappear in 23 years.

Editor: Ben Bond-Lamberty, DOE Pacific Northwest National Laboratory, United States of America

Funding: This work was supported in part by grants from the Forestry Construction of State Forestry Administration of China (No. 2010–240) and the Forestry Construction of Guangdong Province, China (No. 2010–317). The funders had no role in study design, data collection and analysis, decision to publish, or preparation of the manuscript.

Competing Interests: The authors have declared that no competing interests exist.

* E-mail: liuzj@sinicaorchid.org (Z-JL); chenlj@sinicaorchid.org (L-JC)

9 These authors contributed equally to this work.

Introduction

The past century has witnessed an increase of approximately 0.74°C in the global average temperature and extreme weather events. Many predictions have been made about the responses of species to climate change [1–8], such as their non-response or local extinction, disappearance or reduction in suitable distribution ranges, shift of expansion or distribution ranges poleward or toward higher elevations to inhabit areas within their metabolic temperature tolerances, and variations in phenology and behavior, even in genes. Recent studies have shown that species have shifted to low-elevation areas from high ones because the rise in temperature has increased rainfall [9]. As these outcomes challenge the hypothesis about the perceived responses of species

to climate change, the capability of wild species to respond to climate warming, especially to the accompanying extreme weather events, should be investigated.

Orchids usually thrive in climax communities and are very sensitive to environmental changes [10]. *Cephalantheropsis obcordata* belongs to the family Orchidaceae and has special requirements for habitation. It exists in South Fujian, Guangdong, Hainan, Taiwan, and the southern and southeastern parts of Yunnan in China, as well as in Japan, India, Indonesia, Laos, Malaysia, Myanmar, the Philippines, Thailand, and Vietnam [11]. In China, it grows in moist sloping fields where stones have humus soil beside streams in climax subtropical forest communities with higher canopy density [11]. Climate changes directly influence the existence and development of *C. obcordata*, rendering the effects of global warming on the population growth and decline of this species an important area of study.

Much research into the population development trends of endangered species, the establishment of conservation strategy and measures, and the reasonable utilization of current resources by analyzing population quantities has been carried out [12–16], but data on how climate change specifically affects orchid plants, particularly the trends in their population development, are limited [17,18]. With the use of data on the biological characteristics and pollination biology of *C. obcordata*, this current study predicted the dynamic population trends of the species based on its static life and reproduction tables, survival curve, and age pyramid. It also estimated changes in those trends using the Leslie matrix model and the Levins model. This report discusses the spatial pattern, age pyramid, breeding strategy, and population size trends of *C. obcordata* in light of its extinction vulnerability from global warming. The data highlight the need to protect this species while integrating climate change considerations.

Results

Biological characteristics of *C. obcordata*

Growth characteristics and space–time mark. *C. obcordata* is a proximate hygrophyte that grows on shadowly wooded slopes with deep loose humus soil often soaked with water in mountain valleys (Figure 1). Its genet has rhizomes that can grow a number of ramets. New ramets grow from the base of the youngest ramet in early May each year, and a chain-like plant forms year after year. Seedlings need 3 years of vegetative growth to blossom, which usually occurs in the fourth year, followed by sexual reproduction. Floral bud differentiation begins in August, florescence starts in late September and ends in December, and fruiting occurs between November and March. The fruit will split and scatter seeds at maturity, each seed will produce protocorm in humus soil and grow a short stem in the coming year, and a bud will grow from the stem base and then become a new ramet by the following year. Each genet grows only one new ramet annually, representing an obvious space–time mark. The flowering rate of sexually mature genets was $30.24\% \pm 9.61\%$ in this study ($n = 6$).

Spatial pattern. In sample grids of *C. obcordata*, the occurrence frequencies of a square with very few individuals and that with many individuals were higher than the expected values in Poisson distribution, that is, $S^2/\overline{m} = 1.35$, $[\overline{m} = \sum f(x)/n, S^2 = \left(\sum (f(x))^2 - ((\sum f(x))^2/n)\right)/(n-1)]$, which was clearly higher than 1. The spatial structure of *C. obcordata* was thus of clumped distribution [19]. Such spatial pattern is correlated with its habitat adaptability and seed scattering. *Habitat adaptability* refers to the species' requirement for waterlogging in its habitat — such habitat exists only in the buffer section of streams and hills with water seepage, which characterize the spatial distribution pattern of *C. obcordata*; the species hence usually has concentrated growth in narrow places and mutually forms different maculated populations (Figure 1A) that are spatially separated. On the other hand, *seed scattering*, as the term itself suggests, refers to the scattering of seeds that burgeon into seedlings and observe a clumped distribution near the limited habitat of the species.

Age pyramid. Statistical data on each genet by age class revealed the following: the genets from each age class were asymmetrically distributed; the number of genets from each age class that advanced into an upper class greatly differed from the number in lower classes; the proportion of young genets was smaller, and the proportion of aged ones was the smallest, although the proportion of adult genets was very high; the death rate was higher than the birth rate; and the age pyramid was kettle-shaped, indicating that *C. obcordata* population is descending (Figure 2) [19].

Figure 1. Growth condition and spatial structure of *C. obcordata*. (**A**) Porphyritic wild populations. (**B**) An inflorescence. (**C**) A bee *Apis cerana* visits the flower (arrow). (**D**) A hoverfly *Betasyrphus serarius* visits the flower (arrow). (**E**) Fruited plant.

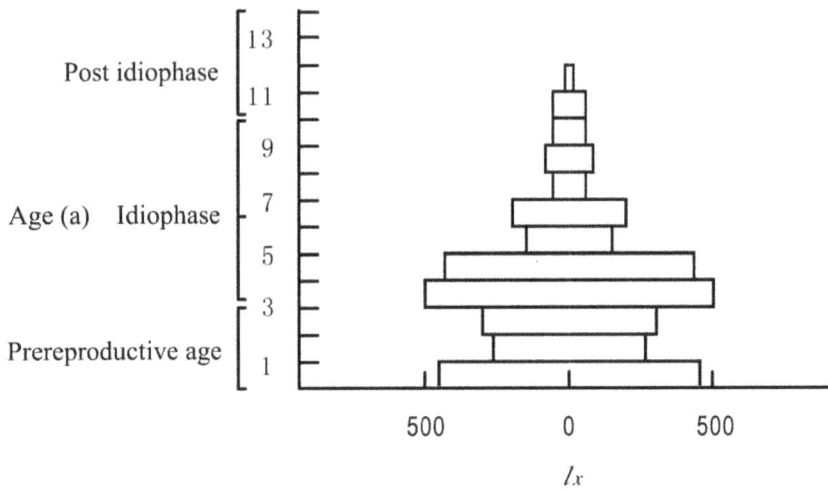

Figure 2. The age pyramid of the *C. obcordata* population.

Reproductive mechanism.   Due to the clumped distribution of *C. obcordata*, the comparatively concentrated flowers had increased the efficiency of pollination (Figure 1B). Bee *Apis cerana* (Figure 1C) and hoverfly *Betasyrphus serarius* (Figure 1D) were found to visit flowers during the entire observation period; 28.32%±15.94% ($n = 10$) flowers produced fruits, with each fruit able to produce 4782±1327 ($n = 10$) seeds (Figure 1E). The fruiting rate of bagged flowers was zero, indicating that *C. obcordata* did not automatically self-pollinate and produce asexual seeds. The fruiting rates of artificial pollination and cross-pollination were both 100%, revealing that the self-pollination and cross-pollination of *C. obcordata* had affinity and that its sexual reproduction was not without obstacles.

Population trend and quantity analysis

Static life table.   The life table of the *C. obcordata* population was formatted based on six subpopulations and the 1-year age class (phenological cycle). Table 1 shows the static life table for this population, with the results demonstrating that the death rate of seedlings from the 1-year age class was as high as 40%, thereby indicating that more of those seedlings died when turning into 2-

year age class seedlings. The death rates of seedlings from the 2- and 3-year age classes were negative, revealing that the seedlings failed to grow into the next age class and that the shortage of seedling storage was severe, as the death rate and disappearance rate (K_x) of the population were rather high before the seedlings (from the 2- and 3-year age classes) reached sexual maturity. The population had trended toward degeneration because of the intense filtration of the environment. Its descending trend was consistent with its kettle-shaped age pyramid. The death rate of sexually mature plants (4-year age class) was 13.64%, and the genet rate after sexual maturity was 63.16%; the death rates of genets in subsequent age classes were very high, except for the negative rate at the 6- and 8-year age classes and the zero rate at the 10-year age class. On one hand, genets of this population had died after sexual reproduction and left the next generation with living space, indicating that individuals' demand for nutrition space had constantly increased after the reproduction phase and that mortality was very high because of the enhanced external fluffing action of the population as influenced by water and other ecological factors. On the other hand, some individuals could grow up to the oldest age class after the physiological senescence phase,

Table 1. Static life table of the *C. obcordata* population.

Year	X	a_x	l_x	d_x	q_x	L_x	T_x	e_x	$\ln l_x$	K_x
2010	1	20	909.09	363.64	400.00	727.27	4863.64	5.35	6.81	0.51
2009	2	12	545.45	−90.91	−166.67	590.91	4136.36	7.58	6.30	−0.15
2008	3	14	636.36	−363.64	−571.43	818.18	3545.45	5.57	6.46	−0.45
2007	4	22	1000.00	136.36	136.36	931.82	2727.27	2.73	6.91	0.15
2006	5	19	863.64	545.45	631.58	590.91	1795.45	2.08	6.76	1.00
2005	6	7	318.18	−90.91	−285.71	363.64	1204.55	3.79	5.76	−0.25
2004	7	9	409.09	272.73	666.67	272.73	840.91	2.06	6.01	1.10
2003	8	3	136.36	−45.45	−333.33	159.09	568.18	4.17	4.92	−0.29
2002	9	4	181.82	45.45	250.00	159.09	409.09	2.25	5.20	0.29
2001	10	3	136.36	0.00	0.00	136.36	250.00	1.83	4.92	0.00
2000	11	3	136.36	90.91	666.67	90.91	113.64	0.83	4.92	1.10
1999	12	1	45.45	45.45	1000.00	22.73	22.73	0.50	3.82	3.82

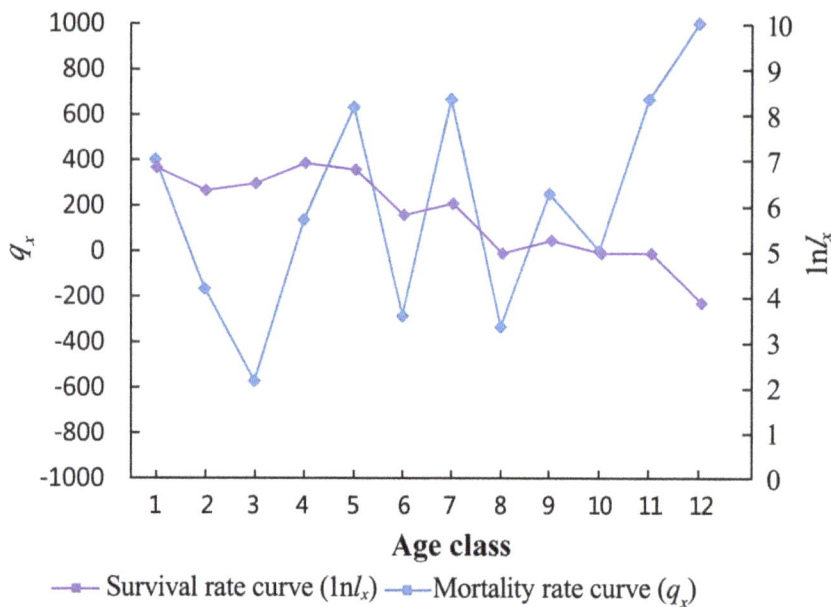

Figure 3. Survival rate and mortality rate curves of the *C. obcordata* population.

occupying environmental conditions (in places with better water conditions) that could better satisfy their growth needs, and the death rate of old individuals was lower, until their physiological senescence.

K-factor. Table 1 shows that the dead genets of *C. obcordata* were mainly seedlings from the 1-year age class and that the genets reached sexual maturity. Field observations revealed that the 1-year age class seedlings and genets died immediately after fructification. In the artificial drought test, death appeared earliest in young ramets and genets after fructification. Drought possibly caused the death (k_{alevin}) of seedlings and plants after reproduction, which is closely correlated with the total death rate (k_{total}). The death of genets after reproduction invalidated their multireproductive ability, which directly reduced the produce of seedlings. Moreover, the change in the death rate of *C. obcordata* at the seedling phase would yield fluctuations in the total death rate and the population size.

Survival curve. The survival curve and death rate curve of *C. obcordata* are illustrated in Figure 3. Figure 3 shows a Deevey type I survival curve, indicating that the survival rate of the 1-year age class seedlings was not high but that the death rate after sexual reproduction was. Figure 3 shows that the population structure was unstable and that the death rates at different age phases varied greatly. Individuals of the 1-year age class and those that were sexually mature had high death rates; however, although the surviving population had a high living quality, the individual number was very small, marking a descending trend. The survival curve and death rate curve reveal that a mass of young individuals could not enter the sexual phase successfully and that those sexually reproductive individuals could not produce a high number of young individuals during their thriving breeding period and instead die in mass, such that stability could not be sustained for the population.

Reproduction table and parameters. The reproduction table of *C. obcordata* is shown as Table 2. The net reproductive rate $[R_0 = \sum l_x m_x, \overline{m} = \sum f(x_1)/\sum (f(x_{4-12})) = 20/71 = 28.16\%]$ was 0.909, indicating that each generation could multiply 0.909 times. The intrinsic rate of increase ($r_m = \ln R_0/T$) was -1.884, with $r_m < 0$

demonstrating that the instantaneous birth rate was lower than the instantaneous death rate. The finite rate of increase ($\lambda = e^r = e^{-1.884}$) was 0.152, which entails that the population had decreased geometrically at a rate of 0.152; the generation span $[T = (\sum Xl_x m_x)/R_0]$ was 5.966a (year), yielding the average age of 5.966a for genets in the generative phase. The result of parameters $R_0 < 1$, $r_m < 0$, $\lambda < 1$ revealed that the population of *C. obcordata* cannot completely self-renew and has been descending [20–22].

Leslie matrix model and predicted model of dynamic quantity. Each sexually reproductive genet from the six subpopulations was tested. The average number of filial seedlings produced ($m_x = 0.282$) represented reproductive ability in calculating the Leslie matrix model and predicting the changes of population size and age structure for 20a (Table 3).

C. obcordata is a perennial plant, and its sexual reproduction can reach the physiological time limit. However, it has weak fecundity and scarce capacity for seedling supplementation, such that the

Table 2. Fecundity schedule of the *C. obcordata* population.

x	l_x	m_x	$l_x m_x$	$Xl_x m_x$
1	0.909	–	–	–
2	0.545	–	–	–
3	0.636	–	–	–
4	1.000	0.282	0.282	1.128
5	0.864	0.282	0.244	1.218
6	0.318	0.282	0.090	0.538
7	0.409	0.282	0.115	0.808
8	0.136	0.282	0.038	0.308
9	0.182	0.282	0.051	0.461
10	0.136	0.282	0.038	0.385
11	0.136	0.282	0.038	0.423
12	0.045	0.282	0.013	0.154

Table 3. Leslie matrix model of the *C. obcordata* population.

0	0	0	0.179	0.174	0.212	0.165	0.282	0.242	0.188	0.071	0.000
0.813											
	1.385										
		1.139									
			0.634								
				0.615							
					0.750						0
						0.583					
							1.000				
								0.857			
									0.667		
		0								0.250	
											0.000

population presents a descending trend. The dynamic quantity model is given as $N_t = N_{t-1} - N_{t-1} e^{-1.884}$, and its predicted results are shown in Table 4.

Levins model of population dynamics and the prediction of dynamic quantity. The extinction of a local population and the establishment of a new one are the two basic processes that determine the trend of a metapopulation [20]. Therefore, the Levins model of dynamic metapopulation of *C. obcordata* can be expressed as follows:

$$dp/dt = mp(1-p) - ep \quad (1)$$

where "m" and "e" are the parameters of misappropriation and extinction, respectively, and 'P = 0 represents extinction. When the population is in a balanced state, given as $dp/dt = 0$, 'P = 1 − e/m (i.e., m<e), the metapopulation will ultimately trend to extinction; conversely, when e<m, the metapopulation will persistently exist [20] if the extinction rate of a local population is less than a certain marginal value. Equation (1) can be restated as follows:

$$dp/dt = (m-e)p[1 - p/(1 - e/m)] \quad (2)$$

The D value of (m–e) in this expression can be identified as an intrinsic rate of increase [20]. The value for *C. obcordata* was −1.884, giving m<e, indicating that its population would ultimately trend to extinction.

Tables 3 and 4 indicate that *C. obcordata* is a diminishing population based on the simulation of the Leslie matrix model and the continuous descending (negative growth) model, which is consistent with data from the reproduction table and biological characteristics of the population. The quantity of each age class presented a successively descending trend — that is, the degradation trend appeared in all age classes. Further predictions based on $N_t = N_{t-1} - N_{t-1} e^{-1.884}$ indicate that the population size of *C. obcordata* would decrease from $N_1 = 117$ genets per six subpopulations to $N_{23} = 0$ in approximately 23 years (Figure 4).

Table 4. Dynamic preestimates of population size and continuous descending model of the *C. obcordata* population.

Age class	N_0	N_1	N_2	N_3	N_4	N_5	N_6	N_7	N_8	N_9	N_{10}	N_{11}	N_{12}	N_{13}	N_{14}	N_{15}	N_{16}	N_{17}	N_{18}	N_{19}	N_{20}	N_{21}	N_{22}	N_{23}
1	20	17	14	12	10	9	7	6	5	5	4	3	3	2	2	2	1	1	1	1	1	1	1	0
2	12	10	9	7	6	5	4	4	3	3	2	2	2	1	1	1	1	1	1	1	0	0	0	0
3	14	12	10	9	7	6	5	4	4	3	3	2	2	2	1	1	1	1	1	1	1	0	0	0
4	22	19	16	13	11	10	8	7	6	5	4	4	3	3	2	2	2	1	1	1	1	1	1	0
5	19	16	14	12	10	8	7	6	5	4	4	3	3	2	2	2	1	1	1	1	1	1	1	0
6	7	6	5	4	4	3	3	2	2	2	1	1	1	1	1	1	1	0	0	0	0	0	0	0
7	9	8	6	5	5	4	3	3	2	2	2	1	1	1	1	1	1	1	0	0	0	0	0	0
8	3	3	2	2	2	1	1	1	1	1	1	0	0	0	0	0	0	0	0	0	0	0	0	0
9	4	3	3	2	2	2	1	1	1	1	1	1	1	0	0	0	0	0	0	0	0	0	0	0
10	3	3	2	2	2	1	1	1	1	1	1	0	0	0	0	0	0	0	0	0	0	0	0	0
11	3	3	2	2	2	1	1	1	1	1	1	0	0	0	0	0	0	0	0	0	0	0	0	0
12	1	1	1	1	1	0	0	0	0	0	0	0	0	0	0	0	0	0	0	0	0	0	0	0
Total	117	101	84	71	62	50	41	36	31	28	24	17	16	12	10	10	8	6	5	4	4	3	3	0

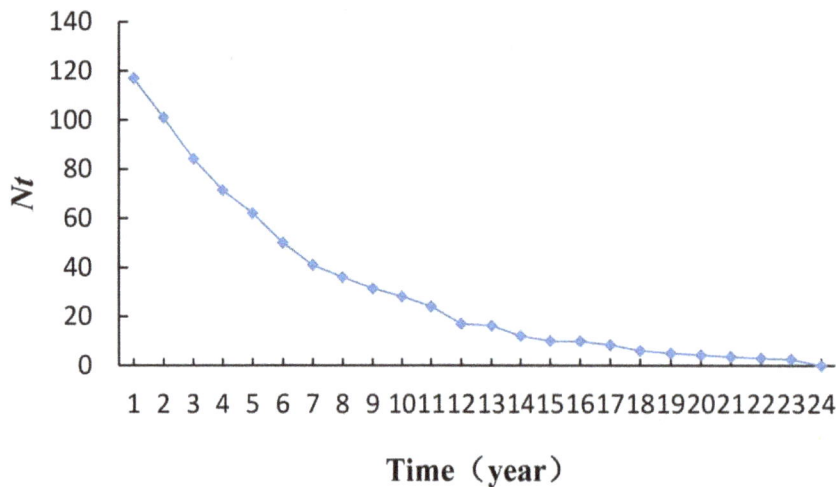

Figure 4. Negative growth model curve of the *C. obcordata* population.

Meteorological data and survival analysis of C. obcordata

Response to rainfall change. Surface flow was introduced after raining and stopped after 20 hours in the artificial sloping field. The genets under such artificial water recharge all survived as well as flowered and fruited normally, producing 47 seedlings, whereas those plants in the natural condition died during the successive episodes of drought from autumn to winter. *C. obcordata* that grew on the poolside also survived, flowered, and fruited, producing 53 seedlings. These results suggest that preserving enough water in the habitat is the key to normal *C. obcordata* growth.

Influence of climate change on the survival of *C. obcordata*. The average temperature from the period of 1980–2010 has increased by 1.1°C compared with the 1970s (21.68±0.24 vs. 22.69±0.24°C), and the average rainfall from the same period has decreased by 11.49 mm compared with the 1970s (1773.67 vs. 1762.18 mm) (see Table S1, data from the Guangdong Weather Archives). The change in weather data over 40 years suggests that the hydrological cycle has accelerated along with climate warming, and the severely uneven temporal and spatial distribution of water has brought about increasingly more severe drought — in particular, seasonal and local drought has become extraordinary. The temperature has gradually ascended, high temperature weather lasts for a long time, the rainfall pattern is frequently fluctuating, relative humidity is decreasing, and evaporation is advancing, further yielding successive episodes of drought in most years from autumn to winter, even extending to spring. Such serious changes in weather have resulted in significant fluctuations in the capacity of stream water in the mountains. Although temperature rise is known to increase rainfall, this study found that rainfall decreased as temperature increased, which was correlated with diminished relative humidity and increased evaporation; these factors increased the frequency of drought ("Report on Dealing with Climate Change and Strengthening Weather Disaster Prevention and Reduction Ability," Guangdong Provincial Government). Such weather pattern caused brook blanking as well as flashfloods; in addition, it has directly changed the habitat as well as threatened the subsistence and development of *C. obcordata*. The survival curve of each age class in the life table agreed with the change in weather data, particularly with the rainfall change and especially in the pre-reproduction and reproduction phases. The survival rate was higher during the years with more rainfall and

lower in those with less. The survival rate fluctuated along with rainfall. The survival rates of all age classes suggest that the population had experienced distinct weather events per age class before entering its current age class.

Discussion

The age pyramid and population growth trend of *C. obcordata* can both be deduced using its life table. The Leslie matrix model can predict the dynamic changes in quantity and structure of a population well; it is an effective means of predicting trends, especially for a descending population such as *C. obcordata*. The continuous descending model $N_t = N_{t-1} - N_{t-1}e^{-1.884}$ can be used to predict the dynamic trend of a population, and its results were congruent to the findings from the spatial pattern test, age pyramid, and Leslie matrix model of this species. After calculation, m<e from the Levins model indicated that the metapopulation of *C. obcordata* would ultimately disappear from Luofu Mountain.

The dynamic quantity trend of the plant population is the product of the interaction between the survival capability of individual plants and the outer environment [21]. The age structure, life table, and survival curve of a plant population under different habitats not only reflect the actual status of the population but also exhibit the resistant relationship between the plant population and the environment [22]. Following this, the age structure, life table, and survival curve of *C. obcordata* reflect its actual status, and the predicted population size trend suggests that the fact that its reproductive seedlings and plants are unable to withstand drought and adapt to weather changes is responsible for its resistant relationship with the environment. Climate warming results in rainfall imbalance, which is an irreversible trend over the short term. As it is difficult to stabilize the growth of *C. obcordata* in conditions with unstable water currents, stabilizing this population forfeits its general conditions.

Due to the rapid development of modern industry, large amounts of CO_2 have been released into the environment, thereby increasing surface temperature, causing the greenhouse effect, and consequently triggering global warming. Temperature rise has changed the long-term adaptability of *C. obcordata* to temperature. In response to high temperature, the plant needs to reduce its cell water capacity, increase the density of its sugar or salt content, slow down its metabolic speed, and engender vigorous transpiration to avoid overheating. However, a reduction in rainfall and extended drought caused by the significantly irregular fluctuations of rainfall both within a year and over time follow the high temperature, which prevents *C. obcordata* from withstanding drought and high temperature, especially due to the undeveloped defense mechanism of its seedlings or newborn leaves, its underdeveloped root system, and its inefficient water imbibition. Moreover, the plant consumes a considerable amount of nutrients after fruition, thereby diminishing its capacity to resist drought. Previous research has shown that the seedlings and plants of *C. obcordata* collectively died after fruiting under the dual impact of reduced water content in the environment and increased inter-water consumption [13]. This species requires a high number of seedlings for replenishment to enter the next age class successfully. The change in rainfall patterns caused by weather warming has been found to kill seedlings and reduce the viability of mature genets. As a result, seed production, seedling growth, and population replenishment prove to be difficult, leaving the population to trend toward senescence.

Vegetation and climate have a remarkable coupling relationship [23]. Climate change is the primary long-term force behind regional vegetation, which will be principally affected by temperature and rainfall. A sudden increase and change in these factors over a short period will significantly affect the species inhabiting such vegetation. An increase in temperature is known to increase water consumption, which in turn causes drought [24]; such water stress triggers the pores of *C. obcordata* to shut, causing the rates of transpiration and photosynthesis to decrease so as to avoid leaf dehydration and reduce the accumulation of dry matter [25]. The abovementioned factors could negatively influence each age-class genet of *C. obcordata* and disturb its long-term adaptability to the environment during its entire life cycle. This study has determined that *C. obcordata* is extremely sensitive to water change, and the findings have confirmed that the rainfall imbalance caused by climate warming may expedite the extinction of the species.

Methods

Biological characteristics

Wild populations of *C. obcordata* in Luofu Mountain (114°05′E, 22°34′N; elevation, 100–500 m) in Huizhou, Guangdong, and artificial populations in the National Orchid Conservation Center of China (Shenzhen,114°10′E, 22°35′N; elevation, 50 m) were evaluated from March 2010 to May 2011 (Figure S1). Six natural subpopulations of *C. obcordata* were observed, and the growth mode of each genet, the quantity of persistent stems, and the growth as well as the flowering and fruiting status of leafed ramets were recorded. Genet age was confirmed via the quantity of stems because each genet can only grow one new ramet annually. The flowering quantity of each age-class genet, the number of flowers on each inflorescence, and the fruiting quantity were calculated. The genet reached sexual maturity, and statistical data on the seedlings in the 1-year age class were confirmed to calculate the offspring quantity of the genet [16].

All necessary permits were obtained for our field studies. The locations for our field studies are not private lands but protected areas, controlled by the Forestry Bureau of Guangdong Province. We have obtained a valid permit from this authoritative organization (see Permit S1). The tested genets of *C. obcordata* were all artificially cultured, and none of them was collected from the field. Our field observations did not collect any plant specimen nor animal or even insect. Although *C. obcordata* is not a rare plant (no endangered plant), it is facing threatened and needs to be protected.

Detection of reproductive mechanism

Biological pollination of *C. obcordata* was carried out in fields [16,26] until all flowers faded or fruited, and the fruiting rate of natural pollination was calculated statistically. The mating system was investigated as described below at the National Orchid Conservation

Center of China. During the flowering period of *C. obcordata*, 40 sample sites were set up and 10 flowers from each sample site were tested under the following treatments: (1) flowers were bagged before blooming until all faded or fruited and the fruiting rate was calculated; (2) flowers were bagged before blooming, pollinia were placed on the stigmata after blooming, the flowers were bagged again until all flowers faded or fruited, and the fruiting rate was calculated; or (3) flowers were bagged before blooming, pollinium was placed on the stigmata of flowers from different genets and bagged again until all flowers faded or fruited, and the fruiting rate was calculated [26].

Dynamic analysis of population size

Formation of the static life table. Based on the trait of single catenulate growth with the obvious space–time mark of *C. obcordata*, the static life table was prepared using the space-to-time method [12,27–29]. It includes the following: X, the age class; l_x (survival rate), the standard survival quantity (1000) at the beginning of X age class; d_x (death quantity), the standard death quantity from X age class to $X+1$ age class; q_x (death rate), the death rate of the genets from X age class, $q_x = d_x/l_x \times 1000$; L_x, the average quantity of the genet from X age class to $X+1$ age class, $L_x = (l_x+l_{x+1})/2$; T_x, the total quantity of the genet from X age class onward, $T_x = L_x+L_{x+1}+L_{x+2}+,...$; e_x (life expectation), the life expectation of the genet that entered X age class, $e_x = T_x/l_x$; a_x (survival quantity), the actual survival quantity at the beginning of X age class (genet number/500 m^2); and K_x, the disappearance rate of the population, $K_x = \ln l_x - \ln l_{x+1}$.

Formation of the survival curve and death rate curve. These curves were plotted via the individual quantity of each age class compared with time to describe the death rate at a specific age class. The death rate curve was drawn with death rate in the y-coordinate and age class in the x-coordinate, whereas the survival curve was drawn with the log value of survival quantity (log value of l_x) in the y-coordinate and age class in the x-coordinate [18].

Formation of the population reproduction table. The population reproduction table [12] includes the following: X, the age class; l_x, the survival rate at X age class; and m_x, the average quantity of offspring of the genet from X age class (based on actual tested quantities of mature genet and 1-year seedlings). The net reproductive rate of the population, $R_o = \Sigma l_x m_x$, the intrinsic rate of increase, $r_m = \ln R_o/T$, the finite rate of increase, $\lambda = e^r$, and the generation span, $T = \Sigma X l_x m_x / \Sigma l_x m_x$ [12], were also calculated.

Structure of the Leslie matrix model and prediction of dynamic quantity. The total survival rate P_x (from X age class to $X+1$ age class) was calculated using the survival rate of the life table, $P_x = L_{x+1}/L_x = (l_{x+1}+l_{x+2})/(l_x+l_{x+1})$. The average quantity of offspring (f_x) generated at X age class and that which survived at $X+1$ age class were calculated using the reproduction value, $f_x = P_x \times m_x$. Lastly, the quantity and age distribution of the population after unit time interval were calculated from the quantity and distribution of the population: $N_{t+1} = M \cdot N_t = M^{(t+1)} \cdot N_o$, where M is the Population Projection Matrix [12,30]:

$$M = \begin{bmatrix} f_0 & f_1 & f_2 & \cdots & \cdots & f_{19} & f_{20} \\ p_0 & 0 & 0 & \cdots & \cdots & 0 & 0 \\ 0 & p_1 & 0 & \cdots & \cdots & \vdots & \vdots \\ \vdots & \vdots & p_2 & \cdots & \cdots & \vdots & \vdots \\ \vdots & \vdots & \vdots & \ddots & & \vdots & \vdots \\ \vdots & \vdots & \vdots & & \ddots & \vdots & \vdots \\ 0 & 0 & 0 & \cdots & \cdots & p_{19} & 0 \end{bmatrix}$$

The Levins model of single metapopulation dynamics. Due to the spatial segregation of the population caused by its growth characteristics and existing habitat, the interaction dynamics of *C. obcordata* requires that the individuals be spread out. Based on the current fragmental situation of its habitat, the Levins model of metapopulation dynamics was utilized to make qualitative and quantitative predictions [20].

Climate change and growth of *C. obcordata*

Analysis of weather data. The weather data used in this study were obtained from the Weather Archives of Guangdong Province. The time series of collected data corresponded with the largest survival age of the genet in population of *C. obcordata*, and the estimated period was from 1998 to 2010. The trend of weather change and change in weather circumstances in the distribution of *C. obcordata* were studied using weather data, including annual average temperature and rainfall. The relationship between the change in rainfall and the quantity trend of the population was analyzed in integration with the static life table.

Detection of influence of water-flow variation on the growth of *C. obcordata*. Artificial simulation was carried out at the National Orchid Conservation Center of China in March 2010 to test the effects of variation in water flow on *C. obcordata* growth: 40 sample sites were set up in an artificial sloping field of 85% shadow where surface flow could occur, with each sample site comprising five genets (all 4–5 years old), and 20 sample sites were set up along the poolside of a stable water storage system, with each sample likewise composed of five genets (all 4–5 years old). After normal growth under artificial management, 20 sample sites from the artificial sloping field were continuously watering; the

others sites were left to grow under natural conditions. The growth statuses in different water conditions were observed and recorded.

Supporting Information

Figure S1 Study area in Guangdong province, China. (A) Site of *C. obcordata* in Luofu Mountain. **(B)** Site of control test of *C. obcordata* in Shenzhen.

Permit S1 The permit for our field studies from Forestry Bureau of Guangdong province, China.

Acknowledgments

We thank Xu-Hui Chen, Wei-Rong Liu, Wen-Hui Rao, Xiu-Dong Xu, and Guo-Hui Huang for helping with the field work.

Author Contributions

Conceived and designed the experiments: Z-JL L-JC X-JX K-WL. Performed the experiments: X-JX K-WL Y-YZ Z-JL G-QZ. Analyzed the data: Z-JL X-JX K-WL L-JC W-CT Y-YH G-QZ. Contributed reagents/materials/analysis tools: Z-JL L-JC X-JX K-WL Y-TZ. Wrote the paper: Z-JL K-WL X-JX L-JC Y-YZ. Discussed the findings and commented on the manuscript: Z-JL X-JX K-WL Y-YZ Y-TZ W-CT Y-YH G-QZ L-JC.

References

1. Melillo JM, Mcguire AD, Kicklighter DW, Moore B, Vorosmarty CJ, et al. (1993) Global climate change and terrestrial net primary production. Nature 363: 234–240.
2. Janack J (1997) Stomatal limitation of photosynthesis as affected by water stress and CO₂ concentration. Photosynthetica 34(4): 473–476.
3. Meier M, Fuhrer J (1997) Effect of elevated CO₂ on orchard grass and red clover grown in mixture at two levels of nitrogen or water supply. Environmental and Experimental Botany 38: 251–262.
4. Root TL, Price JT, Hall KR, Schneider SH, Rosenzweig C, et al. (2003) Fingerprints of global warming on wild animals and plants. Nature 421: 57–60.
5. Liu ZJ, Chen IJ, Liu KW, Li LQ, Zhang YT, et al. (2009) Climate warming brings about extinction tendency in wild population of Cymbidium sinense. Acta Ecologica Sinica 29(7): 3443–3455.
6. Erasmus BFN, Vanjaarsveld SA, Chown LS, Kshatrlya M, Wessels KJ (2002) Vulnerability of South African animal taxa to climate change. Glob Chang Biol 8: 679–693.
7. Midgley GF, Hannah L, Millar D, Thuiller W, Booth A (2003) Developing regional and species-level assessment of climate change impacts on biodiversity in the cape floristic region. Biological Conservation 112: 87–97.
8. Peterson AT, Ortega-Huerta MA, Bartley J, Sánchez-Cordero V, Soberón J, et al. (2002) Future projections for Mexican faunas under global climate change scenarios. Nature 416: 626–629.
9. Crimmins SM, Dobrowski SZ, Greenberg JA, Abatzoglou J, Mynsberge AR (2011) Changes in climatic water balance drive downhill shifts in plant species' optimum elevations. Science 331: 324–327.
10. Liu ZJ, Chen SC, Ru ZZ (2006) The Genus Cymbidium in China. Beijing: Science Press.
11. Chen SC, Gale SW, Cribb PJ (2009) Cephalantheropsis. in Wu zRavan, Hong D, eds. Flora of China 25: 288–289. Beijing: Science Press and St. Louis: Missour Botanical Garden Press.
12. Yue CL, Jiang H, Zhu YM (2002) Analysis on numeric dynamic of population of Cimicifuga nanchuanensis, an endangered plant. Acta Ecologica Sinica 22: 793–796.
13. Zhang WH, Zu YG (1999) Study on population life table and survivorship curves of Adenophora lobophylla, an endangered species, compared with A. potaninii, a widespread species. Acta Phytoecologica Sinica 23: 76–86.
14. Li XK, Su ZM, Xiang WS, Ning SJ, Tang RQ, et al. (2002) Study on the structure and spatial pattern of the endangered plant population of Abies yuanbaoshanensis. Acta Ecologica Sinica 22: 2246–2253.
15. Hu YJ, Wang SS (1988) A matrix model of population growth of dominant tropical rain forest species Vatica hainanensis in hainan island. Acta Ecologica Sinica 8: 104–110.
16. Liu ZJ, Chen IJ, Rao WH, Li LQ, Zhang YT (2008) Correlation between numeric dynamics and reproductive behaviour in Cypripedium lentiginosum. Acta Ecologica Sinica 28: 111–121.
17. Chen SC, Luo YB (2003) Advances in some plant group in China I. a retrospect and prospect of Orchidology in China. Acta Botanica Sinica 45: 2–20.
18. Luo YB, Jia JS, Wang CL (2003) A general review of the conservation status of Chinese orchids. Biodiversity Science 11: 70–77.
19. Pan RC, Ye QS (2005) The physiology of Cymbidium. Beijing: Science Press.
20. Sun RY, Li QF, Niu CJ, Lou AR (2003) Basic Ecology. Beijing: Higher Education Press.
21. Crawley MJ (1986) Plant Ecology. London: Blackwell Scientific Publications.
22. Manuel C, Molles J (2002) Ecology, Concept and Applications (2nd edn.). New York: McGraw-Hill Companies.
23. Hao YP, Chen YF (1998) Progress in estimation of net primary productivity and its responses to climate change. Advances in Earth Science 13: 564–571.
24. Liang N, Maruyama K (1995) Interactive effects of CO₂ enrichment and drought stress on gas exchange and water-use efficiency in Alnus firma. Environmental and Experimental Botany 35: 353–361.
25. Janacek J (1997) Stomatal limitation of photosynthesis as affected by water stress and CO₂ concentration. Photosynthetica 34: 473–476.
26. Liu KW, Liu ZJ, Huang LQ, Li LQ, Chen IJ, et al. (2006) Self-fertilization strategy in an orchid. Nature 441: 945–946.
27. Jiang H (1992) Study on Population Ecology of Picea asperata. Beijing: China Forestry Publishing House.
28. Harper JL (1997) The Population Biology of Plant. London: Academic Press.
29. Harcombe PA (1987) Tree life tables. Bioscience 37: 557–565.
30. Liu JF, Hong W (1999) A study on forecast of population dynamics of Castanopsis kawakamii. Chinese Journal of Applied and Environmental Biology 5: 247–253.

5

Malaria Control and the Intensity of *Plasmodium falciparum* Transmission in Namibia 1969–1992

Abdisalan M. Noor[1,2]*, **Victor A. Alegana**[1], **Richard N. Kamwi**[3], **Clifford F. Hansford**[4], **Benson Ntomwa**[5], **Stark Katokele**[5], **Robert W. Snow**[1,2]

1 Malaria Public Health Department, Kenya Medical Research Institute-Wellcome Trust Research Programme, Nairobi, Kenya, 2 Centre for Tropical Medicine, Nuffield Department of Clinical Medicine, University of Oxford, Oxford, United Kingdom, 3 Office of the Minister, Ministry of Health and Social Services, Windhoek, Namibia, 4 National Institute for Tropical Diseases, Tzaneen, South Africa, 5 National Vector-borne Diseases Control Programme, Directorate of Special Programmes, Ministry of Health and Social Services, Windhoek, Namibia

Abstract

Background: Historical evidence of the levels of intervention scale up and its relationships to changing malaria risks provides important contextual information for current ambitions to eliminate malaria in various regions of Africa today.

Methods: Community-based *Plasmodium falciparum* prevalence data from 3,260 geo-coded time-space locations between 1969 and 1992 were assembled from archives covering an examination of 230,174 individuals located in northern Namibia. These data were standardized the age-range 2 to less than 10 years and used within a Bayesian model-based geo-statistical framework to examine the changes of malaria risk in the years 1969, 1974, 1979, 1984 and 1989 at 5×5 km spatial resolution. This changing risk was described against rainfall seasons and the wide-scale use of indoor-residual house-spraying and mass drug administration.

Results: Most areas of Northern Namibia experienced low intensity transmission during a ten-year period of wide-scale control activities between 1969 and 1979. As control efforts waned, flooding occurred, drug resistance emerged and the war for independence intensified the spatial extent of moderate-to-high malaria transmission expanded reaching a peak in the late 1980s.

Conclusions: Targeting vectors and parasite in northern Namibia was likely to have successfully sustained a situation of low intensity transmission, but unraveled quickly to a peak of transmission intensity following a sequence of events by the early 1990s.

Editor: Joshua Yukich, Tulane University School of Public Health and Tropical Medicine, United States of America

Funding: This work was supported by the Wellcome Trust [Intermediate Research Fellowship grant to AMN (#095127) and a Principal Research Fellowship grant (#079080) to RWS]. Programmatic support for this study was also provided through a Wellcome Trust Major Overseas Programme grant to the Kenya Medical Research Institute/Wellcome Trust Research Programme (#092654). The funders had no role in study design, data collection and analysis, decision to publish, or preparation of the manuscript.

* E-mail: anoor@nairobi.kemri-wellcome.org

Introduction

Namibia is one of six sub-Saharan African countries that have formally declared ambitions to eliminate malaria following significant recent declines in clinical incidence [1–3]. Namibia has a long history of vector control and mass drug administration [4–5], however there has been no formal quantification of the epidemiology of malaria during this period to provide the context for current elimination ambitions. To understand the challenges to achieving the country's ambition of malaria elimination by 2020 [2], lessons from the past are valuable to making rational planning decisions today. Evidence on the levels of interventions scale-up, the subsequent rate of epidemiological transition, and levels of rebound of risks where intervention may have been interrupted, all inform the feasibility of malaria elimination in Namibia.

Advances in computing and geostatistical techniques [6] and the increasing availability of data on malaria transmission have increased our ability to define the spatial and temporal risks of malaria endemicity at high spatial resolutions nationally [7–10] and globally [11,12]. These maps, however, are currently available as predictions to contemporary time-points and do not provide information on the levels of historical risks that is required as the baseline upon which any potential transition of malaria risk is measured, and the probable rebound risks if control were interrupted are defined.

Here, we describe the scale and scope of control in Namibia from 1965 to 1992 in targeted areas in northern regions of the country and spatially model *Plasmodium falciparum* transmission intensity every 5 years using Bayesian goestatistics, from 1969 to

1989, to provide an epidemiological and control context and the likely impact of interventions during this period.

Malaria Epidemiology and Control in Namibia

The earliest surveys of malaria were undertaken in a non-random sample of communities across the country from February to June 1950 [13]. Information was collected on spleen and parasite rates and vector species to define zones of transmission across the country. The regions (Figure 1) of Ovambo (present day Omusati, Oshana, Ohangwena and Oshikoto regions), Kavango and Caprivi bordering Angola and parts of Bushmanland (part of present day eastern Otjozondjupa region) were defined as supporting intense, stable malaria transmission. The Southern regions from Erongo, Khomas and southern Omaheke to the Orange River on the border with South Africa were defined as

largely free of transmission or with very focal pockets of occasional transmission [13–14].

Indoor residual spraying (IRS) using dichlorodiphenyltrichloroethane (DDT) started in 1965 in urban residential houses in the north of the country [4,14,15]. This was expanded to include all malarious areas in the northern territories following the appointment of a malaria public health specialist from South Africa to establish a network of malaria health inspectors in the areas of Ovambo and Kavango. The malaria programme for Ovambo and Kavango were managed from Windhoek with regional officers at Oshakati and Rundu respectively. Initially, the East Caprivi programme was supervised from Pietersburg but in 1967 was transferred to Pretoria in South Africa before returning to Pietersburg in 1972. Soon after, the numbers of field staff reduced and coverage with IRS deteriorated. Supervision for the East

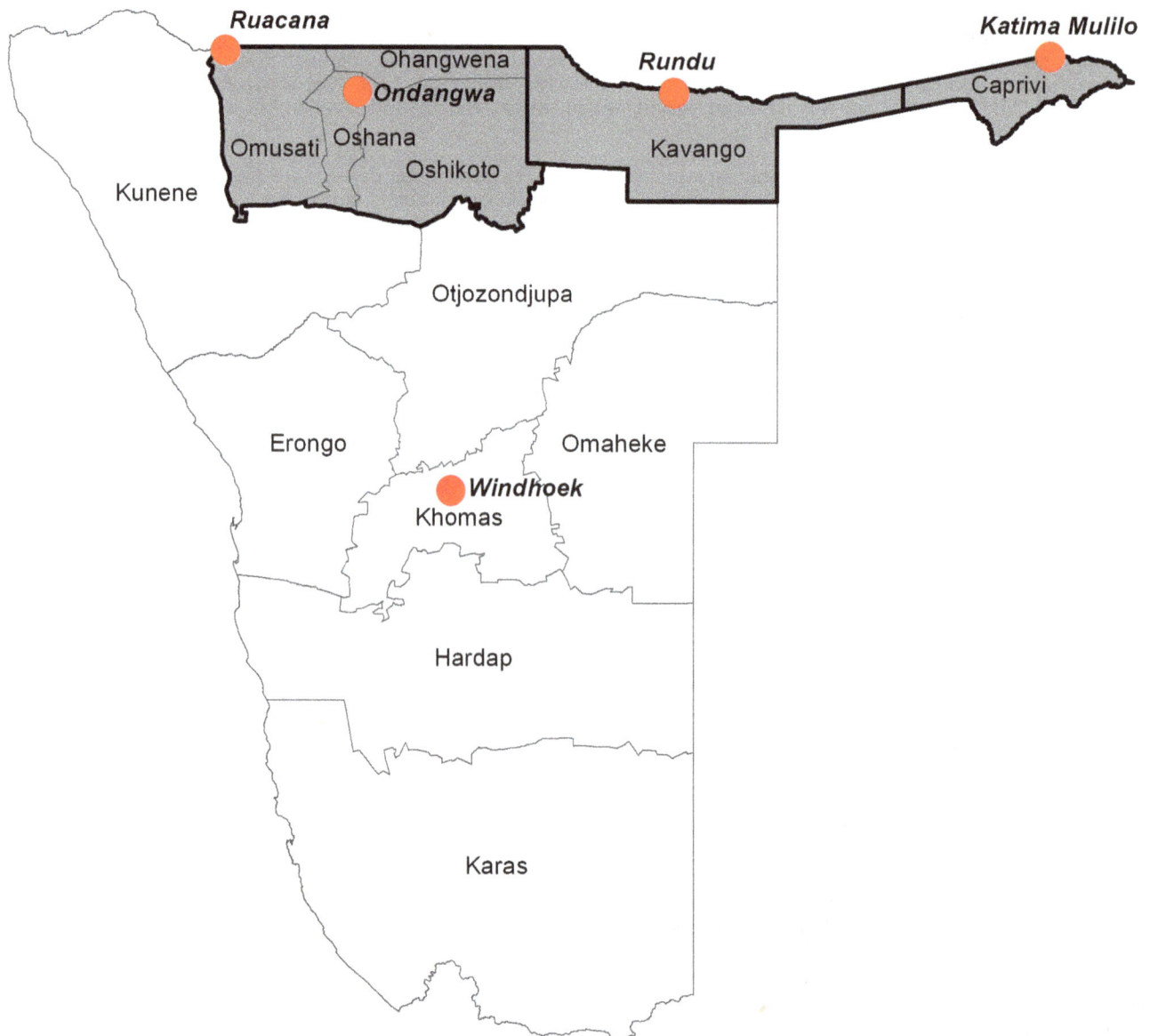

Figure 1. The boundaries of the 13 first level administrative units (regions) of Namibia and locations of the Capital City, Windhoek, the main towns of the north (Rundu, Katima Mulilo, Ruacana and Ondangwa). The grey shaded areas with thick black boundary are the intervention areas of Caprivi, Kavango and Ovambo.

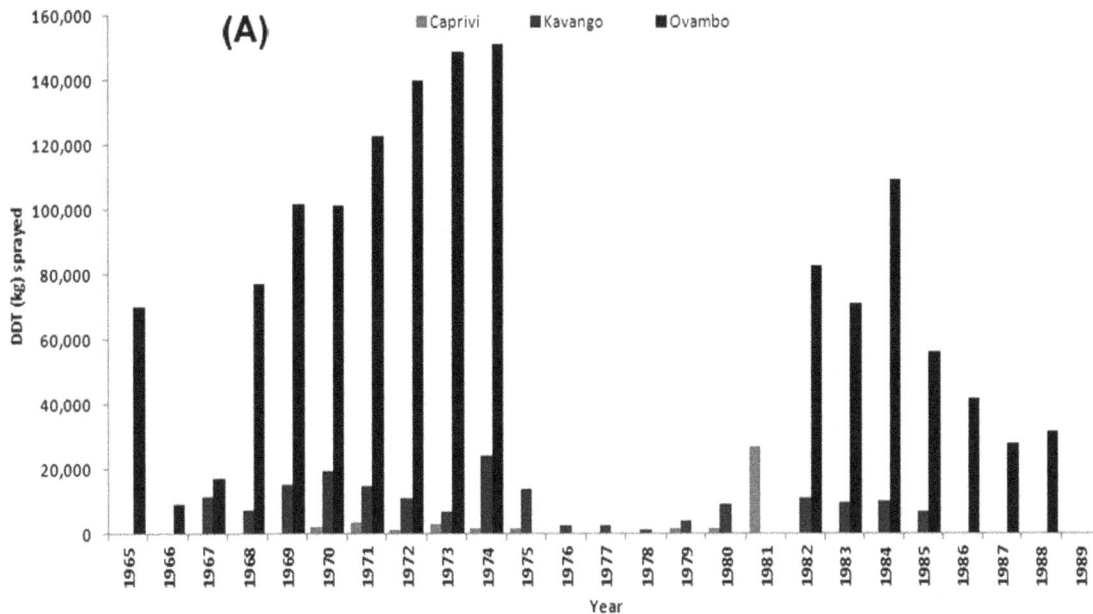

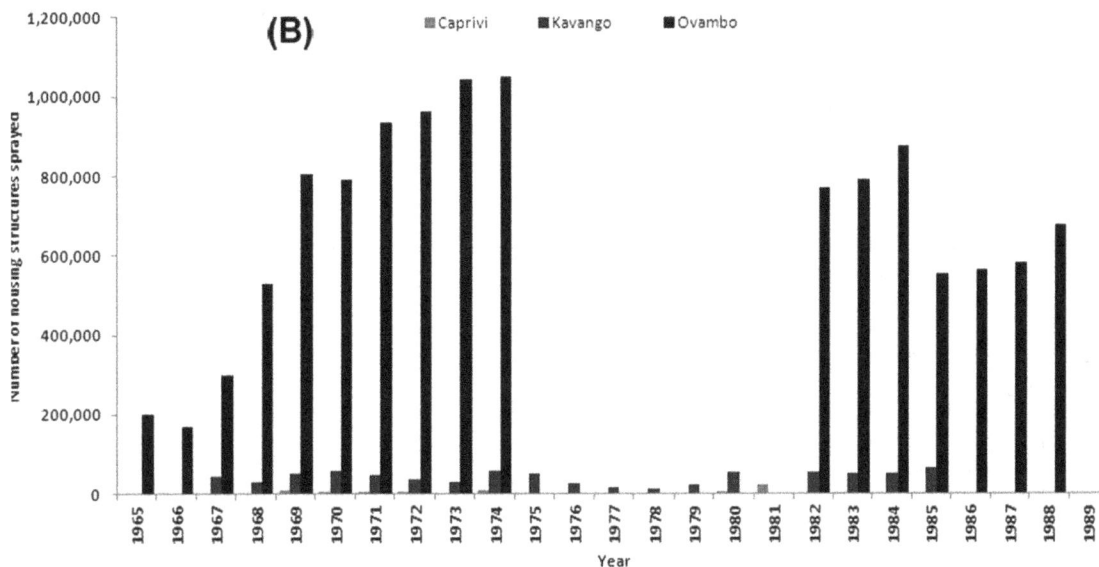

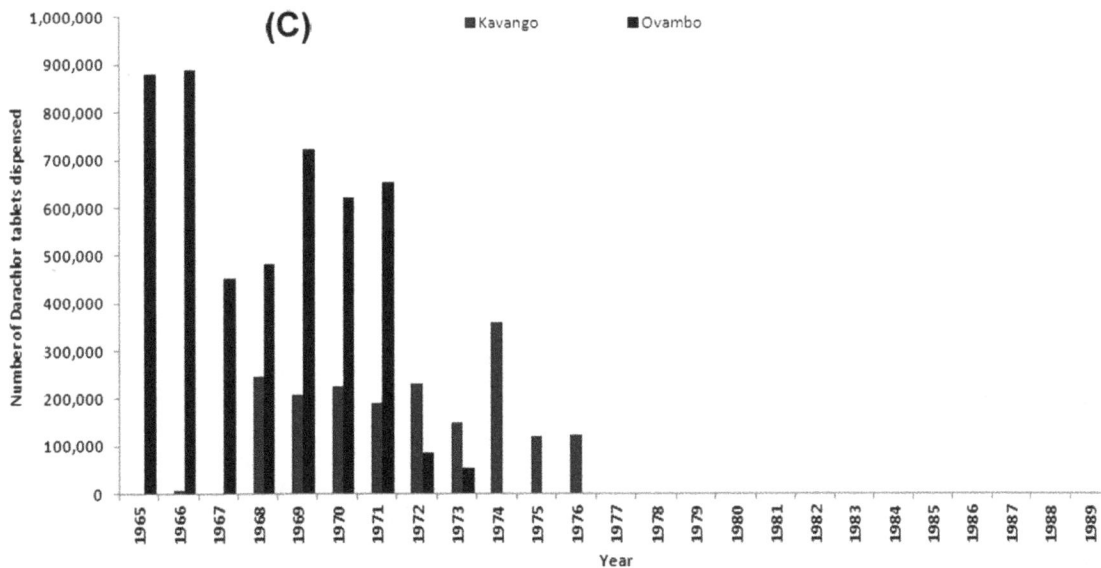

Figure 2. Summary intervention distribution and coverage data assembled from monthly and annual reports archived at Tzaneen for Caprivi, Kavango and Ovambo (present-day Omusati, Oshana, Ohangwena and Oshikoto regions) of: A) the amount of DDT (kilograms) used for IRS; B) the number of house structures sprayed; and C) the number of Darachlor tablets dispensed during the period 1965 to 1989 in A) Kavango and B) Ovambo. Data were unavailable for DDT, housing structures sprayed and Darachor tablets dispensed for the years where data is not shown. Darachlor use was terminated after 1989.

Caprivi was taken over from Windhoek in 1972 with a field station in Katima Mulilo.

Spray operations were undertaken from May to November in Ovambo and twice annually (August to November and January to March) in Kavango and Caprivi [5]. Between 1966 and 1979, over 1.6 million kilograms of DDT was used to spray approximately 12.4 million housing structures in Ovambo, Kavango and Caprivi (Figure 2). In 1980 Bendiocarb was introduced to replace DDT in urban areas. From 1966, approximately 6.7 million tablets of Darachlor (chloroquine+pyrimethamine) were distributed (Figure 2) from January to June each year at dosages of 2 tablets for persons 10 years of age and above and 1 tablet to children below 10 years of age [5].

In 1966, annual mass blood surveys and treatment of febrile cases with Darachlor began in Ovambo and was extended to Kavango and Caprivi in 1967. Technical staff from the National Institute for Tropical Diseases (NITD) in Tzaneen, South Africa, were responsible for these parasitological surveys, which continued through to the late 1980s when large cross-sectional surveys were harder to undertake in part due to the insecurity during the war for independence. A malaria epidemic followed a reduction in both control and survey work in 1990 [16].

Malaria Risk Mapping Process

Assembling _P. falciparum_ parasite prevalence data. Community based _P. falciparum_ parasite rate (_Pf_PR) survey data were compiled from monthly and annual reports available at the archives at the NITD in Tzaneen, South Africa, in 2011. It was possible to transcribe the data at village levels from summaries of the serial mass blood surveys during the years 1967 to 1992 across Ovambo, Kavango, Caprivi, Bushmanland (Otjozondjupa), Hereroland (Omaheke) and Damaraland (Kunene and parts of Erongo). Overall, communities and schools were visited each year at the end of the main transmission season usually between March and June to undertake cross-sectional surveys of infection prevalence. In the majority of surveys individuals of all ages were selected for malaria testing with a few surveys focusing only on children. Thick and thin Giemsa stained blood smears were taken from individuals and transported to the NITD where they were examined using light microscopy by expert microscopists before the results were returned to Namibia to support annual control planning. These active mass blood surveys were suspended after 1992. The longitude and latitude of all survey locations were identified using a variety of approaches including online village databases and gazetteers and a settlement database mapped using Global Positioning Systems (GPS) receivers [17]. Data were assembled as part of routine surveillance system for malaria control. Approval was provided by the Ministries of Health at the time. All data were de-identified and analysed anonymously.

Modelling historical _P. falciparum_ risk within the limits of stable transmission. Three previously described criteria [14] were used to define the limits of stable malaria transmission in Namibia. These were: the suitability of ambient temperature [18]; aridity [19–20]; and medical intelligence [14]. The resulting map classified areas in Namibia into areas that are unsuitable for transmission, those that support unstable transmission and areas of stable transmission (Figure 3, see footnote to this figure for details).

A standard set of ecological and climatic determinants were tested for association with the assembled _P. falciparum_ prevalence data within the stable limits of transmission. These covariates were urbanisation [21], temperature suitability index (TSI) [18], annual average enhanced vegetation index (EVI) [22], annual average precipitation [19] and proximity to main water features [23]. The values of these covariates were extracted to each survey location using ArcGIS 10 Spatial Analyst (ESRI Inc. NY, USA) tool. A total-sets analysis based on a generalized linear regression model and implemented in _bestglm_ package in R [24,25] was then used to select those covariates that were most predictive of _P. falciparum_ prevalence. The best combination of covariates, which was those with the lowest value of the Bayesian Information Criteria (BIC) statistic [26], was selected for the prediction of malaria risk.

A Bayesian model-based geostatistical (MBG) framework was used to produce continuous maps of P. falciparum prevalence at 5×5 km spatial resolution using data from 1967–1992. Individuals participating in each PfPR survey were assumed to be P. falciparum positive with a probability that was the product of a continuous function of the time and location of the survey and an age-correction factor derived from the age range of sampled individuals. A Gaussian random field [27] was used to model the continuous functions of time and space while the age-standardisation factors were modelled using a Bayesian version of the procedure described by [28] to prove predictions within a standard age range 2–10 years (PfPR$_{2-10}$).

Formally, each of the N_i individuals in sample i at a location x and time t was taken to be _P. falciparum_ positive with probability $\widetilde{k}_i P'(x_i,t_i)$, so that the number positive N_i^+ was binomially distributed as follows:

$$N_i^+ | N_i, P'(x_i,t_i) \overset{\text{ind}}{\sim} \text{Bin}\left(N_i, \widetilde{k}_i \, P'(x_i,t_i)\right)$$

Where the coefficients $P'(x_i,t_i)$ were modelled as a Gaussian process and the factor $\widetilde{k}_i$ converted $P'(x_i,t_i)$ to the probability that individuals within the age range for study i at a location x and time t were detected _P. falciparum_ positive, accounting for the influence of age on the probability of detection. The age-standardisation factor $\widetilde{k}_i$ in each population was assumed drawn independently from a distribution $D_{\widetilde{k}}$ whose form is described in the Supplementary Information 1 (SI1), equation SI.2. Skew normal priors were assigned to the square root of the partial sill and the spatial range parameter. A vague but proper prior with an expectation of ten years was used for temporal scale parameter. Uniform priors were assigned to the direction of anisotropy parameter, the square of the "eccentricity" parameter which controls the amount of anisotropy and the temporal parameters for the amplitude of the sinusoidal component and the limiting autocorrelation in the temporal direction. Details of the model specifications are provided in SI1.

The MBG procedure began with an inference stage in which Monte Carlo Markov Chain (MCMC) simulation was used to fit both the geostatistical and age-correction models and a spatial-temporal prediction stage in which samples were generated from the posterior distribution of PfPR$_{2-10}$ at 5×5 km grid within the

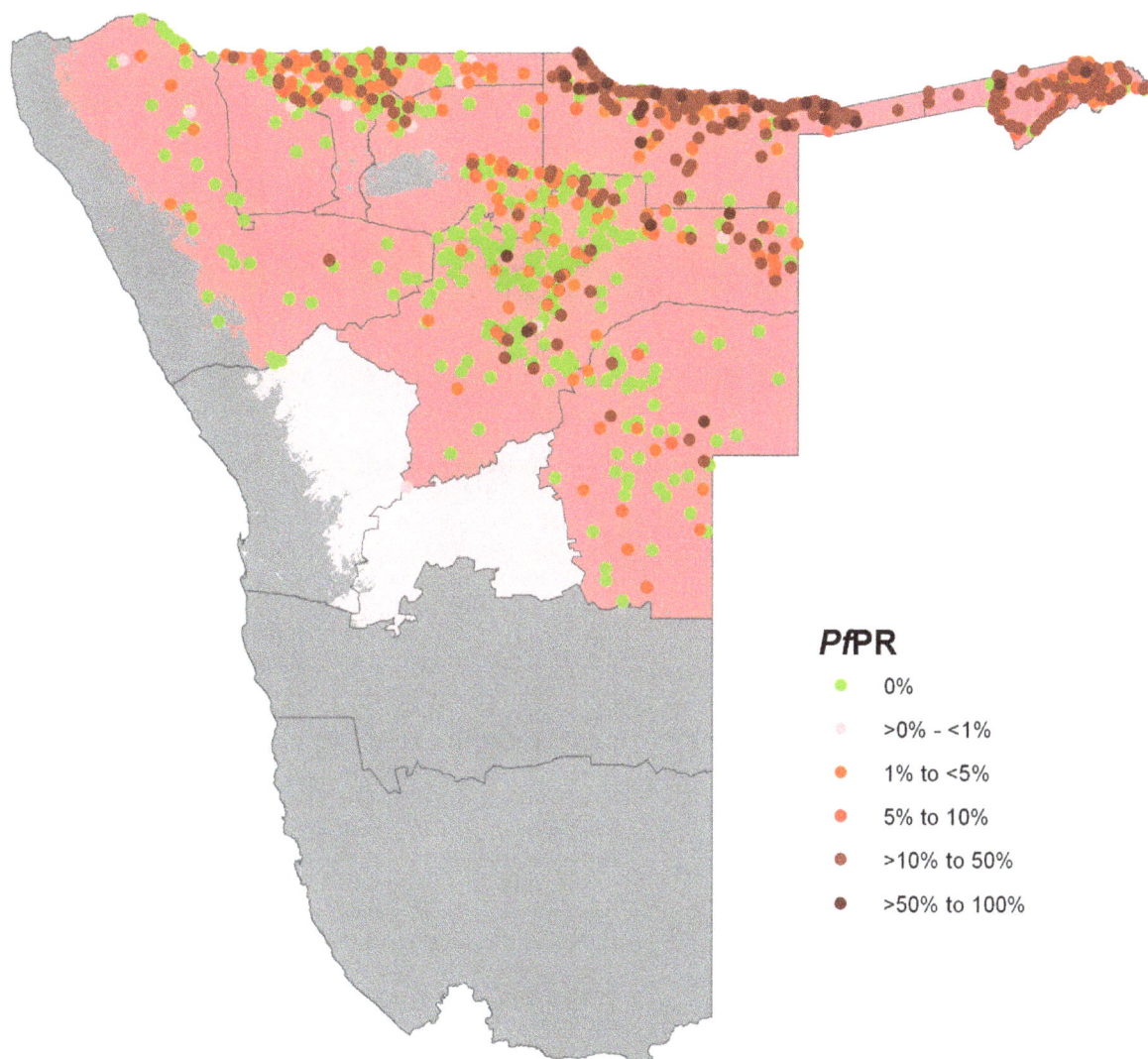

Figure 3. Map of contemporary first level administrative units of Namibia showing the spatial limits of P. falciparum transmission (dark pink) and location of PfPR survey data from the year 1967 to 1992. Where survey data are available for a location in more than one year, the highest PfPR value is displayed top. Light pink areas support unstable transmission and grey areas are malaria free. **Footnote:** A temperature suitability index (TSI) for malaria transmission at 1×1 km spatial resolution [10] was used to delineate areas in Namibia where malaria is unlikely to occur, defined as areas where TSI was equal to zero. TSI was constructed using monthly temperature time series within a biological modelling framework to quantify the effect of ambient temperature on sporogony and vector survivorship and determine the suitability of an area to support transmission globally separately for both P. falciparum and P. vivax. Extreme aridity was defined using synoptic mean monthly enhanced vegetation index (EVI) data [11] to classify into areas unlikely to support transmission, defined as areas where EVI was <0.1 in any two consecutive months of the year [7,12]. In Namibia TSI and aridity identified the Namib Desert on the Atlantic Coast, parts of the Kalahari Desert in the South and the Etosha and other smaller saltpans as areas that were too hot or dry to support malaria transmission. Finally areas defined as operationally risk free based on reported incidences [2,6,7,24] were identified (southern parts of the regions of Kunene and Omaheke and all of Erongo, Hardap, Khomas and Karas). It is hard to eliminate the possibility of any risks and therefore the parts of Erongo, the whole of Khomas and the southern parts of Omaheke were classified to be of unstable transmission where aridity and temperature do not exclude transmission.

limits of stable transmission. The mid-point (in decimal years) between the recorded start and end months was used to temporally reference each survey. For each grid location samples of the annual mean of the full posterior distribution of $PfPR_{2-10}$ for the years 1969, 1974, 1979, 1984 and 1989 were generated. These were then combined to generate a single continuous map of mean $PfPR_{2-10}$. The annual mean $PfPR_{2-10}$ maps were classified into the endemicity classes of $PfPR_{2-10} < 1\%$ (low stable endemic control [29]; $PfPR_{2-10}$ 1% - <5% (hypoendemic 1); $PfPR_{2-10}$ 5% - <10% (hypoendemic 2); $PfPR_{2-10}$ 10% - ≤50% (mesoendemic); $PfPR_{2-10} > 50\%$ (hyper- and holo-endemic).

Model accuracy was estimated by computing the linear correlation, the mean prediction error (MPE) and mean absolute prediction error (MAPE) of the observations and predictions of a 10% hold-out dataset. MPE is a measure of overall model bias while MAPE is a measure model accuracy. The hold-out set was selected using a spatially and temporally declustered algorithm [30] which defined Thiessen polygons around each survey location. Each data point had a probability of selection proportional to the area of its Thiessen polygon so that data located in densely surveyed regions had a lower probability of selection than those in sparsely surveyed regions setting a high threshold for model performance. The Bayesian spatio-temporal geostatistical model was then implemented in full using the remaining 90% of data and predictions were made to the 10% hold-out.

Results

Data assembled comprised 3,497 community PfPR surveys from the period 1967 to 1992 of which 3,260 covering an examination of 230,174 individuals were successfully geocoded and subsequently used in the mapping of risk. The distribution of the PfPR data against the limits of transmission is shown Figure 3. All the tested environmental covariates were separately significant predictors of PfPR but taken together only EVI, precipitation and urbanisation were selected in the best-fit model for the 1967–1992 datasets and subsequently used in the $PfPR_{2-10}$ prediction models. The model MPE was 1.6% while the MAPE of 7.5% indicating good model precision but moderate model bias. The linear correlation of the predicted and observed hold-out set was 0.61.

The annual mean predictions of $PfPR_{2-10}$ reveal small to moderate overall reductions in transmission across the northern regions of Namibia over the period 1969 to 1979, followed by a small rise from 1984 with transmission reaching its maximum over this period by 1989 (Figure 4A–E). In 1969 the highest risk areas were concentrated in Caprivi, Ohangwena and northern Kavango where malaria transmission was largely mesoendemic (>10% to 25% $PfPR_{2-10}$). The southern half of Kavango, northern Oshikoto and an area at the confluence of Otjozondjupa, Oshikoto and Kunene regions had a predicted prevalence of 5% to 10% $PfPR_{2-10}$. Except for areas along the Namib Desert and almost all of Omaheke region where risks were <1% $PfPR_{2-10}$, the rest of country, within the limits of transmission, had predicted prevalence of 1% to <5% $PfPR_{2-10}$.

Between 1969 and 1974 the patterns of risk remained largely unchanged (Figure 4B) but by 1979 the extent of the higher risk classes of 5% to 10% and >10% to 50% $PfPR_{2-10}$ had reduced slightly (Figure 4C). In 1984, however, there appeared to be an expansion of the extent of risk with the high-risk areas (Figure 4D) expanding to margins greater than observed in 1969. This increase continued to 1989 in which the whole of Caprivi, most of Ohangwena and greater areas of northern Kavango were exposed to conditions supporting mesoendemic transmission. By 1989, most of Omaheke region, which was previously exposed to risks of <1% $PfPR_{2-10}$, experienced risks of >1% to <5% $PfPR_{2-10}$ (Figure 4E).

Focusing only on the main intervention areas of Caprivi, Kavango and Ovambo and using 1969 as a reference year, the reduction in risk by 1974 was less 1% overall and by 1979 was between 1% and 2% in the whole of Caprivi and northern Kavango while remaining <1% in the Ovambo and the rest of Kavango (Figure 4F–J). Compared to 1969, transmission had increased by up to about 4% $PfPR_{2-10}$ in Caprivi and most of Kavango and between 4% and 12% in western Ovambo by 1989. The highest mean annual predicted prevalence at anytime from 1969 to 1989 was in Caprivi of around 25% $PfPR_{2-10}$.

Discussion

We have assembled one of the largest long-term, national historical series of malaria infection prevalence in Africa (Figure 3), from surveys undertaken throughout a period of aggressive control targeting both adult vectors and host infection. While data are incomplete it is clear that huge quantities of both DDT and antimalarial drugs were used to control malaria along the northern territories of present day Namibia from the late 1960s through the

1970s (Figure 2). During this period of escalating malaria control large amounts of infection prevalence data were collected to review progress. What was not available to the malaria departments 40 years ago was the ability to model and map changing risk using advanced space-time techniques that interpolate point prevalence data to unsampled locations to accurately estimate risk across a wider area.

We have modelled the changing patterns of malaria transmission intensity between 1969 and 1989 using the assembled data. A striking observation was that the extents of intense transmission declined only marginally throughout the first 5 years of the control efforts, but declined more significantly within the first 10 years by 1979. Thereafter, the margins of higher endemicity risk classes began to expand and by the late 1980s had reached a peak that might be considered comparable to the pre-intervention state (Figures 4 A–E). Focusing only on control priority regions of Caprivi, Kavango and Ovambo, by 1989, the average percentage increase of $PfPR_{2-10}$ from 1969 was between 4% and 12%, with the biggest rebound occurring in eastern Kavango and Caprivi (Figure 4 I–J).

It is not possible here to provide empirical attribution of changing malaria risk to intervention coverage, however, a number of factors may have contributed to a marginal initial decline in risk followed by an expansion to higher risk. First, while the malaria programme maintained concerted efforts to spray houses and provide presumptive treatment this did not cover every household in every region and some regions, notably Caprivi, accessibility was a major obstacle to scale up of interventions. Second, the war for independence, which started in earnest in 1975 and ended in 1988 hampered access in some areas and was associated with large cross-border population movement with Angola probably reducing the likely impact of control on the Namibia side. Third, despite a gradual decline in the use of Darachlor, chloroquine resistance emerged in 1984 and had began to lead to significant clinical failures by the early 1990s [4,2,14]. Forth this region is subject to climatic anomalies that lead to the occurrence of several droughts and above average rainfall across the surveillance period [31]. Finally, human population movement between the higher risk Angola and the northern border regions of Namibia may have played a role in increasing transmission, especially during the war for independence when Namibian fighters engaged South African troops from their Angola bases [32].

To explore the plausibility of these events with changing patterns of observed $PfPR_{2-10}$, Figure 5A–D shows a lowess regression fit in each of the control regions of Caprivi, Kavango and Ovambo against some of the potential modulators of transmission. In Caprivi, the lowess fit of the observed data shows a starting average $PfPR_{2-10}$ of about 20% in 1969, declining to 5% in 1979 and 2% in 1984 before rebounding to 22% in 1989. The period of decline is coincident with the scale-up of DDT indoor residual spraying and Darachlor chemo-suppression. DDT spraying in Caprivi was stopped in 1981, several years earlier than in Kavango and Ovambo, but the decline in observed infection rates appeared to have continued through to 1985, perhaps because of the continued use of Darachlor. The summary of weather anomalies from the 1968/69 to the 1989/90 October-April rainfall season also shows that the first half of the surveillance years were dominated by periods of above average rainfall while in the second half there were more drought events in Caprivi. The year 1979 coincided with the drought of 1978/1979 that also occurred in the 1981/82, 1982/83 and 1984/85 seasons and may have contributed to the very low average infections rates in this region in both 1979 and 1984. In Kavango, average observed infection

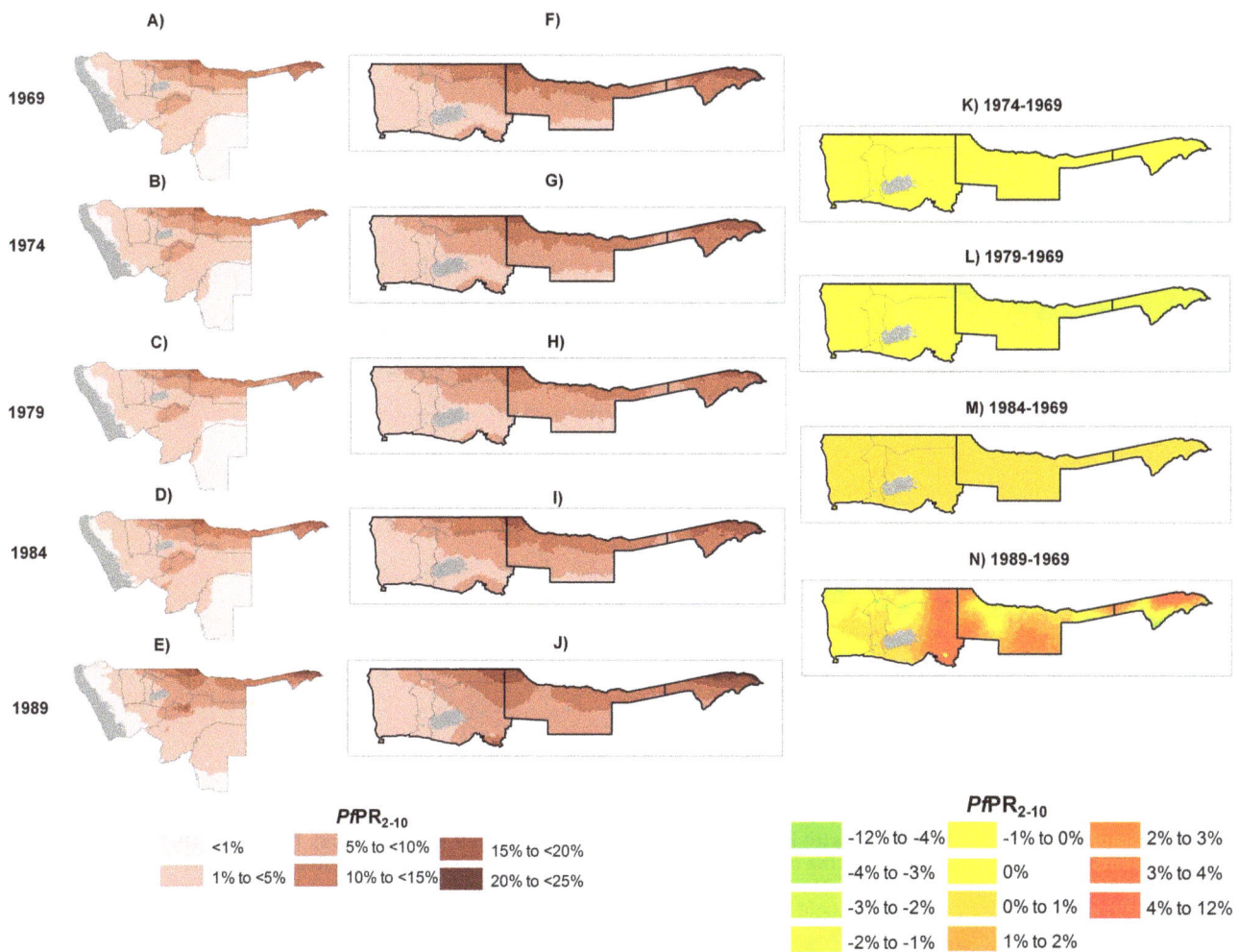

Figure 4. Panel A–D shows the posterior mean distribution of PfPR$_{2-10}$ by the prediction years of 1969–1989 in Namibia. Panel F–J shows enlarged maps of the posterior mean distribution of PfPR$_{2-10}$ by the prediction years of 1969–1989 in the three intervention regions of Caprivi, Kavango and Ovambo. Panel K–N shows the percentage difference from a reference year of 1969 of the PfPR$_{2-10}$ predictions in 1974, 1979, 1984 and 1989. Negative/positive values indicate a decrease/increase in predicted PfPR$_{2-10}$. The grey shade represents areas where transmission is unsuitable due to aridity. All maps are at 5×5 km spatial resolution.

rates were at around 4% in 1969, rising and holding at about 10% from 1974 to 1979, reducing slightly to 8% in 1984 before rising from 1985 to the highest levels of 28% in 1989. Darachlor administration for chemo-suppression appears to have been stopped by 1985 although DDT spraying continued to 1989. Similar to Caprivi, above average rainfall events were common in the period 1969 to 1979 followed by frequent droughts thereafter. Coincident with the PfPR$_{2-10}$ prediction years also were droughts in the 1979/80 and the 1984/85 and extreme rainfall in the 1988/89 season. In Ovambo, average observed prevalence remained low throughout the period 1969 to 1986, at about 2%, before rising to approximately 10% by 1989. Both DDT spraying and Darachlor administration began earlier in Ovambo, several years before the surveillance period, and may explain the very low starting prevalence, although data before 1978 were sparse. DDT spraying and Darachlor administration continued to 1989 and 1988 respectively. The rainfall patterns over the surveillance period were similar to those of Caprivi and Kavango.

In the intervention areas, the years of above average rainfall generally coincided with much higher upper limits of observed *P. falciparum* infections compared to the years when a drought

occurred although this seemed not to dramatically shift the average infections rates (Figure 5). In addition, despite the 13-year period of war starting 1975, a rise in infection rates seems to start only after 1984, although anecdotal information shows that the effect of the war on control may have started earlier. For example in western Caprivi, insecurity affected control efforts during the period 1979 to 1981. Independence fighters stationed in northern Angola also frequently crossed the border into Namibia carrying infections that may have led to localised outbreaks. In parts of Ovambo and Kavango, there was increased reluctance by residents to have their houses sprayed. The emergence of chloroquine resistance in 1984 and its gradual increase through to the 1990s may also have contributed to the observed rise in infection rates during this period across the three intervention areas and the subsequent reported epidemic of 1990 [33].

The intensive efforts and vector control using DDT and parasite suppression using Darachlor in Caprivi, Kavango and Ovambo appeared to be associated, at a glance, with only small to moderate reductions in transmission over the first 15 years of the surveillance period. It is not possible to define a pre-intervention, pre-1960s, endemicity across northern Namibia, although this is likely to have

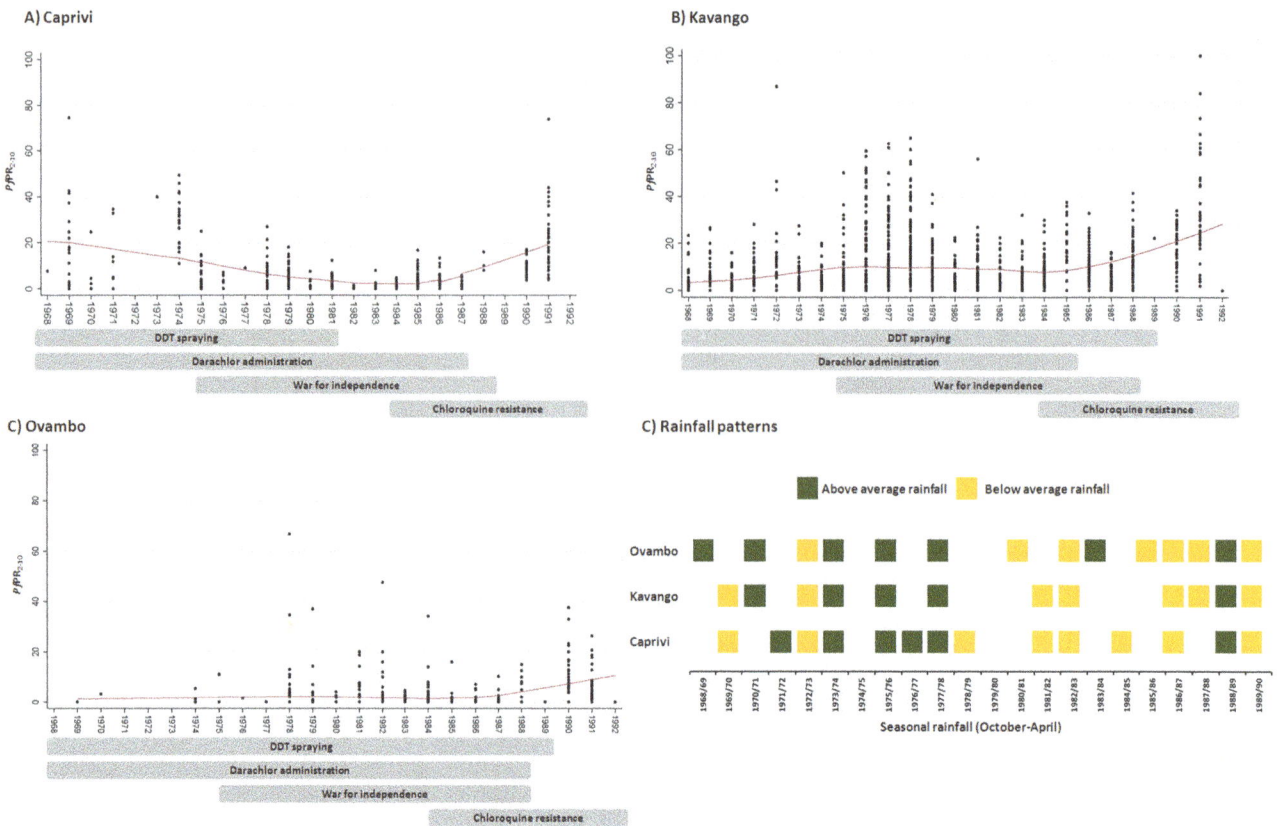

Figure 5. Graphs A–C show scatter plots with lowess regression fit of PfPR$_{2-10}$ by year of survey and the main intervention and political events that may have determined the observed transmission patterns in Caprivi, Kavango and Ovambo respectively. D) is a summary of the droughts (orange) and above average rainfall (green) as measured using rainfall in the October – April rainfall in Caprivi, Kavango and Ovambo. Seasons not designated as one of drought or above average rainfall are considered to have received average rainfall.

been considerably higher than the data shown for 1969 [13]. A reasonable description of the outcome of sustained control between 1969 and 1992 is that up to 1979 a combination of vector and parasite control measures, maintained transmission at a low level, probably negating the effect of the several periods above average rainfall. In the latter years despite declining control the frequent droughts probably helped sustain risks at low levels until the escalating war, the failure of chloroquine and the high rainfall of the 1988/89 season led to the epidemics of 1990 resulting in the expansion of high transmission across the northern territories of Namibia.

Following the epidemic in 1990, the National Malaria Control Programme, later renamed the National Vector-borne Disease Control Programme (NVDCP) was set up by the Namibian government [16]. In March 2009, a meeting of the Elimination Eight (E8) countries of the Southern Africa Development Community (SADC) was held in Windhoek with the aim of eliminating malaria in the four first tier countries (Botswana, Namibia, South Africa and Swaziland) by 2015 while accelerating control with the aim of eventual elimination in the four second tier countries (Angola, Mozambique, Zambia and Zimbabwe) [1]. In April 2010, the Namibian government launched a malaria elimination campaign to move the country to pre-elimination/ elimination in the next five to ten years [34]. Human population movement has been identified as one of the major obstacles to malaria elimination and the historical maps of malaria risk developed in this study will contribute to the quantification of

these risks as they describe the 'probable' innate transmission intensity in the malarious northern regions of Namibia.

In conclusion, the documented efforts directed at controlling malaria in northern Namibia probably sustained a period of low parasite transmission for over ten years, but the combination of climatic anomalies, insecurity, cross-border movement and a declining efficacy and coverage demonstrate how fragile control effects are and when these, harder to manage features occur, malaria transmission increases in its intensity and its spatial extent.

Acknowledgments

The authors would like to thank Christine Theron for her help with assembly of the community survey parasite prevalence data and for facilitation communication with her father C. Frank Hansford. We thank Jacob Ouko, Damaris Kinyoki and Caroline Kabaria for their help with assembly of ancillary spatial data and geo-coding of villages and Anand Patil for his advice on the MBG code. We are most grateful to John Mendelshon for providing spatial databases for Namibia including the location of settlements from which the survey villages were geocoded. The authors are also grateful for comments on earlier drafts of the manuscript from Dr Emelda Okiro.

Author Contributions

Conceived and designed the experiments: AMN RWS CFH RNK. Performed the experiments: AMN VAA RWS. Analyzed the data: AMN VAA RWS CFH. Contributed reagents/materials/analysis tools: SK RNK. Wrote the paper: AMN VAA CFH RNK BN SK RWS.

References

1. The Malaria Elimination 8 Group website. Available: http://www.sadc.int/english/regional-integration/shdsp/sarn/elimination-eight-e8/. Accessed Apr 10 2012.
2. Ministry of Health and Social Services (2010). National Vector Borne Diseases Control Programme. Ministry of Health and Social Services. National Malaria Strategic Plan.
3. Snow RW, Alegana VA, Makomva K, Reich A, Uusiku P, et al. (2010). Estimating the distribution of malaria in Namibia in 2009: assembling the evidence and modeling risk. Ministry of Health & Social Services, Republic of Namibia & Malaria Atlas Project.
4. Hansford CF (1990). Malaria Control in Namibia. National Institute for Tropical Disease. Tzaneen, South Africa: Department of National Health and Population Development, (unpublished document).
5. Kamwi RN (2005). Malaria Situation in Namibia. A Study of Vector Species and Effectiveness of the Past and Current Control Strategies in Selected Parts of Namibia [PhD thesis].Namibia: University of Namibia.
6. Diggle PJ, Moyeed R, Rowlinson B, Thomson M (2002).Childhood malaria in the Gambia: a case-study in model-based geostatistics. Appl Stats;51: 493–506.
7. Kazembe LN, Kleinschmidt I, Holtz TH, Sharp BL (2006). Spatial analysis and mapping of malaria risk in Malawi using point-referenced prevalence of infection data. Int J Health Geogr; 5: e41.
8. Noor AM, Moloney G, Borle M, Fegan GW, Shewshuk T, et al. (2008). The use of mosquito nets and the prevalence of Plasmodium falciparum infection in rural South Central Somalia. PLoS One;3: e2081.
9. Noor AM, Gething PW, Alegana VA, Patil AP, Hay SI, et al. (2009). The risks of Plasmodium falciparum infection in Kenya in 2009. BMC Infect Dis; 9: e180.
10. Gosoniu L, Veta AM, Vounatsou P (2010). Bayesian geostatistical modeling of malaria indicator survey data in Angola. PLoS One; 5: e9322.
11. Hay SI, Guerra CA, Gething PW, Patil AP, Tatem AJ, et al. (2009). A world malaria map: Plasmodium falciparum endemicity in 2007. PLoS Med; 6: e1000048.
12. Gething PW, Patil AP, Smith DL, Guerra CA, Elyazar IR, et al. (2011). A new world malaria map: Plasmodium falciparum endemicity in 2010. Malar J; 10: 378.
13. De Meillon B (1951). Malaria survey of South-West Africa. Bull World Health Org; 4: 333–417.
14. Snow RW, Amratia P, Kabaria CW, Noor AM, Marsh K (2012). The changing limits and incidence of malaria in Africa: 1939–2009. Adv Parasitol; 78: 169–262.
15. Mabaso M, Sharp B, Lengeler C (2004).Historical review of malarial control in southern African with emphasis on the use of indoor residual house-spraying. Trop Med Int Health; 9: 846–56.
16. Ministry of Health and Social Services (1995) National Policy and Strategy for Malaria Control. National Vector Borne Diseases Control Programme.
17. Mendelsohn JM, Jarvis AM, Roberts CS, Robertson T (2002). Atlas of Namibia. A comprehensive atlas and reference work for the country, covering human, physical and biological aspects of Namibia. Compiled for the Ministry of Environment and Tourism, Cape Town.
18. Gething PW, Van Boeckel TP, Smith DL, Guerra CA, Patil AP, et al.(2011). Modelling the global constraints of temperature on transmission of Plasmodium falciparum and P. vivax. Parasite Vectors; 4: 92 doi: 10.1186/1756-3305-4-92.
19. Hijmans RJ, Cameron SE, Parra JL, Jones PG, Jarvis A (2005). Very high resolution interpolated climate surfaces for global land areas. Int J Climatol; 25: 1965–78.
20. Guerra CA, Gikandi PW, Tatem AJ, Noor AM, Smith DL, et al.(2008). The limits and intensity of Plasmodium falciparum transmission: implications for malaria control and elimination worldwide. PLoS Med; 5: e38.
21. AfriPop Project Website. Available: www.afripop.org. Accessed Aug 10 2012.
22. MODIS- EVI data archives Website. Available: ftp://n4ftl01u.ecs.nasa.gov/SAN/MOST/MOD10A2.005/. Accessed Sep 12 2011.
23. Global Lakes and Wetlands Database Website. Available: https://secure.worldwildlife.org/science/data/item1877.html. Accessed Sep 2012.
24. Miller A (2002). Subset Selection in Regression. Boca Raton, FL: Chapman & Hall.
25. Lumley T(2010).leaps: regression subset selection (R package) Version 2.7.
26. Schwarz G(1978). Estimating dimensions of a model. Ann Stat; 6: 461–64.
27. Banerjee S, Carlin BP, Gelfand AE(2004).Hierarchical modeling and analysis for spatial data. Monographs on Statistics and Applied Probability 101. Boca Raton, Florida, U.S.A: Chapman & Hall/CRC Press LLC.
28. Smith DL, Guerra CA, Snow RW, Hay SI (2007). Standardizing estimates of the Plasmodium falciparum parasite rate Malar J; 6: e131.
29. Cohen JM, Moonen B, Snow RW, Smith DL (2010). How absolute is zero? An evaluation of historical and current definitions of malaria elimination. Malar J; 9: 213.
30. Isaacs E, Srivastava R (1989). Applied Geostatistics. Oxford University Press.
31. Botha L. History of droughts in Namibia. Website. Available:http://www.nbri.org.na/old/agricola_files/Agricola1998_99_No10_03_Botha.PDF. Accessed Nov 12 2012.
32. Namibia 1966–1990 Website. Available: http://www.historyofwar.org/articles/wars_namibia.html, Accessed Jun 18 2012.
33. Ministry of Health and Social Services (2005). National Malaria Policy. National Vector Borne Diseases Control Programme.
34. Roll Back Malaria Press Release. Launch of the Namibia Elimination Strategic Plan Website. Available: http://www.rbm.who.int/worldmalariaday/docs/SARNnamibia.pdf. Accessed Dec 2 2012.

6

A *Medicago truncatula* EF-Hand Family Gene, *MtCaMP1*, Is Involved in Drought and Salt Stress Tolerance

Tian-Zuo Wang[1,◐], Jin-Li Zhang[1,◐], Qiu-Ying Tian[1], Min-Gui Zhao[1], Wen-Hao Zhang[1,2]*

1 State Key Laboratory of Vegetation and Environmental Change, Institute of Botany, The Chinese Academy of Sciences, Beijing, P. R. China, **2** Research Network of Global Change Biology, Beijing Institutes of Life Science, The Chinese Academy of Sciences, Beijing, P. R. China

Abstract

Background: Calcium-binding proteins that contain EF-hand motifs have been reported to play important roles in transduction of signals associated with biotic and abiotic stresses. To functionally characterize gens of EF-hand family in response to abiotic stress, an *MtCaMP1* gene belonging to EF-hand family from legume model plant *Medicago truncatula* was isolated and its function in response to drought and salt stress was investigated by expressing *MtCaMP1* in Arabidopsis.

Methodology/Principal Findings: Transgenic Arabidopsis seedlings expressing *MtCaMP1* exhibited higher survival rate than wild-type seedlings under drought and salt stress, suggesting that expression of *MtCaMP1* confers tolerance of Arabidopsis to drought and salt stress. The transgenic plants accumulated greater amounts of Pro due to up-regulation of *P5CS1* and down-regulation of *ProDH* than wild-type plants under drought stress. There was a less accumulation of Na^+ in the transgenic plants than in WT plants due to reduced up-regulation of *AtHKT1* and enhanced regulation of *AtNHX1* in the transgenic plants compared to WT plants under salt stress. There was a reduced accumulation of H_2O_2 and malondialdehyde in the transgenic plants than in WT plants under both drought and salt stress.

Conclusions/Significance: The expression of *MtCaMP1* in Arabidopsis enhanced tolerance of the transgenic plants to drought and salt stress by effective osmo-regulation due to greater accumulation of Pro and by minimizing toxic Na^+ accumulation, respectively. The enhanced accumulation of Pro and reduced accumulation of Na^+ under drought and salt stress would protect plants from water default and Na^+ toxicity, and alleviate the associated oxidative stress. These findings demonstrate that *MtCaMP1* encodes a stress-responsive EF-hand protein that plays a regulatory role in response of plants to drought and salt stress.

Editor: Diane Bassham, Iowa State University, United States of America

Funding: The work was supported by the State Key Lab of Vegetation and Environmental Change. The funders had no role in study design, data collection and analysis, decision to publish, or preparation of the manuscript. The funders had no role in study design, data collection and analysis, decision to publish, or preparation of the manuscript.

Competing Interests: The authors have declared that no competing interests exist.

* E-mail: whzhang@ibcas.ac.cn

◐ These authors contributed equally to this work.

Introduction

Calcium ion ($Ca^{2+)}$ is a fundamental transducer and regulator in many processes associated with growth and development as well as response to biotic and abiotic stresses in plants. In plant cells, Ca^{2+} is stored in cell walls and several vesicular compartments (vacuole, endoplasmic reticulum and mitochondria). Under resting conditions, the cytosolic free calcium concentration ($[Ca^{2+}]_{Cyt}$) is maintained at approx. 100 nM, which is much less than Ca^{2+} concentrations in the "Ca^{2+} stores" [1]. The movement of Ca^{2+} in and out of cells and organelles can be regulated by Ca^{2+} channels and/or pumps [2,3]. Upon perceiving cues associated with abiotic and biotic stresses, a rapid increase in $[Ca^{2+}]_{Cyt}$ will occur by activating Ca^{2+} channels and/or pumps. Signals related to abiotic stress signals including drought, low and high temperatures, salt and oxidative stress often elicit a rapid elevation of $[Ca^{2+}]_{Cyt}$ [4–6]. The elevated $[Ca^{2+}]_{Cyt}$ is recognized by Ca^{2+} sensors, which in turn relay the information to downstream targets, leading to the alteration of gene expression [7]. The Ca^{2+}-dependent proteins regulate physiological and metabolic processes, resulting in phenotypic response to stress [5,8].

Ca^{2+} sensors are Ca^{2+}-binding proteins, and most of the Ca^{2+} sensors contain EF-hand motifs which can bind to Ca^{2+} [9,10]. The major family of Ca^{2+} sensors includes calmodulin (CaM), calcium-dependent protein kinases (CDPK) and calcineurin B-like proteins (CBL) [11,12]. Several Ca^{2+}-binding proteins have been shown to participate in transduction of signals associated with biotic and abiotic stress. CaM is one of the small, conserved Ca^{2+}-binding proteins in eukaryotes [13]. For example, GmCaM4 directly interacts with AtMYB2 in *Arabidopsis thaliana*, and expression of *GmCaM4* in Arabidopsis up-regulates *AtMYB2*-regulated genes, including *P5CS1* (Δ^1-pyrroline-5-carboxylate synthetase 1) that catalyzes Pro biosynthesis, leading to enhanced tolerance to salt stress [14]. AtCML18, a CaM-like protein, can interact with the C-terminus of AtNHX1, a vacuolar Na^+/H^+ antiporter [15]. CDPKs comprise a large family of serine/threonine kinases with kinase domains in the N-terminals in plants [16]. *AtCPK21* is responsive to osmotic stress, whose N-terminal EF-hand pair is a calcium-sensing determinant that is involved in abiotic stress signaling [17]. In addition, CDPKs also play an essential role in response of plants to biotic stress. For instance, silencing *NtCDPK2* renders plants less sensitive to Avr9 virus infection [18]. CBL, a Ca^{2+}-binding protein, forms a complex network with their target CBL-interacting protein kinases (CIPK) to regulate a wide range of physiological processes [19,20]. The expression of *AtCBL1* is markedly up-regulated by drought, cold, and wounding, and is involved in mediation of calcium signaling under certain stress conditions [21]. AtCBL4 (SOS3) can interact with AtCIPK24 (SOS2) by forming a SOS3-SOS2 complex, which in turn activates Na^{+}/H^{+} exchanger SOS1 in the plasma membrane of Arabidopsis to pump Na^{+} out of cells. Mutations in the *SOS3* gene render plants hypersensitive to NaCl [22,23]. Expression of a maize calcineurin B-like protein ZmCBL4 in Arabidopsis confers salt tolerance [24]. In addition, cyclic nucleotide-gated channels (CNGC) may participate in the Ca^{2+}-dependent signaling cascades [25]. There have been a number of studies on the functional characterization of Ca^{2+}-binding proteins in literature so far. However, there have been few studies focusing on the role of Ca^{2+}-binding proteins played in response to abiotic stress in legume species in general and legume model plant *Medicago truncatula* in particular.

In the present study, we isolated a drought-responsive, calcium-binding motif-containing protein gene by the method of suppression subtractive hybridization (SSH), designated *MtCaMP1*, from *Medicago truncatula*, a model plant that has been widely used to study functional genomics of legume plants [26]. We further functionally characterized *MtCaMP1* in response to drought and salt stress by generating transgenic Arabidopsis expressing *MtCaMP1*. Our results demonstrate that expression of *MtCaMP1* in Arabidopsis led to an enhanced tolerance to drought and salt stress. We further explored the physiological mechanisms underlying the enhanced tolerance of the transgenic plants to drought and salt stress.

Materials and Methods

Plant growth and treatments

Seeds of *Medicago truncatula* Jemalong A17 were soaked in concentrated H_2SO_4 solution for approximately 7 min, and then thoroughly rinsed with water. After chilled at 4°C for 2 d, seeds were grown in a pot (diameter 10 cm) filled with vermiculite: peat soil (2:1) under controlled conditions (26°C day/20°C night, 14-h photoperiod, and 50% relative humidity) as described by Wang et al. [27].

The effect of abiotic stress on the expression of *MtCaMP1* was evaluated by treatments of four-week-old *M. truncatula* seedlings with drought, salt and osmotic stress, respectively. Drought stress was initiated by withholding water supply to seedlings for varying periods after seedlings were fully watered. Shoots from seedlings suffering from drought stress were harvested after withholding water for 4, 6, and 8 d. Shoots of M. truncatula seedlings grown under normal watering conditions were harvested and used as a control. Salt and osmotic stresses were achieved by exposing 4-week-old seedlings to 1/2 MS medium supplemented with 200 mM NaCl and 20% PEG6000 for different periods, respectively.

Identification of *MtCaMP1* gene and plant transformation

To identify gene fragments in response to drought stress, suppression subtractive hybridization (SSH) was used to construct a cDNA library [28]. SSH was carried out using a PCR-Select cDNA Subtraction Kit (Clontech) according to the manufacturer's instruction. A gene fragment was identified from this cDNA library. The full-length sequence of *MtCaMP1* cDNA was obtained by BLASTn search [29].

The ORF of *MtCaMP1* was amplified with the primers 5'- GAC GGA TCC ATG TCA TTC CTT TCC ACT CT -3' (*Bam*HI site underlined) and 5'- CAG GAG CTC CTA ACA TAA GAG CAA AAC AC -3' (*Sac*I site underlined). The *Eco*RI/*Bam*HI -digested product was inserted in the downstream of cauliflower mosaic virus 35S (CaMV 35S) promoter of pSN1301 [30]. After pSN1301: *MtCaMP1* was transformed to *Agrobacterium tumefaciens*

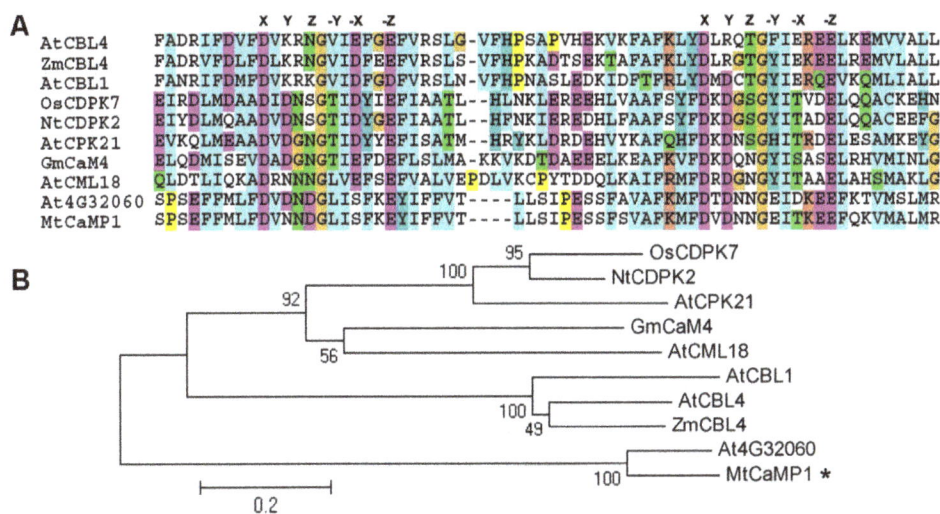

Figure 1. Sequence analysis of EF-hand family proteins. The highly conserved EF-hand motifs of *MtCaMP1* and other known EF-hand family proteins were aligned in panel A. Alignments were performed using the ClustalX2.1 software. X, Y, Z, -Y, -X and -Z represent conserved amino acids of Ca^{2+}-binding loop. Phylogenetic tree of these proteins was constructed by MEGA 5 in panel B. The corresponding ID of AtCBL4 (SOS3), ZmCBL4, AtCBL1, OsCDPK7, AtCPK21, GmCaM4, AtCML18 (CaM15) and At4G32060 are NP_001190377.1, NP_001150076.1, NP_567533.1, BAB16888.1, CAC82998.1, NP_192381.1, NP_001237902.1, NP_186950.1 and NP_194934.1, respectively.

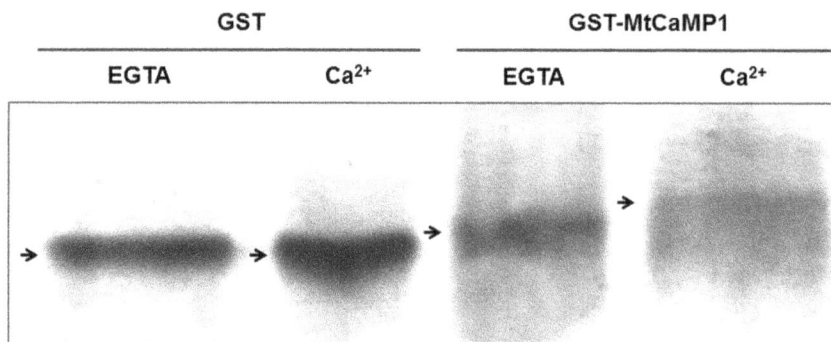

Figure 2. Ca²⁺-dependent electrophoretic mobility shift assay of glutathione S-transferasefucata (GST) and GST-MtCaMP1 recombinant protein. Purified protein was run on a 10% non-denaturing polyacrylamide gel in the presence of Ca²⁺ or EGTA. CaCl2 at 5 mM or EGTA was added into the sample buffer.

GV3101 by electroporation, transformation of *Arabidopsis thaliana* (Col-0) was performed using the *Agrobacterium tumefaciens*- mediated floral dip method [31]. Three independent lines of the T3 generation were randomly chosen for further physiological studies.

Expression and purification of the EF-hand domain of MtCaMP1

To express and purify the EF-hand domain of MtCaMP1 in *Escherichia coli*, the coding region of the EF-hand domain of MtCaMP1 was amplified by PCR using the primer 5′- AGT GGA TCC TTG AGA GGA GAA AGA AG -3′ containing an *Bam*HI site (underlined), and 5′- GTA CTC GAG ATC CCT GTG TTG AAC AC -3′ containing an *Xho*I site (underlined). The PCR products were digested with *Bam*HI/*Xho*I, and ligated to the C-terminal of glutathione S-transferase (GST) in pGEX-4T-1 (GE). The recombinant plasmid was transformed into *E. coil* BL21. Protein expression was induced with 0.5 mM isopropyl thiogalactopyranoside (IPTG) at 30°C. When the optical density at 600 nm of the culture was reached 0.8 IPTG was added. Bacterial cells were harvested after induction for 6 h by centrifuging the culture at 4000 *g* for 10 min. The purification of recombinant protein was carried out by GST Resin (TransGen).

Ca²⁺-dependent electrophoretic mobility shift assay

Ca²⁺-dependent electrophoretic mobility shift assay (EMSA) was carried out according to the method descrbied by Burgess et al. and Takezawa [32,33]. CaCl₂ at 5 mM or ethylene glycol-bis-(β-amino-ethylether) N,N,N',N'-tetra-acetic acid (EGTA) was added into the sample buffer of GST protein and GST-MtCaMP1

recombinant protein. Proteins were analyzed on 10% polyacrylamide gel electrophoresis under non-denaturing conditions, and stained with Coomassie Brilliant Blue R-250.

RNA isolation, semi-quantitative and real-time quantitative PCR

Total RNA was isolated using RNAiso Plus reagent (TaKaRa) and treated with RNase-free DNase I (Promega). The total RNA was reverse-transcribed into first-strand cDNA with PrimeScript® RT reagent Kit (TaKaRa).

The semi-quantitative PCR was used to determine the expression level of *MtCaMP1* in transgenic plants. The primers for transgenic Arabidopsis were 5′-CTC AGC ATC CCA GAA TCA A-3′ and 5′-CAG TTC GCA GTC CAT CCC T-3′. *AtActin11* (accession No. NM_112046) was used as an internal control and was amplified with the following primers: 5′-TGT TCT TTC CCT CTA CGC T-3′ and 5′-CCT TAC GAT TTC ACG CTC T-3′. Real-time quantitative PCR (RT-qPCR) was performed using ABI Stepone Plus instrument. Gene-specific primers used for RT-qPCR were as follows: for *MtCaMP1* (5′-CTC AGC ATC CCA GAA TCA A-3′ and 5′-CAG TTC GCA GTC CAT CCC T-3′); for *AtP5CS1* (5′-CTC GCT TAG TTA TGA CGC-3′ and 5′-CTC CTT TCC ACC CTT TA-3′); for *AtProDH* (5′-ATC TTA CCG TTT ACC CG-3′ and 5′-TCA CCG AAG CGT CCA TA-3′); for *AtHKT1* (5′-CAT CTG GCT CCT AAT CCC T-3′ and 5′-ACC ATA CTC GTC ACG CTT T-3′); for *AtNHX1* (5′-GGT TGC CCT TAT GAT GCT TA-3′ and 5′-CTC ACG GAT CTC CAC TTG TC-3′); *MtActin* gene (accession No. BT141409) and *AtActin11* were used as internal

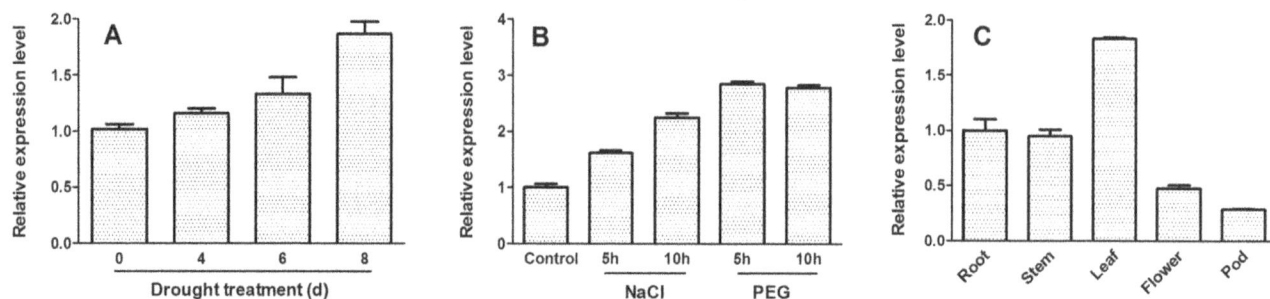

Figure 3. Expression patterns of *MtCaMP1* in *M. truncatula*. Expression of *MtCaMP1* in response to drought stress for varying periods was analyzed in panel A. Effect of 200 mM NaCl and 20% PEG6000 on expression of *MtCaMP1* was shown in panel B. Expression patterns of *MtCaMP1* in different organs were shown in panel C. Data are means±SE with three biological replicates.

controls with primers (5′-ACG AGC GTT TCA GAT G-3′ and 5′-ACC TCC GAT CCA GAC A-3′) and (5′-TGT TCT TTC CCT CTA CGC T-3′ and 5′-CCT TAC GAT TTC ACG CTC T-3′). Primers were designed using the Premier 5 software with the principle to avoid false priming and dimmers. Blasting primers in NCBI and sequencing of PCR produces were used to confirm whether the primers were gene-specific. Each reaction contained 10.0 μL of SYBR Green Master Mix reagent (TOYOBO), 0.8 μL cDNA samples, and 1.2 μL of 10 μM gene-specific primers in a final volume of 20 μL. The thermal cycle used was 95°C for 2 min, 40 cycles of 95°C for 30 s, 55°C for 30 s, and 72°C for 30 s. Three biological and three technological repeats were performed in RT-qPCR. The relative expression level was analyzed by the comparative C_T method using the Microsoft Excel 2010 as described by Livak and Schmittgen [34].

Determination of tolerance to drought and salt stress

Arabidopsis seedlings were planted in pots under the same conditions as described for *M. truncatula*. To determine survival rate after treatment with drought and salt stress, 4-week-old Arabidopsis seedlings were exposed to drought and salt stress by withholding water and irrigating with 200 mM NaCl for 14 days, respectively. After the treatments, plants were transferred to normal growth conditions. The survival rate was determined by scoring the seedlings that failed to grow after recovery from treatments with drought or salt stress. Plants were considered to be dead if all leaves were brown and re-growth did not occur after the plants were transferred to the control growth conditions.

To determinate seed germination rate under osmotic and salt stress, sterile seeds were pointed to 1/2 MS plate (0.8% agar) supplemented with either 300 mM mannitol or 200 mM NaCl at 25°C. Plates that were not supplemented with mannitol or NaCl were used as controls. There were 40 seeds in each plate and the seeds were considered to be germinated at the emergence of the plumule and scored. Seed germination was recorded after 72 h of incubation.

Determination of water loss rate

For measurements of water loss rate, mature leaves from 4-week-old transgenic and wild-type (WT) Arabidopsis plants were weighed immediately. The leaves were then kept on filter paper in an illumination incubator under the conditions of humidity 40% and temperature of 26°C for varying periods (0, 0.5, 1, 2, 3, 4, 5 and 6 h) and weighed again. Water loss rate was determined by measuring the percentage of fresh weight loss relative to the initial plant weights.

Determination of H_2O_2, malondialdehyde (MDA) and proline (Pro) contents

Four-week-old seedlings withheld water or irrigated with 200 mM NaCl for 6 days harvested for determination of H_2O_2

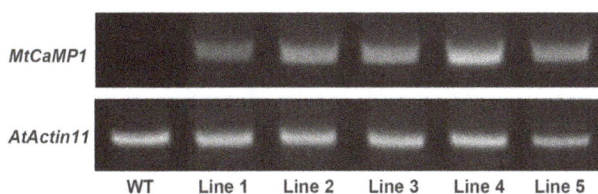

Figure 4. Analysis of *MtCaMP1* expression level in wild-type and transgenic plants. Expression level of *MtCaMP1* in wild-type (WT) and transgenic plants (line1–5) was monitored by RT-PCR.

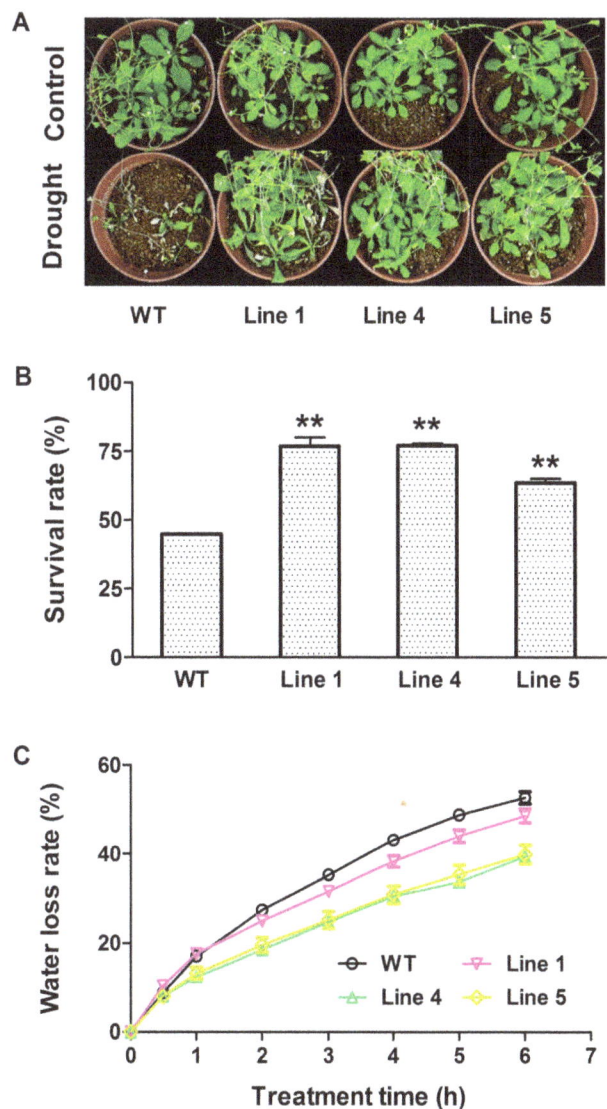

Figure 5. Effect of drought stress on wild-type and transgenic plants. Phenotypes of wild-type and transgenic plants after withholding water for 14 days were shown in panel A. Survival rates were scored after recovery of watering for 7 days (panel B). Water loss rates were determined at 0, 0.5, 1, 2, 3, 4, 5 and 6 h after drought treatment (panel C). Data are mean±SE with three replicates. Asterisks represent statistically significant differences between wild-type and transgenic lines. * $P \leq 0.05$, ** $P \leq 0.01$.

and MDA contents. Seedlings grown in normal conditions were as control. Hydrogen peroxide was measured followed the protocols reported in the literature [35]. MDA content in leaves was determined following the protocol described by Kramer et al. [36]. Pro content in leaves of Arabidopsis was determined as described by Bates et al. [37].

Measurement of Na+ and K+ concentrations

Four-week-old seedlings of transgenic and WT plants irrigated with 1/2 MS solution containing 200 mM NaCl for 6 days were harvested and dried at 80°C, and digested with the mixture of nitric acid and hydrogen peroxide using microwave system (MARS, CEM). The digested samples were used to determine K+ and Na+ contents by ICP-AES (Thermo).

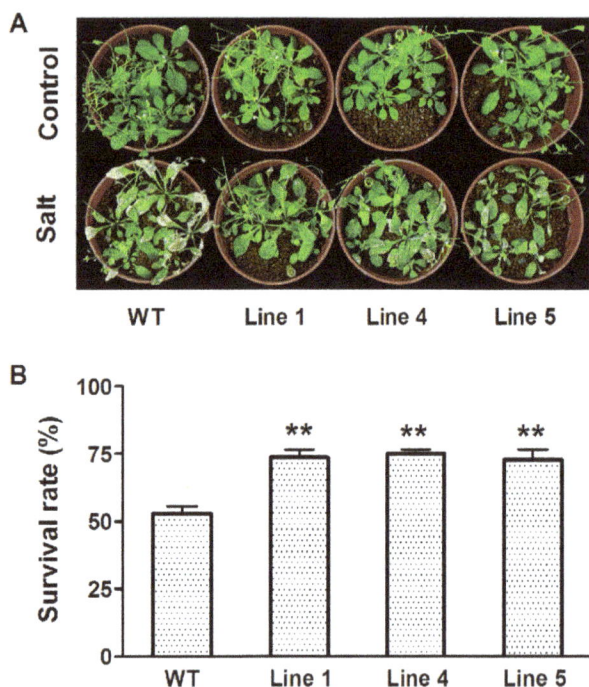

Figure 6. Effect of salt stress on wild-type and transgenic plants. Phenotypes of wild-type and transgenic plants after treatment with NaCl for 14 days were shown in panel A. Survival rates were counted after recovery of watering for 7 days (panel B). Data are mean±SE with three replicates. Asterisks represent statistically significant differences between wild-type and transgenic lines. * $P \leq 0.05$, ** $P \leq 0.01$.

Statistics analyses

All data were analyzed by analysis of variance using the One-way ANOVA method of SPSS17.0 statistics program. Statistical differences are referred to as significant when $P \leq 0.05$.

Results

Isolation and characterization of MtCaMP1 gene

Based on sequences of gene fragments isolated from drought-stress suppression subtractive hybridization library, a full-length sequence of a cDNA was obtained by BLASTn (accession No. BT053074). This gene had a 939 bp open reading frame encoding a protein of 312 amino acid residues, with a calculated molecular mass of about 35.5 kDa. Sequence comparison revealed that the putative protein is a calcium-binding motif-containing protein, and is highly homologous to other EF-hand family proteins (Fig. 1). Therefore, this gene was designated MtCaMP1 (Calcium-binding Motif-containing Protein 1) accordingly.

The EF-hand motif is a helix-loop-helix structure that binds to a single Ca^{2+} ion [9]. This loop consists of 12 residues with the pattern X*Y*Z*-Y*-X**-Z. The residues X, Y, Z, -Y, -X and -Z participate in binding to Ca^{2+} ions, and the intervening residues are represented by asterisk (*). As shown in Figure 1A, MtCaMP1 protein contains two EF-hand motifs which confer EF-hand family proteins the ability of binding Ca^{2+}. Similar motifs have been found in many reported EF-hand family proteins. Because the presence of Ca^{2+} would reduce negative charges and lead to a change in conformation, we observed that the recombinant protein was migrated more slowly to the positive pole (Fig. 2). This result validates that the MtCaMP1 can bind to Ca^{2+}.

Phylogenetic trees based on the full-length amino acid sequences of MtCaMP1 proteins were constructed using the MEGA 5 software (Fig. 1B). The phylogenetic trees show that MtCaMP1 protein had the highest similarity with At4G32060 with unknown function. However, no ortholog of MtCaMP1 was found in Arabidopsis.

Expression patterns of MtCaMP1 gene

A time-dependent increase in MtCaMP1 transcripts was observed upon onset of withholding water (Fig. 3A). In addition to withholding water, expression of MtCaMP1 was also up-regulated by treatments with salt (200 mM NaCl) and osmotic stress (20% PEG6000) (Fig. 3B). Transcripts of MtCaMP1 were detected in roots, stems, leaves, flowers and pods under non-stressed, control conditions, with the expression being greatest in leaves, followed by roots and stems, lowest in pods (Fig. 3C).

Expression of MtCaMP1 gene enhanced tolerance to drought and salt stress

To functionally characterize MtCaMP1, MtCaMP1 was expressed in *Arabidopsis thaliana* (Col-0) under the control of a CaMV 35S promoter. The transgenic lines were confirmed by hygromycin selection, β-glucuronidase (GUS) staining, and RT-qPCR. Compared with the untransformed WT Arabidopsis, the abundance of MtCaMP1 transcript was much higher in the transgenic lines (Fig. 4). Three independent transgenic lines (Line 1, 4 and 5) were used for further physiological studies throughout this paper.

Given that the expression of MtCaMP1 was induced by drought and salt stress in *M. truncatula*, we investigated the role of MtCaMP1 played in response to drought and salt stress by comparing the performance of transgenic plants expressing MtCaMP1 with WT seedlings suffering from drought and salt stress. After Arabidopsis seedlings of 4-week-old of both transgenic plants expressing MtCaMP1 and WT were withheld water for 14 d, transgenic plants exhibited higher survival rate than WT (Fig. 5A, B). In addition, the ability of water retention has been widely used as an indicator for drought tolerance in plants [38]. We found that transgenic plants exhibited lower water loss rate than WT (Fig. 5C), indicating that the expression of MtCaMP1 in Arabidopsis confers the transgenic plants more tolerant to drought stress than WT plants. A similar enhanced tolerance of transgenic plants to salt stress relative to WT plants was also observed (Fig. 6).

Transgenic plants accumulated less H_2O_2 and malondialdehyde (MDA)

Plants suffering from abiotic stress often exhibit symptoms of oxidative stress due to excessive accumulation of reactive oxygen species (ROS) and MDA [39]. There were significant increases in H_2O_2 and MDA contents in both WT and transgenic plants upon exposure to drought and salt stress, and the stress-induced increases in accumulation of H_2O_2 and MDA in the three transgenic lines were significantly less than in WT plants under drought stress (Fig. 7A). A similar less increase in MDA content in the transgenic plants than in WT was also observed when treated with salt stress (Fig. 7B), while expression of MtCaMP1 in Arabidopsis did not affect MDA metabolism under non-stressed, control conditions (Fig. 7B).

Transgenic plants accumulated more Pro under drought stress

Accumulation of free Pro is a common phenomenon for plants suffering from abiotic stress such as cold, drought and salt stress [40]. Pro contents in both WT and transgenic plants were

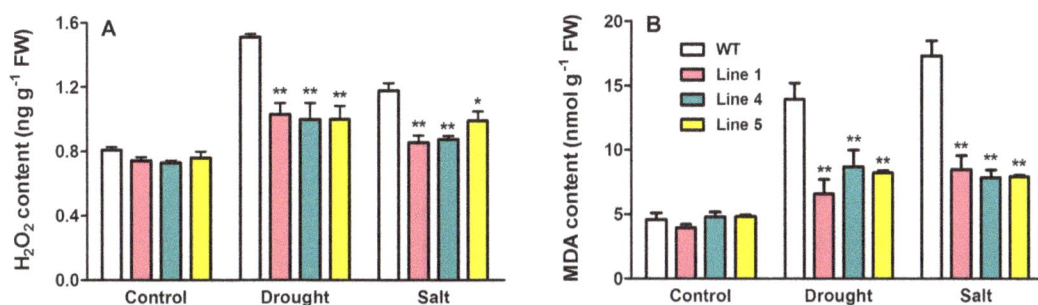

Figure 7. Effect of drought and salt stress on contents of H$_2$O$_2$ and malondialdehyde. H$_2$O$_2$ contents in wild-type and transgenic plants in control and treatment with drought and salt stress (panel A). Malondialdehyde contents in wild-type and transgenic plants in control and treatment with drought and salt stress (panel B). Four-week-old seedlings withheld water or irrigated with 200 mM NaCl for 6 days were used in the experiments. Data are means ± SE of three biological replicates. Asterisks represent statistically significant differences between wild-type and transgenic lines. * $P \leq 0.05$, ** $P \leq 0.01$.

comparable under non-stressed, control conditions (Fig. 8A). A marked increase in Pro contents in both WT and transgenic plants was observed upon challenged by drought and salt stress (Fig. 8A). However, the drought stress-induced increase in Pro accumulation in the three transgenic lines was significantly higher than in WT plants (Fig. 8A). In contrast, the increases in Pro contents in WT and transgenic plants were not significantly different under salt stress (Fig. 8A). To further elucidate the mechanism of the drought-induced Pro accumulation, we monitored the changes in transcripts of *P5CS1* and *ProDH* that encode key enzymes responsible for Pro biosynthesis and degradation, respectively. No differences in the abundance of *P5CS1* and *ProDH* transcripts between WT and transgenic plants were found under control conditions (Fig. 8B, C). There were marked increases in *P5CS1* transcripts in both WT and transgenic plants when treated by drought stress (Fig. 8B). However, the drought stress-induced increases in *P5CS1* transcripts in the transgenic plants were significantly greater than in WT plants (Fig. 8B). In contrast to *P5CS1*, *ProDH* transcripts in WT and transgenic plants differed in their response to drought stress such that drought stress suppressed expression of *ProDH* in the transgenic plants, while it had no effect on *ProDH* transcripts in WT plants (Fig. 8C). In contrast to drought stress, both Pro contents and the expression levels of genes encoding Pro metabolism did not differ between WT and transgenic plants in the present of NaCl in the incubation solution (Fig. 8), suggesting that Pro is not involved in the enhanced tolerance of the transgenic plants to salt stress. These results suggest that expression of *MtCaMP1* facilitates Pro accumulation by stimulating Pro biosynthesis and suppressing Pro degradation at the transcript level under drought stress.

Na$^+$ content of transgenic plants under salt stress

Na$^+$ is toxic to plants when accumulated excessively in the cytosol when exposed to salt stress. To elucidate the physiological mechanisms by which transgenic plants became more tolerant to slat stress than WT plants, the effects of salt stress on Na$^+$ and K$^+$ concentrations in shoots of WT and the transgenic plants were investigated. There were no differences in both Na$^+$ and K$^+$ concentrations in WT and the transgenic plants when they were grown in the control medium in the absence of NaCl (Fig. 9A, B). When the transgenic plants and WT were exposed to solution containing 200 mM NaCl, a marked accumulation of Na$^+$ in both WT and transgenic plants was observed (Fig. 9A). However, Na$^+$ accumulated in the transgenic plants was significantly less than in WT (Fig. 9A). In contrast to Na$^+$, accumulation of K$^+$ in both WT and the transgenic plants was equally inhibited under salt stress (Fig. 9B). This led to an increase in Na$^+$/K$^+$ ratio in both WT and the transgenic lines with the increase being less in the transgenic plants than in WT plants (Fig. 9C). Furthermore, the expression of *AtHKT1* and *AtNHX1* that encode Na$^+$ transporters responsible for Na$^+$ influx into cytosol and into vacuoles were analyzed in the absence and presence of NaCl by RT-qPCR. As shown in Figure 9D, expression of *MtCaMP1* had little effect on *AtHKT1* transcripts under control conditions. Suppression of *AtHKT1* expression by salt stress occurred in both the transgenic and WT plants, but the salt stress-induced down-regulation of *AtHKT1* expression in the transgenic plants was greater than in WT plants (Fig. 9D). Unlike *AtHKT1*, *AtNHX1* transcripts were more abundant in the transgenic plants than in WT plants under control conditions, and salt stress led to significant increases in

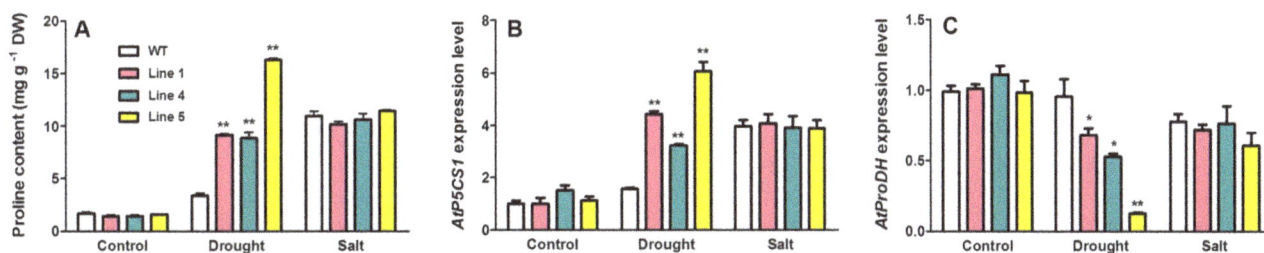

Figure 8. Drought and salt stress-induced changes in Pro contents. Effect of drought and salt stress on contents of Pro was shown in panel A. The expression levels of genes encoding Pro synthetase (*P5CS1*) and dehydrogenase (*ProDH*) under stress were shown in panel B and C, respectively. Data are mean±SE with three replicates. Asterisks represent statistically significant differences between wild-type and transgenic lines. * $P \leq 0.05$, ** $P \leq 0.01$.

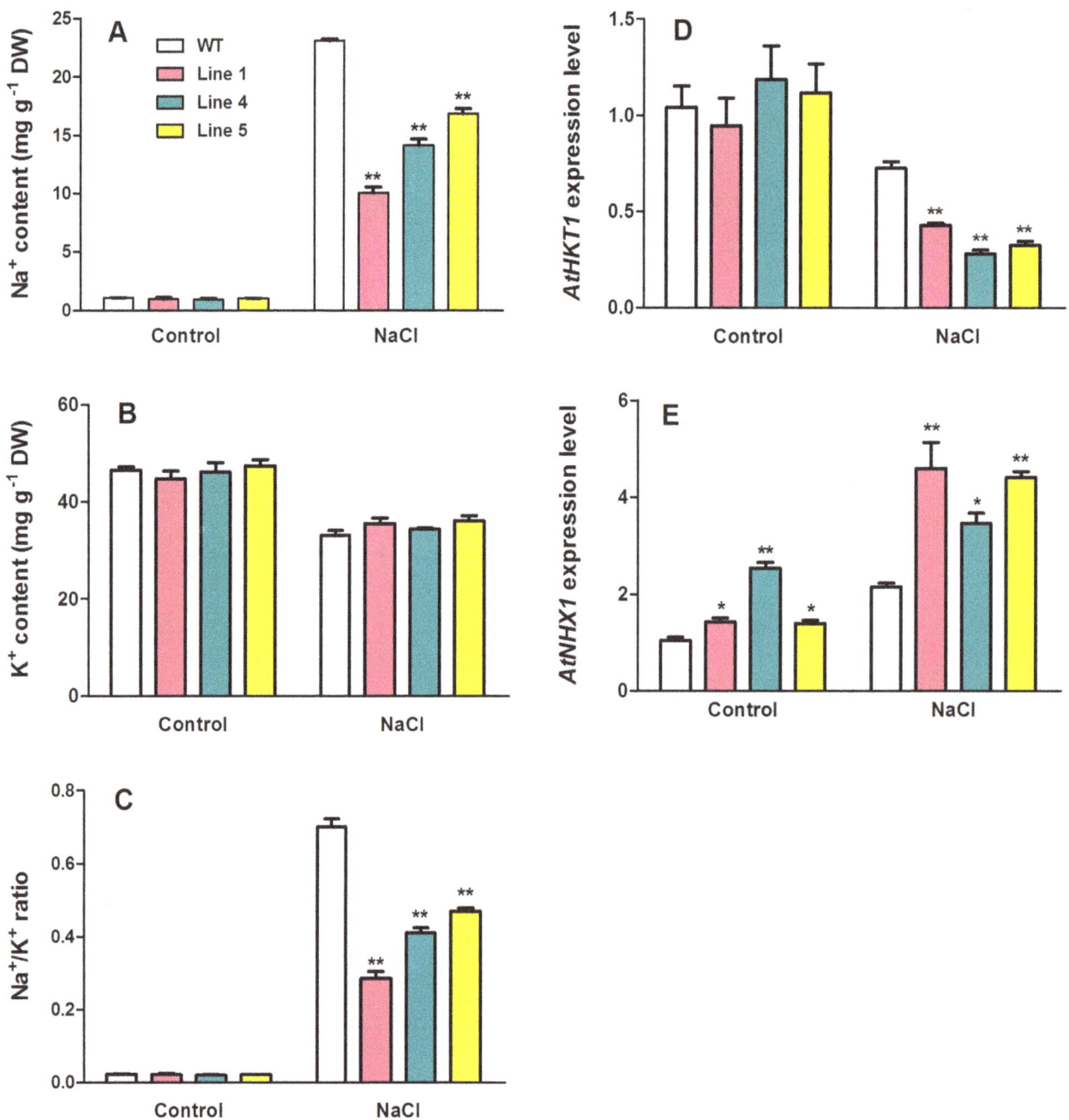

Figure 9. Effect of NaCl on contents of Na$^+$ and K$^+$, and expression of *AtHKT1* and *AtNHX1*. Contents of Na$^+$ and K$^+$ in shoots of wild-type and transgenic plants treated with and without NaCl were shown in panel A and B, respectively. Na$^+$/K$^+$ ratio was shown in panel C. The expression levels of *AtHKT1* and *AtNHX1* were shown in panel D and E, respectively. Four-week-old seedlings irrigated with or without 200 mM NaCl for 6 days were tested in this experiment. Data are mean±SE with three replicates. Asterisks represent statistically significant differences between wild-type and transgenic lines. * $P \leq 0.05$, ** $P \leq 0.01$.

AtNHX1 transcripts in both WT and transgenic plants with the increase greater in the transgenic than in WT plants (Fig. 9E).

Discussion

EF-hand family proteins are involved in perception and transduction of calcium signals by binding to Ca^{2+} ions [10,12]. A number of EF-hand family proteins have been reported to participate in transduction of signals associated with biotic and abiotic stress [5]. For instance, *AtCPK10* plays regulatory roles in

modulation of ABA- and Ca^{2+}-dependent stomatal movements, such that the *cpk10* mutant is more sensitive to drought stress, while the *AtCPK10*-overexpression lines are more tolerant to drought stress than their wild-type counterpart [41]. Furthermore, AtCBL4 has been shown to activate a Na$^+$/H$^+$ exchanger SOS1 in the plasma membrane of Arabidopsis by targeting AtCIPK24, and mutation of *AtCBL4* leads to hypersensitivity to NaCl [22,23]. In the present study, we isolated a gene *MtCaMP1* encoding an EF-hand family protein capable of binding to Ca^{2+} from the legume model plant *M. truncatula*, and evaluated the role of *MtCaMP1* in response to drought and salt stress by generating Arabidopsis plants expressing *MtCaMP1*.

One important finding in the present study is that expression of *MtCaMP1* in Arabidopsis conferred the transgenic plants more tolerant to drought and salt stress. We further explored the physiological mechanisms by which the transgenic plants were more tolerant to drought and salt stress than WT plants. Our results revealed that the transgenic plants accumulated greater amounts of Pro, less amounts of H_2O_2, MDA and Na^+ than WT plants under conditions of drought and salt stress. The accumulated Pro may allow the transgenic plants for more effective osmo-regulation, thus conferring them more tolerant to drought stress by minimizing water loss and maximizing water uptake. The observation that transgenic plants had less water loss rate from leaves than WT plants (Fig. 5C) is consistent with this explanation. There have been reports demonstrating that Pro biosynthesis in plants is regulated by EF-hand family proteins in the literature. For instance, GmCaM4 up-regulates the expression of *P5CS1* by interacting with AtMYB2 [14]. In the present study, we found that expression of *MtCaMP1* in Arabidopsis stimulated Pro biosynthesis and suppressed Pro degradation at the transcriptional level (Fig. 8B, C), thus leading to the enhanced accumulation of Pro under drought stress. In addition to osmo-regulation, Pro has also been suggested to function as a molecular chaperone to stabilize the structure of proteins [42]. Moreover, the less amounts of H_2O_2 and MDA in the transgenic plants than in WT plants under drought stress (Fig. 7) may be accounted for by the higher Pro content in transgenic plants than in WT since Pro may protect the transgenic plants from oxidative damage by acting as an antioxidant [42,43]. Therefore, the greater accumulation of Pro regulated possibly by calcium signals in the transgenic plants under drought stress would confer the transgenic plants to tolerate the drought stress by maintaining favorable water potential and suffering less from oxidative damage.

In contrast to drought stress, accumulation of Pro under salt stress did not differ between the transgenic and WT plants (Fig. 8), suggesting that accumulation is unlikely to account for the differential tolerance to salt stress between WT and the transgenic plants. In this context, we found that expression of *MtCaMP1* in Arabidopsis rendered the transgenic plants accumulated less amounts of Na^+ than WT under salt stress (Fig. 9). Excessive accumulation of toxic Na^+ in plant cells, particularly in the cytosol, disrupts K^+ homeostasis, leading to dysfunction of plant cells. One of the common mechanisms for plants to tolerate salt stress is to minimize Na^+ influx into cytosol and/or maximize Na^+ influx into vacuoles [39]. Our results showed that expression of *MtCaMP1* in Arabidopsis led to a marked reduction in Na^+ accumulation, thus maintaining a higher K^+/Na^+ ratio than in WT plants under salt stress (Fig. 9A–C). The lower Na^+/K^+ ratio is beneficial for plants to maintain physiological processes under salt stress, thus

contributing to the enhanced tolerance of the transgenic plants to salt stress. There are several transporters involved in mediation of Na^+ flux into and out of cytosol in plants, including SOS1 [44], HKT1 [45] and NHX1 [46]. Mutations in the *AtHKT1* gene suppressed *sos3* mutant phenotypes, and analysis of ion contents in the *sos3hkt1* mutant demonstrated that AtHKT1 is involved in mediation of Na^+ influx into plant cells [45]. In addition to preventing Na^+ influx into cytosol, sequestration of toxic Na^+ in the vacuole by NHX in the tonoplast can also contribute to the enhanced tolerance of plants to salt stress by minimizing accumulation of toxic Na^+ in the cytosol [46]. In this context, there are reports demonstrating that EF-hand family proteins are involved in regulation of Na^+ transport mediated by HKT and NHX proteins in the literature. For example, AtCBL4 has been suggested to mediate Na^+ influx into the cytosol and the vacuole by targeting HKT and NHX, respectively [39]. Conversely, it has been reported that Na^+/H^+ exchange activity of AtNHX1 is reduced by AtCML18 [15]. In the present study, we found that the expression levels of *AtHKT1* and *AtNHX1* were significantly reduced and enhanced in the *MtCaMP1*-expressing transgenic plants respectively compared to those in WT plants under the condition of salt stress (Fig. 9D, E). Given that EF-hand family proteins can regulate HKT and NHX by calcium, we speculate that the changes in expression levels of *AtHKT1* and *AtNHX1* in the transgenic plants expressing *MtCaMP1* may also occur in a calcium-dependent manner. It is expected that the down-regulation of *AtHKT1* by expressing *MtCaMP1* would minimize AtHKT1-mediated Na^+ influx into root cells, while the up-regulation of *AtNHX1* in the transgenic plants would facilitate Na^+ accumulation in the vacuole. These changes in patterns of Na^+ transporters at the transcriptional level would allow the transgenic plants to accumulate less toxic Na^+ in the cytosol, thus conferring transgenic plants more tolerance to salt stress. The reduced accumulation of toxic Na^+ in the transgenic plants may also account for the observed less accumulation of H_2O_2 and MDA in the transgenic plants under salt stress.

In conclusion, we demonstrate that expression of *MtCaMP1*, a gene from *M. truncatula* in Arabidopsis conferred the transgenic seedlings tolerant to drought and salt stress. The physiological mechanisms responsible for the enhanced tolerance to drought and salt stress can be accounted for by enhanced accumulation of Pro due to stimulation of Pro biosynthesis and suppression of its degradation as well as inhibition of Na^+ accumulation by up-regulating *AtNHX1* and down-regulating *AtHKT1*.

Author Contributions

Conceived and designed the experiments: TZW WHZ. Performed the experiments: TZW JLZ. Analyzed the data: TZW JLZ QYT MGZ WHZ. Wrote the paper: TZW WHZ.

References

1. Bush DS (1995) Calcium regulation in plant cells and its role in signaling. In: Jones RL, editor. Annual Review of Plant Physiology and Plant Molecular Biology. 95–122.
2. Sanders D, Pelloux J, Brownlee C, Harper JF (2002) Calcium at the crossroads of signaling. Plant Cell 14: S401–S417.
3. Tuteja N (2007) Mechanisms of high salinity tolerance in plants. Methods Enzymol 428: 419–438.
4. Takano M, Takahashi H, Suge H (1997) Calcium requirement for the induction of hydrotropism and enhancement of calcium-induced curvature by water stress in primary roots of pea, *Pisum sativum* L. Plant Cell Physiol 38: 385–391.
5. Knight H (2000) Calcium signaling during abiotic stress in plants. Int Rev Cytol 195: 269–324.
6. Lecourieux D, Ranjeva R, Pugin A (2006) Calcium in plant defence-signalling pathways. New Phytol 171: 249–269.
7. Sanders D, Brownlee C, Harper JF (1999) Communicating with calcium. Plant Cell 11: 691–706.
8. Tuteja N, Mahajan S (2007) Calcium signaling network in plants: an overview. Plant Signal Behav 2: 79–85.
9. Kretsinger RH, Nockolds CE (1973) Carp muscle calcium-binding protein. II. Structure determination and general description. J Biol Chem 248: 3313–3326.
10. Grabarek Z (2006) Structural basis for diversity of the EF-hand calcium-binding proteins. J Mol Biol 359: 509–525.
11. Poovaiah BW, Reddy AS (1993) Calcium and signal transduction in plants. CRC Crit Rev Plant Sci 12: 185–211.
12. Day IS, Reddy VS, Shad Ali G, Reddy AS (2002) Analysis of EF-hand-containing proteins in Arabidopsis. Genome Biol 3: Research0056.
13. Yang T, Poovaiah BW (2003) Calcium/calmodulin-mediated signal network in plants. Trends Plant Sci 8: 505–512.

14. Yoo JH, Park CY, Kim JC, Heo WD, Cheong MS, et al. (2005) Direct interaction of a divergent CaM isoform and the transcription factor, MYB2, enhances salt tolerance in arabidopsis. J Biol Chem 280: 3697–3706.
15. Yamaguchi T, Aharon GS, Sottosanto JB, Blumwald E (2005) Vacuolar Na$^+$/H$^+$ antiporter cation selectivity is regulated by calmodulin from within the vacuole in a Ca^{2+}- and pH-dependent manner. Proc Natl Acad Sci U S A 102: 16107–16112.
16. Cheng SH, Willmann MR, Chen HC, Sheen J (2002) Calcium signaling through protein kinases. The Arabidopsis calcium-dependent protein kinase gene family. Plant Physiol 129: 469–485.
17. Franz S, Ehlert B, Liese A, Kurth J, Cazale AC, et al. (2011) Calcium-dependent protein kinase CPK21 functions in abiotic stress response in *Arabidopsis thaliana*. Mol Plant 4: 83–96.
18. Romeis T, Ludwig AA, Martin R, Jones JD (2001) Calcium-dependent protein kinases play an essential role in a plant defence response. EMBO J 20: 5556–5567.
19. Kolukisaoglu U, Weinl S, Blazevic D, Batistic O, Kudla J (2004) Calcium sensors and their interacting protein kinases: Genomics of the Arabidopsis and rice CBL-CIPK signaling networks. Plant Physiol 134: 43–58.
20. Luan S (2009) The CBL-CIPK network in plant calcium signaling. Trend Plant Sci 14: 37–42.
21. Kudla J, Xu Q, Harter K, Gruissem W, Luan S (1999) Genes for calcineurin B-like proteins in Arabidopsis are differentially regulated by stress signals. Proc Natl Acad Sci U S A 96: 4718–4723.
22. Liu J, Zhu JK (1998) A calcium sensor homolog required for plant salt tolerance. Science 280: 1943–1945.
23. Qiu QS, Guo Y, Dietrich MA, Schumaker KS, Zhu JK (2002) Regulation of SOS1, a plasma membrane Na$^+$/H$^+$ exchanger in Arabidopsis thaliana, by SOS2 and SOS3. Proc Natl Acad Sci U S A 99: 8436–8441.
24. Wang M, Gu D, Liu T, Wang Z, Guo X, et al. (2007) Overexpression of a putative maize calcineurin B-like protein in Arabidopsis confers salt tolerance. Plant Mol Biol 65: 733–746.
25. Talke IN, Blaudez D, Maathuis FJ, Sanders D (2003) CNGCs: prime targets of plant cyclic nucleotide signalling? Trends Plant Sci 8: 286–293.
26. Cook DR (1999) *Medicago truncatula*-a model in the making! Curr Opin Plant Biol 2: 301–304.
27. Wang TZ, Chen L, Zhao MG, Tian QY, Zhang WH (2011) Identification of drought-responsive microRNAs in *Medicago truncatula* by genome-wide high-throughput sequencing. BMC Genomics 12: 367.
28. Wang TZ, Zhao MG, Zhang WH (2012) Construction and analyses of two suppression subtractive hybridization libraries of *Medicago falcata* and *Medicago truncatula* under drought stress. Acta Prata Sin 21: 175–181.
29. Altschul SF, Gish W, Miller W, Myers EW, Lipman DJ (1990) Basic local alignment search tool. J Mol Biol 215: 403–410.
30. Li CJ, Liang Y, Chen CB, Li JH, Xu YY, et al. (2006) Cloning and expression analysis of TSK1, a wheat SKP1 homologue, and functional comparison with Arabidopsis ASK1 in male meiosis and auxin signalling. Funct Plant Biol 33: 381–390.
31. Zhang X, Henriques R, Lin SS, Niu QW, Chua NH (2006) Agrobacterium-mediated transformation of *Arabidopsis thaliana* using the floral dip method. Nat Protoc 1: 641–646.
32. Burgess WH, Jemiolo DK, Kretsinger RH (1980) Interaction of calcium and calmodulin in the presence of sodium dodecyl sulfate. BBA-Protein Structure 623: 257–270.
33. Takezawa D (2000) A rapid induction by elicitors of the mRNA encoding CCD-1, a 14 kDa Ca^{2+}-binding protein in wheat cultured cells. Plant Mol Biol 42: 807–817.
34. Livak KJ, Schmittgen TD (2001) Analysis of relative gene expression data using real-time quantitative PCR and the $2^{-\Delta\Delta Ct}$ method. Methods 25: 402–408.
35. Alexieva V, Sergiev I, Mapelli S, Karanov E (2001) The effect of drought and ultraviolet radiation on growth and stress markers in pea and wheat. Plant Cell Environ 24: 1337–1344.
36. Kramer GF, Norman HA, Krizek DT, Mirecki RM (1991) Influence of UV-B radiation on polyamines, lipid peroxidation and membrane lipids in cucumber. Phytochemistry 30: 2101–2108.
37. Bates LS, Waldren RP, Teare ID (1973) Rapid determination of free proline for water-stress studies. Plant Soil 39: 205–207.
38. Dhanda SS, Sethi GS (1998) Inheritance of excised-leaf water loss and relative water content in bread wheat (*Triticum aestivum*). Euphytica 104: 39–47.
39. Zhu JK (2002) Salt and drought stress signal transduction in plants. Annu Rev Plant Biol 53: 247–273.
40. Ashraf M, Foolad MR (2007) Roles of glycine betaine and proline in improving plant abiotic stress resistance. Environ Exp Bot 59: 206–216.
41. Zou JJ, Wei FJ, Wang C, Wu JJ, Ratnasekera D, et al. (2010) Arabidopsis calcium-dependent protein kinase CPK10 functions in abscisic acid- and Ca^{2+}-mediated stomatal regulation in response to drought stress. Plant Physiol 154: 1232–1243.
42. Szekely G, Abraham E, Cseplo A, Rigo G, Zsigmond L, et al. (2008) Duplicated *P5CS* genes of Arabidopsis play distinct roles in stress regulation and developmental control of proline biosynthesis. Plant J 53: 11–28.
43. Hong Z, Lakkineni K, Zhang Z, Verma DP (2000) Removal of feedback inhibition of Δ^1-pyrroline-5-carboxylate synthetase results in increased proline accumulation and protection of plants from osmotic stress. Plant Physiol 122: 1129–1136.
44. Shi H, Ishitani M, Kim C, Zhu JK (2000) The *Arabidopsis thaliana* salt tolerance gene *SOS1* encodes a putative Na$^+$/H$^+$ antiporter. Proc Natl Acad Sci U S A 97: 6896–6901.
45. Rus A, Yokoi S, Sharkhuu A, Reddy M, Lee BH, et al. (2001) AtHKT1 is a salt tolerance determinant that controls Na$^+$ entry into plant roots. Proc Natl Acad Sci U S A 98: 14150–14155.
46. Yokoi S, Quintero FJ, Cubero B, Ruiz MT, Bressan RA, et al. (2002) Differential expression and function of *Arabidopsis thaliana* NHX Na$^+$/H$^+$ antiporters in the salt stress response. Plant J 30: 529–539.

7

Sensitivity of Temperate Desert Steppe Carbon Exchange to Seasonal Droughts and Precipitation Variations in Inner Mongolia, China

Fulin Yang[1,2], Guangsheng Zhou[3,2]*

1 Key Laboratory of Arid Climatic Change and Reducing Disaster of Gansu Province, Key Open Laboratory of Arid Climatic Change and Disaster Reduction of China Meteorological Administration (CMA), Institute of Arid Meteorology, CMA, Lanzhou, China, 2 State Key Laboratory of Vegetation and Environmental Change, Institute of Botany, Chinese Academy of Sciences, 20 Nanxincun, Xiangshan, Haidian District, Beijing, China, 3 Chinese Academy of Meteorological Sciences, Haidian District, Beijing, China

Abstract

Arid grassland ecosystems have significant interannual variation in carbon exchange; however, it is unclear how environmental factors influence carbon exchange in different hydrological years. In this study, the eddy covariance technique was used to investigate the seasonal and interannual variability of CO_2 flux over a temperate desert steppe in Inner Mongolia, China from 2008 to 2010. The amounts and times of precipitation varied significantly throughout the study period. The precipitation in 2009 (186.4 mm) was close to the long-term average (183.9$\pm$47.6 mm), while the precipitation in 2008 (136.3 mm) and 2010 (141.3 mm) was approximately a quarter below the long-term average. The temperate desert steppe showed carbon neutrality for atmospheric CO_2 throughout the study period, with a net ecosystem carbon dioxide exchange (NEE) of -7.2, -22.9, and 26.0 g C m^{-2} yr^{-1} in 2008, 2009, and 2010, not significantly different from zero. The ecosystem gained more carbon in 2009 compared to other two relatively dry years, while there was significant difference in carbon uptake between 2008 and 2010, although both years recorded similar annual precipitation. The results suggest that summer precipitation is a key factor determining annual NEE. The apparent quantum yield and saturation value of NEE (NEE$_{sat}$) and the temperature sensitivity coefficient of ecosystem respiration (R$_{eco}$) exhibited significant variations. The values of NEE$_{sat}$ were -2.6, -2.9, and -1.4 μmol CO_2 m^{-2} s^{-1} in 2008, 2009, and 2010, respectively. Drought suppressed both the gross primary production (GPP) and R$_{eco}$, and the drought sensitivity of GPP was greater than that of R$_{eco}$. The soil water content sensitivity of GPP was high during the dry year of 2008 with limited soil moisture availability. Our results suggest the carbon balance of this temperate desert steppe was not only sensitive to total annual precipitation, but also to its seasonal distribution.

Editor: Ben Bond-Lamberty, DOE Pacific Northwest National Laboratory, United States of America

Funding: This research was jointly supported by the National Basic Research Program of China (2010CB951303), National Natural Science Foundation of China (40830957), China Postdoctoral Science Foundation (2012M512044), and Natural Science Foundation of Gansu Province (1208RJYA025). The funders had no role in study design, data collection and analysis, decision to publish, or preparation of the manuscript.

Competing Interests: The authors have declared that no competing interests exist.

* E-mail: gszhou@cams.cma.gov.cn

Introduction

Grassland ecosystems comprise approximately 32% of the global natural vegetation [1], making them important to the global carbon balance. Global climate changes are expected to alter precipitation regimes in grassland biomes [2], where the carbon cycle is particularly sensitive to the amount and timing of precipitation [3]. A recent study shows that grassland ecosystem productivity is sensitive to climate change [4]. Changing precipitation regimes and drought can have a profound impact on carbon fluxes in grassland ecosystems, especially in arid and semi-arid regions characterized by limited water [5,6]. In temperate grasslands, interannual variability in total precipitation is the primary climatic factor that causes fluctuations in net annual primary production [7–9] and net ecosystem carbon dioxide exchange (NEE) [10]. Studies on various grassland ecosystems extensively supported the positive relationship between annual NEE and total annual precipitation [11–16]. Depending on the amount of annual precipitation, a grassland ecosystem can switch from being a carbon sink in the wet or normal years to a net carbon source in the drought years [12,13]. However, other studies have shown that the interannual precipitation distribution can alter carbon uptake and release regardless of the total annual precipitation [8,16,17].

Drought is a common factor that limits vegetation growth and ecosystem carbon uptake in semi-arid grasslands. Droughts are related to lower annual rainfall and to different rainfall distributions [18]. The asymmetric distribution of seasonal precipitation can lead to intermittent droughts. Drought spells can substantially modify the seasonal development of leaf area and change plant physiology [19], making them have a large impact on the ecosystem as sources or sinks of atmospheric CO_2 [20]. An extreme drought in Europe in 2003 caused many ecosystems to lose carbon [21]. Aires *et al.* [11] found that winter and early spring droughts suppressed grass production and canopy development, consequently decreasing the maximum daily NEE of the

Mediterranean C3/C4 grassland in southern Portugal significantly during the dry year. Hussain *et al.* [22] showed that the leaf area index (LAI) reduction caused by a summer drought decreased the gross primary production (GPP) and the ecosystem respiration (R_{eco}) in a temperate grassland in Germany. Flanagan *et al.* [12] reported that fall droughts may accelerate leaf fall and shorten the growing season, consequently decreasing the seasonal cumulative GPP. Scott *et al.* [23] suggested that severe droughts can lead to a change in plant community structure, and that NEE was suppressed during the drought years in a semi-desert grassland in the USA. Drought conditions affect the terrestrial carbon balance by modifying the rates of and the coupling between the carbon uptake by photosynthesis (GPP) and release by R_{eco} [24,25]. Reichstein *et al.* [18] suggested that drought conditions may have different effects on plant assimilation and ecosystem respiration, and that short-term drought will suppress R_{eco} before affecting GPP.

Photosynthesis is an important factor in regulating R_{eco} in diurnal [26], daily [11,17], seasonal, and yearly timescales [27]. Davidson *et al.* [28] suggested that gross photosynthesis can control the substrate availability of autotrophic and heterotrophic respiration through root exudates. However, Yan *et al.* [29] reported that water can regulate the effects of the photosynthetic substrate supply on soil respiration in a semi-arid steppe. Insufficient information is available about whether a strong correlation between GPP and R_{eco} holds true in a desert steppe ecosystem, where soil water availability is considered as the factor that limits vegetation growth the most.

Desert steppes with annual precipitation between 150 mm and 250 mm are the most arid grassland ecosystems [30] located in the transitional zones between steppes and deserts [31]. The 17.5 million ha temperate desert steppe area in China provided 0.066 Pg of carbon storage in its biomass [32]. Recent studies have shown that the annual mean temperature and interannual precipitation variability increased and the spring and winter precipitation decreased over the past 40 years [33]. The low soil moisture availability in desert steppes associated with temperature increase may intensify water limitation effects on carbon flux.

In this study, eddy covariance technology was used to continuously measure NEE over the Inner Mongolian temperate desert steppe from 2008 to 2010. The objectives of this study were as follows: (1) to examine the seasonal and interannual variability in GPP, R_{eco}, and NEE; (2) to elucidate environmental and physiological regulations on carbon flux components; and (3) to evaluate the seasonal distribution and the total amount of precipitation that affect the carbon balance over a temperate desert steppe in Inner Mongolia, China.

Materials and Methods

1.1. Ethics Statement

All observational and field studies at the desert steppe were undertaken with relevant permissions from the owners of private land: Mr. L.S. Chai. The field studies did not involve endangered or protected species, and the location was not protected in any way during the study period.

1.2. Study Site

The study site is located north of the Sunitezuoqi County, Inner Mongolia Autonomous Region, China (44°05′N, 113°34′E, 970 m a.s.l.), and it is classified as temperate desert steppe. The plant community is dominated by the bunch grass *Stipa klemenzii* and the herb *Allium polyrrhizum*. The grass canopy is 0.20 m to 0.35 m tall during mid-summer. The study site was fenced in

August 2007 to prevent grazing and other disturbances. The soil in the area was classified as brown calcic with an average bulk density of 1,630 kg m^{-3}. The study site has a mean air temperature of 3.2°C and mean annual precipitation of 183.9±47.6 mm (from 1965 to 2004, Sunitezuoqi Weather Station). Most of the precipitation (85%) falls between May to September. The area has an arid to semi-arid temperate continental climate.

1.3. Eddy Covariance Measurements

An open-path eddy covariance (EC) system was installed at the study site to measure the net ecosystem exchange of CO_2 and latent heat flux at a measurement height of 2.0 m. The EC system included a 3D sonic anemometer–thermometer (CSAT-3, Campbell Scientific Inc., Logan, UT, USA) and an open path infrared gas (CO_2/H_2O) analyzer (LI-7500, LI-COR Inc., Lincoln, NE, USA). Raw signals were recorded at 10 Hz using a data logger (CR5000, Campbell Scientific Inc., Logan, UT, USA).

1.4. Meteorology, Soil and Biotic Factor Measurements

A meteorological tower was situated near the EC tower to measure environmental variables. Air temperature (Ta) and relative humidity (RH) were measured at a height of 2.0 m (HMP45C, Vaisala, Helsinki, Finland). A horizontal wind speed sensor (014A, Campbell Scientific Inc., Logan, UT, USA) was situated at a height of 2.0 m to measure horizontal wind speed (Ws). Photosynthetically active radiation (PAR) and net radiation (Rn) were measured at 2.4 m above the ground by using a quantum sensor (LI-190SB, LI-COR Inc., Lincoln, NE, USA) and a four-component net radiometer (CNR-1, Kipp & Zonen, Delft, Netherlands), respectively. Precipitation was measured above the canopy by using a tipping bucket rain gauge (Model 52203, RM Young Inc., Traverse City, MI, USA). Soil temperature (Ts) at 0.05 m underground was measured using a thermistor (107L, Campbell Scientific Inc., Logan, UT, USA). The soil water content (SWC) at a depth of 0.10 m was measured using time-domain reflectometry probes (CS616, Campbell Scientific Inc., Logan, UT, USA). Two soil heat plates (HFP01, Hukeflux Inc., Delft, Netherlands) were used to measure the soil heat flux (G) at 0.08 m below the soil surface in separate locations. All meteorological and soil sensors were sampled every 2 s, and were stored as a half-hour means by a data logger (CR23X, Campbell Scientific Inc., Logan, UT, USA).

Biomass sampling was conducted monthly in the growing season by clipping eight 1 m ×1 m quadrats. In each quadrat, all the plants were cut at the ground surface and then oven dried at 65°C to a constant weight. The dominant species leaf area ratio (m^2 g^{-1}) was also measured. The total LAI (m^2 m^{-2}) was estimated by means of specific leaf area (m^2 g^{-1}) multiplied by biomass (g m^{-2}) [31]. The monthly samplings of biomass and LAI determination were performed from 2008 to 2009.

1.5. Data Processing and Flux Computation

Raw data from the eddy-covariance measurements were obtained using the EdiRe software (www.geos.ed.ac.uk/abs/research/micromet/EdiRe: developed by University of Edinburgh, UK). The CO_2 fluxes were determined by the eddy covariance method as the mean covariance between fluctuations in vertical wind speed (w') and the carbon dioxide concentration (c') on a half-hourly basis (Eq. 1) [10].

$$NEE = \overline{w'c'} \tag{1}$$

Negative NEE denotes the net carbon uptake of the ecosystem. Prior to the scalar flux computation, spikes exclusion, two-dimension coordinate rotation and air density fluctuations correction [34] were performed [31]. The CO_2 storage term was not included in NEE computation because CO_2 concentration profile was not measured.

Half-hourly flux data were rejected following these criteria: (1) incomplete half-hourly measurement mainly caused by power failure, IRGA calibration; (2) rain events; (3) outliers [35]; and (4) low-turbulence conditions. The moving point test (MPT) was used to determine the friction velocity threshold ($u*_c$) for nighttime CO_2 flux under stable atmospheric conditions [36,37], suggesting the $u*_c$ was 0.11 m s^{-1} for this desert steppe ecosystem. The nighttime NEE data with lower than $u*_c$ was discarded. Negative nighttime CO_2 fluxes occurred prevalently during the non-growing season in this study, which may be due to the instrument LI-7500 surface heating effect. We tried to conduct Burbra correction for the non-growing season data [38]. However, the corrected data showed large noise and unrealistic flux values. The Burba correction did not work in this study. We discarded the negative CO_2 fluxes during 1 January to 30 April and 16 October to 31 December, likely to offset the non performance of the self-heating correction, which anyway did not affect the reliability of the EC flux measurement [39,40].

Due to missing and discarded data, the data gaps during the whole growing period were 36.5% of total data. To derive a continuous time series of NEE, the data gaps of less than 2 h were filled using linear interpolation, and the other data gaps were filled using the look-up table method. If meteorological data was absent, mean daily variation gap-fillings were used [41,42].

The energy balance ratio (EBR) was used to assess the accuracy of the eddy covariance measurements. For a short-statured canopy, EBR can be calculated using the following equation based on the half-hourly dataset after quality controls:

$$EBR = \frac{\sum (LE + H)}{\sum (Rn - G - S)} \quad (2)$$

where LE and H are the latent and sensible heat fluxes (W m^{-2}); Rn is net radiation (W m^{-2}); G is the soil heat fluxes (W m^{-2}); and S is the soil heat storage (W m^{-2}) which can be estimated by Ts and SWC [43]. EBR was 0.9 for the entire observation period, indicating that (H+LE) was close to (Rn-G-S). The flux measurement performance of eddy covariance system was acceptable. So we assumed that the advective losses of energy and CO_2 were small, and energy balance closure correction to CO_2 flux was neglected.

As we had only one eddy covariance tower, the daily-differencing approach [44] was used to estimate the uncertainty in annually integrated NEE. The identical environmental conditions (mean half-hourly PAR within 50 μmol photons m^{-2} s^{-1}, Ta within 2°C, and SWC within 0.5%) were adopted to calculate the standard deviation of the difference.

The partitioning of NEE into GPP and R_{eco} was based on stepwise procedures and algorithms of Reichstein et al. [45]. The equation of Lloyd and Taylor [46] was used to describe the response of half-hourly nighttime R_{eco} to soil temperature (Eq. 3):

$$R_{eco} = R_{ref} \exp^{E_0(1/(T_{ref} - T_0) - 1/(Ts - T_0))} \quad (3)$$

where E_0 is the activation energy (K); T_{ref} is set to 10°C; R_{ref}, reference ecosystem respiration at T_{ref}; T_0 is a constant (−46.02°C) and Ts is the soil temperature. Consistent temperature

sensitivity between daytime and nighttime was assumed for daytime R_{eco} calculations. The daily R_{eco} was taken as the sum of daytime and nighttime R_{eco}. GPP was calculated as follows:

$$GPP = R_{eco} - NEE \quad (4)$$

1.6. Canopy Surface Conductance

Based on the inverted Penman-Montieth equation, the half-hourly canopy surface conductance (g_c, m s^{-1}) was calculated as follows [47]:

$$g_c = \frac{\gamma LE g_a}{\Delta (Rn - G) + \rho c_p g_a VPD - LE(\Delta + \gamma)} \quad (5)$$

where γ is the psychrometric constant (kPa °C^{-1}), LE is the latent heat flux (W m^{-2}), Δ is the slope of the saturation water vapor pressure over the air temperature curve (kPa °C^{-1}), Rn is the net radiation (W m^{-2}), G is the soil heat flux (W m^{-2}), ρ is the air density (kg m^{-3}), C_p is the specific heat of air (J kg^{-1}°C^{-1}), and VPD is the air vapor pressure deficit (kPa), and g_a is the aerodynamic conductance of the air layer between the canopy and the flux measurement height (m s^{-1}). Using the Monteith-Unsworth equation, g_a was obtained as follows [48]:

$$g_a = \left(\frac{\bar{u}}{u_*^2} + 6.2 u_*^{-0.67} \right)^{-1} \quad (6)$$

where $\bar{u}$ is the mean wind speed (m s^{-1}) at a height of 2 m and $u*$ is the friction velocity (m s^{-1}). gc was excluded based on the following criteria: (1) rain events, (2) nighttime data (with incident solar radiation values less than 20 W m^{-2}), (3) low turbulence ($u* < 0.2$ m s^{-1}), and (4) anomalies.

1.7. Data Analysis

1.7.1. Carbon flux response to PAR and Ts. The relationship between the daytime (with incident solar radiation values greater than 20 W m^{-2}) NEE (μmol CO_2 m^{-2} s^{-1}) and PAR (μmol photons m^{-2} s^{-1}) was assessed using a Michaelis-Menten rectangular hyperbola equation [41]:

$$NEE_{daytime} = \frac{\alpha NEE_{sat} PAR}{\alpha PAR + NEE_{sat}} + R_{eco} \quad (7)$$

where α is the apparent quantum yield or the initial slope of the light response curve (μmol CO_2 μmol^{-1} photons), NEE_{sat} is the value of the NEE at a saturating light level, and R_{eco} is a bulk estimate of the ecosystem respiration.

The relationship between the nighttime NEE or R_{eco} (μmol CO_2 m^{-2} s^{-1}) and soil temperature at a 5 cm depth (Ts, °C) was calculated using the Van't Hoff equation [11,31]:

$$NEE_{nighttime} = a \exp(bTs) \quad (8)$$

where a and b are the regression parameters. The temperature sensitivity coefficient (Q_{10}) of R_{eco} was determined as follows:

$$Q_{10} = \exp(10b) \quad (9)$$

1.7.2. Statistical analysis. Regression analysis and model parameter fitting were performed using the statistical package SigmaPlot 10.0 software (Systat Software Inc. San Jose, CA, USA).

Other data computation and analysis were conducted using MATLAB 7.9 software (MathWorks Inc., Natick, MA, USA).

Results

2.1. Environmental Variables

The variations in weather conditions from 2008 to 2010 (Figure 1) recorded by the Sunitezuoqi weather station (1965 to 2004) are summarized in Table 1. The annual precipitation in 2009 (186.4 mm) was close to the average precipitation (183.9 mm) recorded by the Sunitezuoqi weather station, whereas the annual precipitations in 2008 (136.3 mm) and 2010 (141.3 mm) were approximately a quarter below the average. The amount of precipitation received during the growing seasons (May to September) was close among the three study years. However, the timing of the rainfall events differed significantly among the three years. Over 65% of the total annual precipitation fell in May and June 2010, while only 44.6% and 47.5% fell during the same periods in 2008 and 2009, respectively. The amount of precipitation in July and August of 2008, 2009, and 2010 were 73.9%, 52.3%, and 23.9% of the long-term average in the same period, respectively. The temperate desert steppe experienced much drought during the critical period, July and August, for plant development and growth in the measured years, especially in 2010, which had the longest severely dry period. Less than 5 mm of precipitation was experienced in September 2008 and 2009. On the other hand, September 2010 experienced six times more precipitation than September 2008 and 2009.

Daily Ta varied dramatically across the study period, and ranged from −33.4°C to 32.9°C. The annual temperature in 2008 was close to the long-term average temperature, whereas the annual temperatures in 2009 and 2010 were below the long-term average temperature, due to the relatively low temperature during non-growing seasons. However, in June and July 2010, a distinct hot spell occurred during the growing season, resulting in the highest VPD in June and July as compared to the VPD during the same period in 2008 and 2009.

The SWC was related to the precipitation pattern during the growing seasons and ranged from 3.8% (27 May 2008) to 20.1% (19 June 2009). Peak daily SWC values were obtained approximately a day after rain events. The SWC in 2009 was the highest among the three study years. In 2008, the temperate desert steppe experienced two significant dry periods. Only 9.6 mm of precipitation was recorded from 17 May to 1 June (16 days), with an average SWC of 4.1%. The next dry period occurred from 25 July to 8 August (15 days), with a low average SWC of 4.5%. In 2010, the seasonal severe drought started from 24 July to 16 August (24 days), when only 0.2 mm of precipitation was recorded, with a low averaged SWC of 4.2%. The dry period occurred exactly at the normal vegetation peak growth stage. Soil drying continued for approximately a month, with a minimum SWC of 3.9% on 14 August 2010.

The maximum LAIs were 0.38 and 0.50 m^2 m^{-2} in 2008 and 2009, respectively, and occurred during the peak growth period. The LAI in every growth period of 2009 was generally higher than that in 2008 (Figure 2).

2.2. Seasonal Variations in the Cumulative GPP, R_{eco}, and NEE

Strong seasonal variations in GPP, R_{eco}, and NEE of the temperate desert steppe ecosystem were observed. The seasonal variations in 2008 were larger compared with those in 2009 and

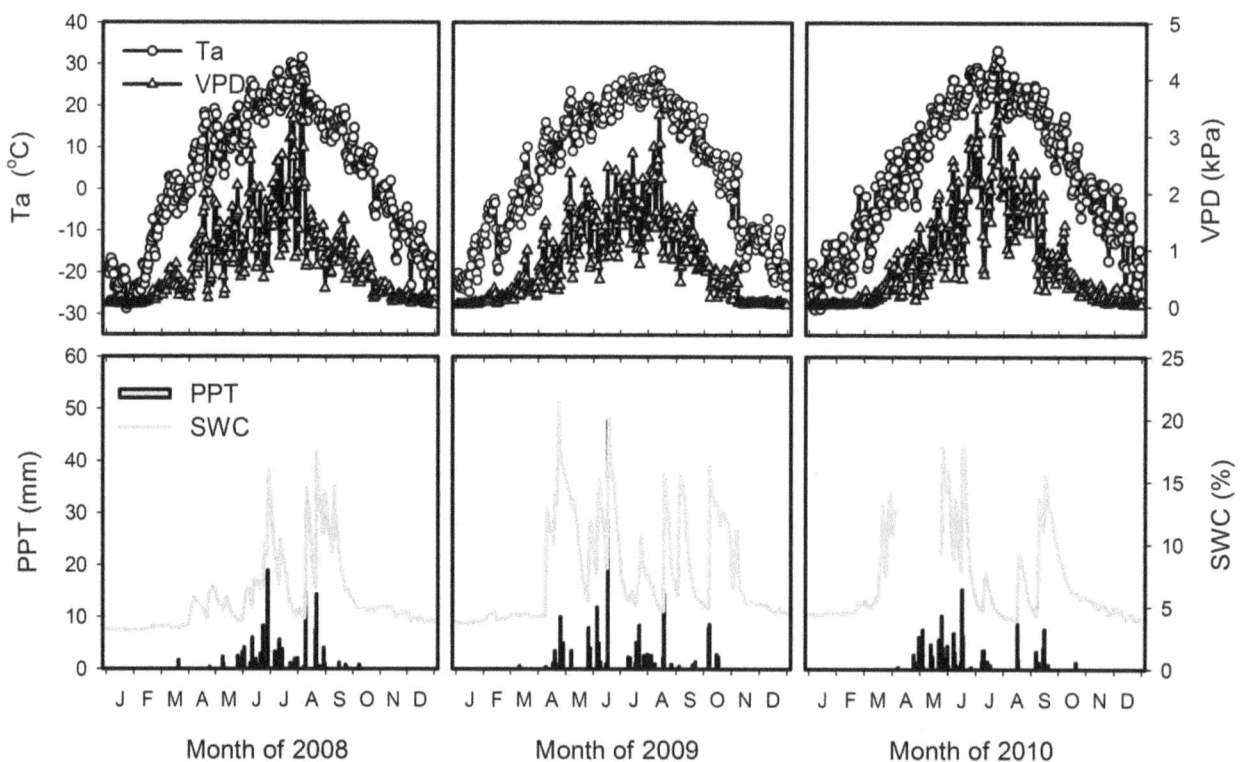

Figure 1. Seasonal variation in daily average temperature (Ta) and vapor pressure deficit (VPD), and daily precipitation (PPT) and daily average soil water content (SWC) throughout the study period. Soil water content data from 6 April to 23 May were omitted because of a malfunction in the connecting cable between the soil moisture sensor and the data logger.

Table 1. Comparison of environmental conditions in temperate desert steppe, Inner Mongolia, during 2008 to 2010.

	Year	May	June	July	August	September	May to September	Annual
PPT	2008	12.6	48.3	23.6	45.0	2.7	132.2	136.3
	2009	18.4	70.2	23.5	25.1	3.2	140.4	186.4
	2010	51.2	40.7	11.2	11.0	20.4	134.5	141.3
	Mean	27.4	53.1	19.4	27.0	8.8	135.7	154.7
	1965 to 2004	16.2	28.4	48.3	44.5	18.6	155.9	183.9
Ta	2008	11.6	19.4	23.9	20.5	13.6	17.8	3.1
	2009	15.1	17.9	22.3	21.4	13.3	18.0	2.4
	2010	13.5	22.1	25.6	20.6	14.8	19.3	2.0
	Mean	13.4	19.8	23.9	20.8	13.9	18.4	2.5
	1965 to 2004	13.7	19.4	22.1	20.1	13.3	17.7	3.2
SWC	2008	4.8	7.6	7.7	9.7	8.9	7.7	5.6
	2009	10.8	12.4	8.6	7.8	8.8	9.7	7.7
	2010	13.7	11.3	5.5	5.6	10.2	8.4	6.8
	Mean	9.8	10.4	7.3	7.7	9.3	8.6	6.7

PPT, precipitation (mm); Ta, air temperature (°C); SWC, soil water content (%).

2010 (Figure 3). The temperate desert steppe continuously lost carbon via soil respiration at low rates during the non-growing seasons of the three study years. The net carbon uptake began in early May (1 May 2008; 5 May 2009 and 2010), when the daily air temperature at a height of 2 m exceeded ca. 10°C. The seasonal variation in carbon fluxes during the growing seasons varied throughout the three study years. Only 19% of the days in 2010 (70 days) had a net sequestration of carbon, while 31% of the days in 2008 (115 days) and 2009 (113 days) had carbon gain. The maximum daily rates of GPP and R_{eco} in 2009 were 12.4 and 6.9 g CO_2 m^{-2} day^{-1}, respectively, and were higher compared with those in the dry years of 2008 and 2010. The maximum daily NEEs in 2008 and 2009 were both approximately -6.0 g CO_2 m^{-2} day^{-1}, whereas that in 2010 was only -4.8 g CO_2 m^{-2} day^{-1}.

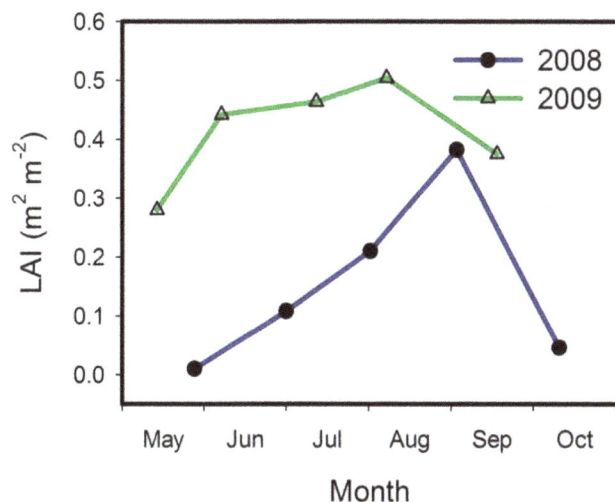

Figure 2. Seasonal dynamics of leaf area index (LAI) in 2008 and 2009.

In terms of the season dynamics, significant differences were evident in the CO_2 exchange of the three study years (Figure 3 and Table 2). In 2008, the CO_2 fluxes over the temperate desert steppe tracked closely with rain events and soil moisture. The ecosystem switched from being a net carbon sink to a source and then went back to a sink for time periods linked to precipitation processes in the growing season. The ecosystem became a daily carbon sink (with a negative NEE) in early May. However, both GPP and R_{eco} significantly decreased during the three drought spells that occurred in the growing season. GPP was lower than the R_{eco} and resulted in positive NEEs during the drought periods. The maximum net CO_2 uptake period occurred in the early growth season of 2009, and not in 2008, because of the higher amounts of rainfall (88.6 mm during May and June 2009) that was experienced during the period. The ecosystem experienced a severe drought from late July to mid-August 2009. GPP decreased at an even higher rate and turned the ecosystem into a weak carbon source (8.3 g CO_2 m^{-2} carbon release during 29 June to 18 August). Although the ecosystem shifted to carbon sink again across the severe drought period, notable carbon uptake characteristics that are similar to those evident in 2008 were not exhibited. Compared to 2008 and 2009, the periods of negative NEE in 2010 were smaller in magnitude and spanned shorter durations because of the severe summer drought, in which the carbon uptake was focused in May and June. The ecosystem suffered considerable seasonal droughts from July to August. The severe water stress (daily SWC lower than 4.5%) resulted in a GPP that was close to zero. The temperate desert steppe ecosystem became a carbon source during the dry period, except for some parts of September when carbon uptake occurred because of the spark rainfall.

Figure 4 shows the seasonal evolutions of the cumulative GPP, R_{eco}, and NEE over the three study years. The desert steppe ecosystem exhibited a neutrality (-4.1 g C m^{-2}) throughout the study period (from January 2008 to December 2010). The uncertainty in annual NEE measurements inferred from daily-differencing approach was ± 12.4, ± 11.9, and ± 10.0 g C m^{-2} yr^{-1} in 2008, 2009, and 2010, respectively. Compared with the two relatively dry years of 2008 and 2010, the

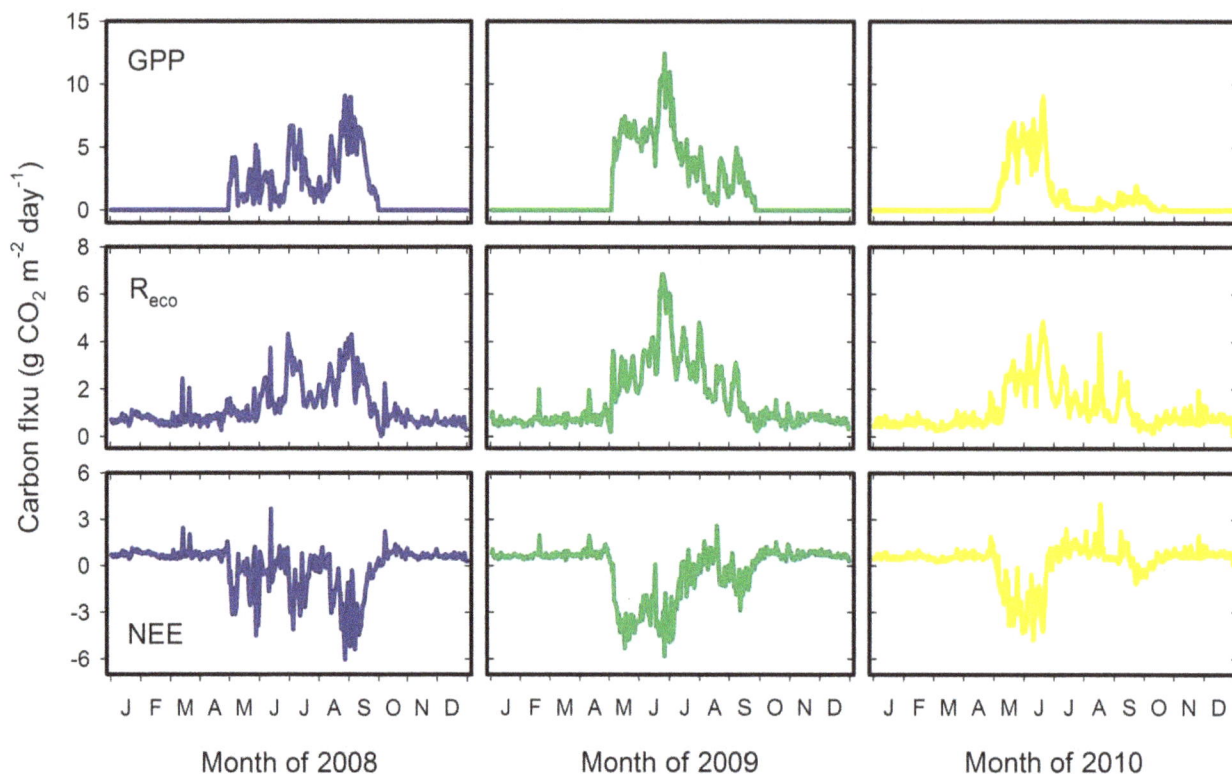

Figure 3. Seasonal variations in daily integrated net ecosystem carbon exchange (NEE), gross primary production (GPP), and ecosystem respiration (R_{eco}) over the course of the study. Negative NEE denotes the net carbon uptake of the ecosystem.

Table 2. Comparison of carbon fluxes in temperate desert steppe, Inner Mongolia, during 2008 to 2010.

	Year	May	June	July	August	September	Annual
GPP	2008	19.1	14.0	29.3	33.8	33.9	130.0
	2009	42.9	58.5	42.5	18.1	16.2	178.3
	2010	29.9	40.9	5.0	2.4	7.1	87.7
	Mean	30.6	37.8	25.6	18.1	19.1	132.0
R_{eco}	2008	8.2	14.9	18.9	20.9	17.5	122.7
	2009	18.7	33.0	30.1	19.7	11.2	155.4
	2010	16.0	24.5	14.1	11.3	9.9	113.8
	Mean	14.3	24.1	21.0	17.3	12.9	130.6
NEE	2008	−10.8	0.9	−10.3	−12.9	−16.3	−7.2
	2009	−24.2	−25.5	−12.5	1.5	−5.1	−22.9
	2010	−13.9	−16.4	9.1	8.9	2.8	26.0
	Mean	−16.3	−13.7	−4.6	−0.8	−6.2	−1.4
R_{eco}/GPP	2008	0.43	1.06	0.65	0.62	0.52	0.94
	2009	0.44	0.56	0.71	1.09	0.69	0.87
	2010	0.54	0.60	2.82	4.71	1.39	1.30
	Mean	0.47	0.74	1.39	2.14	0.87	0.99

GPP, the gross primary production (g C m^{-2}); R_{eco}, the ecosystem respiration (g C m^{-2}); NEE, the net ecosystem carbon exchange (g C m^{-2}); R_{eco}/GPP, the ratio of R_{eco} to GPP.

ecosystem fixed more carbon during 2009 (Figure 4a), but with greater R_{eco} (Figure 4b). Based on the cumulative annual NEE data, the temperate desert steppe was a weak carbon sink during 2008 and 2009, while it was a weak carbon source during 2010 (Figure 4c). Although similar amounts of rainfall were recorded during the dry years of 2008 and 2010, the ecosystem fixed more carbon with 33.2 g C m^{-2} in 2008 than in 2010. The differences in carbon budget were primarily caused by the smaller GPP value from July to September 2010. Moreover, the respiration in 2010 was less than that in 2008 (Table 2).

2.3. Response of Carbon Flux to Environmental Variables

2.3.1. Response of NEE to PAR. The relationship between the daytime NEE and PAR from May to September of the three study years was described using the Micheslis-Menten model (Eq. 7) based on half-hourly data. The rectangular hyperbolic (Eq. 7) produced a good fit to the data when PAR was less than 1,600 μmol photons m^{-2} s^{-1} (approximately the light saturation point). The values of α and NEE$_{sat}$ in 2009 were larger compared with those in 2008 and 2010, with the lowest value observed in 2010. The α value in 2009 was thrice of that in 2010, and the NEE$_{sat}$ value in 2009 was twice of that in 2010. The NEE$_{sat}$ and α values in 2008 were also larger compared with those in 2010, although the averaged SWC in 2010 was slightly higher than that in 2008. The NEE$_{sat}$ and α values for the three integrated study years were −2.3 (±0.01) μmol CO$_2$ m^{-2} s^{-1} and −0.006 (±0.001) μmol CO$_2$ μmol^{-1} photons, respectively (Table 3).

2.3.2. Response of GPP to g_c. The daily g_c values at the study site were consistently low and ranged between 0.06 mm s^{-1} (17 August 2010) and 8.37 mm s^{-1} (9 September 2010) from May

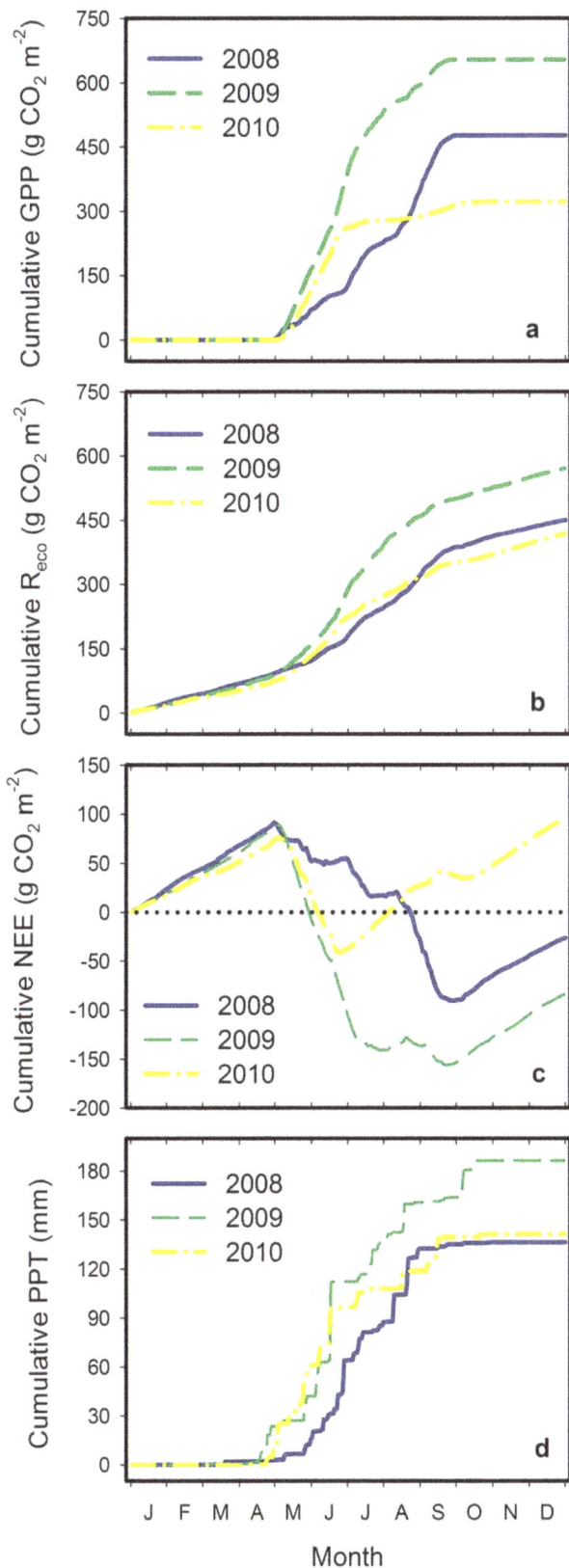

Figure 4. Yearly cumulative net ecosystem carbon exchange (NEE), gross primary production (GPP), ecosystem respiration (R$_{eco}$), and precipitation (PPT) in the Inner Mongolia temperate desert steppe from 2008 to 2010. Negative NEE indicates that the ecosystem is gaining carbon.

Table 3. Mean soil water content and the parameters used to describe the rectangular hyperbolic response of daytime net ecosystem CO_2 exchange to photosynthetically active radiation during May to September of the three measured years as described in Eq. 7.

Years	SWC	α	NEE$_{sat}$	R$_{eco}$	n	R^2	P-value
2008	7.8	−0.006 (±0.001)	−2.6 (±0.1)	0.5 (±0.1)	32	0.94	<0.0001
2009	9.5	−0.010 (±0.003)	−2.9 (±0.2)	0.8 (±0.2)	32	0.94	<0.0001
2010	8.5	−0.003 (±0.002)	−1.4 (±0.1)	0.3 (±0.2)	32	0.75	<0.0001
Three years	8.6	−0.006 (±0.001)	−2.3 (±0.1)	0.5 (±0.1)	32	0.95	<0.0001

SWC, soil water content (%); α, the apparent quantum yield (μmol CO_2 μmol^{-1} photons); NEE$_{sat}$, the saturation value of NEE at an infinite light level (μmol CO_2 m^{-2} s^{-1}); R$_{eco}$, the model-derived bulk ecosystem respiration (μmol CO_2 m^{-2} s^{-1}); n, the number of samples.

to September in the three study years (Figure 5). The season variance of g_c was relatively small in 2008, except for the low g_c value in July. g_c during the early growth season (May and June) was higher than that in the later growth season (September) of 2009, whereas the g_c during the early growth season (May and June) and later growth season (September) was higher than that in the mid growth season (July and August) of 2010. The seasonal variance of g_c was related to the vegetation development and the differences in soil moisture or precipitation. The g_c value fluctuated closely with the drying or wetting of the surface soil and decreased with the depletion of soil moisture. Daily mean g_c values were significantly correlated with the SWC in an exponential manner ($P<0.01$) from May to September of the three study years (Figure 6).

g_c strongly controlled GPP over the temperate desert steppe, and GPP increased linearly with increasing g_c ($P<0.01$), as shown in Figure 7. Slopes of GPP-g_c relationships during the growing seasons significantly varied among the three study years. The slope (2.9±0.3) of GPP-g_c regression line in the 2009 growing season was the highest, while GPP-g_c regression line in 2010 had the lowest slope value (1.9±0.2) among the three study years. These results suggest that the g_c sensitivity of GPP in 2009 was higher than that in 2010, whereas the g_c sensitivity of GPP in 2008 was in the middle range.

2.3.3. Response of GPP and R$_{eco}$ to soil moisture. Figure 8 illustrates the responses of daily GPP and R$_{eco}$ to SWC during May to September of the three study years. Daily GPP and R$_{eco}$ values decreased linearly with increasing soil moisture stress. For each observation year, the SWC sensitivity of R$_{eco}$ was less than the SWC sensitivity of GPP, whereas the slope of R$_{eco}$-SWC regression lines was much lower than that of GPP-SWC responding lines (Table 4). Throughout the study period, the SWC sensitivity of GPP in 2009 was less than that in 2008 and 2010, while the SWC sensitivity of R$_{eco}$ in 2008 was greater than that in 2009 and 2010 (Table 4).

2.3.4. Response to R$_{eco}$ to soil temperature and GPP. The response of half-hourly nighttime NEE (R$_{eco}$) to soil temperatures less than 25°C from May to September of the three study years was analyzed, based on the Van't Hoff equation. The NEE data were averaged with Ts bins of 1°C. The regression coefficients of the fitted curves for the different years are presented in Table 5. For the integrated three growing seasons of the measured years, the temperature sensitivity coefficient (Q$_{10}$) was 2.1, with variance in the different years. In this study, high Q$_{10}$ value (2.3) was observed in 2009 when the desert steppe received

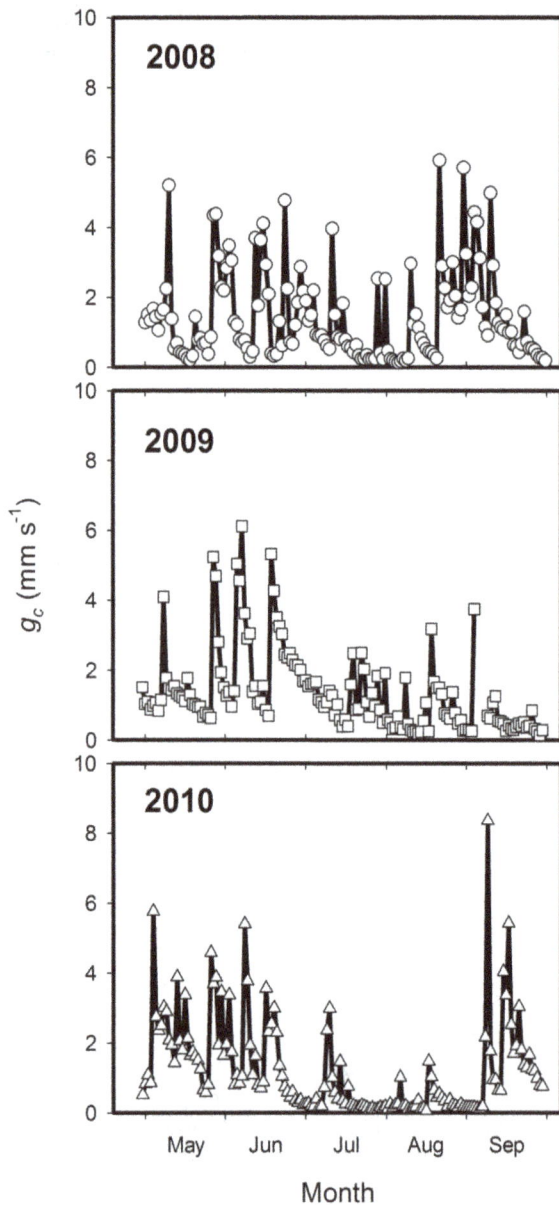

Figure 5. Temporal variations in canopy surface conductance (g_c) from May to September.

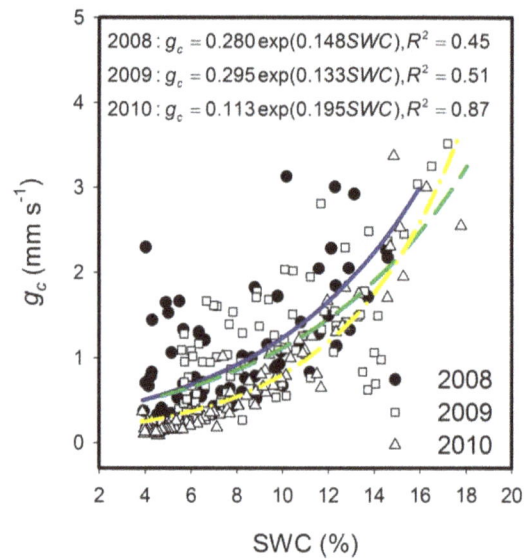

Figure 6. Exponential relation between canopy surface conductance (g_c and soil water content (SWC). Data were obtained from May to September of each of the three study years. Rainy days were excluded from the analysis.

(Figure 10). The relationship between NEE and LAI yielded similar results. Relatively high LAI was observed in July and August 2009 despite a severe drought in the temperate desert steppe (Figure 10).

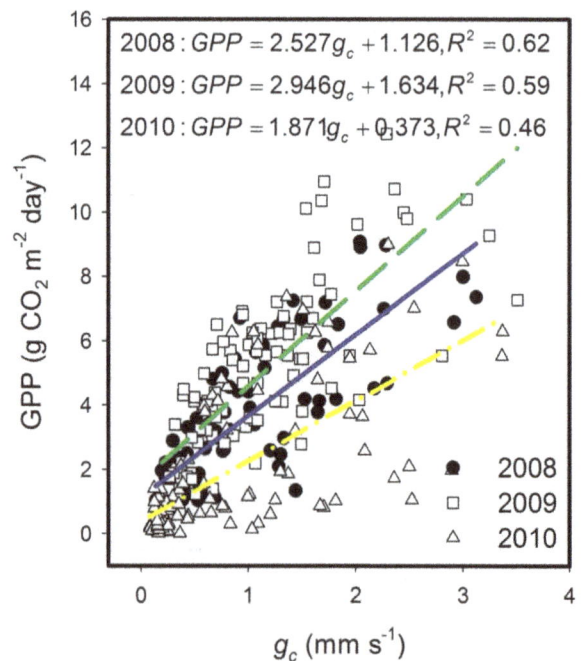

Figure 7. Linear relation between daily gross primary production (GPP) and canopy surface conductance (g_c). Data were obtained from May to September of each of the three study years. Rainy days were excluded from the analysis.

the most rainfall, whereas low Q_{10} value (1.8) occurred in 2010 with the long drought season (Table 5).

The daily R_{eco} values were positively linearly correlated with the daily GPP values, and more than 50% of the variations in R_{eco} can be explained by GPP in the three growing seasons ($P<0.01$, Figure 9). The coefficient of determination (R^2) between R_{eco} and GPP in 2008 (dry soil water conditions) was obviously lower than that in 2009 and 2010 (wet soil water conditions).

2.4. Response of Carbon Fluxes to LAI

The LAI in the growing seasons of 2008 and 2009 were measured. In 2008, changes in LAI explained 45% of the variations in GPP, with the increase in GPP per unit LAI of 6.58 g CO_2 m^{-2} day^{-1}. On the other hand, the linear relationship between GPP and LAI in 2009 changed into a negative relation

Table 4. The regression equations between gross primary productivity and soil water content, and between ecosystem respiration and soil water content in the three years.

Year	Equations	n	R^2	P-value
2008	GPP = 0.45 (±0.11) SWC +0.02 (±1.13)	12	0.61	P = 0.003
	R_{eco} = 0.23 (±0.03) SWC +0.12 (±0.29)	12	0.87	P<0.001
2009	GPP = 0.36 (±0.08) SWC +1.37 (±0.94)	13	0.65	P<0.001
	R_{eco} = 0.16 (±0.03) SWC +1.35 (±0.39)	13	0.68	P<0.001
2010	GPP = 0.41 (±0.05) SWC − 1.85 (±0.47)	11	0.89	P<0.001
	R_{eco} = 0.10 (±0.02) SWC +0.69 (±0.19)	11	0.72	P<0.001

GPP, gross primary productivity (g CO_2 m^{-2} day^{-1}); SWC, soil water content (%); R_{eco}, ecosystem respiration (g CO_2 m^{-2} day^{-1}).

Discussion

3.1. Magnitude of Carbon Flux Compared with Other Grassland Ecosystems

The temperate desert steppe ecosystem accumulated a total of 4.2 g C m^{-2} from 2008 to 2010, averaging 1.4 (±25.0, S.D.) g C m^{-2} yr^{-1} annually, suggesting that it is neutral. Large interannual variability in NEE can also be found in literature on grassland ecosystems in the Inner Mongolia and the Mongolian Plateau (Table S1). For example, net carbon loss was reported in the Inner Mongolia *Stipa Krylovii* steppe [49] and the Inner Mongolia *Leymus Chinensis* steppe [27], while a significant carbon sink was reported for the grazed typical steppe in Central Mongolia [50]. In terms of other grassland ecosystem types, some studies reported a carbon sink in the alpine meadow-steppe in the Qinghai-Tibet Plateau [51,52], while Fu *et al.* [27] found the opposite. For European grasslands, Hussain *et al.* [22] recently reported that the managed grassland in Germany was a carbon sink. Based on the EUROGRASSFLUX data, Gilmanov *et al.* [53] found that four out of nineteen cases of grasslands did not perform as annual net CO_2 sinks. For Mediterranean climate grasslands [5,11,17], the ecosystem could be a carbon source or a carbon sink depending on the precipitation quantity and the timing of rain events. Such alternations between carbon sink and carbon source have been reported on a Canadian temperate grassland [12], a semi-desert grassland in the USA [23]. Due to the seasonal drought and overgrazing, a tropical pasture in Panama was a strong carbon source [54]. Some North American prairies had characteristics similar to a consistent carbon sink [55]. However, fire burning in the spring could result in net carbon loss in the Oklahoma native tallgrass prairie [16]. In general, as shown through the literature cited above, the annual NEE of grassland ecosystems can be a carbon sink or carbon source among different hydrological years. The lower than normal precipitation or drought and management could be the primary factors that lead to the carbon source for grassland ecosystems.

3.2. Environmental Regulation of Carbon Fluxes

3.2.1. Effects of drought on GPP and R_{eco}. Reichstein *et al.* [18] hypothesized that short-term drought would suppress R_{eco} more than GPP because litter and upper soil layers that dry first are the locations of most heterotrophic respiration, whereas photosynthesis could be supported by moisture that is accessible to the roots in deeper soil layers. In the temperate desert steppe studied here, the drought suppressed GPP more than R_{eco} during the three study years. The root distribution of these grasslands

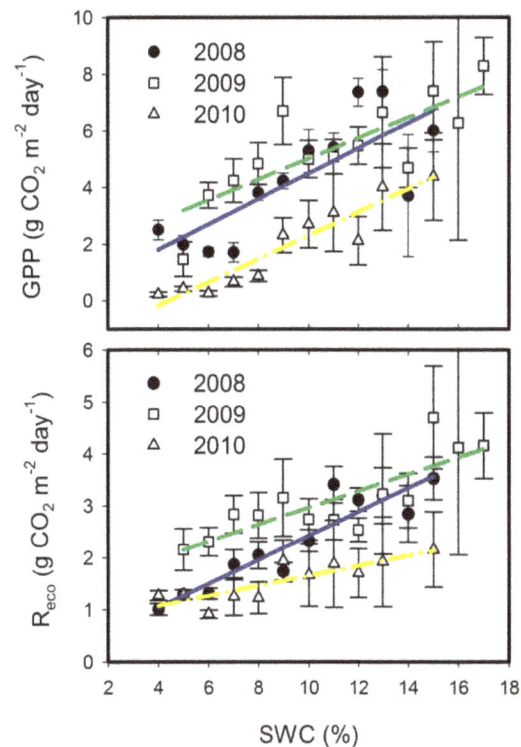

Figure 8. Responses of gross primary productivity (GPP) and ecosystem respiration (R_{eco}) to soil water content (SWC). GPP and R_{eco} data from May to September were averaged with a bin width of 1% for SWC. Error bars represent one standard error. Rainy days were excluded from the analysis.

plants was mainly in the soil layer less than 0.30 m depth [56]. Yang *et al.* [57] reported that SWC deeper than 0.40 m always remained at a near-constant value, and deeper ground water discharged for assimilation was very limited in the temperate desert steppe. Consequently, the impact of the drought might have lasted longer for assimilation than respiration in this ecosystem.

3.2.2. Effects of environmental variables on R_{eco}. Q_{10} in 2010 was lower than that in 2008, although the soil water content in 2010 was higher than that in 2008. The SWC sensitivity of R_{eco} for different hydrological years in the temperate desert steppe was also found to be not consistent with the hypotheses that SWC would constrain R_{eco} more during a dry year than during a wet year [18]. The discrepancy indicated that beyond soil temperature and moisture which is widely accepted [58], some others may make also important influence on R_{eco}, such as biotic factors

Table 5. Regression coefficients as described in Eqs. 8–9, lower than 25°C.

	SWC	a	b	R^2	Q_{10}	P-value
2008	7.7	0.005	0.07	0.81	2.1 (±0.2)	<0.001
2009	10.1	0.005	0.08	0.91	2.3 (±0.2)	<0.001
2010	9.8	0.007	0.06	0.58	1.8 (±0.2)	<0.001
Three years	9.0	0.006	0.07	0.84	2.1 (±0.2)	<0.001

SWC, soil water content (%); Q_{10}, the temperature sensitivity coefficient of ecosystem respiration.

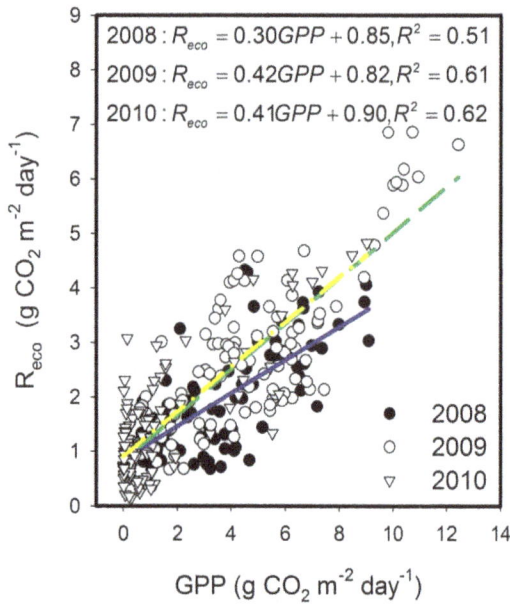

Figure 9. Ecosystem respiration (R_{eco}) responses to gross primary production (GPP) in a linear manner.

(photosynthesis or substrate availability) [59–61]. The data of the current study suggest that GPP may have mediated R_{eco}, which responds more to GPP under wet soil conditions than under dry soil conditions. Based on the results derived from the water addition field experiment, Yan *et al.* [29] also demonstrated that photosynthetic substrate supply was an important factor in regulating soil respiration in both daily and seasonal timescales in semiarid steppe ecosystems and that its effect on R_{eco} increased with increasing water availability. A strong linear relationship between R_{eco} and GPP was also reported in the Mediterranean C3/C4 grassland [11,17] and oak–grass savannah [62]. However, recent results have shown that respiration might be partly driven by GPP, and the respiration components, including autotrophic and heterotrophic maybe partly dependent on the respective substrate availability [8]. Bahn *et al.* [26] and Yan *et al.* [29] reported that the response of R_{eco} to the changes in photosynthesis has a significant time lag of zero to three hours in the grassland ecosystem. Although it may be challenging, the modulation of photosynthesis on respiration should be incorporated as a physiological process to mechanism-based carbon model [63].

3.2.3. Effects of LAI and SWC on carbon Flux. The linear relationship between carbon fluxes and LAI has been demonstrated in the grassland in central Japan [64], Mediterranean climate grassland [5,11,17], Canadian temperate grassland [12], and Inner Mongolia *Leymus chinensis* steppe [27]. The low LAI explanation for the variance of NEE (26%) or GPP (53%) for the

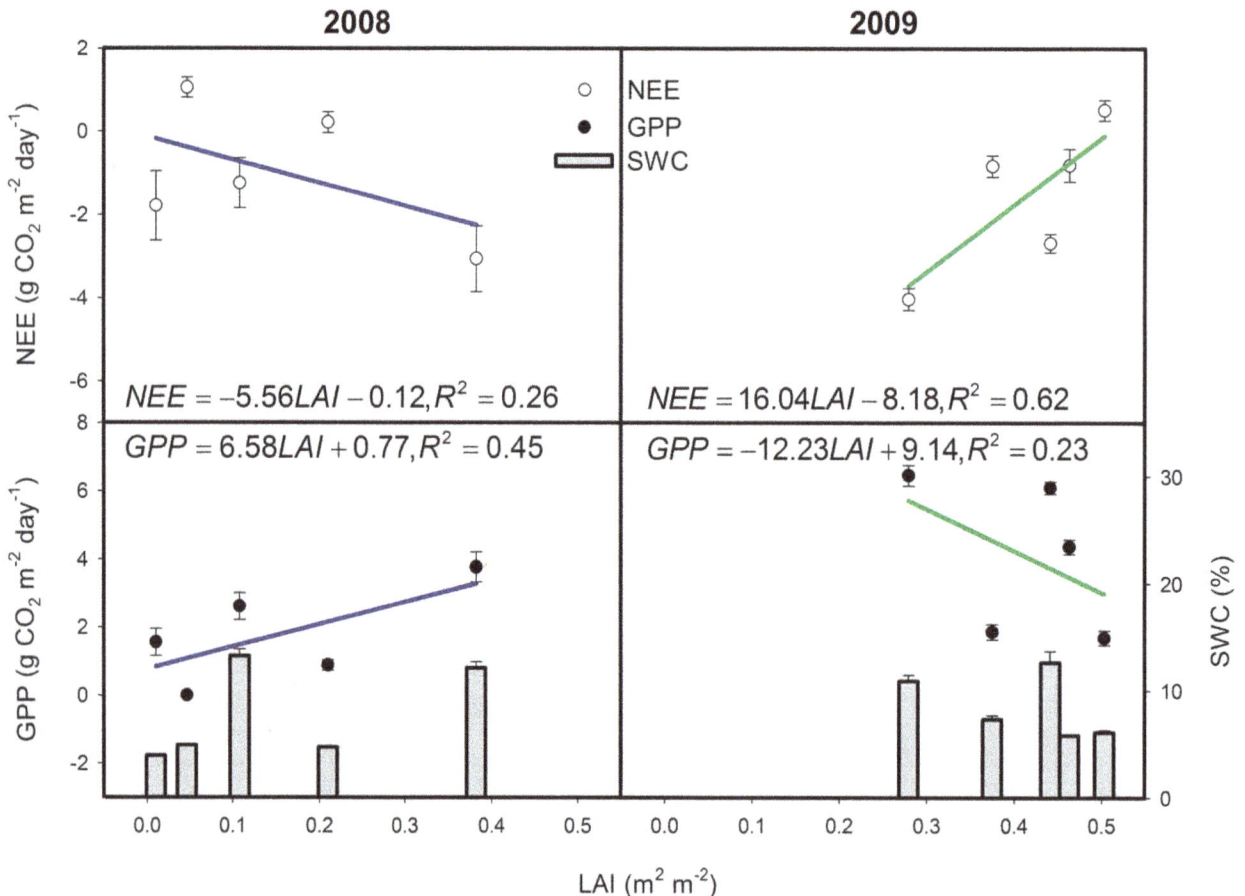

Figure 10. Relationships between daily net ecosystem CO_2 exchange (NEE), daily gross primary production (GPP), and leaf area index (LAI). NEE, GPP, and soil water content (SWC) data represent the seven-day mean that is centered on the day samples for LAI. Error bars represent one standard error.

Mongolia typical steppe [50], and the poor relationship between carbon flux and LAI for the Inner Mongolia typical *Stipa krylovii* steppe have also been reported [49]. Obvious discrepancy in the relationship between carbon flux and LAI in the temperate desert steppe was observed for the two precipitation pattern years. The SWC variance might be the primary factor that affects the response of carbon flux to LAI. A good relationship between carbon flux and LAI can be achieved under sufficient water conditions. However, the vegetation function structural processes, such as LAI, respond to drought stress later than the physiological or biochemical processes, such as vegetation photosynthesis and microbial respiration, which were strongly related to the ecosystem carbon flux. The relationship between carbon flux and LAI may be weak under drought stress conditions, even the possibility of high LAI value. Thus, we suggest that SWC might have the same importance as LAI on the ecosystem carbon uptake in the temperate desert steppe characterized by drought, and it might become dominant under specific conditions.

3.2.4. Effects of seasonal precipitation pattern on carbon balance.
Rainfall seasonal variability alters ecosystem carbon dynamics, regardless of the amount of total annual precipitation [8,65]. In terms of temperate desert steppe, the annual precipitation in 2010 was similar to that in 2008, whereas the precipitation in July and August was significantly lower than that in 2008, leading to a drop in SWC, a decline in photosynthetic activity, and a lower absolute cumulative NEE in 2010. High precipitation in September 2010 could not offset the reductions caused by the severe summer drought. Jongen *et al.* [5] also found that the lack of precipitation at the peak growth could result in a decrease in annual carbon sequestration in the Mediterranean grassland.

Precipitation variance has also been found to influence the timing and duration of canopy development and, therefore, regulate the biomass production on northern mixed-grass prairies [66–68]. Functional changes in grass vegetation could affect the biological and ecological processes, such as stomatal conductance, photosynthesis, and respiration. The long summer drought period in 2010 could have induced the functional change in vegetation in the temperate desert steppe, resulting in the shorter length of the carbon sink days of 2010. The shallow roots of grass species were mostly distributed lower than 50 cm, and soil water availability was dependent on the atmospheric rainfall rather than on the deep groundwater supply [69]. Plant community structure may be sensitive to rainfall variability, especially to seasonal drought stress, which may have an indirect effect on carbon flux [8].

Xu *et al.* [17] reported that the timing of rain events had more impact than the total amount of precipitation on the ecosystem

NEE for the Mediterranean grassland. Similar results were also found for the savannah ecosystem [62] and alpine shrub lands in Qinghai-Tibetan plateau [52]. However, the NEE in the temperate desert steppe was much larger in 2009, with more annual precipitation, than in 2008 and 2010, which had low annual precipitation. These results suggest that carbon sequestration in the temperate desert steppe primarily depends on the amount of annual precipitation and the summer precipitation quantity. Hence, the summer drought could have a significant influence on the annual carbon balance.

Conclusions

The temperate desert steppe ecosystem was close to carbon neutrality during 2008 to 2010. The seasonal variation in NEE was correlated with rain events and soil moisture. This study demonstrated that the carbon sequestration capacity of the temperate desert steppe is sensitive to changes in the precipitation seasonal distribution rather than just the annual precipitation quantity. Summer drought stress could influence the annual carbon balance.

Furthermore, the drought suppressed GPP more than R_{eco} and the SWC sensitivity of GPP was greater during the drought year. The photosynthetic substrate supply displayed an important role in regulating respiration on the daily timescale but the magnitude of this effect became less apparent during the drought year, when there was limited soil moisture availability.

Acknowledgments

We would like to appreciate Drs. Feng Zhang, Fengyu Wang, Fang Bao, and Xiaoyan Ping for their contributions during the experiment, and Drs. Xinghua Sui, Yu Wang, and Li Zhou for their help in data processing. We are grateful to Dr. John E. Hunt for his constructive discussions, and acknowledge the editor and three anonymous reviewers for their valuable comments in the manuscript.

Author Contributions

Conceived and designed the experiments: GZ FY. Performed the experiments: FY GZ. Analyzed the data: FY GZ. Wrote the paper: FY GZ.

References

1. Parton WJ, Scurlock JMO, Ojima DS, Schimel DS, Hall DO (1995) Impact of climate change on grassland production and soil carbon worldwide. Global Change Biol 1: 13–22.

2. Karl TR, Trenberth KE (2003) Modern global climate change. Science 302: 1719.

3. Chou WW, Silver WL, Jackson RD, Thompson AW, Allen-Diaz B (2008) The sensitivity of annual grassland carbon cycling to the quantity and timing of rainfall. Global Change Biol 14: 1382–1394.

4. Baldocchi D (2011) Global change: The grass response. Nature 476: 160–161.

5. Jongen M, Santos Pereira J, Aires LMI, Pio CA (2011) The effects of drought and timing of precipitation on the inter-annual variation in ecosystem-atmosphere exchange in a Mediterranean grassland. Agric Forest Meteorol 151: 595–606.

6. Niu SL, Wu MY, Han Y, Xia JY, Li LH, et al. (2008) Water-mediated responses of ecosystem carbon fluxes to climatic change in a temperate steppe. New Phytol 177: 209–219.

7. Bai Y, Han X, Wu J, Chen Z, Li L (2004) Ecosystem stability and compensatory effects in the Inner Mongolia grassland. Nature 431: 181–184.

8. Klumpp K, Tallec T, Guix N, Soussana JF (2011) Long-term impacts of agricultural practises and climatic variability on carbon storage in a permanent pasture. Global Change Biol 17: 3534–3545.

9. Sala OE, Parton WJ, Joyce LA, Lauenroth WK (1988) Primary production of the central grassland region of the United States. Ecology 69: 40–45.

10. Baldocchi D (2008) Turner review No. 15. 'Breathing' of the terrestrial biosphere: lessons learned from a global network of carbon dioxide flux measurement systems. Aust J Bot 56: 1–26.

11. Aires LMI, Pio CA, Pereira JS (2008) Carbon dioxide exchange above a Mediterranean C3/C4 grassland during two climatologically contrasting years. Global Change Biol 14: 539–555.

12. Flanagan LB, Wever LA, Carlson PJ (2002) Seasonal and interannual variation in carbon dioxide exchange and carbon balance in a northern temperate grassland. Global Change Biol 8: 599–615.

13. Meyers TP (2001) A comparison of summertime water and CO_2 fluxes over rangeland for well watered and drought conditions. Agric Forest Meteorol 106: 205–214.

14. Nagy Z, Pintér K, Czóbel S, Balogh J, Horváth L, et al. (2007). The carbon budget of semi-arid grassland in a wet and a dry year in Hungary. Agric Ecosyst Environ 121: 21–29.

15. Sims PL, Bradford JA (2001) Carbon dioxide fluxes in a southern plains prairie. Agric Forest Meteorol 109: 117–134.

16. Suyker AE, Verma SB, Burba GG (2003) Interannual variability in net CO₂ exchange of a native tallgrass prairie. Global Change Biol 9: 255–265.

17. Xu LK, Baldocchi DD (2004) Seasonal variation in carbon dioxide exchange over a Mediterranean annual grassland in California. Agric Forest Meteorol 123: 79–96.

18. Reichstein M, Tenhunen JD, Roupsard O, Ourcival JM, Rambal S, et al. (2002) Severe drought effects on ecosystem CO₂ and H₂O fluxes at three Mediterranean evergreen sites: revision of current hypotheses? Global Change Biol 8: 999–1017.

19. Hunt JE, Kelliher FM, McSeveny TM, Byers JN (2002) Evaporation and carbon dioxide exchange between the atmosphere and a tussock grassland during a summer drought. Agric Forest Meteorol 111: 65–82.

20. Wang YF, Cui XY, Hao YB, Mei XR, Yu GR, et al. (2011) The fluxes of CO₂ from grazed and fenced temperate steppe during two drought years on the Inner Mongolia Plateau, China. Sci Total Environ 410: 182–189.

21. Ciais P, Reichstein M, Viovy N, Granier A, Ogee J, et al. (2005) Europe-wide reduction in primary productivity caused by the heat and drought in 2003. Nature 437: 529–533.

22. Hussain MZ, Grünwaldb T, Tenhunen JD, Li YL, Mirzae H, et al. (2011) Summer drought influence on CO₂ and water fluxes of extensively managed grassland in Germany. Agric Ecosyst Environ 141: 67–76.

23. Scott RL, Hamerlynck EP, Jenerette GD, Moran MS, Barron-Gafford GA (2010) Carbon dioxide exchange in a semidesert grassland through drought-induced vegetation change. J Geophys Res 115: doi:10.1029/2010JG001348.

24. Meir P, Metcalfe DB, Costa ACL. Fisher RA (2008) The fate of assimilated carbon during drought: impacts on respiration in Amazon rainforests. Philos Trans R Soc Lond B Biol Sci 363: 1849–1855.

25. van der Molen MK, Dolman AJ, Ciais P, Eglin T, Gobron N, Law BE, et al. (2011) Drought and ecosystem carbon cycling. Agric Forest Meteorol 151, 765–773.

26. Bahn M, Schmitt M, Siegwolf R, Richter A, Brüggemann N (2009) Does photosynthesis affect grassland soil-respired CO₂ and its carbon isotope composition on a diurnal timescale? New Phytol 182: 451–460.

27. Fu Y, Zheng Z, Yu G, Hu Z, Sun X, et al. (2009) Environmental influences on carbon dioxide fluxes over three grassland ecosystems in China. Biogeosciences 6: 2879–2893.

28. Davidson EA, Janssens IA, Luo Y (2006) On the variability of respiration in terrestrial ecosystems: moving beyond Q10. Global Change Biol 12, 154–164.

29. Yan L, Chen S, Huang J, Lin G (2011) Water regulated effects of photosynthetic substrate supply on soil respiration in a semiarid steppe. Global Change Biol 17: 1990–2001.

30. Sun HL (2005) Ecosystems of China. Beijing: Science Press. (in Chinese)

31. Yang FL, Zhou GS, Hunt JE, Zhang F (2011) Biophysical regulation of net ecosystem carbon dioxide exchange over a temperate desert steppe in Inner Mongolia, China. Agric Ecosyst Environ 142: 318–328.

32. Fan JW, Zhong HP, Harris W, Yu GR, Wang SQ, et al. (2008) Carbon storage in the grasslands of China based on field measurements of above- and below-ground biomass. Clim Change 86: 375–396.

33. Li XB, Chen YH, Zhang YX, Fan YD, Zhou T, et al. (2002) Impact of climate change on desert steppe in northern china. Adv Earth Sci 17: 254–261. (in Chinese with English abstract)

34. Webb EK, Pearman GI, Leuning R (1980) Correction of flux measurements for density effects due to heat and water vapour transfer. Q J Roy Meteorol Soc 106: 85–100.

35. Papale D, Reichstein M, Aubinet M, Canfora E, Bernhofer C, et al. (2006) Towards a standardized processing of net ecosystem exchange measured with eddy covariance technique: algorithms and uncertainty estimation. Biogeosciences 3: 571–583.

36. Gu LH, Falge EM, Boden T, Baldocchi DD, Black TA, et al. (2005) Objective threshold determination for nighttime eddy flux filtering. Agric Forest Meteorol 128: 179–197.

37. Zhu Z, Sun X, Wen X, Zhou Y, Tian J, et al. (2006) Study on the processing method of nighttime CO₂ eddy covariance flux data in ChinaFLUX. Sci China Ser D 49: 36–46.

38. Burba GG, McDermitt DK, Grelle A, Anderson DJ, Xu LK (2008) Addressing the influence of instrument surface heat exchange on the measurements of CO₂ flux from open-path gas analyzers. Global Change Biol 14: 1854–1876.

39. Koehler AK, Sottocornola M, Kiely G (2011) How strong is the current carbon sequestration of an Atlantic blanket bog? Global Change Biol 17: 309–319.

40. Sottocornola M, Kiely G (2010) Hydro-meteorological controls on the CO₂ exchange variation in an Irish blanket bog. Agric Forest Meteorol 150: 287–297.

41. Falge E, Baldocchi D, Olson R, Anthoni P, Aubinet M, et al. (2001) Gap filling strategies for long term energy flux data sets. Agric Forest Meteorol 107: 71–77.

42. Falge E, Baldocchi D, Olson R, Anthoni P, Aubinet M, et al. (2001) Gap filling strategies for defensible annual sums of net ecosystem exchange. Agric Forest Meteorol 107: 43–69.

43. Oliphant AJ, Grimmond CSB, Zutter HN, Schmid HP, Su HB, et al. (2004) Heat storage and energy balance fluxes for a temperate deciduous forest. Agric Forest Meteorol 126: 185–201.

44. Richardson AD, Hollinger DY, Burba GG, Davis KJ, Flanagan LB, et al. (2006) A multi-site analysis of random error in tower-based measurements of carbon and energy fluxes. Agric Forest Meteorol 136: 1–18.

45. Reichstein M, Falge E, Baldocchi D, Papale D, Aubinet M, et al. (2005) On the separation of net ecosystem exchange into assimilation and ecosystem respiration: review and improved algorithm. Global Change Biol 11: 1424–1439.

46. Lloyd J, Taylor JA (1994) On the temperature dependence of soil respiration. Funct Ecol 8: 315–323.

47. Allen RG, Pereira LS, Raes D, Smith M (1998) Crop evapotranspiration. Guidelines for computing crop water requirements. Rome: FAO. 19 P.

48. Monteith JL, Unsworth MH (1990) Principles of environmental physics. London: Edward Arnold. 291 P.

49. Wang Y, Zhou G (2008) Environmental effects on net ecosystem CO₂ exchange at half-hour and month scales over Stipa krylovii steppe in northern China. Agric Forest Meteorol 148: 714–722.

50. Li SG, Asanuma J, Eugster W, Kotani A, Liu JJ, et al. (2005). Net ecosystem carbon dioxide exchange over grazed steppe in central Mongolia. Global Change Biol 11: 1941–1955.

51. Kato T, Tang Y, Gu S, Hirota M, Du M, et al. (2006) Temperature and biomass influences on interannual changes in CO₂ exchange in an alpine meadow on the Qinghai-Tibetan Plateau. Global Change Biol 12: 1285–1298.

52. Zhao L, Li YN, Xu SX, Zhou HK, Gu S, et al. (2006) Diurnal, seasonal and annual variation in net ecosystem CO₂ exchange of an alpine shrubland on Qinghai-Tibetan plateau. Global Change Biol 12: 1940–1953.

53. Gilmanov TG, Soussana JE, Aires L, Allard V. Ammann C, et al. (2007) Partitioning European grassland net ecosystem CO₂ exchange into gross primary productivity and ecosystem respiration using light response function analysis. Agric Ecosyst Environ 121: 93–120.

54. Wolf S, Eugster W, Potvin C, Turner BL, Buchmann N (2011) Carbon sequestration potential of tropical pasture compared with afforestation in Panama. Global Change Biol 17: 2763–2780.

55. Suyker AE, Verma SB (2001) Year-round observations of the net ecosystem exchange of carbon dioxide in a native tallgrass prairie. Global Change Biol 7: 279–289.

56. Ping X, Zhou G, Zhuang Q, Wang Y, Zuo W, et al. (2010) Effects of sample size and position from monolith and core methods on the estimation of total root biomass in a temperate grassland ecosystem in Inner Mongolia. Geoderma 155: 262–268.

57. Yang FL, Zhou GS (2011) Characteristics and modeling of evapotranspiration over a temperate desert steppe in Inner Mongolia, China. J Hydrol 396: 139–147.

58. Raich JW, Tufekciogul A (2000) Vegetation and soil respiration: correlations and controls. Biogeochemistry 48: 71–90.

59. Craine JM, Wedin DA, Chapin FS (1999) Predominance of ecophysiological controls on soil CO₂ flux in a Minnesota grassland. Plant Soil 207: 77–86.

60. Thomey ML, Collins SL, Vargas R, Johnson JE, Brown RF, et al. (2011) Effect of precipitation variability on net primary production and soil respiration in a Chihuahuan Desert grassland. Global Change Biol 17: 1505–1515.

61. Wan S, Luo Y (2003) Substrate regulation of soil respiration in a tallgrass prairie: results of a clipping and shading experiment. Global Biogeochem Cycles 17: doi:10.1029/2002GB001971.

62. Ma S, Baldocchi DD, Xu L, Hehn T (2007) Inter-annual variability in carbon dioxide exchange of an oak/grass savanna and open grassland in California. Agric Forest Meteorol 147: 157–171.

63. Tang J, Baldocchi DD, Xu L (2005) Tree photosynthesis modulates soil respiration on a diurnal time scale. Global Change Biol 11: 1298–1304.

64. Saigusa N, Oikawa T, Liu S (1998) Seasonal variations of the exchange of CO₂ and H₂O between a grassland and the atmosphere: an experimental study. Agric Forest Meteorol 89: 131–139.

65. Harper CW, Blair JM, Fay PA, Knapp AK, Carlisle JD (2005) Increased rainfall variability and reduced rainfall amount decreases soil CO₂ flux in a grassland ecosystem. Global Change Biol 11: 322–334.

66. Smart AJ, Dunn BH, Johnson PS, Xu L, Gates RN (2007) Using weather data to explain herbage yield on three Great Plains plant communities. Rangeland Ecol Manag 60: 146–153.

67. Wayne Polley H, Frank AB, Sanabria J, Phillips RL (2008) Interannual variability in carbon dioxide fluxes and flux-climate relationships on grazed and ungrazed northern mixed-grass prairie. Global Change Biol 14: 1620–1632.

68. Krishnan P, Meyers TP, Scott RL, Kennedy L, Heuer M (2012) Energy exchange and evapotranspiration over two temperate semi-arid grasslands in North America. Agric Forest Meteorol 153: 31–44.

69. Shen W, Reynolds JF, Hui D (2009) Responses of dryland soil respiration and soil carbon pool size to abrupt vs. gradual and individual vs. combined changes in soil temperature, precipitation, and atmospheric [CO₂]: a simulation analysis. Global Change Biol 15: 2274–2294.

8

Comparative Demography of an At-Risk African Elephant Population

George Wittemyer[1,2,3]*, **David Daballen**[3], **Iain Douglas-Hamilton**[3,4]

1 Department of Fish, Wildlife and Conservation Biology, Colorado State University, Fort Collins, Colorado, United States of America, **2** Graduate Degree Program in Ecology, Colorado State University, Fort Collins, Colorado, United States of America, **3** Save The Elephants, Nairobi, Kenya, **4** Department of Zoology, University of Oxford, Oxford, United Kingdom

Abstract

Knowledge of population processes across various ecological and management settings offers important insights for species conservation and life history. In regard to its ecological role, charisma and threats from human impacts, African elephants are of high conservation concern and, as a result, are the focus of numerous studies across various contexts. Here, demographic data from an individually based study of 934 African elephants in Samburu, Kenya were summarized, providing detailed inspection of the population processes experienced by the population over a fourteen year period (including the repercussions of recent increases in illegal killing). These data were compared with those from populations inhabiting a spectrum of xeric to mesic ecosystems with variable human impacts. In relation to variability in climate and human impacts (causing up to 50% of recorded deaths among adults), annual mortality in Samburu fluctuated between 1 and 14% and, unrelatedly, natality between 2 and 14% driving annual population increases and decreases. Survivorship in Samburu was significantly lower than other populations with age-specific data even during periods of low illegal killing by humans, resulting in relatively low life expectancy of males (18.9 years) and females (21.8 years). Fecundity (primiparous age and inter-calf interval) were similar to those reported in other human impacted or recovering populations, and significantly greater than that of comparable stable populations. This suggests reproductive effort of African savanna elephants increases in relation to increased mortality (and resulting ecological ramifications) as predicted by life history theory. Further comparison across populations indicated that elongated inter-calf intervals and older ages of reproductive onset were related to age structure and density, and likely influenced by ecological conditions. This study provides detailed empirical data on elephant population dynamics strongly influenced by human impacts (laying the foundation for modeling approaches), supporting predictions of evolutionary theory regarding demographic responses to ecological processes.

Editor: Matt Hayward, Australian Wildlife Conservancy, Australia

Funding: Funding for this study included the National Science Foundation GRFP and IRFP OISE-0502340, Save the Elephants private donors and the Escape Foundation (http://www.escapefoundation.org). The funders had no role in study design, data collection and analysis, decision to publish, or preparation of the manuscript.

Competing Interests: The authors have declared that no competing interests exist.

* E-mail: g.wittemyer@colostate.edu

Introduction

Detailed demographic data are available for only a handful of wildlife species of high economic or conservation value [1]. For the limited number of species that have been studied, data from multiple populations are typically not available, with existing information usually compiled from a single population [2]. As such, we often lack information on life history strategies across species and ecological systems, limiting our capacity to assess demographic responses to human pressure and ecological changes [3,4]. Understanding species specific responses is of particular interest for threatened species or those that are economically salient to their range states.

One of the better studied tropical, large ungulates is the African elephant, for which multiple studies using individual based monitoring or culling data provide demographic parameters. The African elephant plays a keystone role [5] in the diversity of habitats it occupies (from deserts to rainforests) by influencing canopy cover [6], affecting species distribution [7] and dispersing seeds [8]. The state of elephant populations thus is critical to the integrity of the ecosystems it inhabits [9]. Further, this species is of high economic value in terms of commercial trade (for ivory) and tourism as well as being a flagship species. The conservation status of African elephant populations vary broadly across the continent with recent eradications in parts of Central and West Africa [10] and nearly a quarter of the species inhabiting one region in Botswana [11]. As such, its designated conservation status, trade in its products [12] and management of the species [13] are contentious [14]. As a species of high conservation concern and extreme life history strategy, with the longest mammalian gestation period (as well as long reproductive life), understanding the demographic status and response to different human pressures and ecological conditions is invaluable for theoretical and practical applications. Most of the existing data on elephants, however, were compiled from well protected populations with stable or increasing populations at the time of the study. Unfortunately, these data do not represent the status of many elephant populations across the continent [10,11,15], and it is imperative conservation and management bodies have comparative data on populations experiencing greater pressure to determine differences in population status and response to human impacts.

Here we summarize the demographic data derived from 14 years of continuous, individual based monitoring of the Samburu elephant population in northern Kenya [16]. This area is of high conservation interest as one of the Convention on International Trade in Endangered Species (CITES) Monitoring of Illegal Killing of Elephants (MIKE) sites, a program to assess the relationship between legal ivory trade and illegal killing of elephants [17,18]. During the study period, this population was subject to consistent human pressure and predation [19,20] the impact of which recently spiked [21]. Life table metrics from the Samburu population are presented and its demography during periods of relatively low and high human induced mortality compared and contrasted with information from published demographic studies of African elephants in different ecosystems. The demographic processes related to varying degrees of human pressure and different ecological zones are discussed in relation to the conservation status of this species.

Methods

Study System

Individually based demographic data on the Samburu elephant population were collected through an individual identification study of all elephants inhabiting the semiarid, 220 km² Samburu and Buffalo Springs national reserves in northern Kenya (0.3–0.8° N, 37–38°E) (Fig. 1). These reserves are centered on the Ewaso N'giro River, the only permanent water in the area, and are a focal area for wildlife and tourism. Rainfall in the region averages approximately 350 mm per year (IQR: 242 mm to 401 mm) and occurs during biannual rainy seasons generally taking place in April/May and November/December. Due to the rainfall pattern in the system, demographic data were collated for the twelve month period between Oct. 1st and Sept. 30th in relation to the date of consistent separation between wet and dry periods in the ecosystem. Intermittent droughts (defined as years during which one of the wet seasons received no precipitation or when total annual rainfall was below the 25th quartile) affect the system periodically.

Individual identification using natural markings (protocols established in other populations [22,23]) of all elephants of the reserves began in 1997 [16], since which time the elephants have been closely monitored. As a result of heavy tourist use of the parks, the reserve elephants are habituated to vehicles, enabling easy observation. The area used by these elephants is large (>3500 km²) and all elephants rely substantially on areas outside the protected areas [20]. Outside the protected areas, the elephants occur in largely unpatrolled, communally and government owned lands (with the exception of patrolled community conservation areas), where human elephant conflict and illegal killing occurs [24] and has significant impacts on the population demography [21].

Demographic Data

The data presented in this study were collected from November 1997 through September 2011 from resident elephants (as defined in [16]) in the national reserves, numbering between 408–558 individuals during the 14 year study (Fig. 2). Analysis was conducted on 509 individual females (accounting for 4628 live female years) and 425 males (accounting for 2914 live male years). The presence or absence of individual elephants, location, and time were recorded during weekly travel along 5 established transects (approximately 20 km long) in the protected areas [25],

from which mortalities and births were deduced (see below). The study elephants are not always present in the national reserves [26], therefore sampling was opportunistic along these transects. During the 14 year study, 494 births and 340 deaths were recorded among these resident, focal elephants (Fig. 2) and 156 were thought to have dispersed (see description below). The average age at the first observation of new born elephants was 23 days (S.D. = 45). New adult females (a single family group) were identified in the population on one occasion in 2005, but were excluded from this analysis. New males periodically entered the population and when regularly observed were added to the studied cohort (n = 34).

Mortalities were identified in one of three ways as described in [19]: (1) known individuals were found dead; (2) repeated (>3) observations established individuals were missing from their family group; and (3) no observations for >3 years of individuals that previously had been observed annually by the monitoring team. None of the elephants recorded as dead were subsequently observed. Carcasses of the majority of registered mortalities were never found, therefore, the actual dates of death for these mortalities were unknown and year of death was assigned by adding its average inter-observation interval to the last date of observation as described in [19]. Carcasses were found during daily monitoring within the reserves (often initially sighted by tourists), but its identity was not always possible to assign depending on the age of the carcass, scavenger impacts, and cause of death (e.g. predation, disease, poaching). Researchers also investigated any reports of dead elephants in the wider ecosystem (approximately a 10 km buffer around the reserves) to assess if a carcass was a known individual. When the carcass of a known individual was found outside the reserves, relevant information (location, date and cause of death) were recorded and added to the standard monitoring data described previously. The cause of death of all carcasses found within the reserves and those of known individuals found outside the reserves were collated to derive the proportion of illegally killed elephants ($PIKE = \frac{\# Illegally\ Killed}{Total\ Carcasses}$) in the study area, the standard monitoring metric for the MIKE programme. In contrast to other demographic data presented here, PIKE figures were collated over the calendar year (Jan-Dec) to match continental protocol [17].

Because subadult males (3–18 years old) regularly leave their family groups and appear to range widely, they were recorded as dead only when their carcasses were observed or death could be inferred by other indicators (e.g. the presence of wounded family members). When subadult males transitioned from being integrated with the family, to trailing the family, to not being associated with the family (e.g. were no longer seen with their families but not known or suspected to have died), they were defined as dispersed and subsequently removed from the analyses (n = 156). As such, some individuals that died were likely treated as having dispersed. In the latter stages of the study when human induced mortality was high (2009–2011), some young females (orphans) from poached families were lost and could not be assumed dead since they were inconspicuous (e.g. indistinguishable from their age mates) and their regular associates were killed. These lost females were treated as dispersed, though it is likely some died. Due to these lost individuals, presented mortality levels should be considered minimums.

Of the 940 elephants in this study, the age of 596 individuals (63%) were known (i.e. they were observed within 1.5 years of the estimated date of birth) with the rest estimated. Visual characteristics established from elephants of known age [27] were used to estimate the age of individuals and these age estimates were validated in the study population by comparing visual estimates of

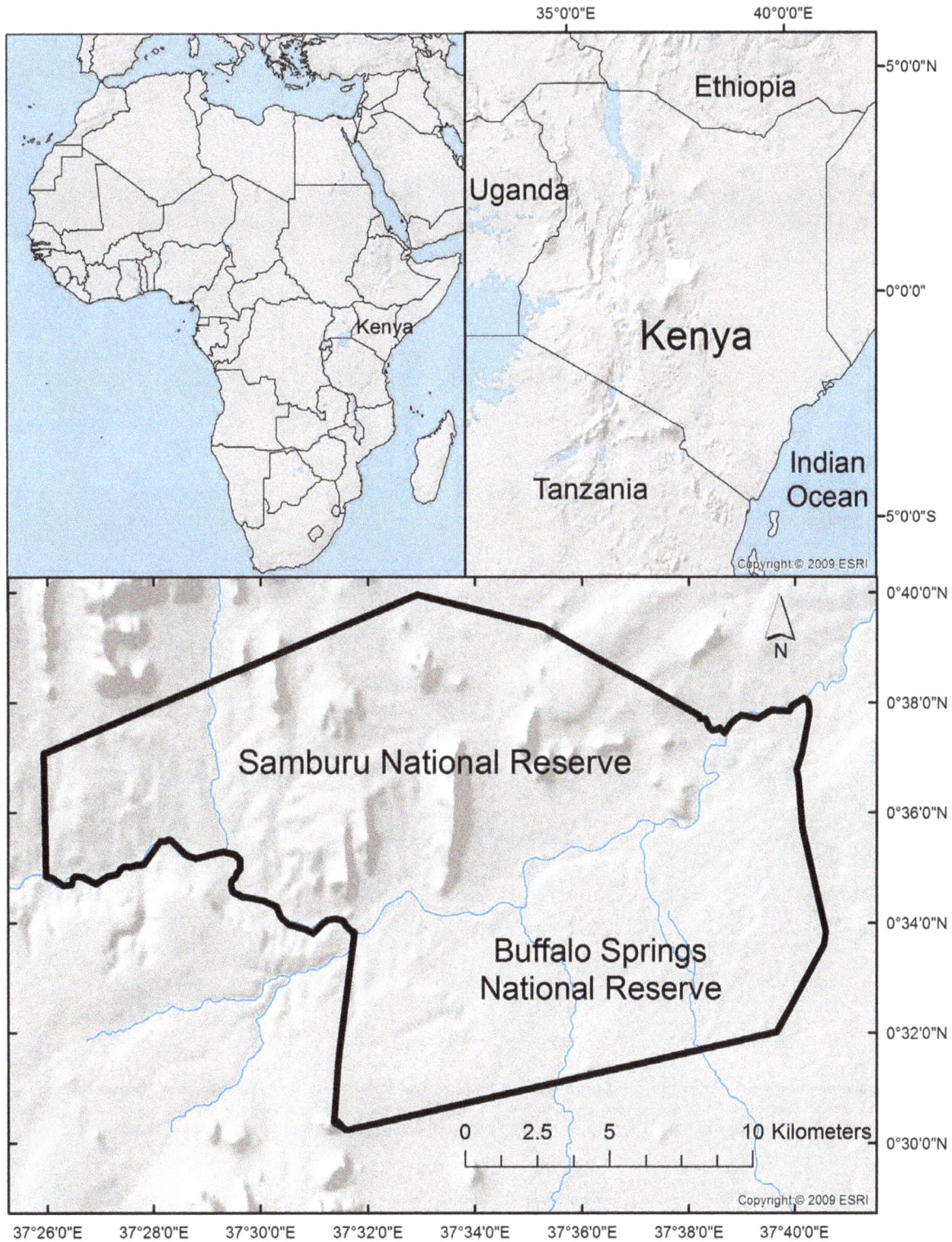

Figure 1. Map of the Samburu and Buffalo Springs National Reserves in Kenya, East Africa.

age with ages of dead or anesthetized individuals determined from dentition [28]. Age estimates of mature individuals based on physical appearance were within ± 3 years of the age based on molar progression for 80% of the elephants [28].

Analysis

Annual counts of live, dead and new born elephants were collated and used to derive population trends and time specific age structure in the population. Standard life table analyses were conducted to derive age specific mortality, life expectancy,

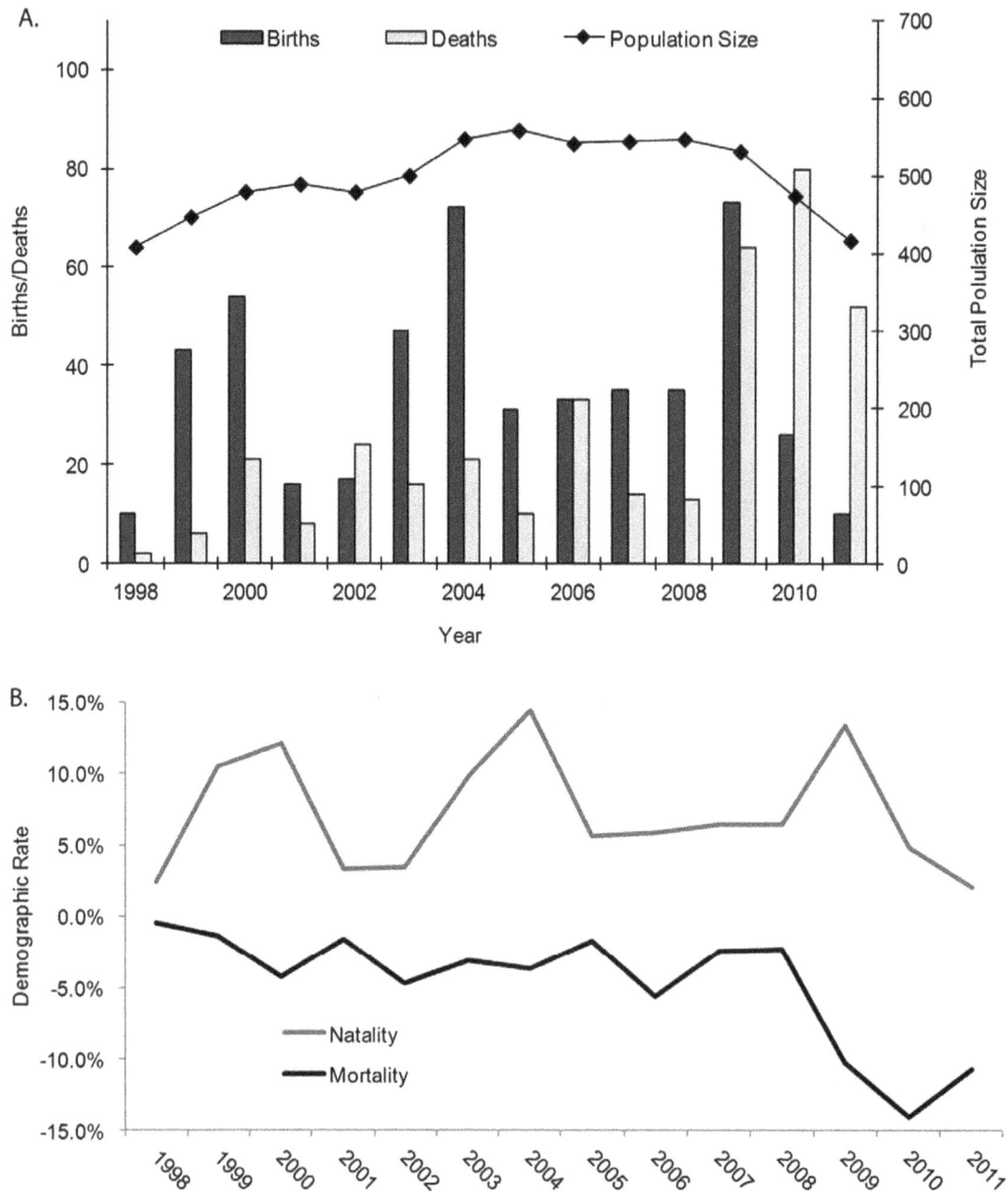

Figure 2. Annual Samburu population demography and population trend: (a) The number of births (black bars), deaths (gray bars), and the population size at the end of each year. (b) The population mortality and natality varied annually the duration of the study. Natality pulses every 3–4 years are a function of the prolonged (22 month) gestation period of elephant resulting in multi-year inter-calving intervals.

reproductive number, and generation time [29]. The survival functions for the right censored and left truncated data on males and females were calculated as $\hat{S} = \prod_{t_i \leq t} \left[1 - \frac{d_i}{Y_i} \right]$, where d_i is the number of individuals that died at time t_i and Y_i is the number of individuals at risk at time t_i [30]. Sex specific survival curves were derived for the low (1998–2008) and high (2009–2011) mortality periods, a division based on the year when the population began to decline. Survival functions were used to assess age specific reproductive onset among females (data on reproductive viability were not available for males) and age specific natal group dispersal among males (females do not disperse from their natal group). Dispersed individuals lost during the study (i.e. not observed post

dispersal) were removed from successive analyses as of the last year of definitive identification. A log-rank test of Kaplan-Meier survivorship [30] was used to compare survival across sexes and between populations (derived from individual based monitoring: Addo Elephant Park [31], Amboseli National Park [27,32], Tarangire National Park [33,34]; data from culling: Kruger National Park [35,36]) and hazard functions of reproductive onset between populations, as implemented in R 2.14.1 OIsurv library [37]. The compilation of demographic parameters derived for Samburu were only available for the Amboseli N. P. study, which was conducted using the same individual based approach [27,32]. Therefore, comparison was focused on these two populations and included the other populations when possible. Only known age individuals, i.e. those identified when estimated to be 18 months or younger, were used in analysis of male dispersal and only females estimated to be under the age of 5 at identification were used in analysis of reproductive onset.

Analysis of demographic data from 12 different populations across Africa [22,27,31,33,36,38,39,40,41,42] was conducted to assess the relationship between the response variables of age at first calving, inter-calving interval and annual mortality and covariate data on population density estimates and average annual rainfall in the study ecosystem (the only dependent and independent variables available across all populations). Information theoretic approaches (bias-adjusted Akaike's information criterion (AIC$_c$)) [43] were used to compare performance of generalized linear models with a Gaussian link function and results from the top model were presented.

Results

This known sample of the elephant population of Samburu fluctuated from an initial 410 individuals at the beginning of 1998 to a peak of 558 in 2005 (the population was relatively stable between 2004–2008), and declining to 417 at the end of 2011 (Fig. 2a). Annual population growth (including the effects of immigration and emigration) averaged 0.17% over the fourteen year study, but was highly variable increasing during the early years of the study (with a maximum annual increase of 9.1% in 1999) and decreasing during the latter half of the study (with a maximum annual decline of 12.8% in 2011). Net change through migration (immigration plus emigration) averaged −1.4% (S.D. = 1.5%) per year, predominantly in the form of young male dispersal from their natal groups. Excluding migration, annual population growth averaged 2.9% (S.D. = 7.33%) per annum. Annual mortality varied broadly during the fourteen year study, averaging 4.71% (S.D. = 4.09%) per annum with a peak of 14.1% in 2010 at the end of a major drought spanning 2009–2010 (Fig. 2b). Prior to the onset of the major drought in 2009, mortality averaged 2.82% (S.D. = 1.57%) per annum. Similarly, annual natality was highly variable averaging 7.21% (S.D. = 4.10%) per annum, with a maximum of 14.4% in 2004 and a minimum of 2.1% in 2011 (Fig. 2b). The low recruitment in 1998 and 2011 (Fig. 1a) is the result of drought induced lows in fecundity [44].

Female and male survival to the age of 10 years was 70% and 64% respectively over the fourteen year study, with the majority of deaths occurring between 2009–2011 during which survival to 10 years was estimated at 34% and 27% respectively (Fig. 3a; see Table S1). Male survival was significantly lower than female survival in Samburu over the fourteen year study ($\chi^2 = 114$, d.f. = 1, p<0.001; Fig. 3a). Age specific survival among the Samburu elephants was the lowest recorded in an intensively studied population, with Samburu female survival between 1998–2008 (mean = 22.3±1.59 years) significantly lower than the females of Amboseli ($\chi^2 = 27.5$, d.f. = 1, p<0.001) and Tarangire ($\chi^2 = 9.3$, d.f. = 1, p = 0.002) and Samburu male survival (mean = 19.3±1.35 years) during this period significantly lower than the males of Amboseli ($\chi^2 = 872$, d.f. = 1, p<0.001) (Fig. 3b (females) and 3c (males)). Maximum life span in Samburu for females and males was estimated at 64 and 54 years respectively, with life expectancy at birth estimated at 21.81 years for females (1998–2008 = 33.54 years; 2009–2011 = 8.01 years) and 18.85 years for males (1998–2008 = 22.96 years; 2009–2011 = 9.42 years) in contrast to female life expectancy of 40 years and male life expectancy of 24 years in Amboseli [27].

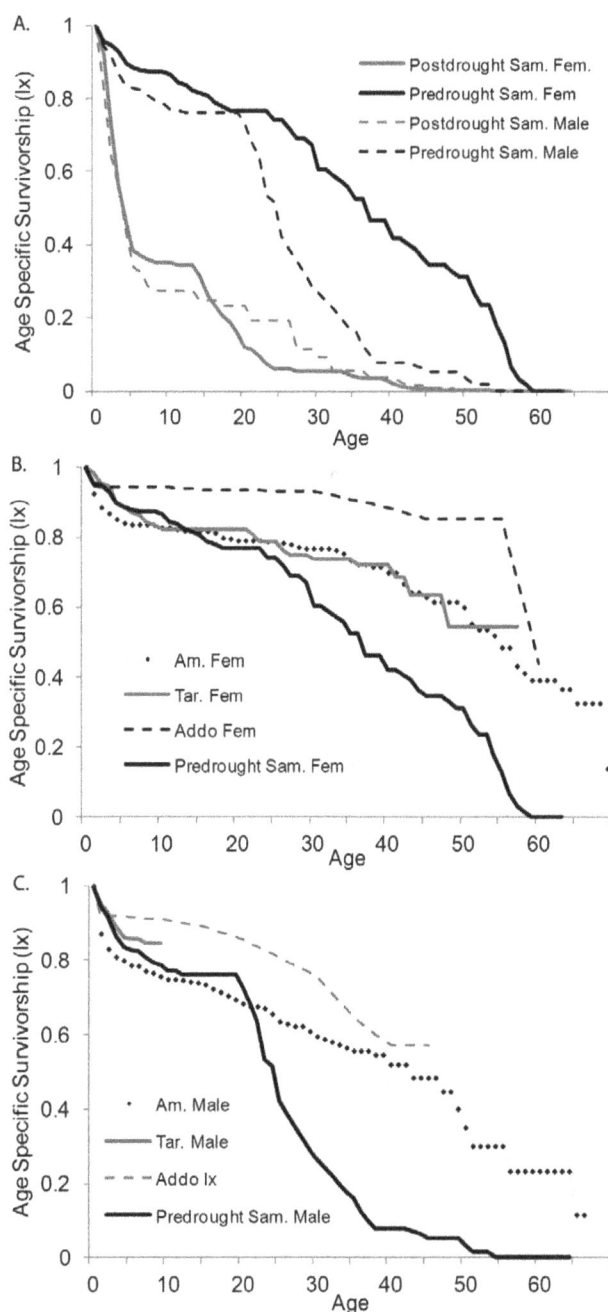

Figure 3. Comparative survivorship: (a) Survivorship curves for Samburu females (solid lines) and males (dashed lines) before (black) and after (gray) 2009 demonstrate the ramification of morality increases associated with the extreme drought of 2009 and successive increased illegal killing. (b) Survivorship among females in Samburu predrought (black line) was lower than that of females from Amboseli N. P. (gray line black dots), Tarangire N. P. (gray line), and Addo Elephant Park (dashed line). (c) Survivorship among males in Samburu predrought (black line) was lower than that of males from Amboseli N. P. (gray line black dots) and Addo Elephant Park (dashed line). Male survivorship was recorded only to the age of 8 years in Tarangire N.P. (gray line).

Illegal killing of elephants over the study period accounted for 32% of the known causes of death (i.e. those diagnosed from carcasses). Among elephants over the age of 9 years (approximate age of puberty), over half of the known deaths were from illegal

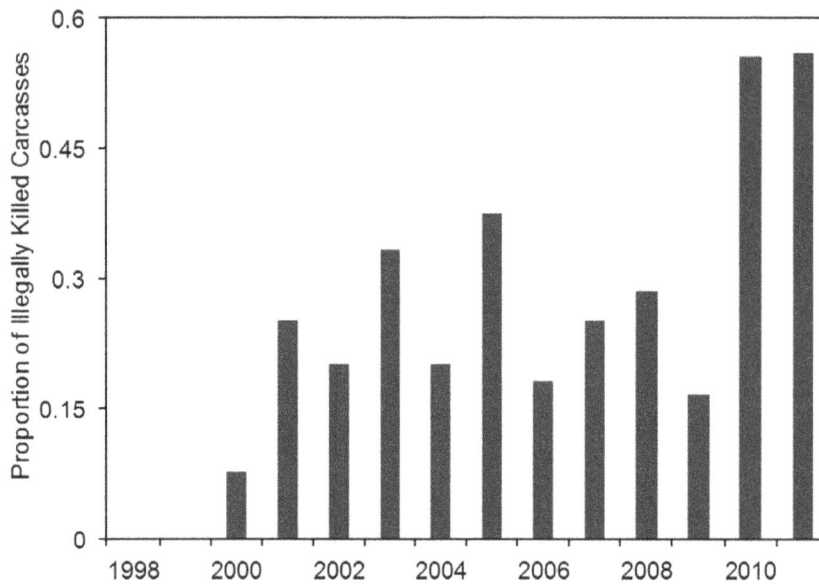

Figure 4. Annual proportions of illegally killed elephants (PIKE): The rate among found carcasses of illegally killing was greatest during the last three years of the study.

killing (Table 1). The PIKE doubled in the last three years of the study, averaging 0.43 between 2009–2011 in contrast to 0.20 between 1998–2008, with the highest PIKE recorded in 2011 at 0.56 (Fig. 4).

As calculated by standard life table approaches, the net reproductive number, R_0, in the Samburu population was 1.59 and the mean generation time was 24.1 years. The youngest known primiparous age was 8.5 years and the oldest estimated age of first reproduction was 19 years. Analysis of primiparous females 8 years or older (n = 52) demonstrated the average age of primiparity in the population was 11.34 years (S.E. = 0.08) (Fig. 5). This was significantly lower than the average age of primiparity among Amboseli females (i.e. the Samburu mean was outside the Amboseli 95% C.I.) and historic populations with relatively high densities (Lake Manyara and Queen Elizabeth N.P.), and close to the ages of reproductive onset reported in the culled population of Kruger National Park and recovering Tarangire populations (Douglas-Hamilton 1972; Laws et al. 1975; Moss 2001; Foley and Faust 2010; Freeman et al. 2009).

Table 1. Causes of mortality in the Samburu population assessed from carcasses.

Sex	Natural		Illegally Killed		Unknown		% Illegally Killed
	M	F	M	F	M	F	
0–2	15	4	0	0	0	0	0%
3–9	17	13	2	2	4	1	10%
10–18	2	5	1	4	3	4	26%
19–30	6	3	3	10	1	1	54%
31–50	3	5	7	9	1	2	59%
50+	0	2	1	2	0	0	60%
Total	75		41		17		31%

Age specific fecundity was 0.1 female calves per year (Fig. 6a), calculated as the number of female births per female year (female year is defined as the number of females alive at age x in the population). Fecundity was relatively constant between early and middle age, with age specific fecundity of 0.096 for females from 9–18 years (age span of primiparous females) and 0.092 for females 19–30 years. Fecundity was slightly higher for mature females between 30 and 50 years at 0.13 female calves per year, as a result of an increase in fecundity between 30 and 40 years of age. Similar to Amboseli N.P. [27], fecundity decreased after the age of 40 years. Only three of 20 females were observed to give birth in their 50's with age specific fecundity at 0.032 females per year between 50 and 60 years. No female was observed to give birth after the estimated age of 54 years (Fig. 6a). Reproductive value declined linearly after its peak at age 10 years (Fig. 6b). Inter-calf intervals among females with calves identified within 30 days of birth (n = 264) averaged 4.01 years (S.D. = 0.94). This interval was reduced by approximately one year when a female lost her previous calf within the first year (n = 14), as found in Amboseli N.P. (Moss 2001).

Reproductive success of all known males was not possible to assess, but previous work indicates fecundity is skewed towards older age classes [45]. The youngest known male to disperse from his natal family unit was 5 years old and the oldest was 18 years old (Fig. 5). Analysis of male dispersal age (n = 111) in the population demonstrated a mean dispersal age of 9.66 years (S.E. = 0.07). Approximately one fifth (21%) of these dispersals occurred after the death of the male juveniles mother. Males as young as 5 years were observed to disperse from their natal group without any obvious disruptive inducement (e.g. mother death).

The number of elephants over the age of 30 years in the known population declined over the course of the study. At its peak in 2000, 38 known males were estimated to be over the age of 30 years, but by the end of 2011 only 12 constituted this age class (7 of which were recruited into the age class after 2000, i.e. only 5 of the original 38 survived to 2011). Similarly, known females over the age of 30 peaked in 2002 and 2006 at

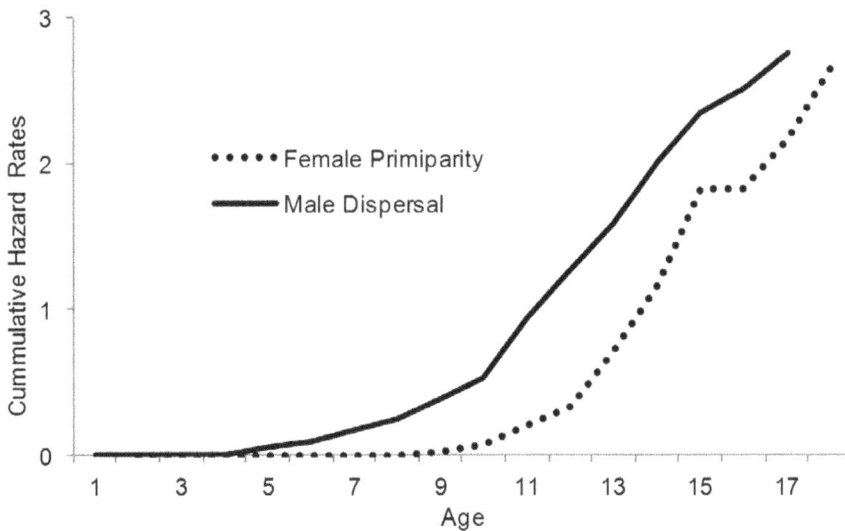

Figure 5. Indices of the age of reproductive onset: (a) Cumulative hazard rate for female primparous age (dotted line) and male dispersal (solid line) in the Samburu population.

59 individuals, but numbers were nearly halved between 2009 and 2011 with 32 remaining at the end of 2011 (Fig. S1). While some of this mortality was due to the serious 2009 drought, at least half is thought to have been caused by illegal killing (Table 1). This strong mortality altered the age structure and age-related social organization. The known population sex ratio has grown more skewed across the years with significantly more females than males across six age classes in 1998 ($\chi^2 = 80.38$, d.f. = 5, p<0.001) when 42% of the elephants were male, 2004 ($\chi^2 = 71.81$, d.f. = 5, p<0.001) when 41.6% were male, and 2011 ($\chi^2 = 335.51$, d.f. = 5, p<0.001) when 32% were male (Fig. S2). Social groups as defined in [25] have also changed dramatically, with 10 of the 50 quantitatively defined groups being extirpated (no known breeding females surviving) and 13 with no breeding female over the age of 25 years (Fig. S3).

The best supported model in an analysis of primiparous age (average = 13.4 years) across 12 populations (Table 2) demonstrated a significant positive correlation with the interaction term rainfall and density ($t = 2.80$, $p = 0.021$). The best supported model of inter-calving interval (average = 5 years) demonstrated a positive relationship with rainfall ($t = 2.871$, $p = 0.015$), excluding density ($t = -0.197$, $p = 0.847$) as a predictive variable. The best supported model of mortality (average = 4.1% per year across the 11 populations with reported values) showed a non-significant negative relationship with density ($t = -0.708$, $p = 0.497$).

Discussion

In addition to providing detailed population level demographic data, the Samburu study offers the first individually based demographic analysis of the impacts of illegal killing, offering detailed insight to demographic parameters (survival, fecundity, and population level metrics) in the face of moderate to high human pressures. Such pressure increased as the study progressed, allowing contrasts between periods of increase/stability and decline in a single population. Unfortunately, illegal killing and related population decline is increasingly common across Africa [17,18], therefore the results from this study are directly relevant to understanding the conservation status of the species.

Demographic Ramifications of Illegal Killing and Drought Induced Mortality

While survivorship was particularly low during the latter three years of the study, a period marked by relatively high illegal killing and intense drought, survivorship was significantly lower than the less impacted population of Amboseli even during the initial period (1998–2008) of increase/stability (Fig. 3b and 3c). Over the fourteen year study, life expectancy in Samburu among females was approximately half and males approximately a quarter of the less impacted population in Amboseli N. P. As predicted by life history theory [46,47], reproductive effort was greater (earlier age of primiparity and shorter inter-calf interval) in the Samburu relative to the Amboseli population. These metrics in the Kruger elephant population, where high levels of mortality were induced by management based culling, were similar to those found in Samburu [35,36], indicating savanna elephants may demonstrate a compensatory reproductive response to lower survival across systems. A similar response was proposed as the cause for the high reproductive surge in the Tarangire population, recorded during a period of recovery after extensive poaching impacted the population [33,34], though survival during the period of data collection was similar to that in Amboseli (Fig. 3b). This may indicate a lag in the purported compensatory reproductive response relative to survival rates, potentially associated with changes in the age structure of the population (see below).

Below 25th percentile rainfall (drought) years repeatedly caused declines in the survival of the younger age classes (see mortality in 2000, 2006 and 2010). During the acute drought of 2009–2010, survival of mature individuals declined as a result of both natural drought and a surge in illegal killing (with life expectancy dropping approximately 20 years during the latter portion of the study). The reduction of adult survival being partially a function of drought was unexpected given the drought resistant biology of megaherbivores [48] and results from previous analyses from this population showing that ecological conditions poorly predicted adult mortality [19]. The impact of drought induced mortality was compounded by the related but lagged slowing of reproduction (as demonstrated in [44,49]), which resulted in low recruitment in 2011 (10 total births) because of the long (22 month) gestation period of elephants. Therefore the impact of droughts on

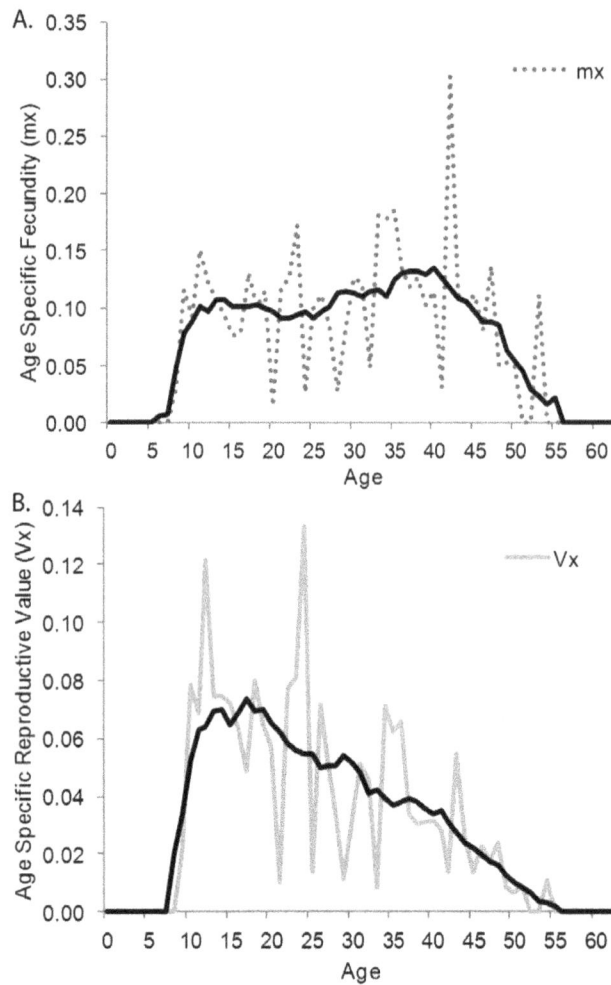

Figure 6. Summary of age class fecundity: (a) Age specific fecundity (m_x) (dashed line = actual, and solid line = 5-year running average) peaked when the Samburu females were in their mid 40s. (b) Age specific reproductive value (v_x) (dashed line = actual, and solid line = 5-year running average) declined after the peak between 15 and 20 years of age.

population growth was typically carried out over 2–3 years. The regulatory importance of natural ecological fluctuations to this species has been discussed in relation to management concerns [50], but the strength of the impact of a single year major drought (killing close to a sixth of the population and impacting the population for multiple years) may not have been recognized (though see [42,51]).

Illegal human killing caused over half the recorded mortality in the Samburu elephants over the age of 9 (and indirectly caused the deaths of all victim's dependent calves under 2 years). The high illegal killing in the latter part of the study had serious ramifications for the structure and organization of the population. In contrast to representative age structure culling [36], the illegal killing appeared to select adult individuals in Samburu and particularly males resulting in increasing skew in the sex ratio over the course of the fourteen year study (Fig. S1). Social disruption also resulted, with numerous well known and stable family groups being completely lost (i.e. no surviving breeding females) causing increased numbers of unaffiliated juveniles (orphans) [21]. These orphans take different strategies in relation to their social context,

remaining solitary, clustering with other orphans or joining other groups. Such processes were first noted by genetic studies of the population [52], but the number of orphaned individuals has risen sharply. In addition, the relative density of individuals in older age classes has declined compared with densities during the study onset. It is critical to follow this population in the future to understand if such alterations impact demographic processes.

Comparative Demography Across Elephant Populations

Demographic data on African elephants have been compiled from a variety of ecological settings from semi-arid to highly mesic savannas and span a gradient from high to low densities, offering an opportunity to investigate environmental correlates with different population parameters (Table 2). As identified by [41], higher densities generally were associated with slower reproduction (i.e. greater age of reproductive onset), though calving interval showed no relationship to density among the populations assessed. Annual mortality was not correlated with either density or rainfall. Most demographic data available from high density populations happen to be located in mesic environments, however, reducing the ability to draw inference on the interaction between density and ecological conditions at this scale of analysis. The interaction between density and fecundity was not linearly predictable, with Lake Manyara N.P. [22] demonstrating relatively fast reproduction among the high density populations though slower reproduction than the low density, human impacted populations. The slower reproduction in Amboseli relative to Samburu, Tarangire, and Kruger was not attributable to density (with all at relatively low densities); an older age structure in Amboseli may be the most significant difference [27], but how this influences fecundity remains unclear. Evidence of density dependence was not found in demographic studies of the highly managed Addo system [31], a population which periodically reached high densities, but for which the impacts of density were mitigated through park expansion and resource provisioning that dampened ecological processes that influence elephant populations [53]. Within the Samburu population, inter annual fluctuations in survival and fecundity associated with rainfall can be interpreted as resulting from density dependent effects, with high mortality induced by droughts (lower carrying capacity) and birth pulses resulting from mesic years (high carrying capacity).

Management Implications of Samburu Demography

The contrast between the conservation status of Amboseli and Samburu despite broadly similar management contexts is particularly salient to assessment of this species' status. Both populations inhabit similar semi-arid ecosystems (average annual rainfall in both is ~350 mm) and are located in Kenyan protected areas that cover a small proportion of their home ranges, for which security is maintained by the Kenya Wildlife Service (though Amboseli is a national park in contrast to the national reserve status in Samburu) [24,27]. Rather than ecological or protective institutional differences, the differences in human pressure across these populations is related to the local political context and instability in the ecosystems, with the diverse ethnic landscape in Samburu being prone to more frequent unrest relative to the homogeneous ethnic context around Amboseli [24,32]. The critical ramifications of local political and economic contexts on population conservation status [19] highlights potential weaknesses in analyses focused predominantly on national scale indicators to understand human drivers of demographic processes [17].

Finally, basic life table metrics derived from the Samburu data provide information required for red list assessments as specified by the IUCN [54]. One of the key demographic parameters

Table 2. African elephant demographic parameters from 12 wild populations (one sampled twice) collected through individual registration, radio tracking or culling.

	Youngest Conception Age	Avg. Primiparous Age (years)	Mean Calving Interval (years)	Annual Mortality	Density ele/mile^2	Yearly Rain (mm)	Citation
Addo N.P. South Africa	9	12.5	3.3	1.4%	9.6	392	Gough and Kerley 2006
Amboseli N.P. Kenya	7.1	13.7	4.5	4.15%	<1	350	Lee and Moss pers comms. 2012, Moss 2001
Etosha N.P. Namibia	–	11.1	3.8	5.0%	<1	250–450	Lindeque 1988
Kruger N.P. South Africa	8	13	3.8	3.2%	<1	450–700	Whyte et al. 1998, Freeman et al. 2009
Lake Manyara N.P. Tanzania	8	13	4.6	3.5% (2.7–3.9%)	6.2	650	Douglas-Hamilton 1972
Luangwa N.P. Zambia	12	16	3.4	–	3–6	800	Hanks 1972
Mkomazi N.P. Tanzania	–	12.2	4.2	5.0%	2.5	400	Laws et al. 1975
Murchison Falls N.P. Uganda	7	11	8.6	–	1.7	1,085	Buss and Smith 1966
Murchison North Uganda	–	16.3	9.1	5.1–6.6%	3–4	1,085	Laws et al. 1975
Murchison South Uganda	–	17.8	5.6		6–7	1,085	Laws et al. 1975
Samburu N.R. Kenya	6.7	11.5	4.0	2.8% pre 4.7% total	<1	350	This study
Tarangire N.P. Tanzania	6.9	11.2	3.3	1.9%	<1	656	Foley and Faust 2010
Tsavo N.P. Kenya	–	14.5	6.8	3.8–5.0%	3	200–700	Laws et al. 1975
Captive	12	16.3	–	–			Dale 2010

applied in red list assessments is generation time, on which the period considered (3x the generation time) when assessing changes in numbers or distribution are based [55]. The generation time in Samburu (24.1 years) was approximately equal to the 25 years calculated for Kruger, but the combined decreased survivorship and increased fecundity (early age of primiparity and shorter inter-calving interval) found in these human impacted populations potentially decrease generation time relative to that of an unperturbed population. The 17.4 year generation time reported for Amboseli was calculated using both sexes [27], not strictly females. Therefore, the standard generation time for Amboseli would be higher. Despite the faster reproduction, age specific fecundity and reproductive value were broadly similar to those reported for Amboseli (Amboseli values included both males and females in their assessment, but due to their reported 1:1 sex ratio can be halved for comparison), with reproductive output peaking in mid-aged females and declining in older age classes indicating senescence [56].

These data from Samburu and their comparison to other populations provide a more comprehensive understanding of the demographic processes occurring across populations in this species. As such, these data can facilitate more accurate modeling of elephant population dynamics across the spectrum of conditions they encounter. In modeling approaches, however, it is important to recognize site specific factors structure demographic processes, limiting inference provided by extrapolating from weak correlations between demographic parameters and coarse ecological indices.

Supporting Information

Figure S1 Change in number of older age class individuals: The number of mature adult (30 years or older) males (gray line) declined consistently between 2000 and 2011, while mature females (black line) declined rapidly during the last three years of the study.

Figure S2 Change in population sex ratio: The sex ratio among the closely monitored elephants has increasingly become skewed with fewer males.

Figure S3 Change in population social structure: High mortality in the latter half of the study caused severe social disruption for nearly half of the intensively monitored social units. Disrupted groups had no breeding females over the age of 25 years and extirpated groups had no remaining breeding females during the specified year.

Acknowledgments

We thank the Kenyan Office of the President; Kenya Wildlife Service; the Samburu and Buffalo Springs national reserves' county councils, wardens, and rangers; and C. Leadisimo, D. Lentipo, and G. Sabinga and the Save the Elephants team.

Author Contributions

Conceived and designed the experiments: GW IDH. Performed the experiments: GW DD. Analyzed the data: GW. Contributed reagents/materials/analysis tools: GW IDH. Wrote the paper: GW.

References

1. Gaillard JM, Festa-Bianchet M, Yoccoz NG (1998) Populations dynamics of large herbivores: variable recruitment with constant adult survival. Trends in Ecology & Evolution 13: 58–63.

2. Gaillard JM, Festa-Bianchet M, Yoccoz NG, Loison A, Toigo C (2000) Temporal variation in fitness components and population dynamics of large herbivores. Annual Review of Ecology and Systematics 31: 367–393.

3. Stockwell CA, Hendry AP, Kinnison MT (2003) Contemporary evolution meets conservation biology. Trends in Ecology & Evolution 18: 94–101.

4. Owen-Smith N, Mason DR, Ogutu JO (2005) Correlates of survival rates for 10 African ungulate populations: density, rainfall and predation. Journal of Animal Ecology 74: 774–788.

5. Power ME, Tilman D, Estes JA, Menge BA, Bond WJ, et al. (1996) Challenges in the quest for keystones. Bioscience 46: 609–620.

6. Dublin HT, Sinclair ARE, McGlade J (1990) Elephants and fire as cuases of multiple stable states in the Serengeti Mara woodlands. Journal of Animal Ecology 59: 1147–1164.

7. Pringle RM (2008) Elephants as agents of habitat creation for small vertebrates at the patch scale. Ecology 89: 26–33.

8. Blake S, Deem SL, Mossimbo E, Maisels F, Walsh P (2009) Forest Elephants: Tree Planters of the Congo. Biotropica 41: 459–468.

9. Laws RM (1970) Elephants as Agents of Habitat and Landscape Change in East-Africa. Oikos 21: 1-&.

10. Bouche P, Douglas-Hamilton I, Wittemyer G, Nianogo AJ, Doucet JL, et al. (2011) Will Elephants Soon Disappear from West African Savannahs? Plos One 6.

11. Blanc JJ, Barnes RFW, Craig CG, Dublin HT, Thouless CR, et al. (2007) African Elephant Status Report 2007: An update from the African elephant database. Glands, Switzerland: IUCN.

12. Douglas-Hamilton I (1987) African elephants: population trends and their causes. Oryx 21: 11–24.

13. Owen-Smith N, Kerley GIH, Page B, Slotow R, van Aarde RJ (2006) A scientific perspective on the management of elephants in the Kruger National Park and elsewhere. South African Journal of Science 102: 389–394.

14. Wasser S, Poole J, Lee P, Lindsay K, Dobson A, et al. (2010) Elephants, Ivory, and Trade. Science 327: 1331–1332.

15. Maisels F, Strindberg S, Blake S, Wittemyer G, Hart J, et al. (in review) Devastating decline of forest elephants in Central Africa. PLoS One.

16. Wittemyer G (2001) The elephant population of Samburu and Buffalo Springs National Reserves, Kenya. African Journal of Ecology 39: 357–365.

17. Burn RW, Underwood FM, Blanc J (2011) Global Trends and Factors Associated with the Illegal Killing of Elephants: A Hierarchical Bayesian Analysis of Carcass Encounter Data. PLoS One 6.

18. Douglas-Hamilton I, Wittemyer G, Ihwagi F (2010) Levels of illegal killing of elephants in the Laikipia-Samburu MIKE site. Doha, Qatr: Convention on the International Trade in Endangered Species of Wild Fauna and Flora.

19. Wittemyer G (2011) Effects of Economic Downturns on Mortality of Wild African Elephants. Conservation Biology 25: 1002–1009.

20. Wittemyer G, Daballen DK, Rasmussen HB, Kahindi O, Douglas-Hamilton I (2005) Demographic Status of elephants in the Samburu and Buffalo Springs National Reserves, Kenya. African Journal of Ecology 43: 44–47.

21. Wittemyer G, Daballen D, Douglas-Hamilton I (2011) Rising ivory prices threaten elephants. Nature 476: 282–283.

22. Douglas-Hamilton I (1972) On the ecology and behaviour of the African elephant: Elephants of Lake Manyara [D.Phil.]. Oxford: Oxford University.

23. Moss CJ (1996) Getting to know a population. In: Kangwana K, editor. Studying Elephants. Nairobi, Kenya: African Wildlife Foundation. 58–74.

24. Kahindi O, Wittemyer G, King J, Ihwagi F, Omondi P, et al. (2010) Employing participatory surveys to monitor the illegal killing of elephants across diverse land uses in Laikipia-Samburu, Kenya. African Journal of Ecology 48: 972–983.

25. Wittemyer G, Douglas-Hamilton I, Getz WM (2005) The socioecology of elephants: analysis of the processes creating multitiered social structures. Animal Behaviour 69: 1357–1371.

26. Wittemyer G, Getz WM, Vollrath F, Douglas-Hamilton I (2007) Social dominance, seasonal movements, and spatial segregation in African elephants: a contribution to conservation behavior. Behavioral Ecology and Sociobiology 61: 1919–1931.

27. Moss CJ (2001) The demography of an African elephant (Loxodonta africana) population in Amboseli, Kenya. Journal of Zoology 255: 145–156.

28. Rasmussen HB, Wittemyer G, Douglas-Hamilton I (2005) Estimating age of immobilized elephants from teeth impression using dental acrylic. African Journal of Ecology 43: 215–219.

29. Caughley G (1977) Analysis of Vertebrate Populations. London: Wiley Interscience.

30. Klein JP, Moeschberger ML (2003) Survival Analysis: Techniques for Censored and Truncated Data. New York: Springer.

31. Gough KF, Kerley GIH (2006) Demography and population dynamics in the elephants Loxodonta africana of Addo Elephant National Park, South Africa: is there evidence of density dependent regulation? Oryx 40: 434–441.

32. Moss CJ, Croze H, Lee PC (2011) The Amboseli Elephants: a long-term perspective on a long-lived mammal. Chicago: The University of Chicago Press.

33. Foley CAH, Faust LJ (2010) Rapid population growth in an elephant Loxodonta africana population recovering from poaching in Tarangire National Park, Tanzania. Oryx 44: 205–212.

34. Foley CAH, Pettorelli N, Foley L (2008) Severe drought and calf survival in elephants. Biology Letters 4: 541–544.

35. Freeman EW, Whyte I, Brown JL (2009) Reproductive evaluation of elephants culled in Kruger National Park, South Africa between 1975 and 1995. African Journal of Ecology 47: 192–201.

36. Whyte I, van Aarde R, Pimm SL (1998) Managing the elephants of Kruger National Park. Animal Conservation 1: 77–83.

37. R Development Core Team (2012) R: A Language and Environment for Statistical Computing. Vienna, Austria.

38. Buss IO, Smith NS (1966) Observations on reproduction and breeding behavior of the African elephant. Journal of Wildlife Management 30: 375–388.

39. Hanks J (1972) Reproduction of the elephant (Loxodonta africana) in the Luangwa Valley, Zambia. Journal of Reproduction and Fertility 30: 13–26.

40. Laws RM (1969) Aspects of reproduction in the African elephant (Loxodonta africana). Journal of Reproductive Fertility Supplement 6: 193–217.

41. Laws RM, Parker ISC, Johnstone RCB (1975) Elephants and Their Habitats: The Ecology of Elephant in North Bunyoro, Uganda. Oxford, U.K.: Calrendon Press.

42. Lindeque M (1988) Population Dynamics of Elephants in Etosha National Park, S.W.A./Namibia. Stellenbosch, South Africa: University of Stellenbosch.

43. Burnham KP, Andersen DR (1998) Model Selection and Inference: a practical theoretic approach. New York: Springer.

44. Wittemyer G, Rasmussen HB, Douglas-Hamilton I (2007) Breeding phenology in relation to NDVI variability in free-ranging African elephant. Ecography 30: 42–50.

45. Rasmussen HB, Okello JBA, Wittemyer G, Siegismund HR, Arctander P, et al. (2008) Age- and tactic-related paternity success in male African elephants. Behavioral Ecology 19: 9–15.

46. Stearns SC (1992) The Evolution of Life Histories. Oxford, U.K.: Oxford University Press.

47. Roff DA (1992) The Evolution of LIfe Histories: Theory and Analysis. New York: Chapman & Hall, Inc.

48. Owen-Smith N (1988) Megaherbivores The Influence of Very Large Body Size on Ecology. Cambridge, U.K.: Cambridge University Press.

49. Wittemyer G, Ganswindt A, Hodges K (2007) The impact of ecological variability on the reproductive endocrinology of wild female African elephants. Hormones and Behavior 51: 346–354.

50. van Aarde RJ, Jackson TP (2007) Megaparks for metapopulations: Addressing the causes of locally high elephant numbers in southern Africa. Biological Conservation 134: 289–297.

51. Dudley JP, Criag GC, Gibson DS, Haynes G, Klimowicz J (2001) Drought mortality of bush elephants in Hwange National Park, Zimbabwe. African Journal of Ecology 39: 187–194.

52. Wittemyer G, Okello JBA, Rasmussen HB, Arctander P, Nyakaana S, et al. (2009) Where sociality and relatedness diverge: the genetic basis for hierarchical social organization in African elephants. Proceedings of the Royal Society B-Biological Sciences 276: 3513–3521.

53. Chamaille-Jammes S, Valeix M, Fritz H (2007) Managing heterogeneity in elephant distribution: interactions between elephant population density and surface-water availability. Journal of Applied Ecology 44: 625–633.

54. Mace GM, Collar NJ, Gaston KJ, Hilton-Taylor C, Akcakaya HR, et al. (2008) Quantification of Extinction Risk: IUCN's System for Classifying Threatened Species. Conservation Biology 22: 1424–1442.

55. IUCN (2012) The IUCN Red List of Threatened Species. IUCN, Geneva.

56. Packer C, Tatar M, Collins A (1998) Reproductive cessation in female mammals. Nature 392: 807–811.

9

Effects of Landscape-Scale Environmental Variation on Greater Sage-Grouse Chick Survival

Michael R. Guttery[1]*, David K. Dahlgren[2], Terry A. Messmer[3], John W. Connelly[4], Kerry P. Reese[5], Pat A. Terletzky[3], Nathan Burkepile[6], David N. Koons[7]

1 Department of Forest and Wildlife Ecology, University of Wisconsin, Madison, Wisconsin, United States of America, 2 Kansas Department of Wildlife and Parks, Hays, Kansas, United States of America, 3 Department of Wildland Resources, Utah State University, Logan, Utah, United States of America, 4 Idaho Department of Fish and Game, Blackfoot, Idaho, United States of America, 5 Department of Fish and Wildlife Sciences, University of Idaho, Moscow, Idaho, United States of America, 6 Northland Fish and Game, Whangarei, New Zealand, 7 Department of Wildland Resources and the Ecology Center, Utah State University, Logan, Utah, United States of America

Abstract

Effective long-term wildlife conservation planning for a species must be guided by information about population vital rates at multiple scales. Greater sage-grouse (*Centrocercus urophasianus*) populations declined substantially during the twentieth century, largely as a result of habitat loss and fragmentation. In addition to the importance of conserving large tracts of suitable habitat, successful conservation of this species will require detailed information about factors affecting vital rates at both the population and range-wide scales. Research has shown that sage-grouse population growth rates are particularly sensitive to hen and chick survival rates. While considerable information on hen survival exists, there is limited information about chick survival at the population level, and currently there are no published reports of factors affecting chick survival across large spatial and temporal scales. We analyzed greater sage-grouse chick survival rates from 2 geographically distinct populations across 9 years. The effects of 3 groups of related landscape-scale covariates (climate, drought, and phenology of vegetation greenness) were evaluated. Models with phenological change in greenness (NDVI) performed poorly, possibly due to highly variable production of forbs and grasses being masked by sagebrush canopy. The top drought model resulted in substantial improvement in model fit relative to the base model and indicated that chick survival was negatively associated with winter drought. Our overall top model included effects of chick age, hen age, minimum temperature in May, and precipitation in July. Our results provide important insights into the possible effects of climate variability on sage-grouse chick survival.

Editor: Mark S. Boyce, University of Alberta, Canada

Funding: This research was supported by the Idaho Department of Fish and Game Federal Aide in Wildlife Restoration Project W-160-R, the United States Natural Resources Conservation Service, U.S. Bureau of Land Management, Utah State University Quinney Professorship for Wildlife Conflict Management, Jack H. Berryman Institute of Wildlife Damage Management, Utah State University Extension Service. The funders had no role in study design, data collection and analysis, decision to publish, or preparation of the manuscript.

Competing Interests: The authors have declared that no competing interests exist.

* E-mail: micrgutt@gmail.com

Introduction

Selective pressures result in the evolution of a life history conducive to species persistence under the environmental conditions encountered throughout the species' evolutionary history. Environmental conditions are not static, but rather experience climatic, geological, and successional changes through time. While such changes continue to occur naturally, anthropogenic disturbances have critically altered many of these processes, resulting in environments changing at rates that exceed the ability of some species to adapt [1]. The impact of rapidly changing environments may be particularly severe for species with limited dispersal opportunities (i.e., those existing in highly fragmented habitats; [2]). Efforts to conserve such species must focus on identifying the key demographic rates that are limiting population growth and the environmental factors that affect these rates [3].

During the 20th century, greater sage-grouse (*Centrocercus urophasianus*; hereafter sage-grouse) populations experienced precipitous declines as a result of anthropogenic habitat destruction, degradation, conversion, and fragmentation [4,5]. In response to

declining populations and increasing threats to remaining habitat, the Canadian Committee on the Status of Endangered Wildlife in Canada declared sage-grouse to be an endangered species in 1998 [6]. The United States Fish and Wildlife Service (USFWS) designated the sage-grouse as a candidate for protection under the Endangered Species Act in 2010 [7].

Sage-grouse are endemic to sagebrush (*Artemisia* sp.) dominated habitats of western North America, which have historically been very stable given that sagebrush is a long-lived and persistent plant. As such, sage-grouse evolved to use sagebrush for food and cover throughout the majority of their annual cycle. However, sage-grouse chicks do not consume sagebrush during their early development but instead require forbs and their associated arthropod communities. These components of the sagebrush ecosystem are highly dependent upon precipitation levels and therefore may exhibit high interannual variability. Thus, sage-grouse evolved a life history characterized by high annual adult survival but relatively low and variable reproductive rates compared to most other tetronids [8,9].

Recently, researchers have applied life cycle models to gain a better understanding of factors affecting greater sage-grouse at the population [10] and range-wide scales [9]. Although both studies found sage-grouse population growth rates to be most sensitive to variability in adult female survival, they also found chick survival to have the second largest impact on population growth. While numerous studies have evaluated factors which influence survival rates of adult female sage-grouse [11,12], little is known about factors affecting chick survival. Generally, demographic rates to which the population growth rate is highly sensitive have low temporal variability [13,14]. Thus, chick survival should exhibit greater inter-annual variability and could therefore contribute more to spatio-temporal variability in population growth rate [15] even though sage-grouse populations are more sensitive to hen survival.

Previously published studies of factors affecting sage-grouse chick survival [16,17,18] have focused on micro-scale habitat factors such as percent coverage and height of forbs and grasses and availability of arthropods at chick location sites. These studies follow logically from previous research on sage-grouse brood habitat selection [19,20,21,22] and chick diets [23,24,25,26]. Collectively, these studies clearly demonstrate that broods typically select relatively mesic habitats with abundant forbs and arthropods and that these choices are related to chick survival. However, existing studies have not investigated the impacts of large-scale environmental processes (drought, temperature, etc.) on sage-grouse chick survival.

Landscape-scale environmental factors such as habitat condition, drought, and climate may be correlated with chick survival. Normalized Difference Vegetation Index (NDVI) is a commonly used index of plant production and habitat quality [27,28,29,30], with higher values of the index corresponding to increased levels of "greenness". Despite being less sensitive to plant phenology in sagebrush steppe ecosystems [29] and potential biases due to image quality, NDVI has been shown to be positively related to sage-grouse recruitment and population growth [27]. Drought and climatic variables can work independently and in concert to affect habitat parameters and can be reflected in NDVI values. For example, measures of drought, precipitation, and temperature can be correlated to winter snow pack which is known to be a major driver of vegetation dynamics throughout much of the mountainous regions of western North America [31]. However, climatic variables may affect sage-grouse chick survival in ways other than through their influence on habitat quality. Young grouse may be susceptible to exposure mortality during periods of extreme temperatures [32]. Additionally, numerous studies have documented increased nest and chick predation rates following precipitation events (i.e., moisture facilitated predation hypothesis; [33,34,35]). This effect is typically attributed to increased scent production resulting from increased bacterial growth when skin and feathers are wet [36]. Although the assumptions underlying the moisture facilitated predation hypothesis have not been thoroughly evaluated in the context of the hypothesis, the processes of moisture facilitating microbial activity and increased microbial activity resulting in increased scent production have been well documented in other fields of study [37,38,39,40].

The objective of our study was to model the effects of landscape scale biotic (habitat greenness) and abiotic (climate and drought) factors on sage-grouse chick survival. We demonstrate the utility of data that can be readily obtained for virtually any geographic region or temporal period via web-based resources for predicting sage-grouse chick survival.

Materials and Methods

Study areas

Data were collected as part of 2 larger studies conducted in Idaho and Utah (Fig. 1). The Idaho study was conducted from 1999–2002 in sagebrush-grassland habitats of the Upper Snake River Plain in southeastern Idaho (44°13′N, 112°38′W). This area was characterized by relatively low topographic relief with elevation across the site ranging from 1300–2500 m. Approximately 50% of the area was privately owned, with the remainder being public lands administered by the U.S. Bureau of Land Management (BLM). Annual precipitation varied by elevation with low elevation areas receiving 17.5 to 30.0 cm of precipitation. Most sage-grouse habitat at lower elevations was dominated by Wyoming big sagebrush (*A. tridentata wyomingensis*). At higher elevations precipitation ranged from 30.5 to 45.5 cm annually and the habitat was dominated by mountain big sagebrush (*A. t. vaseyena*). Livestock grazing and cropland agriculture were the dominant land uses across the area [41].

The Utah study area was located on Parker Mountain in south-central Utah (38°17′N, 111°51′W). Research on sage-grouse chick survival was conducted on this site from 2005 through 2009. The area encompassed 107,478 ha and was administered by the Utah School and Institutional Trust Lands Administration (40.8%), United States Forest Service (20.2%), BLM (33.9%), and private ownership (5.1%). Parker Mountain is a sagebrush-dominated plateau at the southern edge of the sage-grouse range. It is one of the few areas remaining in Utah with relatively stable numbers of sage-grouse, and it includes some of the largest contiguous tracts of sagebrush in the state [42]. Grazing by domestic livestock is the predominant land use practice across the study site. The area receives between 40 and 51 cm of precipitation annually, which generally exhibits a bi-modal pattern, occurring either as rain during the seasonal monsoonal period from late summer and early autumn or as snow during winter.

Field methods

We captured female sage-grouse on and around leks using spotlights, binoculars and long handled nets [43,44] during early spring (March–April). Captured hens were classified as being either second-year (SY) or after second-year (ASY) birds based on wing characteristics as described by Beck et al. [45]. Birds were fitted with 15–19 g necklace style radio-transmitters (Advanced Telemetry Systems, Isanti, MN, USA; Holohil Systems, Carp, Ontario, Canada) and released at the capture location.

Marked hens were monitored during April and May to determine if they initiated a nest. Nesting was confirmed visually, but hens were never intentionally flushed from their nest due to the tendency of female sage-grouse to abandon their nest if disturbed [46,47]. Nesting hens were visually monitored every 2–3 days to determine nest fate. Nests were monitored daily as the anticipated hatch date approached.

We captured chicks by using telemetry equipment to locate radio-marked hens. During capture events, the brood hen was flushed and chicks were captured by hand and placed in an insulated container to help maintain body temperature. We captured most broods within 48 hours of hatching with all broods being captured within 1 week of hatching. Captured chicks were weighed to the nearest gram and marked with a ≤1.5 g backpack-style radio-transmitter (Advanced Telemetry Systems, Insanti, MN in 1999–2001 and 2005, Holohil Systems, Carp, Ontario, Canada in 2006–2008, and American Wildlife Enterprises, Monticello, FL in 2009) attached with 2 sutures [48]. For the Idaho study site, 2–3 chicks per brood were selected at random to receive radio-

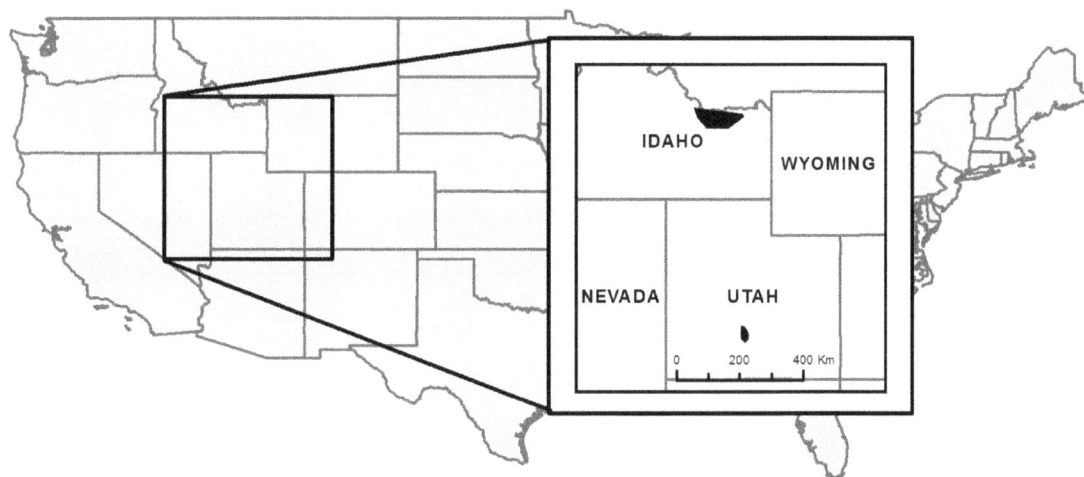

Figure 1. Map of study areas in Idaho and Utah.

transmitters. At the Utah study area, we marked all captured chicks except in 2006 when 3 chicks from each brood were randomly selected to receive transmitters. Chicks found dead in the immediate vicinity of the capture site were considered to have died as a result of handling and were excluded from subsequent analyses. Broods were typically checked within 12 hours of being marked and all chicks classified as capture mortalities were found intact within a few meters of the release site, indicating that their death was directly attributable to the capture event. Our decision to exclude chicks classified as capture mortalities from our analysis may have inflated survival estimates if some of these mortalities were in fact not related to capture. However, we do not believe this was a common occurrence if it occurred at all.

Marked chicks were located every 1–2 days until they reached 42 days of age. Monitoring intervals did occasionally exceed 2 days due to inclement weather events or difficulties locating broods following large movements. Extensive efforts were made to find any chicks missing from a brood. We occasionally recovered chick transmitters with no chick remains or signs of predation. These recoveries were classified as mortalities although it is possible that transmitters may have been lost for reasons other than chick death. Alternatively, we could have right-censored these specific events. While this would have been a valid option, we chose to treat the events as mortalities to ensure that our survival estimates were conservative. Due to the difficulty of distinguishing predation from scavenging, we did not assign specific causes of mortality.

All necessary permits were obtained for the described field studies. Permission to capture and mark sage-grouse in Idaho was obtained from the Idaho Department of Fish and Game and from the Utah Division of Wildlife Resources for the Utah study site. Grouse capture and transmitter attachment procedures were approved under the Utah State University Institutional Animal Care and Use Committee (IACUC) permit #945R and #942 and University of Idaho IACUC permit #2000-7.

Covariate data

We compiled year and site specific covariate data pertaining to drought, landscape greenness, and climate. We included seasonal (preceding winter and current summer) and monthly (May–July) Palmer Drought Severity Index (PDSI) and Palmer Z-Index (PZI) values. For climate and drought covariates, we defined winter as the period from 1 November to 30 April because precipitation

would likely fall as snow on both study sites during these months. Summer was defined as the period from 1 May through 31 July. We did not include August because very few broods were monitored beyond July.

While both the PDSI and PZI indices are measures of drought and their values interpreted similarly (negative values correspond to drought conditions while positive values indicate wet conditions), the PDSI is most appropriate for measuring conditions across long time periods (several months) while the PZI is designed to measure conditions across shorter time periods (several weeks to a few months, [49]). Although drought is often thought about in terms of the presence or absence of precipitation, PDSI and PZI also account for site specific rates of evapotranspiration, soil moisture recharge, runoff, and moisture loss [50]. Additionally, both drought indices are calculated relative to the long-term average drought conditions at a specific site. As such, values of each index are standardized to have a common interpretation across locations [50]. All drought data were downloaded from the National Oceanic and Atmospheric Administration's National Climate Data Center (http://www.ncdc.noaa.gov/temp-and-precip/time-series/index.php).

Climate variables of interest included total precipitation, minimum temperature, and maximum temperature for the same seasonal and monthly periods described above. Unlike drought covariates, climate covariates were not adjusted to account for other physical processes or long-term site-specific averages. Because complete and representative weather station data were not available for both study sites, we used the Parameter-elevation Regressions on Independent Slopes Model (PRISM; http://www.prism.oregonstate.edu/) to estimate climatic data for both sites. PRISM is a knowledge-based climate analysis system capable of generating gridded predictions of climate data from known point climate data and a digital elevation map [51]. We used ArcMap10 to generate minimum convex polygons around all chick locations at both sites to define our study sites. We then extracted climate variable data from the corresponding PRISM layer.

Phenological change in landscape greenness was measured using NDVI for each study area. We generated NDVI values using Landsat 4–5 satellite images obtained from the United States Geological Survey EarthExplorer website (http://earthexplorer.usgs.gov/). We selected images captured between 1 May and 31 August with minimal cloud coverage. Due to variability in image quality (i.e., cloud cover) and capture date, we were not able to use images taken on identical dates across years. Images were processed using the ERDAS Imagine remote sensing image analysis software (Intergraph, Madison, AL, USA) to apply radiometric corrections that eliminated background noise while retaining temporal variance in vegetative reflectance [52]. We used ERDAS Imagine to calculate NDVI on a pixel-by-pixel basis for each image based on the ratio of red to near-infared reflectance [53]. For each site, we fit our observed NDVI values to a linear model:

$$NDVI = Year + Date + Date^2 + Year * Date + Year * Date^2 + \varepsilon,$$

to estimate daily NDVI values where $\varepsilon \approx N(0,\sigma^2)$. This model provided a good fit to the data ($R^2 > 0.80$, $F_{11} > 4.50$, $P < 0.001$). We used predicted values to estimate mean and maximum NDVI values for May, June, July, and summer (as defined above). We also estimated the mean NDVI value at the date of hatching, 15 days before hatch, date of each survival observation, and 5 and 10 days prior to the date of each observation. Finally, because variability in NDVI was low due to sagebrush obscuring the phenological progression of forbs and grasses, we adjusted all NDVI values by subtracting out the year and site specific NDVI value on 1 May. This linear transformation effectively removed baseline site and year variation, thereby allowing our analysis to focus more directly on the effect of within-year plant phenology at a site.

Analysis

Missing chicks whose fate could not be determined were removed (i.e., right-censored) from the data set at the time of their last confirmed detection. Failure to locate chicks may have been the result of transmitter failure, the chick being removed from the study site by a predator, or long distance movements that exceeded the range of the transmitters. On a few rare occasions, chicks were found alive several weeks after going missing. The flexibility of our model allowed us to reintroduce these chicks back into the data set once rediscovered. Alternatively, we could have assumed that missing chicks were either dead or alive but our approach likely provides the most realistic estimate of chick survival because only chicks with known fate were allowed to influence daily survival rates [17].

We modeled sage-grouse chick daily survival rates from hatch to 42-days of age using the known-fate maximum likelihood estimator developed by Manly and Schmutz [54] and extended by Fondell et al. [55]. This model assumes a piecewise survival function such that the survival rate from age t to age $t+1$ is:

$$\Phi_t = \exp(-\alpha_i)$$

where $\alpha_i \geq 0$, for $t_{i-1} \leq t < t_i$, with $t_0 = 0$, and i = 1,2,...,p. Therefore the daily survival rate (DSR) for ages 0 to t_1 days is assumed to be $\exp(-\alpha_1)$, the DSR for ages t_1 to t_2 days is assumed to be $\exp(-\alpha_2)$, and so forth, with p survival intervals. If N_a chicks are observed in a brood at age a then the number of survivors at age $b > a$, N_b, has a binomial distribution with mean:

$$E(N_b|N_a) = N_a \Phi_a \Phi_{a+1} \cdots \Phi_{b-1}.$$

To account for extra-binomial variance, the variance term is:

$$V_1(N_b|N_a) = D * V(N_b|N_a),$$

Where D is a constant and $V(N_b|N_a)$ is the binomal variance given by:

$$V(N_b|N_a) = N_a \Phi_a \Phi_{a+1} \cdots \Phi_{b-1}(1 - \Phi_a \Phi_{a+1} \cdots \Phi_{b-1}).$$

This variance formulation assumes that most extra-binomial variation is the result of lack of independence in the fates of chicks within broods. Given this formulation and assumptions, the log-likelihood function for the observed number n_b at the end of a survival period is derived from the normal density function and takes the form:

$$L(\alpha_1, \alpha_2, \cdots, \alpha_p, D) =$$
$$\sum \left[-0.5 \log_e\{2\pi V_1(N_b|N_a)\} - \frac{0.5\{n_b - E(N_b|N_a)\}^2}{V_1(N_b|N_a)} \right],$$

where the summation is over all the instances in the data set where a brood size is observed at time a and then observed next at time b [54].

This generalized linear model is appropriate because it allows for variable observation intervals, changes in brood size due to missing chicks, and accounts for lack of independence in fates among chicks within a brood by using a quasi-likelihood approach [17,54,55]. Values of D near 1 indicate minimal dependence in the fates of brood mates whereas larger values correspond to decreasing independence among brood mates [54]. Covariates were modeled using a logit-link. Maximum likelihood estimates for all parameters were estimated using the 'OPTIM' function in R 2.14.1 [56].

To examine processes affecting chick survival in our populations, we first developed models that included alternative parameterizations of chick age. For example, we created models with categorical age classes wherein the categories were based on biological development of chicks, such as pre- versus post-flight ages or early ages when the diet consists primarily of insects versus later ages when forbs become important. We also considered linear and quadratic models of age treated as a continuous variable. Competing models of the various chick age parameterizations were ranked using the quasi-likelihood version of Akaike's Information Criterion adjusted for sample size (QAIC$_c$: [57,58]). Models with ΔQAIC$_c \leq 2$ were considered to be equally supported by the data, and when this occurred we applied the principle of parsimony and based our inference on the model with the fewest parameters [59]. Upon identifying the best parameterization of chick age, we next considered the addition of hen age (SY or ASY) and hatch date effects, as both have been shown to be important predictors of sage-grouse chick survival [17,60]. Year and site effects were not modeled explicitly because all covariates of interest were site and year specific (i.e., site and year effects were modeled implicitly). The validity of the approach was assessed by adding year and site effects to our final top model and monitoring the change in QAIC$_c$.

We then developed candidate model sets for each of the 3 covariate groups. Covariates within each group tended to be correlated. To insure the interpretability of parameter estimates (i.e., to avoid multicollinearity), covariates with a Pearson correlation coefficient (ρ) greater than 0.50 were not included in the same model. To determine which of the 3 groups of covariates had the greatest impact on chick survival, we did not include covariates from different groups in the same model. These restrictions limited the complexity of models we considered. Upon identifying the top model for each of the 3 covariate groups, we obtained 95% bootstrap confidence intervals for model parameters using 5,000 samples with replacement from our dataset [61]. All continuous covariates were Z-standardized prior to analyses. We calculated the proportional reduction in deviance [62] for each model relative to the null model:

$$D_I = 1 - (dev_I / dev_N)$$

where D_I is the Zheng-score for the model of interest, dev_I is the deviance for the model of interest, and dev_N is the deviance for the null model (unless otherwise noted, an intercept-only model) and deviance was calculated as -2*quasi-log-likelihood. The Zheng-score is a goodness-of-fit measure for generalized linear models of longitudinal data and can be interpreted similarly to a standard coefficient of determination, R^2, in a linear model [62]. We then further assessed model fit by calculating the ratio of the Zheng-score for the model of interest relative to the spatially and temporally saturated model [63]:

$$R_I = D_I / D_{FS}$$

where D_I is the Zheng-score for the model of interest and D_{FS} is the Zheng-score for the fully spatial and temporally saturated model. Values of R close to zero indicate little improvement in model fit over the null model, whereas values of R that approach one indicate model fit similar to the fully saturated model.

Results

Chick statistics

Most hens had a single brood during the course of our multi-year study; however, 24 of the 142 hens had broods during more than one year of the study. Peak hatch date ranged from 25 May to 7 June at the Utah study area and from 19 May to 30 May for the Idaho area. During the 9 years of study we attached radio transmitters to 518 chicks from 142 broods, resulting in 11,188 chick exposure days (Table 1). Chick age at the time of capture ranged from 1 to 8 days. A total of 18 chicks were determined to have died as a result of capture, and were excluded from analyses.

Table 1. Capture statistics for greater sage-grouse chicks marked in Idaho (1999–2002) and Utah (2005–2009).

Year	Broods[1]	Chicks[2]	Hen Ages[3]	Marked[4]
1999	13	30	SY = 3, ASY = 10	2.31
2000	15	42	SY = 4, ASY = 11	2.80
2001	14	40	SY = 1, ASY = 13	2.86
2002	24	71	SY = 5, ASY = 19	2.96
2005	21	89	SY = 11, ASY = 10	4.24
2006	21	61	SY = 0, ASY = 21	2.90
2007	12	55	SY = 4, ASY = 8	4.58
2008	11	66	SY = 2, ASY = 9	6.00
2009	11	64	SY = 1, ASY = 10	5.82
Total	142	518	SY = 31, ASY = 111	3.65

[1]Number of broods captured.
[2]Total number of chicks marked with radio-transmitters.
[3]SY = second year hen (hatched the previous year), ASY = after second year hen (hatched ≥2 years earlier).
[4]Average number of chicks marked per brood.

We censored an additional 159 missing chicks from the dataset after the last date of telemetry observation.

Base null model

Our best intra-annual model of chick survival included linear and quadratic effects of chick age (Table S1). This model clearly out-performed all other intra-annual temporal models in terms of QAICc value. This model was then used as the base model for evaluation of the main effects of hen age and hatch date. Comparison of QAICc scores for these models (Table S2) shows that the model including only hen age and the additive effects of hen age and hatch date were both competitive (Δ QAICc<2). Because the model containing only the effect of hen age was more parsimonious, we chose to retain this model as the base null model for comparison of climate, drought, and greenness phenology covariates [64].

NDVI models

All NDVI covariates considered were highly correlated (all ρ>0.62). Thus, we did not construct models containing multiple NDVI covariates. All 13 single NDVI-effect models produced positive beta estimates (Table S3). Five models were equally supported by the data (ΔQAICc<2.0, Table S3). However, none of the 13 models, including the top 5 models, resulted in a meaningful increase in model fit (as measured by the R-score) relative to the base model. Additionally, the 95% confidence intervals for the effect of the average NDVI in July relative to May 1 for a given site and year (the model with lowest QAIC$_c$) were symmetrical around zero, indicating a weak and imprecise effect (Table S4).

Drought models

As with our NDVI covariates, all drought covariates were highly correlated (all ρ>0.77) so only single-effect models were considered. Of the 10 drought models considered, the model including the effect of the PZI for the preceding winter performed best (Table S5). The addition of winter PZI to the base model resulted in an approximate 40% increase in the R-score, indicating a substantial improvement in model fit. Further, the 95% confidence interval for winter PZI indicated a significant positive effect of the covariate on chick survival (Fig. 2, Table S6).

Climate models

Several combinations of climatic covariates had correlation coefficients below our critical value. Additive effects of multiple climatic covariates were modeled if the correlation between all covariates was less than 0.5 and the model was deemed to be ecologically meaningful. These conditions resulted in the construction of 18 models (Table S7). The top climate model (minimum temperature in May+total precipitation in July) fit the data well (R-score = 0.766) and was the overall top model (Table 2). Both climatic effects in the top model were negatively associated with chick survival (Figs. 3–4). Despite the model fitting the data well, only the effect of July precipitation was significantly different from zero (Table S8). To ensure that the effects in our top model were robust and were not confounded by underlying effects of site and/or year, we added site and year effects to our top climate model (Table S9). Models containing year effects did not converge and models including additive and interactive site effects were not supported by the data (based on ΔQAIC$_c$), indicating that our results are robust across the 2 study sites. Allowing daily survival to change with chick age and holding all other covariate values at the sample mean, predicted values from our top model yielded a 42-

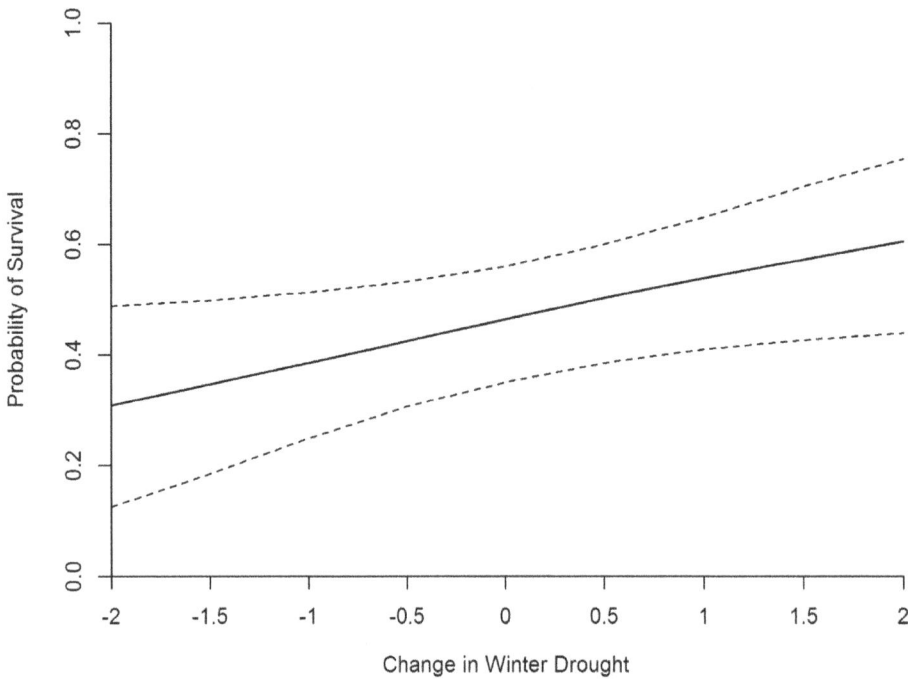

Figure 2. Effects of changes in winter drought severity (PZI) on the probability of greater sage-grouse chick survival to 42 days of age. Dashed lines indicate 95% confidence intervals. Negative values correspond to increasingly severe drought conditions. A change of 0.0 is equal to the mean Winter Palmer Z-Index score observed during the extent of this study. Palmer (1965) stated that a drought score of -2 was indicative of moderate drought.

day survival probability of 0.475 (95% CI = 0.375 to 0.566). Estimates of D from the 3 top models all indicate that dependence in the fates among brood mates was low (1.6149 to 1.7085) but non-negligible (Tables S4, S6, S8).

Discussion

Studies of avian survival are often short-term and conducted on a single study area. While such studies provide important

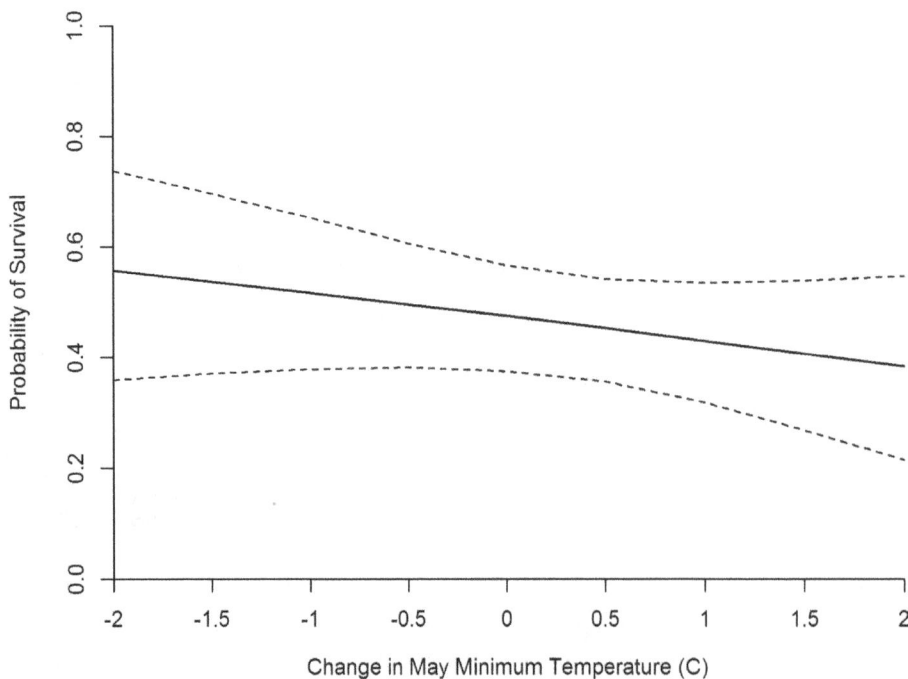

Figure 3. Effects of May minimum temperature on the probability of greater sage-grouse chick survival to 42 days of age. Dashed lines indicate 95% confidence intervals. A change of 0.0 is equal to the mean May minimum temperature observed during the extent of this study.

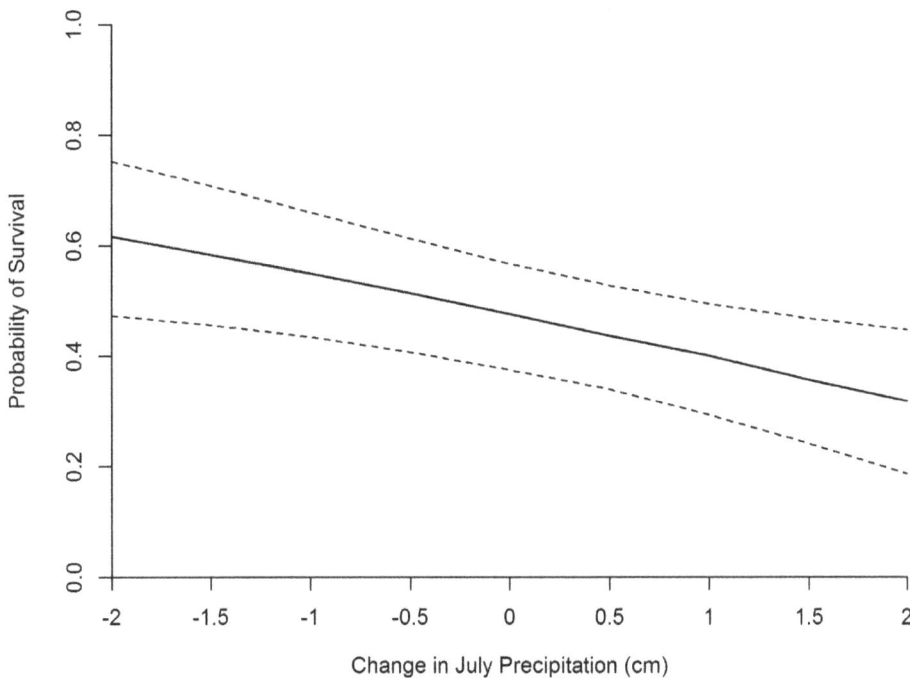

Figure 4. Effects of July precipitation on the probability of greater sage-grouse chick survival to 42 days of age. Dashed lines indicate 95% confidence intervals. A change of 0.0 is equal to the mean July precipitation observed during the extent of this study.

information, for many species there is a lack of knowledge concerning general large-scale factors which influence dynamics across space and time. An understanding of large-scale population drivers is essential for effective wildlife conservation planning and provides a baseline for developing meaningful hypotheses about specific local factors affecting populations at smaller spatial and temporal scales. Our study is the first to attempt to establish this

Table 2. Comparison of top chick survival models among the landscape-scale covariate groups.

Model	K	QAICc	ΔQAICc	w_i	R-score
Base+Saturated Model	13	−121.45	0.00	0.999	1.000
Base+May Min Temp+July Precip (−,−)	7	−58.30	63.15	0.000	0.766
Base+Winter PZI (+)	6	58.21	179.66	0.000	0.396
Base+July Mean NDVI (+)	6	178.53	299.98	0.000	0.022
Quadratic Chick Age+Hen Age (Base)	5	183.48	304.93	0.000	0.000

All models contain the base effects of quadratic chick age and hen age. Models were evaluated using the Quasi-Akaike's Information Criterion (QAIC). K = number of parameters. w_i = model weight (i.e. the likelihood of a particular model being the best model). R-score = percent reduction of deviance relative to the base model (Quadratic Chick Age+Hen Age). The saturated model contains effects for site (1 parameter) and each year (7 parameters). Typically 8 parameters would be required to model the effects of 9 years. However, because years did not overlap between the 2 sites we were able to fully specify year effects with only 7 parameters.

baseline for the survival of greater sage-grouse chicks across multiple populations.

Independence of brood mates

Our modeling approach allowed simultaneous incorporation of commonly collected demographic information (hatch date, chick age, hen age) as well as publically accessible landscape level biotic (NDVI) and abiotic (temperature, precipitation, drought) information into survival models implemented in R [56]. Additionally, our approach allowed us to account for the potential lack of independence among chicks from the same brood [55]. Estimates of D from the top 3 models (Tables S4, S6, S8) ranged from 1.6149 to 1.7085 and, in all cases, confidence intervals did not include one or the mean number of chicks marked per brood (3.65, Table 1). This finding indicates that, while not highly dependent, fates were not independent among brood mates. This supports our decision to use the Manly and Schmutz [54] survival estimator rather than traditional known-fate survival estimators that assume fates of individuals are independent.

Survival rate

Overall, chick survival during this study was relatively high. Our top model produced an average 42-day survival probability of 0.475 (95% CI = 0.375 to 0.566). This is similar to the 42-day survival rate of 0.50 reported for sage-grouse chicks by Dahlgren et al. [17] and considerably higher than the 28-day survival rate of 0.392 reported by Gregg and Crawford [18]. However, comparison of our observed survival rate to those of Gregg and Crawford [18] are potentially confounded by the use of different transmitter attachment methods (suture attachment versus subcutaneous implant). Gregg and Crawford [18] report a total of 32 chick mortalities attributable to capture compared to only 18 in our study despite a similar total number of chicks being marked in both studies. It is possible that our survival rates may be inflated if some chicks treated as capture mortalities were incorrectly classified as such. However, the low incidence of chicks being classified as capture mortalities makes it unlikely that any misclassifications would significantly influence our findings.

Our analysis supported previous research that has shown both chick age and hen age to be important predictors of sage-grouse chick survival [17,60] (Tables S1 and S2). Interestingly, our models indicate that chicks hatched to second-year hens experience higher survival rates than chicks hatched to older hens. This effect has been previously reported for sage-grouse [17,60] although the mechanism underlying it has not been thoroughly explained. Despite being poorly estimated throughout (Tables S4, S6, S8), we chose to retain chick and hen age covariates in all models to minimize bias in estimates of the effects of interest.

Effects of NDVI

In recent years, NDVI has proven to be a useful tool for understanding various aspects of animal ecology [65]. We found positive relationships between all of our measures of NDVI phenology and chick survival (Table S3). However, none of the NDVI measures resulted in substantial improvements in model fit, as measured by the R-score, relative to the base (chick age+hen age) model, nor were the effects significant. Blomberg et al. [27] similarly found that NDVI was positively associated with sage-grouse recruitment and population growth, but that NDVI provided weak predictive power relative to other predictors such as precipitation.

Given the importance of invertebrates and herbaceous vegetation in the diet of sage-grouse chicks [23,18,25], the poor predictive power of NDVI for sage-grouse chick survival is somewhat surprising because NDVI is a well-established index of net primary production [65,66,67], and invertebrate production is positively related to plant production [68]. We suggest that the extensive coverage of sagebrush across both study sites resulted in phenological measures of NDVI being less sensitive to changes in coverage of forbs and grasses, thereby diminishing the ability of NDVI to measure changes of direct relevance to sage-grouse chicks. Correspondingly, Paruelo and Lauenroth [29] found a generally smaller range of NDVI values in sagebrush-steppe ecosystems than in grasslands where plant phenological changes are likely easier to detect.

Effects of drought

Although local availability and abundance of specific invertebrates and forbs is proximally related to sage-grouse chick survival [17,18], survival is likely under the primary influence of physical factors such as precipitation (amount and timing), temperature, and drought. Accordingly, our analysis indicated that abiotic factors were better predictors of sage-grouse chick survival than phenology of NDVI. Our top drought model (Table S5) indicated the presence of a significant relationship between winter drought and chick survival (Table S6). Since smaller PZI (and PDSI) values correspond to increasingly severe drought conditions, the positive parameter estimate associated with the winter drought effect implies that winter droughts lead to reduced chick survival (Fig. 2). Unfortunately, our data do not allow us to identify the true causal mechanism(s) underlying this relationship. Schwinning et al. [69] found that winter drought, even more so than summer drought, affects plant production during the following summer. Therefore, winter drought may affect sage-grouse chick survival via its influence on brood habitat quality. Additionally, winter drought may influence chick survival by affecting resource provisioning during egg formation. Forb abundance during the pre-nesting period is positively associated with hen nutrition [70], and hen

nutrition prior to nest initiation is positively related to reproductive investment [71]. Thus, we suggest that either or both of these effects may be the mechanism behind the relationship between winter drought and chick survival that we observed.

Effects of climate

Blomberg et al. [27] reported relatively stable survival rates for adult sage-grouse but found that recruitment was variable and strongly influenced by annual climatic variation. These findings led the authors to conclude that stability of sage-grouse populations is dependent upon stable annual survival rates and occasional large inputs of new individuals into the population when climatic conditions are amenable to chick and juvenile survival. Our results support this assertion that climatic variables play a primary role in determining sage-grouse reproductive success. Of the 3 groups of predictors of chick survival we considered, models containing climatic effects clearly outperformed all other models (Table S7 versus Tables S5 and S3).

Our top climatic model fit the data well (Table 2, R-score = 0.766). In addition to the effects of chick and hen age, the top model included the minimum temperature in May (MMT) and precipitation in July, both producing negative parameter estimates (Table S8). We initially hypothesized that MMT could have either a positive or negative association with chick survival. Specifically, we predicted that MMT could be positive if higher minimum temperature resulted in fewer chicks dying due to exposure. Alternatively, we predicted that a negative effect of MMT would be attributable to high minimum temperatures leading to early snow melt and thus lower soil moisture and poor habitat quality throughout the brood-rearing period. We conclude that the latter is the case. Although particularly cold temperatures in late May could potentially result in increased exposure mortality, consideration of our peak hatch dates (see Section 3.1) reveals that it is unlikely that many chicks would be hatched early enough to be exposed to extreme low temperatures likely occurring in the first half of May. We also note that the minimum temperature in June is positively associated with chick survival, possibly indicating that exposure mortality does increase as temperature decreases during this timeframe.

Our interpretation of the negative effect of MMT on chick survival does raise concerns about the impact of projected climate on future sage-grouse reproductive success. Significant temperature increases have been documented across western North America in recent decades, and climate models consistently predict that temperatures will continue to increase into the foreseeable future [72]. Observed and projected warming trends have also been connected to observed and projected transitions from winter precipitation falling as rain rather than snow and consequently reduced spring snow pack [72,73,74].

The trend in warming temperatures could impact sage-grouse population dynamics as a result of phenological asynchrony [75], increased spread of exotic species such as red fox (*Vulpes vulpes*; [76]) and cheatgrass (*Bromus tectorum*; [5]), and increased frequency and severity of wildfires [77]. Blomberg et al. [27] concluded that projected climate change could result in reduced recruitment of sage-grouse. Our results support this conclusion. Figure 3 shows model-derived chick survival estimates across a range of MMT. According to our results, a 2°C increase in mean MMT, well within the range projected by most climate models [72], will result in an approximate 10% reduction in sage-grouse chick survival. This effect could be mitigated if sage-grouse are able to adjust their hatch dates to correspond with earlier snow melt and advanced plant phenology. A simple linear regression of our observed median hatch dates on MMT shows a significant correlation

($p = 0.0171$, $R^2 = 0.58$). This demonstrates that sage-grouse may be capable of synchronizing the timing of nesting with MMT for at least the range of MMT observed during our study. However, it is not clear if the level of plasticity in breeding phenology is sufficient to compensate for future climatic changes. Additionally, if warming results in a shift in the form of winter precipitation from snow to rain [73], chick survival may still be negatively affected by poor habitat quality, even if hens are able to adjust nest initiation to correspond with early snowmelt.

We initially hypothesized that July precipitation (JP) would have a positive effect on sage-grouse chick survival due to a moisture associated increase in plant and arthropod forage. However, our analysis showed a significant negative effect of JP (Table S8). While this result may be less intuitive, we conclude that it is real and meaningful. At least 2 mechanisms may underlie the relationship between chick survival and JP. First, chicks may be susceptible to exposure mortality in July. As noted above, the Utah study area is located in a monsoonal zone and receives a substantial proportion of the annual precipitation during late summer (primarily July and August). Monsoonal storms across the Utah study area often build quickly and result in significant temperature reductions followed by rain, hail, or both. By July, chicks are larger in size and are more independent of the brood hen [78]. If chicks are too large to be effectively brooded, severe monsoonal storms may result in chicks becoming soaked by rain and/or losing body temperature due to low temperatures or hail. While the Idaho study area is not in a monsoonal zone, it is possible that occasional severe July storms could produce similar effects. However, by July chicks should be more capable of thermoregulation relative to the early development period in June. Thus, we conclude that if exposure were a major source of chick mortality, models including the effect of June precipitation or the minimum temperature in June would have performed better.

Alternatively, JP may negatively affect sage-grouse chick survival through an interaction between increased moisture/humidity and predator search efficiency (i.e., moisture facilitated predation hypothesis). Moisture on skin and feathers increases bacterial activity, subsequently increasing scent production [36,37,38,39,40]. Mammalian predators have been shown to respond rapidly to the presence of prey odor [79] and increased scent production may lead to enhanced prey detection rates. A number of studies have found increased nest predation rates following precipitation events for greater sage-grouse and other gallinaceous birds [33,80,81], and the phenomenon of moisture-facilitated predation may apply to chicks and adult birds as well [34,82]. We do not present observed chick predation rates due to concerns about correctly distinguishing between predation and scavenging. However, both of our study sites were inhabited by a suite of potential mammalian predators. Both study sites received predator management to reduce coyote (*Canis latrans*) predation on livestock, but coyotes and other common predators of sage-grouse chicks (red fox, badger [*Taxidea taxus*], weasels [*Mustella* sp.], and rattlesnakes [*Crotalus viridis*]) were present on both sites. In addition to the potential effects of moisture-facilitated predation by olfactory predators, JP may increase predation by avian predators if sage-grouse broods move to areas with less sagebrush cover following precipitation events to expedite drying and/or warming. Although not formally documented, we did observe broods along roadways at a higher frequency following precipitation events than at other times.

Effects of climate change on precipitation are less clear than the effects on temperature [72]. Climate models are inconclusive as to the sign of the effect on precipitation [83], and effects may vary by season [84]. In the absence of a consensus about effects of climate

change on summer precipitation, anticipating the effect of changing precipitation on ecological communities and populations is difficult. Our analysis indicates that a 2 cm change in JP (positive or negative) would result in an approximate 15% change in sage-grouse chick survival (Fig. 4).

Sage-grouse are a species of great conservation concern in western North America. Chick survival has been shown to be an important determinant of population growth rates [9,10], yet relatively little is known about climatic or other large-scale environmental factors affecting survival rates. Previous studies have identified specific habitat characteristics that influence survival [17,18]. These studies have led to a proliferation of efforts to manage brood-rearing habitats without thorough consideration of the abiotic factors influencing both habitat quality and chick survival. Our study clearly demonstrates that large-scale abiotic factors such as drought, temperature and precipitation have significant effects on chick survival. These factors are beyond the control of state and federal wildlife management agencies and highlight the importance of considering current and future climatic conditions when developing policy and conservation strategies for this species. However, the effects we observed were measured for populations inhabiting large intact tracts of sagebrush habitat. The availability of adequate amounts of suitable habitat is a prerequisite that must be met for the effects of the abiotic factors we studied to be relevant.

Supporting Information

Table S1 Models for effect of age on greater sage-grouse chick survival.

Table S2 Models for effect of hen age and hatch date on greater sage-grouse chick survival. 'Age' is the top age varying model from Table S1. Signs in parentheses indicate the direction of respective covariate effects excluding chick age.

Table S3 Models for the effects of habitat greenness as measured by the Normalized Difference Vegetation Index (NDVI) on greater sage-grouse chick survival. Signs in parentheses indicate the direction of respective covariate effects excluding chick age. All models (except the intercept only model) contain the base effects of quadratic chick age and hen age. Models were evaluated using the Quasi-Akaike's Information Criterion (QAIC). K = number of parameters. w_i = model weight (i.e. the likelihood of a particular model being the best model). R-score = percent reduction of deviance relative to the base model (Quadratic Chick Age+Hen Age).

Table S4 Parameter estimates with 95% confidence intervals for the top model of the effects of Normalized Difference Vegetation Index (NDVI) on greater sage-grouse chick survival. Confidence intervals were calculated based on 5,000 bootstraps of the original data set.

Table S5 Models for the effects of drought on greater sage-grouse chick survival. Signs in parentheses indicate the direction of respective covariate effects excluding chick age. All models (except the intercept only model) contain the base effects of quadratic chick age and hen age. Models were evaluated using the Quasi-Akaike's Information Criterion (QAIC). K = number of parameters. w_i = model weight (i.e. the likelihood of a particular model being the best model). R-score = percent reduction of deviance relative to the base model (Quadratic Chick Age+Hen Age).

Table S6 Parameter estimates with 95% confidence intervals for the top model of the effect of drought on greater sage-grouse chick survival. Confidence intervals were calculated based on 5,000 bootstraps of the original data set.

Table S7 Models for the effects of climate on greater sage-grouse chick survival. Signs in parentheses indicate the direction of respective covariate effects excluding chick age. All models (except the intercept only model) contain the base effects of quadratic chick age and hen age. Models were evaluated using the Quasi-Akaike's Information Criterion (QAIC). K = number of parameters. w_i = model weight (i.e. the likelihood of a particular model being the best model). R-score = percent reduction of deviance relative to the base model (Quadratic Chick Age+Hen Age).

Table S8 Parameter estimates with 95% confidence intervals for the top model of the effects of climate on

greater sage-grouse chick survival. Confidence intervals were calculated based on 5,000 bootstraps of the original data set.

Table S9 Evaluation of the effects of site and year effects on the top model from Table S7. All models contain the base effects of quadratic chick age and hen age. Models were evaluated using the Quasi-Akaike's Information Criterion (QAIC). K = number of parameters. w_i = model weight (i.e. the likelihood of a particular model being the best model).

Acknowledgments

We thank the many technicians who assisted with collecting data for this research. D. Menuz, J. Guttery and 2 anonymous reviewers provided valuable comments on earlier versions of this manuscript.

Author Contributions

Conceived and designed the experiments: MRG DKD TAM JWC KPR NB. Performed the experiments: MRG DKD NB. Analyzed the data: MRG PAT DNK. Wrote the paper: MRG.

References

1. Jump AS, Peñuelas J (2005) Running to stand still: adaptation and the response of plants to rapid climate change. Ecol Letters 8: 1010–1020.
2. Thomas CD, Cameron A, Green RE, Bakkenes M, Beaumont LJ, et al. (2004) Extinction risk from climate change. Nature 427: 145–148.
3. Norris K (2004) Managing threatened species: the ecological toolbox, evolutionary theory and declining-population paradigm. J Applied Ecol 41: 413–426.
4. Garton EO, Connelly JW, Horne JS, Hagen CA, Moser A, et al. (2011) Greater sage population dynamics and probability of persistence. In: Knick, ST, Connelly, JW, eds. Greater sage-grouse: ecology and conservation of a landscape species and its habitats. Studies in Avian Biology, Vol. 38. University of California Press, Berkeley, USA. pp 293–382.
5. Miller RF, Knick ST, Pyke DA, Meinke CW, Hanser SE, et al. (2011) Characteristics of sagebrush habitats and limitations to long-term conservation. In: Knick, ST, Connelly, JW, eds. Greater sage-grouse: ecology and conservation of a landscape species and its habitats. Studies in Avian Biology, Vol. 38. University of California Press, Berkeley, USA. pp 145–184.
6. Harris W, Lungle K, Bristol B, Dickinson D, Eslinger D, et al. (2001) Canadian sage grouse recovery strategy. Saskatchewan Environment and Resource Management and Alberta Natural Resources Services.
7. USFWS (2010) Endangered and Threatened Wildlife and Plants; 12-month findings for petitions to list the greater sage-grouse (*Centrocercus urophasianus*) as threatened or endangered. Federal Register 50 CFR Part 17 [FWS-R6-ES-2010-00189]. http://www.fws.gov/mountain-prairie/species/birds/sagegrouse/FR03052010.pdf
8. Connelly JW, Hagen CA, Schroeder MA (2011) Characteristics and dynamics of greater sage populations. In: Knick, ST, Connelly, JW, eds. Greater sage-grouse: ecology and conservation of a landscape species and its habitats. Studies in Avian Biology, Vol. 38. University of California Press, Berkeley, USA. pp 53–68.
9. Taylor RL, Walker BL, Naugle DE, Mills LS (2012) Managing multiple vital rates to maximize greater sage-grouse population growth. J Wildl Manage 76: 336–347.
10. Dahlgren DK (2009) Greater sage-grouse ecology, chick survival, and population dynamics, Parker Mountain, Utah. Utah State University, Logan: Ph.D. Dissertation.
11. Anthony RG, Willis MJ (2009) Survival rates of female greater sage-grouse in autumn and winter in southeastern Oregon. J Wildl Manage 73: 538–545.
12. Moynahan BJ, Lindberg MS, Thomas JW (2006) Factors contributing to process variance in annual survival of female greater sage-grouse in Montana. Ecol Appl 16: 1529–1538.
13. Pfister CA (1998) Patterns of variance in stage-structured populations: evolutionary predictions and ecological implications. Proc Nat Acad Sci 95: 213–218.
14. Gaillard JM, Yoccoz NG (2003) Temporal variation in survival of mammals: a case of environmental canalization? Ecology 84: 3294–3306.
15. Caswell H (2001) Matrix population models: construction, analysis and interpretation. 2nd edition. Sinauer Associates, Sunderland, Massachusetts, USA.
16. Aldridge CL, Boyce MS (2007) Linking occurrence and fitness to persistence: habitat-based approach for endangered greater sage-grouse. Ecol Appl 17: 508–526.

17. Dahlgren DK, Messmer TA, Koons DN (2010) Achieving better estimates of greater sage-grouse chick survival in Utah. J Wildl Manage 74: 1286–1294.
18. Gregg MA, Crawford JA (2009) Survival of greater sage-grouse chicks and broods in the northern Great Basin. J Wildl Manage 73: 904–913.
19. Drut MS, Crawford JA, Gregg MA (1994) Brood habitat use by sage grouse in Oregon. Great Basin Naturalist 54: 170–176.
20. Klebenow DA (1969) Sage grouse nesting and brood habitat in Idaho. J Wildl Manage 33: 649–662.
21. Sveum CM, Crawford JA, Edge WD (1998) Use and selection of brood-rearing habitat by sage grouse in south central Washington. Great Basin Nat 58: 344–351.
22. Wallestad RO (1971) Summer movements and habitat use by sage grouse broods in central Montana. J Wildl Manage 35: 129–136.
23. Drut MS, Pyle WH, Crawford JA (1994) Diets and food selection of sage grouse chicks in Oregon. J Range Manage 47: 90–93.
24. Johnson GO, Boyce MS (1990) Feeding trials with insects in the diet of sage grouse chicks. J Wildl Manage 54: 89–91.
25. Klebenow DA, Gray GM (1968) Food habits of juvenile sage grouse. J Range Manage 21: 80–83.
26. Peterson JG (1970) The food habits and summer distribution of juvenile sage grouse in central Montana. J Wildl Manage 34: 147–155.
27. Blomberg EJ, Sedinger JS, Atamian MT, Nonne DV (2012) Characteristics of climate and landscape disturbance influence the dynamics of greater sage-grouse populations. Ecosphere 3: 55.
28. Box EO, Holben BN, Kalb V (1989) Accuracy of the AVHRR vegetation index as a predictor of biomass, primary production, and new CO_2 flux. Vegetatio 80: 71–89.
29. Paruelo JM, Lauenroth WK (1995) Regional patterns of normalized difference vegetation index in North American shrublands and grasslands. Ecology 76: 1888–1898.
30. Pettorelli N, Vik JO, Mysterud A, Gaillard JM, Tucker CJ, et al. (2005) Using the satellite-derived NDVI to assess ecological responses to environmental change. TREE 20: 503–510.
31. Walker DA, Halfpenny JC, Walker MD, Wessman CA (1993) Long-term studies of snow-vegetation interactions. BioScience 43: 287–301.
32. Hannon SJ, Martin K (2006) Ecology of juvenile grouse during the transition to adulthood. Journal of Zoology 269: 422–433.
33. Herman-Brunson KM, Jensen KC, Kaczor NW, Swanson CC, Rumble MA, et al. (2009) Nesting ecology of greater sage-grouse *Centrocercus urophasianus* at the eastern edge of their historic distribution. Wildl Biol 15: 237–246.
34. Lehman CP, Flake LD, Rumble MA, Thompson DJ (2008) Merriam's turkey poult survival in the Black Hills, South Dakota. Intermountain J Sci 14: 78–88.
35. Roberts SD, Coffey JM, Porter WF (1995) Survival and reproduction of female wild turkeys in New York. J Wildl Manage 59: 437–447.
36. Syrotuck WG (1972) Scent and the scenting dog. Arner Publishers, Rome, New York, USA.
37. Barros N, Gomez-Orellana I, Feijóo S, Balsa R (1995) The effects of soil moisture on soil microbial activity studied by microcalorimetry. Thermochimica Acta 249: 161–168.
38. Bohn HL, Bohn KH (1999) Moisture in biofilters. Enviro Progress 18: 156–161.

39. Schimel JP, Gulledge JM, Clein-Curley JS, Lindstrom JE, Braddock JF (1999) Moisture effects on microbial activity and community structure in decomposing birch litter in the Alaskan taiga. Soil Bio Biochem 31: 831–838.

40. Zavala MAL, Funamizu N (2005) Effects of moisture content on the composting process in a biotoilet system. Compost Sci Util 13: 208–216.

41. Beck JL, Reese KP, Connelly JW, Lucia MB (2006) Movements and survival of juvenile greater sage-grouse in southeastern Idaho. Wildl Soc Bull 34: 1070–1078.

42. Beck JL, Mitchell DL, Maxfield BD (2003) Changes in the distribution and status of sage-grouse in Utah. Western North American Naturalist 63: 203–214.

43. Giesen KM, Schoenberg TJ, Braun CE (1982) Methods for trapping sage grouse in Colorado. Wildl Soc Bull 10: 224–231.

44. Wakkinen WL, Reese KP, Connelly JW, Fischer RA (1992) An improved spotlighting technique for capturing sage grouse. Wildl Soc Bull 20: 425–426.

45. Beck TDI, Gill RB, Braun CE (1975) Sex and age determination of sage grouse from wing characteristics. Colorado Department of Natural Resources Game Information Leaflet 49, Denver, USA.

46. Baxter RJ, Flinders JT, Mitchell DL (2008) Survival, movement, and reproduction of translocated greater sage-grouse in Strawberry Valley, Utah. J Wildl Manage 72: 179–186.

47. Holloran MJ, Heath BJ, Lyon AG, Slater SJ, Kuipers JL, et al. (2005) Greater sage-grouse nesting habitat selection and success in Wyoming. J Wildl Manage 69: 638–649.

48. Burkepile NA, Connelly JW, Stanley DW, Reese KP (2002) Attachment of radiotransmitters to one-day-old sage grouse chicks. Wildl Soc Bull 30: 93–96.

49. Karl TR (1986) The sensitivity of the Palmer drought severity index and Palmer's Z-index to their calibration coefficients including potential evapotranspiration. J Climate Applied Meteorology25: 77–86.

50. Palmer WC (1965) Meteorological drought. U.S. Weather Bureau Research Paper No.45.

51. Daly C, Taylor GH, Gibson WP, Parzybok TW, Johnson GL, et al. (2000) High-quality spatial climate data sets for the United States and beyond. Trans Amer Soc Ag Eng 43: 1957–1962

52. Schroeder TA, Cohen WB, Song C, Canty MJ, Yang Z (2006) Radiometric correction of multi-temporal Landsat data for characterization of early successional forest patterns in western Oregon. Remote Sensing of Environ 103:16–26.

53. Jensen J R (2005) Introductory digital image processing: a remote sensing perspective, 3rd edition. Pearson Prentice Hall.

54. Manly BFJ, Schmutz JA (2001) Estimation of brood and nest survival: comparative methods in the presence of heterogeneity. J Wildl Manage 65: 258–270.

55. Fondell TF, Miller DA, Grand JB, Anthony RM (2008) Survival of dusky Canada goose goslings in relation to weather and annual nest success. J Wildl Manage 72: 1614–1621.

56. R Development Core Team (2011) R: a language and environment for statistical computing. R Foundation for Statistical Computing, Vienna. http://www.R-project.org/

57. Akaike H (1973) Information theory and an extension of the maximum likelihood principle. In: Petrov BN, Csaki BF, eds. Second International Symposium on Information Theory, Academiai Kiado, Budapest. pp 267–281.

58. Burnham KP, Anderson DR (2002) Model selection and multimodel inference: a practical information-theoretic approach. Springer-Verlag, New York, New York, USA.

59. Hamel S, Côté SD, Festa-Bianchet M (2010) Maternal characteristics and environment affect the costs of reproduction in female mountain goats. Ecology 91: 2034–2043.

60. Guttery MR (2011) Ecology and management of a high elevation southern range greater sage-grouse population: vegetation manipulation, early chick survival, and hunter motivations. Utah State University, Logan: Ph.D. Dissertation.

61. Dixon PM (1993) The bootstrap and the jackknife: describing the precision of ecological indices. In: Scheiner SM, Gurevitch J, eds. Design and analysis of ecological experiments. Chapman and Hall, New York, New York, USA. pp 290–318.

62. Zheng B (2000) Summarizing the goodness of fit of generalized linear models for longitudinal data. Statistics in Medicine 19: 1265–1275.

63. Iles DT, Rockwell RF, Matulonis P, Robertson GJ, Abraham KF, et al. (2013) Predators, alternative prey, and climate influence annual breeding success of a long-lived sea duck. J Animal Ecology 82: 683–693.

64. Arnold TW (2010) Uninformative parameters and model selection using Akaike's Information Criterion. J Wildl Manage 74: 1175–1178.

65. Pettorelli N, Ryan S, Mueller T, Bunnefeld N, Jędrzejewska B, et al. (2011) The normalized difference vegetation index (NDVI): unforseen successes in animal ecology. Climate Research 46: 15–27.

66. Fang J, Piao S, Tang Z, Peng C, Ji W (2001) Interannual variability in net primary production and precipitation. Science 293: 1723a.

67. Field CB, Randerson JT, Malmström CM (1995) Global net primary production: combining ecology and remote sensing. Remote Sensing of Environ 51: 74–88.

68. Wenninger EJ, Inouye RS (2008) Insect community response to plant diversity and productivity in a sagebrush-steppe ecosystem. J Arid Environ 72: 24–33.

69. Schwinning S, Starr BI, Ehleringer JR (2005) Summer and winter drought in a cold desert ecosystem (Colorado Plateau) part II: effects on plant carbon assimilation and growth. J Arid Environ 61: 61–78.

70. Gregg MA, Barnett JK, Crawford JA (2008) Temporal variation in diet and nutrition of preincubating greater sage-grouse. Range Ecol Manage 61: 535–542.

71. Dunbar MR, Gregg MA, Crawford JA, Giordano MR, Tornquist SJ (2005) Normal hematologic and biochemical values for prelaying greater sage-grouse (*Centrocercus urophasianus*) and their influence in chick survival. J Zoo and Wildl Medicine 36: 422–429.

72. Karl TR, Melillo JM, Patterson TC, eds(2009) Global climate change impacts in the United States. Cambridge University Press.

73. Knowles N, Dettinger MD, Cyan DR (2006) Trends in snowfall versus rainfall in the western United States. J Climate 19: 4545–4559.

74. Mote PW, Hamlet AF, Clark MP, Lettenmaier DP (2005) Declining mountain snowpack in western North America. Bull American Meteorological Society 86: 39–49.

75. Parmesan C (2006) Ecological and evolutionary responses to recent climate change. Annual Review of Ecology, Evolution, and Systematics 37: 637–669.

76. Walther G, Post E, Convey P, Menzel A, Parmesan C, et al. (2002) Ecological responses to recent climate change. Nature 416: 389–395.

77. Brown TJ, Hall BL, Westerling AL (2004) The impact of twenty-first century climate change in wildland fire danger in the western United States: an applications perspective. Climatic Change 62: 365–388.

78. Schroeder MA, Young JR, Braun CE (1999) Sage grouse (*Centrocercus urophasianus*). In: Poole A, Gill F, eds. The Birds of North America, No. 425. The Academy of Natural Science, Philadelphia, Pennsylvania. pp 1–28.

79. Hughes NK, Price CJ, Banks PB (2010) Predators are attracted to the olfactory signals of prey. Plos ONE 5: e13114.

80. Lehman CP, Rumble MA, Flake LD, Thompson DJ (2008) Merriam's turkey nest survival and factors affecting nest predation by mammals. J Wildl Manage 72: 1765–1774.

81. Webb SL, Olson CV, Dzialak MR, Harlu SM, Winstead JB, et al. (2012) Landscape features and weather influence nest survival of a ground-nesting bird of conservation concern, the greater sage-grouse, in human-altered environments. Ecol Proc 1: 1–15.

82. Hohensee SD, Wallace MC (2001) Nesting and survival of Rio Grande turkeys in northcentral Texas. Proceedings of the National Wild Turkey Symposium 8: 85–91.

83. Chambers JC, Pellant M (2008) Climate change impacts on northwestern and intermountain United States rangelands. Rangelands 30: 29–33.

84. Mote PW (2006) Climate-driven variability and trends in mountain snowpack in western North America. J Climate 19: 6209–6220.

10

Drought Tolerance in Wild Plant Populations: The Case of Common Beans (*Phaseolus vulgaris* L.)

Andrés J. Cortés[1,2*], **Fredy A. Monserrate**[3], **Julián Ramírez-Villegas**[4,5], **Santiago Madriñán**[2], **Matthew W. Blair**[6*]

1 Evolutionary Biology Center, Uppsala University, Uppsala, Sweden, **2** Laboratorio de Botánica y Sistemática, Universidad de los Andes, Bogotá, Colombia, **3** Centro Agropecuario, Servicio de Enseñanza Nacional, Buga, Colombia, **4** School of Earth and Environment, University of Leeds, Leeds, United Kingdom, **5** CGIAR Program on Climate Change, Agriculture and Food Security (CCAFS), Cali, Colombia, **6** Department of Plant Breeding, Cornell University, Ithaca, NY, United States

Abstract

Reliable estimations of drought tolerance in wild plant populations have proved to be challenging and more accessible alternatives are desirable. With that in mind, an ecological diversity study was conducted based on the geographical origin of 104 wild common bean accessions to estimate drought tolerance in their natural habitats. Our wild population sample covered a range of mesic to very dry habitats from Mexico to Argentina. Two potential evapotranspiration models that considered the effects of temperature and radiation were coupled with the precipitation regimes of the last fifty years for each collection site based on geographical information system analysis. We found that wild accessions were distributed among different precipitation regimes following a latitudinal gradient and that habitat ecological diversity of the collection sites was associated with natural sub-populations. We also detected a broader geographic distribution of wild beans across ecologies compared to cultivated common beans in a reference collection of 297 cultivars. Habitat drought stress index based on the Thornthwaite potential evapotranspiration model was equivalent to the Hamon estimator. Both ecological drought stress indexes would be useful together with population structure for the genealogical analysis of gene families in common bean, for genome-wide genetic-environmental associations, and for postulating the evolutionary history and diversification processes that have occurred for the species. Finally, we propose that wild common bean should be taken into account to exploit variation for drought tolerance in cultivated common bean which is generally considered susceptible as a crop to drought stress.

Editor: John P. Hart, New York State Museum, United States of America

Funding: This research was supported by the Generation Challenge Program. The funders had no role in study design, data collection and analysis, decision to publish, or preparation of the manuscript.

Competing Interests: The authors have declared that no competing interests exist.

* E-mail: andres.cortes@ebc.uu.se (AJC); mwb1@cornell.edu (MWB)

Introduction

Common bean (*Phaseolus vulgaris* L.) is a key source of nutrients and dietary protein for over half a billion people in Latin America and Africa and nearly 4 million hectares are grown in zones where drought is severe, such as in northeastern Brazil, coastal Peru, the northern highlands of México and in dry parts of Africa [1]. Therefore, increasing drought tolerance in common bean commercial varieties is highly desirable. A considerable reservoir for this task may be available in the wild and cultivated collections of common bean, as can be suggested by their high genetic diversity and phenotypic variability [2,3,4].

Wild common bean is an annual, viney plant that germinates among small trees and shrubs in forest clearings or in disturbed environments with the onset of seasonal rains [5,6]. Specifically, the growth cycle of the wild common bean is from 8 to 10 months in length. Hence, in tropical bimodal rainfall regions wild common bean is subjected to a mid-term drought, while in sub-tropical unimodal rainfall regions wild beans can be subjected to more prolonged periods of water stress. These drought stresses are characteristic of environments in the inter-Andean valleys of the Andes in South America and in northern parts of Mesoamerica especially the volcanic axis and mountains of Mexico [7,8].

Although wild common bean is promising in terms of drought tolerance, the evaluation of drought physiology traits in wild populations would be impractical due to long growth cycle and seed dehiscence [7]. Consequently, alternative methods should be explored in order to discover potential drought tolerance sources in wild populations based on the characteristics of their natural habitats as done for other species [4,9,10,11].

In this sense, potential evapotranspiration (PET) modeling is a powerful tool to predict drought severity for a geographic site or the accessions' origin, so as to identify sources of drought tolerance in cases in which no phenotypic evaluations are available [12]. PET is a theoretical value that aims to characterize the quantity of water that will flux from the soil-biosphere system toward the atmosphere given the effects of evaporation and transpiration and provided that soil water is enough to supply the demand [13]. PET can be calculated purely with climatic variables provided that the hypothetical effect of each of the variables for evaporation and transpiration is known [14,15].

Calculations of PET consider that transpiration and evaporation are proportional to temperature. Two lines of evidence support this assumption. First, increasing temperatures leads to an increase in the maximum density of water vapor (until the air is fully saturated with water vapor), and coincidentally to a relative decrease in humidity (relative humidity = real density of water vapor/maximum density of water vapor). Relative air humidity is proportional to the water potential gradient between the plant-soil system and the surrounding air. Hence, there would be a net flux of water from the plant toward the surrounding air during drought stress events. In addition, temperature is proportional to the energy transferred and is a necessary variable for understanding the phase change between liquid and gaseous water [15,16]. PET modeling also considers that evaporation and transpiration are proportional to radiation because radiation is proportional to the energy transferred from sunlight to plants. Radiation is thus a necessary variable for understanding phase change. Finally leaf conductance is proportional to radiation, at least for C3 plants due to the modulation of stomata opening [17]. The two common non-intensive methods to calculate PET based on temperature and radiation are the Thornthwaite method [16] and the Hamon method [15]. The former considers the effects of both temperature and radiation explicitly, while the latter is purely based on temperature effects.

To determine net water flux given the effects of temperature and radiation, habitat drought index can be calculated comparing the values for PET and for precipitation (P). In particular, three scenarios can be recognized: PET and P are equivalent, PET is higher than P, and P is higher than PET [18,19]. Two considerations must be taken into account before deciding on the biological meaning of each scenario. First, PET and P estimations in a specific period of time are based on stochastic variables. Second, the period of time considered for the previous calculations and analysis will determine if the plant is actually subjected to significant water stress or not. For instance, PET higher than P implies that drought stress is a constant if PET and P are measured in time scales in which a water deficit impacts the plant's physiology. In contrast, P higher than PET presumes that the plant does not experience water stress during the time scale of measurements. However, it is notable and counter-intuitive that PET equals P does not predict an absolute stress condition because of the stochastic components of both PET and P, and because of the soil water holding capacity [20,21].

The objectives of this research were 1) to evaluate the environmental variability in collection site habitats for a core collection of wild common bean that had been previously fingerprinted for genepool and sub-population structure and 2) to determine through two (Thornthwaite vs. Hamon) methods of PET modeling the extent of drought tolerance, the correlation of drought tolerance with collection site characteristics and the association of drought tolerance estimates based on environmental data for the collection site with genetic population and sub-population structure of the wild bean collection. We also evaluated whether the classification of geographical distribution based on drought stress was dissimilar between wild and cultivated common beans, and if their patterns of geographical variation could be determined by local adaptation to hydrological regimes or by evolutionary inertia.

Materials and Methods

Plant Material

A total of 104 wild common bean accessions were considered in this study. All the genotypes were loaned by the Genetic Resources Unit at the International Center for Tropical Agriculture and are preserved under the treaty for genetic resources from the Food and Agriculture Organization, hereafter abbreviated as the FAO collection. In addition, information on drought tolerance of 297

cultivars of common beans from Pérez et al. [22] were considered to compare distribution of cultivated and wild common beans. These two reference collections were selected to be a representative sample of genepools and races, based on a subset of core collections for wild [23] and cultivated [24] beans. Their analysis with neutral molecular markers has also been previously described [25]. Finally, the definition of wild genetic sub-populations is according to Blair et al. [26] and Broughton et al. [27]. Geographic information was provided for each accession by the Genetic Resource Unit (http://isa.ciat.cgiar.org/urg/main.do). In order to estimate drought tolerance for wild common bean, 19 bioclimatic variables were downloaded from WorldClim (http://www.worldclim.org) and they were recorded for each wild accession point (table 1).

Multivariate Analysis

Visual correlation between population structure previously accessed for the wild collection and drought stress was assessed by overlaying the dot-map distribution of the accessions with the precipitation pattern using ESRI's ArcView (ESRI, Inc.). Furthermore, the variation of precipitation and temperature in different regions along the latitudinal pattern was assessed using DIVA-GIS 7.1.6 [28]. With the aim of determining which variables were useful for estimating the drought stress of each accession habitat, the all 19 bioclimatic variables as well as two subsets of these (one of them including precipitation-related variables and the other one including drought-related variables – see table 1) were subjected to scatter plot, principal components and cluster analysis (Pearson's correlation coefficient (r), Spearman's rank correlation coefficient (ρ) and middle joint method). The variables in the sub-set of drought-related were chosen according to their relation to mean temperature during the warmest period and precipitation during the driest period, which both are associated to drought events. For example, annual precipitation and precipitation of the driest period indicate long- and short-term stress, respectively. Graphics were revised and edited in SigmaPlot (Systat-Software, Inc.). Population structure was considered based on results from Blair et al. (2009) and an analysis of variance was carried out to recognize differences between populations using XLSTAT 7.5.2 (Addinsoft, Inc.) and STATISTIX 8 (Statistix: Analytical Software, Inc.).

Potential Evapotranspiration and Drought Index Calculation

Monthly potential evapotranspiration was calculated for each accession using the bioclimatic information and the Thornthwaite [14,16] and Hamon [15] approximations. Thornthwaite equations considered the effects of temperature and radiation on the calculation of the potential evapotranspiration (PET). In particular, we used the following equation (for the month "j"):

$$PET_{Thornthwaite,\, j} \left\{ \begin{array}{l} 1.6L_j \left(\frac{10T_j}{I}\right)^a \; if \; T_j > 0 \\ 0 \; if \; T_j \leq 0 \end{array} \right\}$$

Where

T_j = monthly mean air temperature (°C)

Table 1. Contribution (%) of each bioclimatic variable to the PCA analysis.

Main Variable	Bioclimatic Variable	Total Variables (T)			Only Precipitation Variables (P)			Drought-related variables (*, S)		
		F1+ 40.41%	F2 30.17%	F3 13.04%	F1 57.16%	F2 26.27%	F3 9.28%	F1 42.02%	F2 34.28%	F3 11.88%
Temperature	P1. Annual Mean Temperature*	4.08	**11.26**	1.19				4.52	**22.73**	0.94
	P2. Mean Diurnal Temp. Range (Mean(period max-min))	7.87	1.11	0.01						
	P3. Isothermality (P2/P7)	4.58	3.23	**11.63**						
	P4. Temperature Seasonality (Coefficient of Variation)*	6.47	1.99	7.25				6.72	4.06	**34.78**
	P5. Max Temperature of Warmest Period*	0.01	**16.38**	0.00				**15.80**	7.48	1.45
	P6. Min Temperature of Coldest Period	*10.17*	2.03	3.23						
	P7. Temperature Annual Range (P5–P6)	**9.07**	2.59	2.77						
	P8. Mean Temperature of Wettest Quarter	0.85	*14.25*	0.02						
	P9. Mean Temperature of Driest Quarter*	6.01	7.67	2.16				1.81	**24.28**	5.31
	P10. Mean Temperature of Warmest Quarter*	0.91	**15.03**	0.00				9.95	**15.02**	1.75
	P11. Mean Temperature of Coldest Quarter	7.68	4.83	4.42						
Precipitation	P12. Annual Precipitation*	**8.58**	0.06	10.41	**18.14**	5.53	2.13	5.96	**12.26**	4.14
	P13. Precipitation of Wettest Period	5.29	0.83	**16.18**	11.32	*21.87*	1.59			
	P14. Precipitation of Driest Period*	6.10	3.75	2.36	**16.73**	7.67	2.59	**15.97**	5.59	2.29
	P15. Precipitation Seasonality (Coefficient of Variation)*	3.51	7.94	2.13	6.97	**24.60**	0.01	**16.48**	0.62	9.38
	P16. Precipitation of Wettest Quarter	5.11	0.85	**16.20**	11.52	*20.38*	3.91			
	P17. Precipitation of Driest Quarter*	6.44	3.70	1.16	**16.15**	9.92	1.65	**15.81**	6.03	1.02
	P18. Precipitation of Warmest Quarter	2.17	1.09	**18.58**	10.39	1.90	**51.51**	6.99	1.94	**38.93**
	P19. Precipitation of Coldest Quarter	5.10	1.39	0.29	8.77	8.12	**36.61**			

+Three categories (F1–F3) are used to analyze the 19 bioclimatic variables. Three main components and the percentage of explained variance are indicated for each category.

*Variables used in the third analysis (selected because of being strictly drought-related variables).

- Bold numbers: Variables with significant contribution in the definition of the respective component and pertinent for drought stress estimation.
- Bold and italic numbers: Variables with significant contribution in the definition of the respective component but not conceptually pertinent for drought stress estimation.

$$I = \text{annual heat index} = \begin{cases} \sum_{i=1}^{12} \left(\frac{T_j}{5}\right)^{1.514} & if\ T_j > 0 \\ 0\ if\ T_j \leq 0 \end{cases}$$

$$a = \text{cubic function of } I = 6.75 * 10^{-7} * I^3 - 7.71 * 10^{-5} * I^2$$
$$+ 1.792 * 10^{-2} * I + 0.49239$$

L_j = value of adjustment of sunlight depending on the latitude

$$= \frac{D_j}{12}$$

$$D_j = \text{day duration (h)} = 24 - \left(\frac{24}{\pi}\right)$$
$$* \cos^{-1}\left(\frac{\sin\left(\frac{0.8333*\pi}{180}\right) + \sin\left(\frac{Latitude*\pi}{180}\right) * \sin(A_j)}{\cos\left(\frac{Latitude*\pi}{180}\right) * \cos(A_j)}\right)$$

$Latitude$ = latitude (+ / − decimal degrees)

$$A_j = \sin^{-1}(0.39795 * \cos(0.2163108 + 2 * \tan^{-1}(0.9671396$$
$$* \tan(0.00860 * (J_j - 186)))))$$

J_j = day of the year (day 15 for month average)

On the other hand, Hamon approximation estimated PET based exclusively on temperature effects. In this sense, we used the following equation (for the month "j"):

$$PET_{Hamon,j} = 13.97 d_j D_j^2 W_{t,j}$$

Where

d_j = number of days in a month

D_j = monthly hours of daylight in units of 12 hrs

(view previous section)

$$W_{t,j} = \text{saturated water vapor density}\left(\frac{g}{m^3}\right) = \frac{4.95 e^{0.062 T_j}}{100}$$

T_j = monthly mean air temperature (°C)

Monthly drought indices were obtained by comparison of these PET estimators with monthly total precipitation (P). The following

drought index (DI) ratio was used (for the month "j" and the PET calculation approximation "i"):

$$DI_{j,i} = 100\left(\frac{PET_{j,i} - P_j}{PET_{j,i}}\right)\ where\ -100 < DI < 100$$

Finally, normalized annual mean drought index ($\overline{DI}_i$) and annual maximum drought index ($\max[DI_{j,i}]$) were determined for each accession following both strategies to calculate the PET. The first pretends to analyze long term theoretical habitat drought, while the second one explores short term and sporadic drought.

Results

Clusters of Environmental Variability

The wild accessions were distributed among different precipitation regimes that followed a latitudinal gradient from North to South America. Thus, accessions from Central America and northwest South America (regions near the equator) were associated with higher precipitation (annual precipitation and precipitation of the wettest quarter), while accessions from northern Mexico and central Andean regions found in Argentina, Bolivia and Peru (near the tropics of Cancer and Capricorn) were associated with lower precipitation (figure 1 and 2). Additionally, wild common beans occupied more geographical regions with extensive drought stress than cultivated accessions (t = 3.21, p-value = 0.0014, n = 399).

Five sub-populations have been recognized within wild common beans: Mesoamerica (Mexican wilds), Guatemala, Colombia, Ecuador and North Peru, and Andean (wilds from Argentina, Bolivia and southern Peru) sub-population per Blair et al. [26] and Broughton et al. [27]. The clustering of genetic groups within each of the wild sub-populations also followed a latitudinal gradient except for a pair of accessions collected in Peru that belonged to the Mesoamerican sub-population and two accessions collected in Mexico that belonged to the Colombian and Andean sub-populations. In these cases, the precipitation regime at the collection sites for these accessions was similar to the precipitation pattern associated with the whole sub-population (figure S1). For example, Mesoamerican and Andean wild populations were generally restricted to low precipitation habitats, whereas Colombia and Guatemala populations were distributed in higher precipitation habitats, although these could be found along wider gradients of total rainfall (table 1 and 2).

Precipitation of the driest and wettest periods, mean and maximum Thornthwaite Drought Index, and mean and maximum Hamon Drought Index of each sub-population in biplot analysis confirmed the separation of the sub-populations into ecological niches (table 2 and figure 3). Each of the main components (F1-3) showed a slightly different pattern of environmental variation in relation with population structure, as is depicted in table 2. Among the three analytical sets which were considered (all bioclimatic variables, precipitation variables and drought-related variables) the first explained population structure the best (see dashed divisions in table 2 and variables in bold). The behavior of the variables in predicting wild accessions sub-grouping was confirmed by the clustering analysis with all bioclimatic variables (figure S2). However, the resolution provided by the PCA and clustering analysis to discern between natural populations was not comparable with that achieved by the use of drought indices. Analysis of variance (table 2) confirmed these observations.

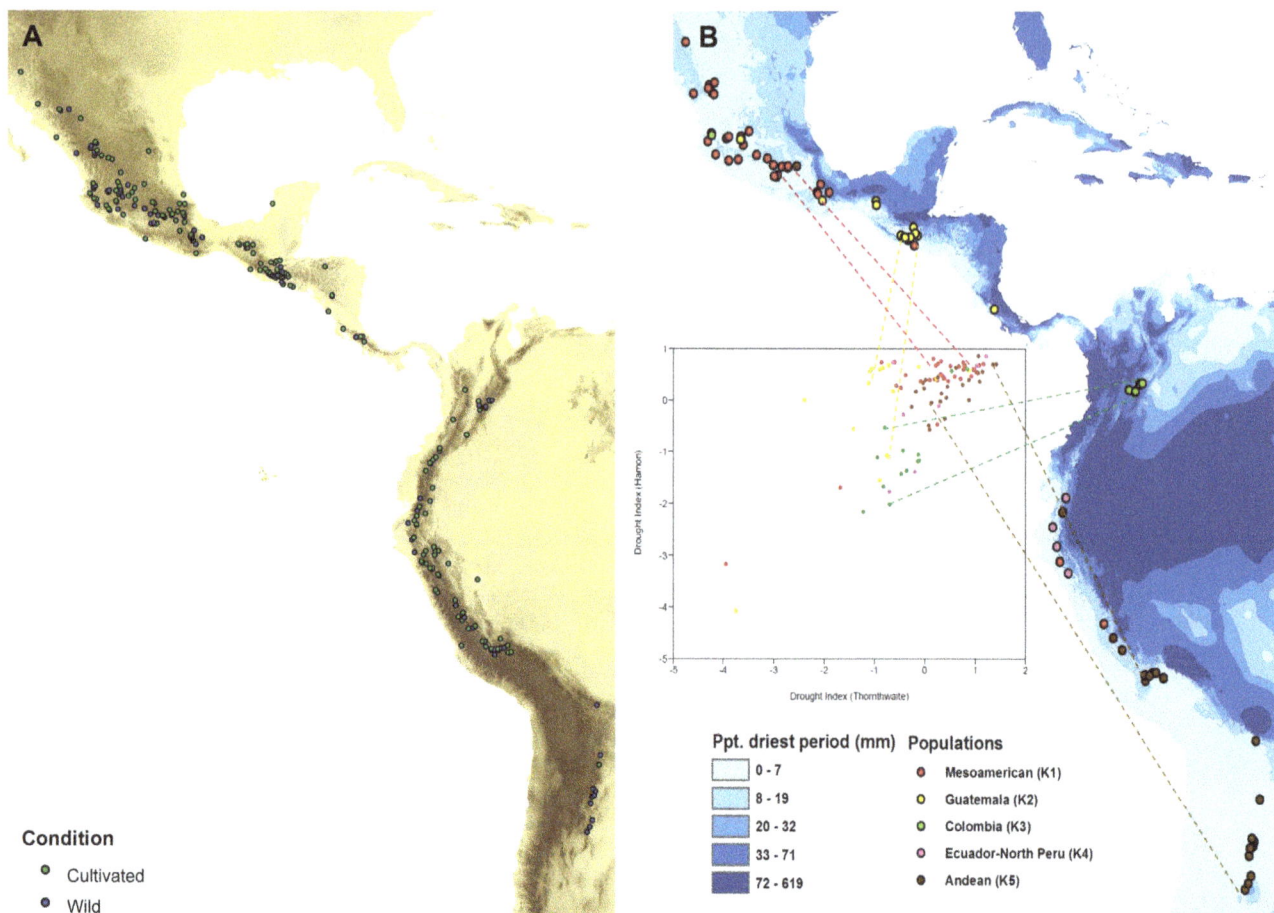

Figure 1. Geographic distribution of wild (104 accessions) and cultivated (297 accessions) common bean accessions (A), and precipitation during the driest period along the geographic range of wild common bean (B). A dispersion diagram between the estimated drought index using the potential evapo-transpiration (PET) of Thornthwaite and the estimated drought index using the PET of Hamon is presented in B. Populations definition as in Blair *et al.* [26] and Broughton et al. [27].

Comparison of Methods to Estimate Drought Tolerance

Among the main components of the PCA analysis for all bioclimatic variables, the first two were not determined by variables conceptually meaningful in the context of drought stress estimation, such as mean temperature of driest quarter and precipitation of warmest quarter (table 1). However, the third component was significantly determined by these variables, although it explained only 13% of total variation. Annual precipitation was present in all the components and therefore

did not offer discriminatory power. Variables related with drought stress contributed considerably to the first components of the PCA analysis for the two additional subsets of bioclimatic variables (precipitation-related variables and strictly drought related variables). The contribution of drought stress relevant variables to the other two components of both subsets was lower. We observed that the first component of the PCA analysis for precipitation variables was a good estimator of long term habitat drought stress, while the second component for the drought-related variables was a good estimator of short term and sporadic drought stress.

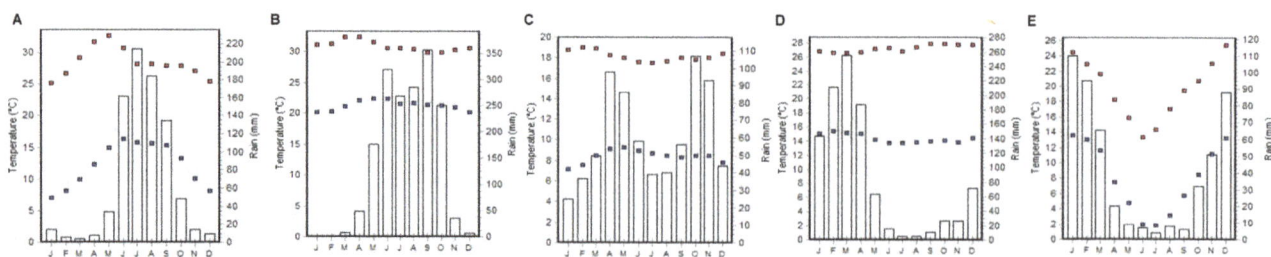

Figure 2. Temporal variation of precipitation (bars), maximum temperature (red squares) and minimum temperature (blue squares) in different representative regions: A. Mexico (−102° latitude, 20° longitude), B. Guatemala (−90°, 14°), C. Colombia (−74°, 4°), D. Ecuador-North Peru (−80°, −4°) and E. Argentina (−65°, −24°).

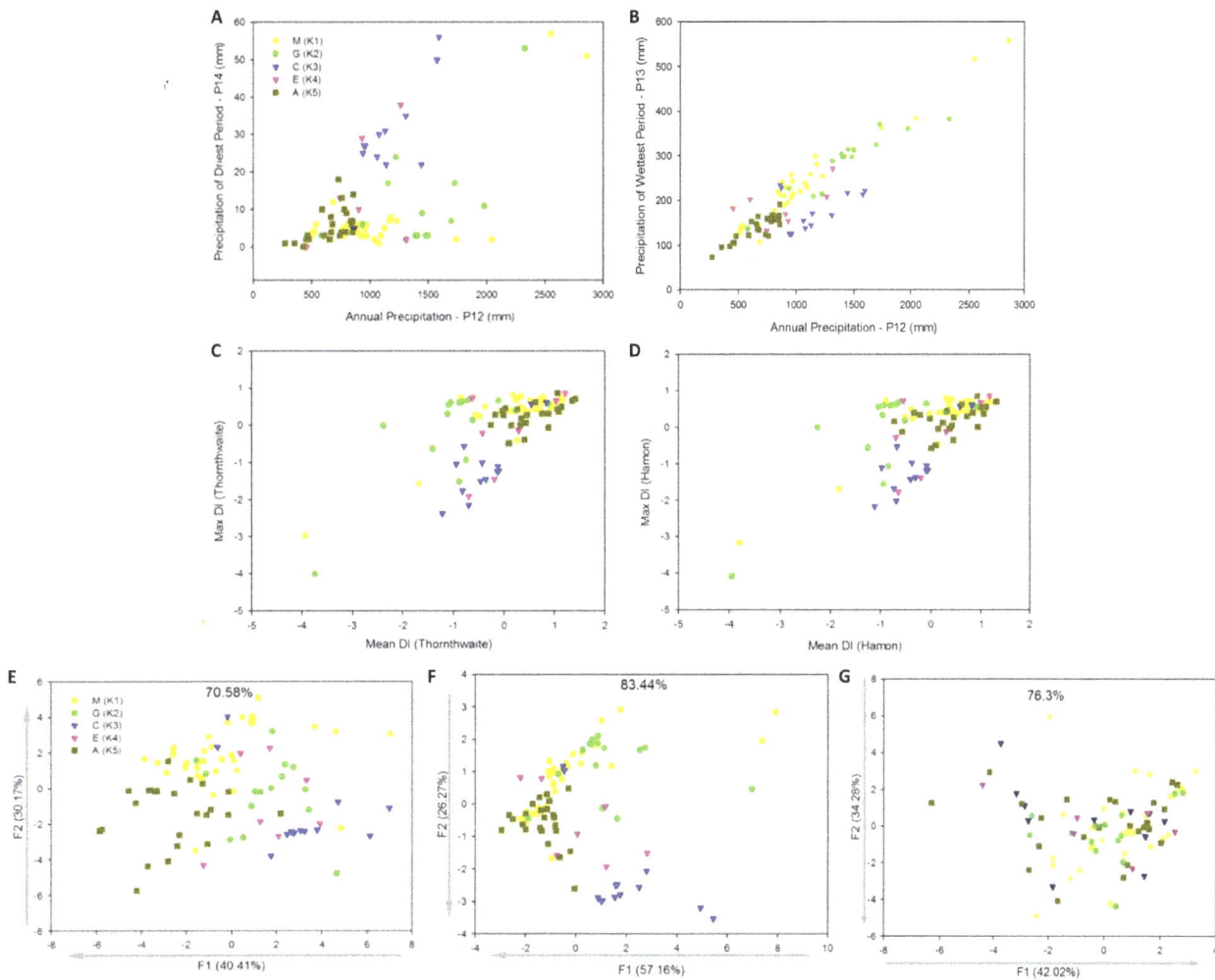

Figure 3. Scatter plots for: A. mean annual precipitation (P12) and precipitation of the driest period (P14), B. mean annual precipitation (P12) and precipitation of the wettest period (P13), C. mean and maximum Thornthwaite Drought Index (DI), D. mean and maximum Hamon DI, E. two main components of the PCA for all bioclimatic variables (P1–P19– table 1), F. two main components of the PCA for precipitation related bioclimatic variables (P12–P19– table 1), and G. two main components of the PCA for drought-related bioclimatic variables (table 1). Arrows indicate the increase in the estimated drought stress for each component. Wild populations: M: Mesoamerican, G: Guatemala, C: Colombia, E: Ecuador-North Peru and A: Andean. Numbers in E, F and G are percentage of explained variation by each component.

Table 2. Pairwise comparisons of significant variation for each bioclimatic variable, component and drought severity estimator in relation with population structure (p-value<0,001).

	P1	P2	P3	P4	P5	P6	P7	P8	P9	P10	P11	P12	P13	P14	P15	P16	P17	P18	P19	DIT	DIT max	DIH	DIH max	F1P	F2P	F1S	F2S
M (K1)	A	A	–C	A	A	–B	A	A	A	A	A	–B	–B	–B	A	AB	–B	–B	–B	AB	A	A	A	–B	A	A	A
G (K2)	A	–B	–BC	–B	–B	AB	–B	–B	A	–BC	A	A	A	–B	–B	A	–B	A	–B	–C	–BC	–B	–BC	A	A	–BC	A
C (K3)	A	–B	AB	–B	–B	A	–B	AB	A	AB	A	AB	–BC	A	–C	–BC	A	AB	A	–BC	–C	AB	–C	A	–C	–C	A
E (K4)	A	–B	A	–B	–B	AB	–B	AB	A	–BC	A	–BC	–BC	AB	AB	–BC	–B	AB	AB	AB	ABC	A	ABC	AB	–B	–BC	A
A (K5)	–B	A	–C	A	–B	–C	A	–B	–B	–C	–B	–C	–C	–B	–B	–C	–B	–B	–B	A	AB	A	AB	–B	–B	–B	–B

Variable abbreviations: P1–P19: main variables as defined in table 1, DIT: Annual Mean Drought Index (Thornthwaite), DIT, Max: Maximum Drought Index (Thornthwaite), DIH: Annual Mean Drought Index (Hamon), DIH, Max: Maximum Drought Index (Hamon), F1P, F2P: Main two factors for bioclimatic precipitation variables (P12–P19), F1S, F2S: Main two factors for drought-related variables.

Kruskal-Wallis tests were applied in all cases except for DI (Drought Index) estimations, where an ANOVA followed by a Tukey's-b post-hoc test was used. A, B and C are different ranks. Populations with more than one letter could not be assigned to a single rank. Mean for each variable for each population: A>B>C.

Bold variables: Drought-related variables. Selected variables for further analysis based on their conceptual power to describe drought tolerance.

This pattern was corroborated by Pearson's correlation (r) and Spearman's rank correlation coefficients (ρ) tests for the significant PCA components with annual mean Drought Index and maximum Drought Index (from Thornthwaite and Hamon), respectively (table 3). The three main components were correlated with temperature and precipitation variables, whereas drought indices based on evapotranspiration were only correlated with precipitation variables, as would be expected. Pearson's and Spearman's coefficients provide the same pattern of correlation, which means that the assumption of normality is not a strong premise for our other analysis.

Finally, in the evaluation of the two indices of potential evapotranspiration obtained from the bioclimatic variables we found both to be similarly predictive. The correspondence between Thornthwaite and Hamon drought index estimators (normalized annual mean and annual maximum) was indicated by the correlation analysis where r-values were of 0.99 and 1.00 showing them to be nearly analogous estimators. Additionally, the value of these indices to detect long and short drought stress for a habitat based on potential evapotranspiration is shown by the correlation with mean and maximum annual precipitation (Table 1 and figure S2).

Discussion

The evaluation of drought physiology traits in wild common bean populations would have been impractical due to their long growth cycle and low biomass. Hence, the ecological analysis of wild bean accessions geographical origin performed here was useful in successfully predicting the drought tolerance of these genotypes in a case where other sources of information were not available. Two particular issues were considered: the way in which ecological variation is structured into natural populations, and the ideal and unequivocal estimator of drought tolerance.

Ecological Diversity is Structured along the Populations of Wild Bean

The analysis disclosed three non-overlapping categories of drought tolerance associated with population structure and extensively correlated with a latitudinal pattern. Correlation between ecology and geographical distance is a common phenomenon in natural populations which responds to isolation of sub-populations [29]. This is a consequence of independent evolution in different subpopulations of a species evolving towards adaptation to specific microclimatic conditions [30,31]. Moreover, random accumulation of genetic variability is uneven along populations because of genetic drift and bottlenecks. Consequently, the genetic resources available in each population to deal with new or old selective forces are dissimilar between groups or demes. The evolution pathway followed by each population is therefore unique [32]. This is the foundation of the adaptive radiation hypothesis according to which meta-population structure will favor adaptive and diversifying selection.

Previous research from Tiranti and Negri [33] demonstrated that selective microenvironmental effects play a role in shaping genetic diversity and structure in common bean wild accessions. In this study we have confirmed that ecological diversity is associated with structuring into natural populations in wild beans. In contrast, cultivated accessions of common bean are mainly structured into races, with less explanation of diversity by latitudinal shifts or gradients. This difference between wild and cultivated common bean might be a consequence of continental level rainfall patterns. Specifically, in tropical environments near the equator with bimodal rainfall a mid-season dry period occurs that can last two to four weeks. In contrast in the sub-tropics, a dry period of three or more months can occur. In response to this mid-cycle drought of the sub-tropics, wild *P. vulgaris* enters a survival mode of slow growth and reduced physiological activity until

Table 3. Pearson's correlation coefficients (r – above the diagonal) and Spearman's rank correlation coefficients (ρ – below the diagonal) among some representative climatic variables, components and drought severity estimators.

	DIT	DIT max	DIH	DIH max	F1T	F2T	F1P	F2P	F1S	F2S	P12	P14	P1	P9
DIT		0.73	0.99	0.74	-0.71	0.4	-0.9	-0.14	0.73	-0.39	-0.91	-0.7	0.03	-0.14
DIT max	0.55		0.73	1	-0.6	0.59	-0.82	0.43	0.88	-0.29	-0.57	-0.96	0.15	0.02
DIH	0.99	0.56		0.75	-0.67	0.45	-0.89	-0.13	0.76	-0.33	-0.89	-0.7	0.09	-0.07
DIH max	0.57	1.00	0.58		-0.59	0.6	-0.83	0.41	0.89	-0.28	-0.58	-0.96	0.17	0.03
F1T	-0.71	-0.33	-0.67	-0.33		0	0.84	-0.01	-0.47	0.84	0.81	0.68	0.56	0.68
F2T	0.41	0.66	0.46	0.68	-0.01		-0.28	0.58	0.85	0.5	-0.06	-0.46	0.8	0.66
F1P	-0.89	-0.60	-0.86	-0.60	0.83	-0.25		0	-0.72	0.6	0.91	0.87	0.18	0.29
F2P	-0.17	0.46	-0.14	0.46	0.15	0.60	0.13		0.47	0.21	0.34	-0.4	0.32	0.25
F1S	0.69	0.82	0.72	0.83	-0.36	0.89	-0.59	0.44		0	-0.5	-0.82	0.44	0.28
F2S	-0.38	-0.01	-0.32	-0.01	0.83	0.47	0.59	0.35	0.12		0.65	0.44	0.88	0.91
P12	-0.89	-0.33	-0.86	-0.34	0.82	-0.03	0.91	0.41	-0.37	0.68		0.63	0.3	0.4
P14	-0.57	-0.94	-0.57	-0.94	0.45	-0.50	0.69	-0.38	-0.70	0.20	0.43		0	0.12
P1	-0.01	0.34	0.05	0.34	0.54	0.77	0.22	0.39	0.49	0.85	0.36	-0.14		0.95
P9	-0.16	0.24	-0.10	0.24	0.66	0.65	0.35	0.35	0.36	0.91	0.48	-0.03	0.96	

Bold: significant values: <0.05 for r or <-0.5 and >0.5 for ρ.
DI_T: Normalized Annual Thornthwaite Drought Index.
DI_H: Normalized Annual Hamon Drought Index.
$DI_{max\ N}$: Normalized Maximum Month Drought Index (Thornthwaite (T) or Hamon (H)).
F#i: Two main components using all bioclimatic variables (i = T), only precipitation variables (i = P), or only drought-related variables (i = S) (table 1).
Original Control Bioclimatic Variables: P12: Annual Precipitation, P14: Precipitation of Driest Period, P1: Annual Mean Temperature, P9: Mean Temperature of Driest Quarter.

rainfall resumes and flowering occurs [7]. Cultivated beans on the other hand are less frequently subjected to these environmental pressures and tend to mature in a shorter length of time.

Interestingly, we observed that wild common bean occupy more geographical regions with extensive drought stress than cultivated accessions. Those regions include the arid areas of Peru, Bolivia and Argentina, and the valleys of northwest Mexico. In addition, it is necessary to emphasize that the correlation between population structure and climatic variability could also be a partial consequence of other correlated latitudinal variation not necessarily driven by day length and temperature. Hence, population structure as well as climatic variability constraints must be taken into account to analyze genetic variation in relation with theoretical drought stress of each habitat.

In summary, we have detected a broad habitat distribution for wild common beans that is useful for drought tolerance. Cultivated common bean is traditionally considered susceptible to drought, but that seems not to be the case for wild common beans. In addition, some differences must exist between the adaptations of wild populations to arid regimes which are reflected in the subpopulations found in different ecologies. Several of them are valuable for plant breeding. Therefore, we propose, as was suggested by Acosta et al. [34], that wild common bean be taken into account to exploit variation for drought tolerance, however care is needed to avoid the reduction in yield associated with the wild bean genotype.

Thornthwaite and Hamon Drought Estimators Perform Similarly

Environmental analysis provided us with a set of non-redundant variables useful to describe long and short-term theoretical habitat drought stress. It was convenient to consider estimators based on potential evapotranspiration because of the conceptual power of this approach. Besides, it was appropriate to include the two main components from the subsets of bioclimatic variables related with precipitation or drought stress, because these emphasize temperature and precipitation, while other estimators only emphasize precipitation. These components allowed us to test their effect on the global analysis. Finally, it was practical to incorporate annual precipitation and precipitation of the driest period because these variables gave us a direct idea of long and short-term drought stress. All these estimators were congruent with visual inspections over precipitation maps for the area of geographical origin of the wild accessions. This is in line with the fact that bioclimatic variables are all different combinations of monthly air temperature and monthly total precipitation.

Some divergence between the Hamon and Thornthwaite models and tests with the different subsets of bioclimatic variables have demonstrated the possible ways to exploit environmental variability in order to infer different aspects of drought stress for the differed habitats. Therefore, the scope of the application will determine which metric is adequate. For example, the Thornthwaite drought index and the first component of the PCA analysis that used only bioclimatic variables directly associated with drought tolerance are good estimators of short term and sporadic drought stress. However, the Hamon drought index and the first component of the PCA analysis that used only precipitation variables are the best estimators of long-term drought stress. On the contrary, the second and third main components of the PCA analysis that included all the bioclimatic variables have low power and specificity to detect any kind of drought stress. Overall, robustness and resolution to discern between subpopulations were more extensive for the Thornthwaite and Hamon estimators than for the other two components. Thus, the former estimators should be preferred.

Some theoretical issues remain in order to guarantee the pertinence of each estimator. First, one must consider the link between habitat/geographical origin drought stress and plant drought tolerance. Two aspects modulate this relationship: 1) abiotic stress is a highly genotype×environment and plant species dependent phenomenon [12,35], and 2) the collection site of a genotype in a semi-arid habitat does not make it necessarily drought tolerant. Several assumptions in the PET modeling must also be considered to access the boundaries in inferences made by the models [12]. Namely, our estimated drought stress is useful in a comparative perspective. It must not be used to make inter-specific comparisons because stress is a plant-specific perception and not a site characteristic, and because we are not including any soil water dynamics (by assuming that all precipitation water is potentially available to the plants). Another assumption is that plant distribution must be in equilibrium with niche requirements and ecological forces [36,37], so that the errant presence of poorly adapted genotypes can be discarded.

A further consideration is the relationship between habitat ecology and drought stress. For example, precipitation patterns could be more related with the incidence of plant pathogens and the consequent biotic stresses than with drought stress [38]. To avoid this limitation we suggest rejecting estimators based on non-drought-related bioclimatic variables. Furthermore, we suggest using model-based estimators that consider the specific ways in which environmental variables can modulate drought stress. Hence, this is another argument to prefer Thornthwaite and Hamon drought indexes over the estimators derived from the PCA analysis.

In terms of selecting the best model, Thornthwaite and Hamon estimators were complementary. The Thornthwaite model takes into account latitudinal variation in addition of radiation and temperature [14,16]. Meanwhile, the Hamon model focused on the latter two [15]. However, given the high correlation that is expected between day length, latitude, and seasonal temperature, the high consistency between both models does not turn surprising. In order to be able to consider both short and long term drought events, we also propose the use of the maximum monthly drought index and the normalized average drought index, keeping in mind the limitations described in the previous paragraphs.

In summary, we have estimated short and long term drought stress for the habitats and geographical origin of wild common bean accessions using multivariate methods and physiological (PET) modeling techniques. The habitat drought stress index based on the Thornthwaite and Hamon PET models are equivalent and are promising as predictors of overall drought tolerance. Recent examples illustrate how this resource should be coupled with considerations about population structure as a way to identify and exploit natural variation [39,40,41,42,43]. This will ultimately facilitate oncoming genealogical analysis and genome-wide genetic-environmental association studies that aim predicting fitness in wild populations [43,44,45].

Supporting Information

Figure S1 Geographic distribution for wild common bean accessions in relation with rainfall. A. Wild common bean populations and precipitation in the driest period (mm) for the entire range of distribution, B. for Mexico, C. and for Peru, Bolivia and Argentina. D. Precipitation in the wettest period (mm) and total annual rainfall (mm).

Figure S2 Dendogram of accessions constructed using the Pearson's correlation coefficient and the middle joint method for all bioclimatic variables (P1–P19,

table 1). Accession names contain: accessions number+population assignation (Mesoamerican (Mexican wilds): K1, Guatemala: K2, Colombia: K3, Peru and Ecuador: K4, Andean (wilds from Argentina, Bolivia and southern Peru): K5), as defined by Blair *et al.* [26] and Broughton et al. [27]+quintiles for habitat drought stress (annual mean Thornthwaite Drought Index (DI), maximum Thornthwaite DI, annual mean Hamon DI, maximum Hamon DI). Branch colors are based on population structure. Red lines indicate groups of accessions with overall high quintiles.

Acknowledgments

The authors wish to thank the personnel of the Genetic Resource Unit for seed collection, cataloguing, characterization and multiplication. We are also grateful with the three anonymous referees for reading and commenting on this manuscript and with the publication fund of Facultad de Ciencias at Universidad de los Andes. The Genomes and Phenotypes Gradschool at Uppsala University is also acknowledged for the scholarship that allowed the first author to go to the ECMTB 2011 meeting (Krakow), were some of the ideas presented in this manuscript were refined.

Author Contributions

Conceived and designed the experiments: AJC MWB. Performed the experiments: AJC FAM JR. Analyzed the data: AJC FAM JR. Contributed reagents/materials/analysis tools: SM MWB. Wrote the paper: AJC MWB.

References

1. Singh SP (2005) Common Bean (*Phaseolus vulgaris* L.). In: Singh RJ, Jauhar PP, editors. Genetic Resources, Chromosome Engineering, and Crop Improvement, Grain Legumes. London: CRC Press.
2. Gepts P, Aragão F, Barros E, Blair M, Brondani R, et al. (2008) Genomics of *Phaseolus* Beans, a Major Source of Dietary Protein and Micronutrients in the Tropics. In: Moore P, Ming R, editors. Genomics of Tropical Crop Plants. New York: Springer.
3. Lane A, Jarvis A (2007) Changes in climate will modify the geography of crop suitability : Agricultural biodiversity can help with adaptation. SAT eJournal 4: 1–11.
4. Vermeulen S, Zougmore R, Wollenberg E, Thornton P, Nelson G, et al. (2012) Climate change, agriculture and food security: A global partnership to link research and action for low-income agricultural producers and consumers. Current Opinion in Environmental Sustainability 4: 128–133.
5. Debouck DG. Biodiversity, ecology and genetic resources of *Phaseolus beans* – Seven answered and unanswered questions. In: Oono K, editor; 2000; Tsukuba, Japan. National Institute of Biological Resources. 95–123.
6. Debouck DG (1999) Diversity in *Phaseolus* species in relation to the common bean. In: Singh SP, editor. Common bean improvement in the twenty first century. Dordrecht, The Netherlands: Kluwer Academic Publishers. 25–52.
7. Beebe S, Rao IM, Cajiao C, Grajales M (2008) Selection for Drought Resistance in Common Bean Also Improves Yield in Phosphorus Limited and Favorable Environments. Crop Science 48: 582–592.
8. Frahm MA, Rosas JC, Mayek-Perez N, Lopez-Salinas E, Acosta-Gallegos JA, et al. (2004) Breeding beans for resistance to terminal drought in the lowland tropics. Euphytica 136: 223–232.
9. Jarvis A, Ferguson ME, Williams DE, Guarino L, Jones PG, et al. (2003) Biogeography of Wild *Arachis*: Assessing Conservation Status and Setting Future Priorities. Crop Science 43: 1100–1108.
10. Jarvis A, Williams D, Hyman G, Vargas I, Williams K, et al. (2002) Spatial analysis of wild peanut distributions and the implications for plant genetic resources conservation. Plant Genetic Resources Newsletter 121: 29–35.
11. Jones P, Jarvis A, Hyman G, Beebe SE, Pachico DH (2007) Climate proofing agricultual research investments. SAT eJournal 4: 1–29.
12. Bartels D, Sunkar R (2005) Drought and salt tolerance in plants. Critical Reviews in Plant Sciences 24: 23–58.
13. Xu C, Singh VP (2002) Cross Comparison of Empirical Equations for Calculating Potential Evapotranspiration with Data from Switzerland. Water Resources Management 16: 197–219.
14. Thornthwaite CW, Mather J, R. (1957) Instructions and Tables for Computing Potential Evapotranspiration and the Water Balance. Climatology 10: 185–311.
15. Hamon WR (1961) Estimating Potential Evapotranspiration. Proceedings of the American Society of Civil Engineers 87: 107–120.
16. Thornthwaite CW, Mather JR (1955) The Water Balance. Climatology 8: 1–104.
17. Amede T, Schubert S (2003) Mechanisms of drought resistance in grain legumes II: Stomatal regulation and root growth. Ethiopian Journal of Science 26: 137–144.
18. Griffiths H, Parry MAJ (2002) Plant responses to water stress. Annals of Botany 89: 801–880.
19. Munns R (2002) Comparative physiology of salt and water stress. Plant, Cell and Environment 25: 239–250.
20. Heyman DP, Sobel MJ (1990) Stochastic models. Amsterdam: North-Holland. 723 p.
21. Knowles LL, Maddison WP (2002) Statistical phylogeography. Molecular Ecology 11: 2623–2635.
22. Pérez JC, Monserrate F, Beebe S, Blair M (2008) Field evaluation of a common bean reference collection for drought tolerance. In: CIAT, editor. Annual Report Outcome Line SBA-1 Product 2: Beans that are more productive in smallholder systems of poor farmers. Palmira, Colombia: CIAT.
23. Tohme J, González O, Beebe S, Duque MC (1996) AFLP Analysis of Gene Pools of a Wild Bean Core Collection. Crop Science 36: 1375–1384.
24. Tohme J, Jones P, Beebe S, Iwanaga M (1995) The combined use of agroecological and characterization data to establish the CIAT *Phaseolus vulgaris* core collection. In: Hodgkin T, Brown AH, van Hintum JL, Morales EA, editors. Core Collections of Plant Genetic Resources. New York: John Wiley & Sons.
25. Blair M, Diaz LM, Buendia HF, Duque MC (2009) Genetic diversity, seed size associations and population structure of a core collection of common beans (*Phaseolus vulgaris* L.). Theoretical and Applied Genetics 119: 955–972.
26. Blair MW, Soler A, Cortés AJ (2012) Diversification and Population Structure in Common Beans (*Phaseolus vulgaris* L.). Plos One 7: e49488.
27. Broughton WJ, Hernandez G, Blair M, Beebe S, Gepts P, et al. (2003) Beans (*Phaseolus spp.*) - model food legumes. Plant and Soil 252: 55–128.
28. Hijmans R, Guarino L, Cruz M, Rojas E (2002) Computer tools for spatial analysis of plant genetic resources data: 1. DIVA-GIS. Plant Genetic Resources Newsletter 127: 15–19.
29. Kotlik P, Markova S, Choleva L, Bogutskaya NG, Ekmekci FG, et al. (2008) Divergence with gene flow between Ponto-Caspian refugia in an anadromous cyprinid *Rutilus frisii* revealed by multiple gene phylogeography. Molecular Ecology 17: 1076–1088.
30. Chen CN, Chiang YC, Ho THD, Schaal BA, Chiang TY (2004) Coalescent processes and relaxation of selective constraints leading to contrasting genetic diversity at paralogs *AtHVA22d* and *AtHVA22e* in *Arabidopsis thaliana*. Molecular Phylogenetics and Evolution 32: 616–626.
31. Gepts P, Papa R (2003) Possible effects of (trans)gene flow from crops on the genetic diversity from landraces and wild relatives. Environmental biosafety research 2: 89–103.
32. van der Sluijs I, Van Dooren TJM, Hofker KD, van Alphen JJM, Stelkens RB, et al. (2008) Female mating preference functions predict sexual selection against hybrids between sibling species of cichlid fish. Philosophical Transactions of the Royal Society B-Biological Sciences 363: 2871–2877.
33. Tiranti B, Negri V (2007) Selective microenvironmental effects play a role in shaping genetic diversity and structure in a *Phaseolus vulgaris* L. landrace: implications for on-farm conservation. Molecular Ecology 16: 4942–4955.
34. Acosta JA, Kelly JD, Gepts P (2007) Prebreeding in common bean and use of genetic diversity from wild germplasm. Crop Science 47: S44–S59.
35. Seki M, Kamei A, Yamaguchi-Shinozaki K, Shinozaki K (2003) Molecular responses to drought, salinity and frost: common and different paths for plant protection. Current Opinion in Biotechnology 14: 194–199.
36. Brady KU, Kruckeberg AR, Bradshaw HD (2005) Evolutionary ecology of plant adaptation to serpentine soils. Annual Review of Ecology Evolution and Systematics 36: 243–266.
37. Valladares F, Niinemets U (2008) Shade Tolerance, a Key Plant Feature of Complex Nature and Consequences. Annual Review of Ecology Evolution and Systematics 39: 237–257.
38. Diévart A, Clark SE (2004) LRR-containing receptors regulating plant development and defense. Development 131: 251–261.
39. Cortés AJ, Chavarro MC, Blair MW (2011) SNP marker diversity in common bean (*Phaseolus vulgaris* L.). Theoretical and Applied Genetics 123: 827–845.
40. Cortés AJ, Chavarro MC, Madriñán S, This D, Blair MW (2012) Molecular ecology and selection in the drought-related *Asr* gene polymorphisms in wild and cultivated common bean (*Phaseolus vulgaris* L.). BMC Genetics 13:58.
41. Cortés AJ, This D, Chavarro C, Madriñán S, Blair MW (2012) Nucleotide diversity patterns at the drought-related *DREB2* encoding genes in wild and cultivated common bean (*Phaseolus vulgaris* L.). Theoretical and Applied Genetics 125: 1069–1085.
42. Galeano CH, Cortés AJ, Fernandez AC, Soler A, Franco-Herrera N, et al. (2012) Gene-Based Single Nucleotide Polymorphism Markers for Genetic and Association Mapping in Common Bean. BMC Genetics 13: 48.
43. Blair MW, Cortés AJ, Penmetsa RV, Farmer A, Carrasquilla-Garcia N, et al. (2013) A high-throughput SNP marker system for parental polymorphism screening, and diversity analysis in common bean (*Phaseolus vulgaris* L.). Theoretical and Applied Genetics 126: 535–548.

44. Hancock AM, Brachi B, Faure N, Horton MW, Jarymowycz LB, et al. (2011) Adaptation to Climate Across the *Arabidopsis thaliana* Genome. Science 334: 83–86.

45. Kelleher CT, Wilkin J, Zhuang J, Cortés AJ, Quintero ÁLP, et al. (2012) SNP discovery, gene diversity, and linkage disequilibrium in wild populations of *Populus tremuloides*. Tree Genetics & Genomes 8: 821–829.

11

Seasonal Exposure to Drought and Air Warming Affects Soil Collembola and Mites

Guo-Liang Xu[1]*, Thomas M. Kuster[2], Madeleine S. Günthardt-Goerg[2], Matthias Dobbertin[3], Mai-He Li[4]

1 Key Laboratory of Vegetation Restoration and Management of Degraded Ecosystems, South China Botanical Garden, Chinese Academy of Sciences, Guangzhou, China, **2** Forest Dynamics, Swiss Federal Research Institute WSL, Birmensdorf, Switzerland, **3** Long-term Forest Ecosystem Research, Swiss Federal Research Institute WSL, Birmensdorf, Switzerland, **4** Tree Physioecology, Swiss Federal Research Institute WSL, Birmensdorf, Switzerland

Abstract

Global environmental changes affect not only the aboveground but also the belowground components of ecosystems. The effects of seasonal drought and air warming on the genus level richness of Collembola, and on the abundance and biomass of the community of Collembola and mites were studied in an acidic and a calcareous forest soil in a model oak-ecosystem experiment (the Querco experiment) at the Swiss Federal Research Institute WSL in Birmensdorf. The experiment included four climate treatments: control, drought with a 60% reduction in rainfall, air warming with a seasonal temperature increase of 1.4°C, and air warming + drought. Soil water content was greatly reduced by drought. Soil surface temperature was slightly increased by both the air warming and the drought treatment. Soil mesofauna samples were taken at the end of the first experimental year. Drought was found to increase the abundance of the microarthropod fauna, but reduce the biomass of the community. The percentage of small mites (body length $\leq$ 0.20 mm) increased, but the percentage of large mites (body length >0.40 mm) decreased under drought. Air warming had only minor effects on the fauna. All climate treatments significantly reduced the richness of Collembola and the biomass of Collembola and mites in acidic soil, but not in calcareous soil. Drought appeared to have a negative impact on soil microarthropod fauna, but the effects of climate change on soil fauna may vary with the soil type.

Editor: Martin Krkosek, University of Otago, New Zealand

Funding: This work was supported by the Knowledge Innovation Program of the Chinese Academy of Sciences (No. KSCX2-EW-J-28) and the National Natural Science Foundation of China (no. 40801096). G.L. Xu's research stay in Switzerland was supported by Sino-Swiss Science and Technology Cooperation Program (project No. EG07-092009) which is gratefully acknowledged. The funders had no role in study design, data collection and analysis, decision to publish, or preparation of the manuscript.

Competing Interests: The authors have declared that no competing interests exist.

* E-mail: xugl@scbg.ac.cn

Introduction

Soil mesofauna exert strong regulatory control over the soil food web and have substantial effects on important soil characteristics, including the distribution of soil particles, the soil's water-holding capacity and water infiltration rate, the lability of organic compounds and mineralization, immobilization, the availability of N and other nutrients, the transport of compounds, and the composition, abundance, dispersal, and activity of bacteria and fungi [1–4]. Soil mesofauna are also particularly sensitive to environmental changes, and therefore thought to be an excellent bioindicator [5–7]. However, few studies on soil mesofauna have been made in the context of environmental change [8], in spite of its important ecosystem functions. This contrasts with the way above-ground effects [9], responses of soil microbial communities [10], litter decomposition [11], and nutrient cycling [12] have been extensively investigated. Environmental changes, such as changes in precipitation and air temperature, are likely to lead to changes in the mesofauna and could alter belowground biological processes, with potential consequences for ecosystem functions.

Warming and changes in precipitation amounts can strongly influence microarthropod reproduction and development rates by altering the soil temperature and moisture [8]. Soil moisture has often been reported to be the most important environmental

variable affecting both the structure and function of the soil fauna community [13–z16]. Generally, it seems that the abundance of all faunal groups decreases with drought. For example, soil mites, one of the most abundant group of mesofauna has been found to be positively related to soil moisture across many ecosystems [17–20]. Drought seems to be a limiting factor for Collembola. For example, drought periods were associated with a reduction (in absolute numbers and diversity) in Collembola species that dwell in the forest litter and the moss layer [6,14,21]. Moisture-induced shifts in the community composition of soil mesofauna may influence community structure, which in turn may affect body size distribution, although to our knowledge, this has not yet been explicitly explored. These changes may have important impacts on ecosystem functions, including changes in decomposition rate under future climatic changes [8].

Warming can alter the soil fauna community by leading to changes in the abundance and composition of soil bacteria and fungi, and influencing plant physiology and community structure. Changes in community structure are directly related to the resource availability and microhabitat conditions of the ecosystem [9,22]. According to several studies, the responses of soil fauna to warming vary with climate [8,23–27]. Wolters [28] found that, for ten consecutive years, the annual mean temperature was significantly correlated with alterations in collembolan density in a beech forest on limestone (Göttingen, Germany). Oribatida and predatory mites (Mesostigmata and Prostigmata), in particular the Oribatida species *Diapterobates notatus*, also tended to increase in number with warming [25]. However, the abundance of most Collembola, including *Hypogastrura tullbergi*, *Lepidocyrtus lignorum* and *Isotoma anglicana*, tended to reduce with warming [25]. Only minor changes in the soil fauna occurred at higher temperatures, even after 6 years of elevated temperature treatment [29]. These contradictory results may be due to the significant side-effects of drought, which have often been found to accompany experimental increases in temperature. Harte et al. [1] found that warming increased microarthropod abundance and biomass only under wet conditions, but not under dry conditions. In fact, increases in temperature are often associated with decreases in moisture, and the indirect effects of changes in soil moisture may be more important for the survival and reproductive ability of soil fauna than the direct effects of warming [30]. Thus targeted experiments and monitoring studies are needed to distinguish between the effects of warming and drought.

The current study was performed as part of the "Querco" experiment [31], in which combined air warming and/or drought treatments were studied in two types of forest soils. The overall goal of the Querco experiment is to understand the effects of climate change on near-natural oak model ecosystems at the Swiss Federal Research Institute WSL [32]. We hypothesized that the abundance and biomass of soil Collembola and mites, and the taxon richness of Collembola, will be decreased by drought but increased by warming without interaction.

Methods and Materials

Design of the "Querco" Experiment

The current study was conducted in 2007 at the Swiss Federal Research Institute WSL in Birmensdorf, Switzerland, as part of the Querco experiment [31]. The experimental set-up included 16 hexagonal model ecosystem chambers (6 m² in area, 3 m in height, and 1.5 m in soil depth), with roofs that close automatically during rainfall. The chambers were arranged in a Latin square with four treatments and four replications each. The treatments were control, air-temperature warming, drought, and air warming + drought. In the control treatment, opened chamber walls provided the same air temperature as at ambient conditions. In the air warming treatment, the opening of the chamber walls was reduced, which increased the air temperature in summer 2007 (June, July and August) passively by 1.4°C, in line with the IPCC A2 scenario for Switzerland [33–34]. Subsequently, soil temperature (−10 cm) was also increased, e.g. by 0.7°C in August 2007.

Irrigation was applied with the same ion composition as the 30-year precipitation mean of the site [31]. The irrigation regime was the same in the air-warming treatment as in the control. Sufficient water (according to the monitoring by TDR and tensiometers) was supplied by means of sprinklers at intervals of 2 to 3 days from May to October. In the drought and air-warming + drought treatment, the irrigation was interrupted several times (Fig. 1). Therefore, in these treatments, the amount of irrigation from April to October was 60% lower than the long-term mean at the site, as it used as a reference to the severe variation in the IPCC A2 scenario. Each chamber was divided into two soil-lysimeter compartments containing one of two forest soils with similar soil texture, namely, acidic loamy sand (haplic alisol, pH 4.1) or calcareous sandy loam (calcaric fluvisol, pH 7.3). The sandy forest soils (for details, see Kuster et al. [31]) were taken in autumn 2005 from sites stocked with adult oaks, and then passively homoge-

nized during transport and filling in the lysimeter compartments. In spring 2006, 24 2-year-old oak seedlings were planted in each soil compartment. Three common European oak species were selected, each with four provenances [32]. The oak seedlings were randomly distributed in each compartment. Treatments started in spring 2007 with the mesofauna samples taken at the end of the season.

Determination of Soil Water Content, Soil Temperature, Foliage and Root Biomass, and Soil Respiration

Volumetric soil water contents were measured using time domain reflectometry (TDR 100, Campbell Scientific Inc., USA) in each soil compartment at 0–25 cm depth at 1-week intervals throughout the growing season from June to December 2007. Soil temperatures were measured hourly below the soil surface at a depth of 1 cm with iButton temperature loggers (Maxim Integrated Products Inc., USA). Foliage biomass was sampled at the end of the growing season in 2007. The root biomass in 2007 could not be measured as the experiment was running afterwards for another two years. We therefore estimated the root biomass in 2007 by using the allometric relationship between foliage and root mass in 2009 and the foliage biomass sampled in 2007. The soil respiration rate was measured at permanent docking cylinders (diameter = 10 cm) in each soil subplot using a 6400-09 soil CO_2 flux chamber connected to an LI-6400 infrared gas analyser (both LI-Cor Bioscience Inc, Lincoln, Nebraska, USA). Each measurement was conducted three times in a row to average out short-term variations.

Sampling, Extraction and Identification of Soil Collembola and Mites

Soil cores (0–5 cm depth, 5 cm diameter) were collected from 22 October to 10 November 2007 with a steel cylinder at three locations in each soil compartment of three of the four replicate chambers. The three soil cores from each soil compartment in each chamber were combined to form a mixed sample. Immediately after collection, the soil samples were transported to the laboratory, and soil Collembola and mites were extracted using Tullgren dry funnels [35] for 48 hours. All specimens were sorted and counted with a dissecting microscope and examined with an Olympus BX41 research microscope. All Collembola were identified to genus level, mainly according to the keys in "Checklist of the collembolan of the world" [36], but also according to the keys in Potapov [37] and Bretfeld [38]. Abundance was expressed as ind. m^{-2}. Soil mites were classified into four groups, namely Mesostigmata, Oribatida, Prostigmata and Astigmata [39].

Biomass Calculations

Individual body length and width was measured at 10–80× magnification with a dissecting microscope equipped with an ocular micrometer with 0.01-mm precision. Dry biomass was calculated from regression equations estimating weights from linear dimensions:

Collembola dry mass: $Y = 0.0024L^{3.676}$ [40]   (1)

where Y is the dry weight (mg) and L is the length (mm) of individuals.

The dry mass of the mites was calculated using the equations of Douce [41]:

Oribatida with $W_G \times W_B > 0.013$: $Y = 156.33(W_G \times W_B) - 1.31$ (2)

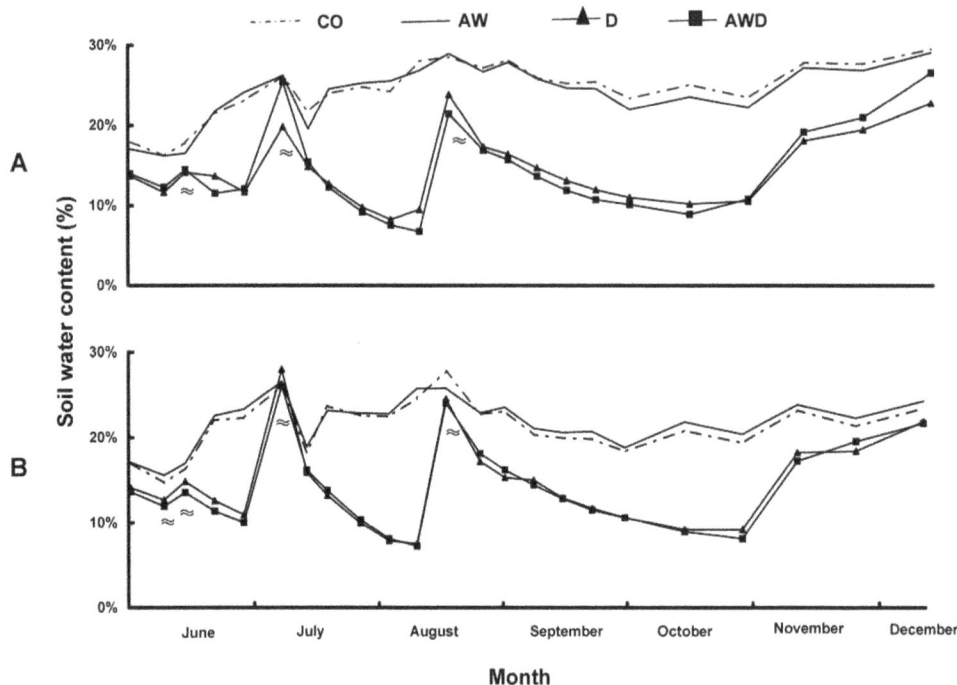

Figure 1. Effects of climate treatments and soil types on soil water content. Soil water content (SWC) at a depth of 0–25 cm in acid (A) and calcareous (B) soils as affected by the treatments: CO = control, AW = air warming, D = drought, and AWD = air warming + drought. Values are means of four plots in 2007. ≈ indicates no significant drought effects between (CO and AW) and (D and AWD) (n = 4).

Oribatida with $W_G \times W_B < 0.013$: $\log_e Y = 1.5\log_e(W_G \times W_B) + 6.11$ (3)

Mesostigmata: $Y = 150.27(L \times W_G) - 2.32$ (4)

Prostigmata: $Y = 19.26(L \times W_B) + 0.04$ (5)

where Y is the dry weight (μg); L is the maximum body length (mm) excluding chelicera; W_G is the maximum gnathosomal width (mm); and W_B is the maximum body width (mm).

A few Astigmata were also found. Their body shape is similar to *Stigmaeus* (Prostigmata), and their biomass was estimated using equation (5).

Statistical Analysis

Our experiment had a split-plot design with two subplots (soil types) within each plot (chamber with climate treatment). The main and interactive effects of the climate treatments and soil types were analyzed with an ANOVA accounting for split plots. The effect of the chambers was tested as a third factor, and climates were accordingly tested against its interaction with the chambers. The effect of the chambers was only kept in the ANOVA model if it was significant [42]. The differences among climate treatments in each soil type were evaluated with an LSD *post hoc* multiplied comparison. Pooling both soil types, the main and interactive effects between drought and warming were analyzed with a two-way ANOVA. The homogeneity of variances was confirmed by Levene's test before analysis. All tests were considered to be significant at $P<0.05$ level. SPSS 13.0 was used for all analyses.

Results

Soil Water Content and Temperature

During the experiment, soil water content (SWC) did not differ between the air warming treatment and the control, and also not between drought and air warming + drought. This confirms that the temperature treatment was not confounded by drought effects (Fig. 1). SWC was substantially lower in soils, both acidic and calcareous, with a drought treatment (drought and air warming + drought) compared to in well-watered soils (control and air warming). After the plots were re-watered in July and August, the differences between the treatments disappeared accordingly (Fig. 1). There was a significant soil type effect on the SWC from mid August until the end of 2007, which indicates that the values in the acidic soil type were higher than in the calcareous soil type, especially in the control and air warming treatments. There was a positive air warming effect on the soil temperature in August, September and October, whereas the drought treatment raised the temperature below the soil surface only in August and September. However, when Tukey HSD pair-wise comparisons were performed for each soil type separately, the only significant difference was between control and air warming + drought in September (Fig. 2). There was no soil type effect on the monthly soil temperature.

Plant Foliage, Root Biomass and Soil Respiration

The foliage biomass at the end of the 2007 growing season was significantly reduced by the drought treatment in the acidic soil type, but not in the calcareous soil type. A significant soil type effect indicated a higher foliage biomass and coarse root biomass in the calcareous than in the acidic soil type. Drought significantly reduced the coarse root biomass more in the acidic soil than in the calcareous soil (Table 1). As an indicator for soil respiration in 2007, the soil respiration at the end of a drought period in 2009 was significantly reduced by the drought treatment.

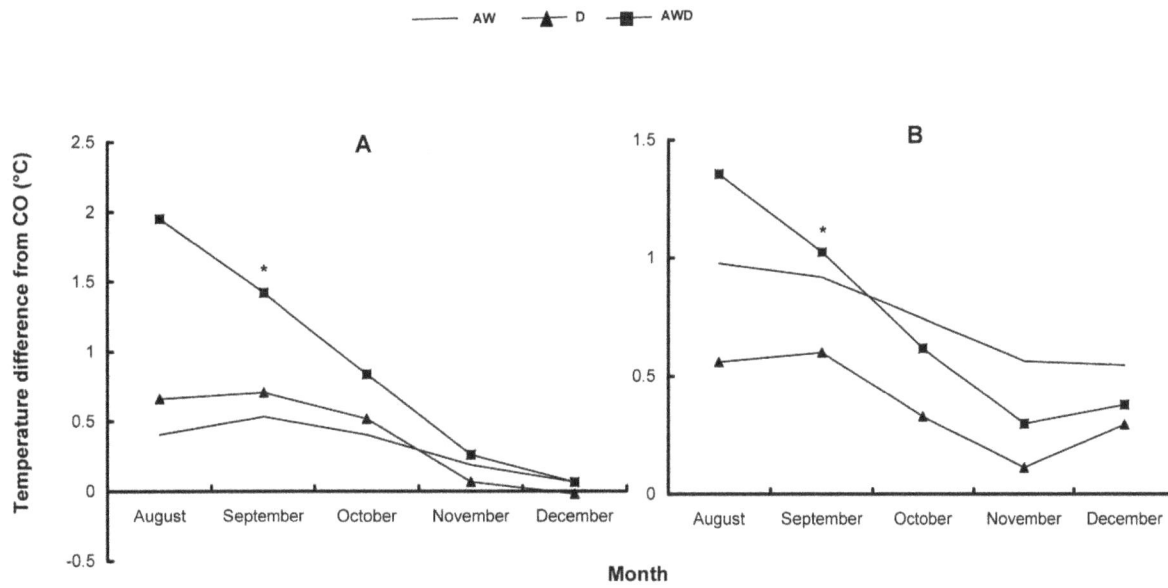

Figure 2. Effects of climate treatments and soil types on soil surface temperature. Mean surface temperature relative to controls (CO) in acid (A) and calcareous (B) soils as affected by the treatments: AW = air warming, D = drought, and AWD = air warming + drought in 2007, where * indicates a significant difference from CO ($P < 0.05$, n = 4).

Abundance, Group Richness, and Biomass of Soil Collembola and Mites

Fourteen genera of Collembola were identified, and the abundance of each Collembola genus and mite group in the four treatments and two soils were determined (Table 2). Only in the acidic soil, Collembola genus level richness was significantly reduced by all climate treatments (Fig. 3). There was a slight trend for Collembola to be richer in the calcareous than in the acidic soil type, with a weak climate-soil interaction (Table 3). Similarly, climate treatments reduced the biomass of the microarthropod fauna only in the acidic soil (Fig. 4B), and also with a weak climate-soil interaction (Table 3). If both soil

Table 1. Effects of climate treatments and soil types on soil respiration, foliage biomass and coarse root biomass.

		DW Foliage 07	DW Root 07	Soil Respiration
A	Control	0.156[a] (0.013)	0.814[a] (0.111)	6.9[a] (0.6)
	Air-warming	0.121[ab] (0.009)	0.531[a] (0.058)	7.8[a] (1.1)
	Drought	0.110[b] (0.005)	* 0.526[a] (0.037)	2.7[b] (0.1)
	AW & D	0.119[ab] (0.006)	0.555[a] (0.036)	2.0[b] (0.1)
B	Control	0.162[a] (0.011)	0.835[a] (0.058)	9.5[a] (0.7)
	Air-warming	0.140[a] (0.004)	0.697[a] (0.021)	7.2[a] (0.7)
	Drought	0.141[a] (0.010)	* 0.822[a] (0.051)	3.0[b] (0.9)
	AW & D	0.126[a] (0.005)	0.600[a] (0.038)	2.8[b] (0.5)

Air-warming (AW) and drought (D) treatment effects on soil respiration on 20.8.09 at the maximum of a drought period ($\mu mol\ CO_2\ m^{-2}\ s^{-1}$), with the foliage biomass and coarse root biomass in 2007 (kg m^{-2}, means ± SE) in the two soils ("A" acidic, "B" calcareous). Different letters indicate significant differences between the respective treatments in the same soil. DW indicates dry weight. An asterisk (*) indicates a significant difference between acidic and calcareous soil for the respective treatment.

types were combined, drought appeared to increase the abundance of microarthropod fauna (Table 4, Fig. 4A), but significantly to decrease their biomass (Table 4). This however, is probably mainly due to the effect of drought observed in the acidic soil (Fig. 4B).

Body Size of Mites and Collembola as Affected by the Treatments

The body size distribution of soil mites was altered by drought (Table 4). Drought increased the percentage of mites with small body size (≤ 0.20 mm), but reduced the percentage of mites with large body size (>0.40 mm). The smaller body size was mainly to do with the effect of drought observed in the acidic soil (Table 5). Furthermore, the small and large mites responded differently to air warming. Smaller mites were generally negatively associated with air warming in both soil types, while larger mites were positively associated with air warming in calcareous soil (Table 5). The percentage of large Collembola (>0.60 mm) was reduced by drought in combination with warming (Table 4), whereas the soil type showed no distinguishable pattern (data not shown).

Discussion

Effects of Drought and Warming on Soil Collembola and Mites

Previous reports concerning the effect of warming on soil collembolan abundance have been inconsistent. This might be due to the inability of the previous experimental designs to distinguish the warming effects from the drought effects inadvertently caused by the warming treatment [6,27–28]. In our study, the experimental design allowed us to separate the effects of warming and drought. The air warming treatment, which in the Querco experiment increased the air and the soil surface temperature slightly, only induced a negative response with respect to the richness of Collembola at the genus level and the biomass of Collembola + mites in the acidic soil (Fig. 3 and 4B). The effects on the soil water content of the two soils were similar (Fig. 1). Air

☐CO ▨AW ☐D ▨AWD

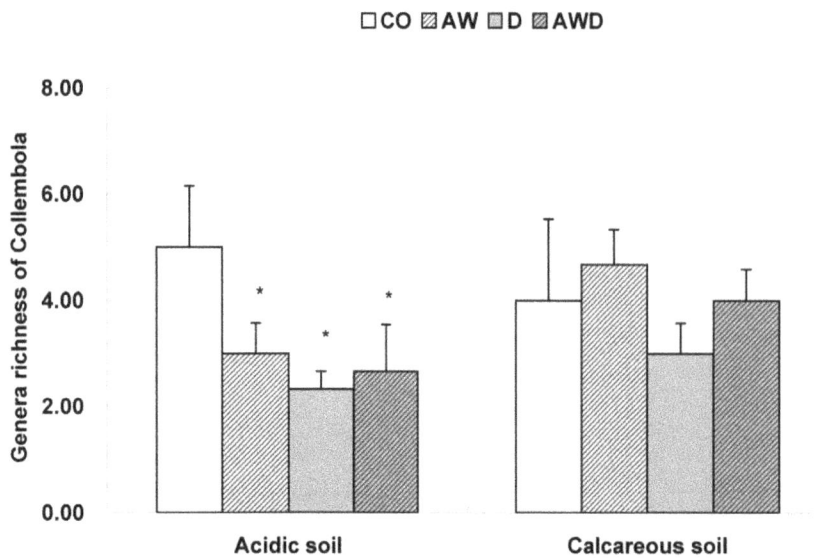

Figure 3. Effects of climate treatments and soil types on Collembola richness. Effects of the treatments (CO = control, AW = air warming, D = drought, and AWD = air warming + drought) on the genus level richness (number of genera) of Collembola in two forest soils (acidic and calcareous) in the Querco experiment at WSL in 2007, with mean values + SE, where * indicates significant differences from CO (P<0.05, n = 3).

warming not only had little effect on microarthropod fauna, apart from on their body size distribution, but it also had little effect on soil respiration, leaf biomass and coarse root biomass (Table 1).

In contrast, drought reduced the microarthropod biomass and Collembola richness, which is consistent with our hypothesis and with previous studies. Moisture conditions generally have a strong impact on Collembola behaviour and survival [13]. A field experiment in a spruce (*Picea abies*) monoculture forest near Giessen in Hesse, Germany, found that drought treatment drastically reduced the size of the Collembola communities in litterbags, and concluded that drought treatment stressed the Collembola, and reduced their abundance and species richness

[14]. In a spruce forest in Sweden, euedaphic and hemiedaphic species of Collembola were distinguished and both were found to be negatively affected by experimental drought [21]. A long-term study (1992–2002) in Scots pine forests in the North Vidzeme Biosphere Reserve (northern Latvia) found that the abundance and diversity of Collembola species in the forest litter and the moss layer were reduced during drought periods [6].

How exactly drought affects Collembola is still not clear. Drought may, for example, directly influence the physiological reactions, resistance to dehydration, development responses, oviposition rate, or fecundity [21]. The indirect effects of drought

☐CO ▨AW ☐D ▨AWD

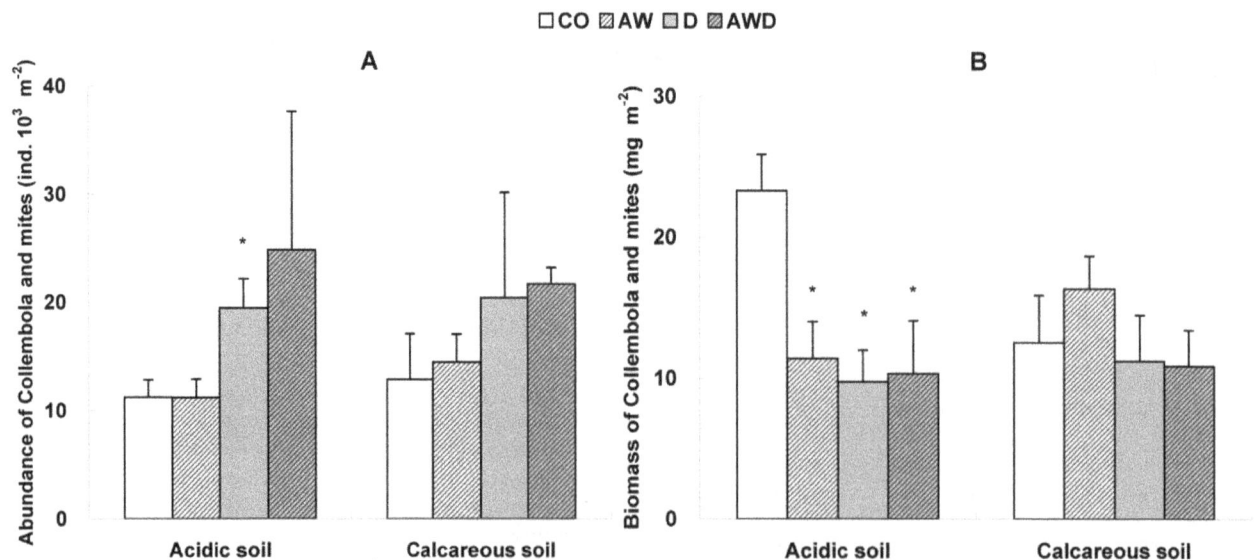

Figure 4. Effects of climate treatments and soil types on the community of Collembola and mites. Effects of the treatments (CO = control, AW = air warming, D = drought, AWD = drought + air warming) on (A) the abundance of Collembola and mites (ind. 10^3 m^{-2}) and (B) the biomass of Collembola and mites (mg m^{-2}) in two forest soils (acidic and calcareous), with mean values + SE, where * indicates significant differences from CO (P<0.05, n = 3).

Table 2. Abundance of soil Collembola and mites sampled in the Querco experiment.

	CO		AW		D		AWD	
	A	**B**	**A**	**B**	**A**	**B**	**A**	**B**
Collembola								
Entomobrya	57 (57)	–	–	–	–	–	–	–
Folsomia	57 (57)	–	57 (57)	–	–	57 (57)	–	–
Isotomiella	226 (150)	226 (226)	113 (113)	–	–	–	57 (57)	–
Isotomodes	–	–	–	57 (57)	113 (113)	113 (57)	–	340 (98)
Lepidocyrtus	–	–	–	–	–	–	113 (113)	–
Marcuzziella	57 (57)	–	–	–	–	–	–	–
Neelides	170 (98)	57 (57)	–	340 (170)	–	–	–	–
Oligaphorura	57 (57)	–	–	113 (113)	–	–	–	113 (57)
Pachyotoma	57 (57)	–	–	–	–	–	–	–
Parisotoma	170 (98)	226 (57)	679 (259)	623 (463)	962 (591)	226 (226)	226 (150)	170 (98)
Proisotoma	–	57 (57)	–	113 (113)	113 (113)	–	57 (57)	–
Protaphorura	–	57 (57)	–	–	–	–	–	–
Thalassaphorura	113	396	226	1189	–	510	–	623
	(113)	(247)	(226)	(490)	–	(259)	–	(396)
Tullbergia	4529	2378	4360	4926	7757	2944	5492	3680
	(1076)	(196)	(1306)	(2247)	(2072)	(1429)	(884)	(932)
Acari								
Astigmata	–	–	–	–	57 (57)	–	396 (204)	–
Mesostigmata	3227	2944	1755	3001	2038	623	2491	1019
	(354)	(1630)	(575)	(1331)	(260)	(150)	(1501)	(98)
Oribatida	1982	5718	3284	2831	2887	4360	5435	8096
	(653)	(2237)	(2362)	(982)	(1348)	(3041)	(3229)	(3906)
Prostigmata	566	849	736	1302	5548	11607	9229	7643
	(113)	(392)	(204)	(558)	(2774)	(8319)	(7967)	(2666)

Mean (SE) abundance of Collembola and mites (ind. m^{-2}) as affected by the treatments (CO = control, AW = air warming, D = drought, and AWD = air warming + drought) in two forest soil types ("A" acidic, "B" calcareous) in the Querco experiment at WSL.

may also play a role in determining microhabitat heterogeneity, fungi biomass and diversity. Soil bacteria and fungi are particularly sensitive components in soils, showing simultaneous responses to decreases in substrate humidity and quality [14]. The species structure of fungi is, for example, known to change with varying moisture levels [43–44].

Most Collembola and Oribatida are mycophagous and may selectively feed on different fungal species [5,45]. For example, five needle-excavating oribatids species may need particular fungi to make oviposition possible [11]. A detailed analysis of the Collembola community structure showed that certain species are highly adapted to specific characteristics of the substrate and thus respond rapidly to changes in microhabitat conditions [14]. In the acidic soil in the Querco experiment, especially, such indirect effects have been partly confirmed by the way drought seems to reduce the fungal abundance (unpublished data) and soil respiration (Table 1). Changes in the soil fauna community also influence the functions of an ecosystem, e.g. decomposition [14] and nitrogen cycling [46].

Table 3. Effects of climate treatments and soil types on Collembola and mites.

	Collembola richness		Community abundance		Community biomass	
	F	*P*	*F*	*P*	*F*	*P*
Climates (*df* = 3)	1.00	0.44	1.45	0.30	2.66	0.12
Soils (*df* = 1)	3.05	0.12	0.03	0.87	0.26	0.62
Climates * Soils (*df* = 3)	2.41	0.14	0.11	0.95	3.18	0.09

F- and *P*- values of the main and interactive effects of treatments (climates) and soils (ANOVA, split-plot design in the Querco experiment at WSL) on Collembola genus level richness (number of genera), and the community (Collembola and mite) abundance and biomass.

Table 4. Effects of drought and air warming on Collembola and mites.

	Community abundance		Community biomass		Collembola of small size (≤0.30 mm)		Collembola of large size (>0.60 mm)		Mite of small size (≤0.20 mm)		Mite of large size (>0.40 mm)	
	F	P	F	P	F	P	F	P	F	P	F	P
Drought (df = 1)	5.62	0.05	6.18	0.04	0.03	0.86	3.76	0.09	20.51	<0.01	17.85	<0.01
Warming (df = 1)	0.26	0.63	0.84	0.39	0.23	0.64	0.12	0.74	1.52	0.25	0.34	0.57
Drought * Warming (df = 1)	0.04	0.85	0.94	0.36	<0.01	0.99	6.92	0.03	1.06	0.33	0.41	0.54

F- and P- values of the main and interactive effects of treatments (drought and warming) on the community (Collembola and mite) abundance and biomass, and on the number of Collembola of small size (≤ 0.30 mm), Collembola of large size (>0.60 mm), mite of small size(≤0.20 mm), and mite of large size(>0.40 mm) by two-way ANOVA in the Querco experiment at WSL.

In our study, we found that drought actually increased the abundance of the microarthropod fauna, which is inconsistent with some other reports [13,20–21]. However, the contribution of mites to this result was difficult to estimate without specific taxonomic identification. For Collembola, we found that the significant reduction of richness caused by drought was mainly due to the absence of some rare genera, such as *Oligaphorura*, *Marcuzziella*, and *Pachyotoma*, and the increase in abundance under drought was dependent on the contribution of two genera, *Parisotoma* and *Tullbergia*, whose abundance increased by 200% and 55%, respectively, compared to that of the control. This might reflect the way soil fauna adopt different adaptation strategies to cope with low humidity, which may be morphological, physiological or reproductive adaptations [47].

Lindberg et al. [19] found some drought-tolerant Collembola species in the drought plots they studied using PCA analysis. A detailed analysis of the Collembola community structure indicated that certain species, such as *Willemia anophthalma* and *Mesaphorura tenuisensillata* seemed to be resistant to drought [14]. It could be that they reproduced through parthenogenesis, which had been shown to be a common strategy to increase the population among small euedaphic Collembola living deeper in the soil and among active colonizers, such as *Mesaphorura machrochaeta* (Familia: Tullbergiidae) and *Parisotoma notabilis* [48]. Our results indicated that the composition of Collembola species under drought tends to shift towards a dominance of drought-resistant species, while drought-sensitive species are at risk of disappearing.

Effects of Drought on Body Size Distribution

Body mass is a fundamental organismal trait and is closely related to an organism's physiology and ecology [49]. The body size distribution of soil mites was markedly changed by drought in this study, with an increase in small mites and a decrease in large mites. Whether this change in body size distribution resulted from a change in species composition or a change in size within species could not be determined. To our knowledge, few previous studies reported a change in the distribution of mesofauna body size in response to environmental changes. It seems, however, to be a general phenomenon in many animals, and reflects intraspecific change in some cases [50–52].

In Australia, light brown apple moths are reported to be smaller during warm, dry months than during cool, wet months [53]. Also in Australia, the growth and size of the wild brush-tailed phascogale is reduced during drought years [54]. Numerous authors have argued that such patterns of morphological variation are evidence of adaptation to environmental variables (referred to by Boyce [55]). Jones [52], for example, proposed that a small body size was advantageous during drought. However, a clear understanding of the adaptive significance of this variation is still missing. In a recent issue of the journal OIKOS, eight papers explored the influence of body size on many processes, ranging from individual biological rates to ecological networks [50]. Changes in body size reported here could imply changes in the soil food web, in the organism's ecosystem functions, e.g. its metabolic or ingestion rate, and in its ecology. For example, its strength in interacting with other species, such as prey-handling

Table 5. Effects of climate treatments and soil types on body length categories of mites.

Length (mm)	Soil	Percentage of soil mites in each length category (%)			
		CO	AW	D	AWD
≤ 0.20	Calcareous	50 (18) ab	31 (14) b	84 (11) a	90 (5) a
	Acidic	40 (13) bc	33 (3) c	73 (2) a	67 (16) ab
0.21–0.40	Calcareous	39 (15) a	41 (23) a	8 (5) a	9 (5) a
	Acidic	8 (8) a	33 (14) a	14 (3) a	23 (12) a
>0.40	Calcareous	10 (3) b	28 (9) a	8 (6) b	1 (0) b
	Acidic	53 (6) a	34 (17) ab	12 (5) b	10 (6) b

The percentages of different body length of mites in each treatment and soil type in the Querco experiment at WSL. Values are means (SE). Values in a row followed by different letters are significantly different (P<0.05).

ability and risk of being attacked by predators, may change, corresponding to changes in body size [49].

Role of Soil Types

Our experiment showed that more changes as a result of the climate treatments occurred in the acidic than in the calcareous soil. The negative effects of drought on Collembola and mites in the acidic soil may be correlated with the presence of aluminum ions. Some soil fauna are reported to survive in stressed and polluted areas by using detoxification mechanisms, such as the activation of metal-binding proteins and the precipitation of metals as intracellular electron-dense granules [56–57]. However, the presence of aluminum ions is generally thought to be toxic. Aluminum ions have been found to be more mobilized in the acidic than in the calcareous soil (221 mg kg^{-1} in the acidic vs. <2 mg kg^{-1} in the calcareous topsoil, Kuster, T.M., unpublished data) and can reach toxic levels for soil biota [12]. Another possible explanation for the strong effect of drought on mesofauna in the acidic than in the calcareous soil may relate to carbon, energy, and nutrient input for the food web. Drought significantly reduced the coarse root biomass in the acidic soil than in the calcareous soil (Table 1), and microbial biomass carbon and microbial biomass nitrogen were significantly lower in the acidic soil (Hu et al., personal communication). Roots, root exudates, decaying organic matter, and the associated microorganisms provide most of the carbon and energy that fuels the soil food web [58–59]. A reduction in these components may lead to a reduction in mesofauna richness.

The sensitivity of acidic soil to climate treatments, especially to drought, was also confirmed by the reduction in other components of the ecosystem, such as the coarse root and foliage biomass. For example, the foliage biomass was found to significantly decrease under drought in the acidic soil but not in the calcareous soil (Table 1). Over a three-year period the trees grew better on acidic than on calcareous soils, therefore consuming more water, these characteristics disappeared when drought was imposed, indicating stronger drought effects on trees grown in acidic than in alkaline soil [31]. Common beans are reported to grow longer roots and extract more soil moisture as an important mechanism to cope with soil water [60]. The roots of drought-tolerant bean species reached a soil depth of 1.3 m under drought stress at Palmira (soil pH 7.7), while the roots extended only to 0.7 m under acid soil conditions at Quilichao (soil pH 5.0) [60]. This suggests that the mechanism to cope with drought was inhibited in acidic soil. In conclusion, it seems that mesofauna are sensitive to air warming and drought, particularly in acidic soils.

Acknowledgments

We thank Dr. Feng Zhang and Prof. Erhard Christian for their help with the identification of Collembola, and Dr. Silvia Dingwall for editing the language.

Author Contributions

Conceived and designed the experiments: MGG MD MHL. Performed the experiments: GLX MHL TK MD. Analyzed the data: GLX MHL TK. Wrote the paper: GLX.

References

1. Harte J, Rawa A, Price V (1996) Effects of manipulated soil microclimate on mesofaunal biomass and diversity. Soil Biology and Biochemistry 28: 313–322.
2. Brussaard L (1997) Biodiversity and ecosystem functioning in soil. Ambio 26: 563–570.
3. Freckman DW, Blackburn TH, Brussaard L, Hutchings P, Palmer MA, et al. (1997) Linking biodiversity and ecosystem functioning of soils and sediments. Ambio 26: 556–562.
4. Rusek J (1998) Biodiversity of Collembola and their functional role in the ecosystem. Biodiversity and Conservation 7: 1207–1219.
5. Hopkin SP (1997) Biology of the Springtails (Insecta, Collembola). Oxford: Oxford University Press. 326p.
6. Jucevica E, Melecis V (2006) Global warming affect Collembola community: A long-term study. Pedobiologia 50: 177–184.
7. Xu GL, Schleppi P, Li MH, Fu SL (2009) Negative responses of Collembola in a forest soil (Alptal) under increased atmospheric N deposition. Environmental Pollution 157: 2030–2036.
8. Kardol P, Reynolds WN, Norby RJ, Classen AT (2011) Climate change effects on soil microarthropod abundance and community structure. Applied Soil Ecology 47: 37–44.
9. Kardol P, Cregger MA, Campany CE, Classen AT (2010) Soil ecosystem functioning under climate change: plant species and community effects. Ecology 91: 767–781.
10. Castro HF, Classen AT, Austin EE, Norby RJ, Schadt CW (2010) Soil microbial community responses to multiple experimental climate change drivers. Applied Environmental Microbiology 76: 999–1007.
11. Hågvar S (1998) Mites (Acari) developing inside decomposing spruce needles: biology and effect on decomposition rate. Pedobiologia 42: 358–377.
12. Matson P, Lohse KA, Hall SJ (2002) The globalization of nitrogen deposition: consequences for terrestrial ecosystems. Ambio 31: 113–119.
13. Verhoef HA, van Selm AJ (1983) Distribution and population dynamics of Collembola in relation to soil moisture. Holarctic Ecology 6: 387–394.
14. Pflug A, Wolters V (2001) Influence of drought and litter age on Collembola communities. European Journal Soil Biology 37: 305–308.
15. Tsiafouli MA, Kallimanis AS, Katana E, Stamou GP, Sgardelis SP (2005) Responses of soil microarthropods to experimental short-term manipulations of soil moisture. Applied Soil Ecology 29: 17–26.
16. Moron-Rios A, Rodriguez MA, Perez-Camacho L, Rebollo S (2010) Effects of seasonal grazing and precipitation regime on the soil macroinvertebrates of a Mediterranean old-field. European Journal of Soil Biology 46: 91–96.
17. Wallwork JA (1983) Oribatid in forest ecosystems. Annual Review of Entomology 28: 109–130.
18. Badejo MA (1990) Seasonal abundance of soil mites (Acarina) in two contrasting environments. Biotropica 22(4): 382–390.
19. Lindberg N, Engtsson JB, Persson T (2002) Effects of experimental irrigation and drought on the composition and diversity of soil fauna in a coniferous stand. Journal of Applied Ecology 39: 924–936.
20. Badejo MA, Akinwole PO (2006) Microenvironmental preferences of oribatid mite species on the floor of a tropical rainforest. Experimental Applied Acarology 40: 145–156.
21. Lindberg N (2003) Soil fauna and global change –Responses to experimental drought, irrigation, fertilisation and soil warming. Dissertation, Swedish University of Agricultural Sciences.
22. Sjursen H, Michelsen A, Jonasson S (2005) Effects of long-term soil warming and fertilization on microarthropod abundances in three sub-arctic ecosystems. Applied Soil Ecology 30: 148–161.
23. Coulson SJ, Hodkinson ID, Wooley C, Webb NR, Block W, et al. (1996) Effects of experimental temperature elevation on high-arctic soil microarthropod populations. Polar Biology 16: 147–153.
24. Huhta V, Hänninen SM (2001) Effects of temperature and moisture fluctuations on an experimental soil microarthropod community. Pedobiology 45: 279–286.
25. Dollery R, Hodkinson ID, Jonsdottir IS (2006) Impact of warming and timing of snow melt on soil microarthropod assemblages associated with Dryas-dominated plant communities on Svalbard. Ecography 29: 111–119.
26. McGeoch MA, Le Roux PC, Hugo EA, Chown S (2006) Species and community responses to short-term climate manipulation: Microarthropods in the sub-Antarctic. Austral Ecology 31: 719–731.
27. Hägvar S, Klanderud K (2009) Effect of simulated environmental change on alpine soil arthropods. Global Change Biology 15: 2972–2980.
28. Wolters V (1998) Long-term dynamics of a collembolan community. Applied Soil Ecology 9: 221–227.
29. Haimi J, Laamanen J, Penttinen R, Räty M, Koponen S, et al. (2005) Impacts of elevated CO_2 and temperature on the soil fauna of boreal forests. Applied Soil Ecology 30: 104–112.
30. Sinclair BJ, Stevens MI (2006) Terrestrial microarthropods of Victoria Land and Queen Maud Mountains, Antarctica: implications of climate change. Soil Biology and Biochemistry 38: 3158–3170.
31. Kuster TM, Arend M, Bleuler P, Günthardt-Goerg MS, Schulin R (2012) Water regime and growth of young oak stands subjected to air warming and drought on two different forest soils in a model ecosystem experiment. Plant Biology 14: 1–10.
32. Arend M, Kuster T, Günthardt-Goerg MS, and Dobbertin M. (2011) Provenance-specific growth responses to drought and air warming in three European oak species (Quercusrobur, Q.petraea and Q.pubescens). Tree Physiology 31: 287–297.
33. IPCC (2007) Climate Change 2007: Synthesis report. Cambridge University Press, Cambridge.

34. CH2011 (2011) Swiss Climate Change Scenarios CH2011, C2SM, MeteoSwiss, ETH, NCCR Climate, and OcCC. CH2011, Zurich, Switzerland, 88.

35. Brady J (1969) Some physical gradients set up in Tullgren funnels during the extraction of mites from poultry litter. Journal of Applied Ecology 6: 391–402.

36. Janssens F (2007) Checklist of the Collembola of the world. http://www.collembola.org/.

37. Potapov M (2001) Synopses on PalaearcticCollembola vol. 3. Isotomidae.Görlitz: State Saxonian Museum of Natural History. 603p.

38. Bretfeld G (1999) Synopses on PalaearcticCollembola. vol. 2.Symphypleona.-Görlitz: State Saxonian Museum of Natural History. 318p.

39. Johnston DE (1982) Acari. In Synopsis and classification of living organisms, S.P. Parker (ed.), p. 111. McGraw-Hill, New York.

40. Hódar JA (1996) The use of regression equations for estimation of arthropod biomass in ecological studies. ActaŒcologica 17: 421–433.

41. Douce GK (1976) Biomass of soil mites (Acari) in Arctic coastal tundra. Oikos 27: 324–330.

42. Chen PY (2005) Statistics Software Tutorial of SPSS Application. People Medical Publishing House, Beijing.

43. Bääth E, Söderström B (1982) Seasonal and spatial variation in fungal biomass in a forest soil. Soil Biology and Biochemistry 14: 353–358.

44. Widden P (1986) Functional relationships between Quebec forest soil microfungi and their environment. Canadian Journal of Botany 64: 1424–1432.

45. Maraun M, Migge S, Schaefer M, Scheu S (1998) Selection of microfungal food by six oribatid mite species (Oribatida, Acari) from two different beech forests. Pedobiologia 42: 232–240.

46. Laakso J, Setälä H (1999) Sensitivity of primary production to changes in the architecture of belowground food webs. Oikos 87: 57–64.

47. Alvarez T, Frampton GK, Goulson D (1999) The effects of drought upon epigeal Collembola from arable soils. Agricultural and Forest Entomology 1: 243–248.

48. Chahartaghi M, Scheu S, Ruess L (2006) Sex ratio and mode of reproduction in Collembola of an oak-beech forest. Pedobiologia 50: 331–340.

49. Digel C, Riede JO, Brose U (2011) Body sizes, cumulative and allometric degree distributions across natural food webs. Oikos 120: 503–509.

50. Blanchard JL (2011) Body size and ecosystem dynamics: an introduction. Oikos 120: 481–482.

51. Petchey OL, Belgrano A (2010) Body-size distributions and size-spectra: universal indicators of ecological status? Biology letters 6: 434–437.

52. Jones G (1987) Selection against large size in the Sand Martin Ripariariparia during a dramatic population crash. Ibis 129: 274–280.

53. Danthanarayana W (1975) Factors determining variation in fecundity of the light brown apple moth, EpiphyasPostvittana (Walker) (Tortricidae). Australian Journal of Zoology 23: 439–451.

54. Rhind SG, Bradley JS (2002) The effect of drought on body size, growth and abundance of wild brush-tailed phascogales (Phascogale tapoatafa) in southwestern Australia. Wildlife Research 29: 235–245.

55. Boyce MS (1978) Climatic variability and body size variation in the muskrats (Ondatrazibethicus) of North America. Oecologia 36: 1–19.

56. Dallinger R (1996) Metallothionein research in terrestrial invertebrates: synopsis and prospectives. Comp. Biochem. Physiol. 113 C: 125–133.

57. Pigino G, Miglliorini M, Paccagnini E, Bernini F (2006) Localisation of heavy metals in the midgut epithelial cells of Xenillus tegeocranus (Hermann, 1804) (Acari: Oribatida). Ecotoxicology and Environmental Safety 64: 257–263.

58. Norby RJ, Jackson RB (2000) Root dynamics and global change: Seeking an ecosystem perspective. New Phytologist 147: 3–12.

59. Pollierer MM, Langel R, Körner C, Maraun M, Scheu S (2007) The underestimated importance of belowground carbon input for forest soil animal food webs. Ecology Letters 10: 729–736.

60. Sponchiado BN, White JW, Castillo JA, Jones PG (1989) Root growth of four common bean cultivars in relation to drought tolerance in environments with contrasting soil types. Experimental Agriculture 25: 249–257.

12

Leaf Area Index Drives Soil Water Availability and Extreme Drought-Related Mortality under Elevated CO$_2$ in a Temperate Grassland Model System

Anthony Manea*, Michelle R. Leishman

Department of Biological Sciences, Macquarie University, North Ryde, NSW, Australia

Abstract

The magnitude and frequency of climatic extremes, such as drought, are predicted to increase under future climate change conditions. However, little is known about how other factors such as CO$_2$ concentration will modify plant community responses to these extreme climatic events, even though such modifications are highly likely. We asked whether the response of grasslands to repeat extreme drought events is modified by elevated CO$_2$, and if so, what are the underlying mechanisms? We grew grassland mesocosms consisting of 10 co-occurring grass species common to the Cumberland Plain Woodland of western Sydney under ambient and elevated CO$_2$ and subjected them to repeated extreme drought treatments. The 10 species included a mix of C$_3$, C$_4$, native and exotic species. We hypothesized that a reduction in the stomatal conductance of the grasses under elevated CO$_2$ would be offset by increases in the leaf area index thus the retention of soil water and the consequent vulnerability of the grasses to extreme drought would not differ between the CO$_2$ treatments. Our results did not support this hypothesis: soil water content was significantly lower in the mesocosms grown under elevated CO$_2$ and extreme drought-related mortality of the grasses was greater. The C$_4$ and native grasses had significantly higher leaf area index under elevated CO$_2$ levels. This offset the reduction in the stomatal conductance of the exotic grasses as well as increased rainfall interception, resulting in reduced soil water content in the elevated CO$_2$ mesocosms. Our results suggest that projected increases in net primary productivity globally of grasslands in a high CO$_2$ world may be limited by reduced soil water availability in the future.

Editor: Gerrit T. S. Beemster, University of Antwerp, Belgium

Funding: This research was funded by a Macquarie University Research Excellence and National Climate Change Adaptation Research Flagship scholarship to AM. The funders had no role in study design, data collection and analysis, decision to publish, or preparation of the manuscript.

Competing Interests: The authors have declared that no competing interests exist.

* E-mail: anthony.manea@students.mq.edu.au

Introduction

A major driver that shapes the physiology, ecology and evolution of terrestrial plants is climatic extremes [1]. It is widely acknowledged that the magnitude and frequency of climatic extremes, such as drought, are likely to increase under future climate change conditions [2]. The IPCC [2] defines a climatic extreme, such as extreme drought, as an event that occurs once every 20 years, on average. The potential for climatic extremes to alter the structural and functional dynamics of ecological communities [2], [3], coupled with their increasing magnitude and frequency in the future, suggests research on climatic extremes should be a high priority.

Grass-dominated systems (grasslands, savannas and open grassy woodlands) occupy more than 30% of the global terrestrial landscape [4] and play an important role in the global carbon cycle [5]. The productivity of grass-dominated systems (referred to as grasslands from now on) is strongly mediated by soil water availability [4], [6], [7]. For example, Fay et al [8] increased the mean annual rainfall to an experimental grassland community in Kansas by 250% and found that both soil water content (SWC) and aboveground net primary productivity significantly increased (see exception [9]). Consequently it is likely that soil water availability prior to an extreme drought event will be a major driver in the response of the grassland to the event.

It has rarely been considered how CO$_2$ concentration will alter soil water availability in grasslands and thus modify grassland responses to extreme drought. One of the major drivers of soil water availability in grasslands is canopy transpiration [6]. The amount of water that is lost through canopy transpiration depends on the stomatal conductance and leaf area index (LAI) of the grasses [10]. Reduction in stomatal conductance of grasses under elevated CO$_2$ has been well-documented [6], [11], [12]. For example, Morgan et al [7] found that in the Wyoming mixed-grass prairies in the United States, the annual SWC increased on average by 17.3% over a three year period due partly to reductions in stomatal conductance under elevated CO$_2$ levels. The physiology (C$_3$ or C$_4$) and origin (native or exotic) of the grasses play an important role in this reduction of stomatal conductance under elevated CO$_2$ levels. C$_4$ plants evolved in a low CO$_2$ environment, allowing high rates of photosynthesis at low stomatal conductance [13]. This suggests that C$_4$ plants should have greater reductions in stomatal conductance compared to C$_3$ plants under elevated CO$_2$ levels. In addition, meta-analysis studies have shown that at the leaf-level natives tend to have lower stomatal conductance than exotics [14].

In contrast to reductions in stomatal conductance, grasses often have higher leaf area index (LAI) under elevated compared to ambient CO_2 levels, resulting in increased canopy transpiration. For example, the LAI of the C_4 bunchgrass *Pleuraphis rigida* significantly increased under elevated CO_2 in the 10 year Mojave Desert FACE experiment [15]. In a meta-analysis of semi-wild and wild C_3 and C_4 grasses, leaf area (related to LAI) increased under elevated CO_2 by 15% and 25% respectively [16]. Therefore reductions in the stomatal conductance of grasses under elevated CO_2 may be offset by increases in their LAI [17].

In this study we asked whether the response of grasslands to repeat extreme drought events is modified by elevated CO_2, and if so, what are the underlying mechanisms? Experimental mesocosms containing common co-occurring native and exotic grass species of the Cumberland Plain Woodland of western Sydney, Australia were grown under ambient and elevated CO_2 levels and exposed to repeated one in 20 year extreme drought events. We hypothesized that a reduction in the stomatal conductance of the grasses, particularly the C_4 and native grasses, grown under elevated CO_2 levels would be offset by increases in the LAI of the grasses. That is, the total canopy transpiration of the grasses would not differ between the CO_2 treatments. Therefore the retention of soil water in the mesocosms and consequently the vulnerability of the grasses to extreme drought would not differ between the CO_2 treatments.

Methods

Species selection

We selected five native and five exotic perennial grass species which commonly co-occur in a grassy open woodland community known as Cumberland Plain Woodland that occurs in western Sydney, Australia. All the exotic species are considered to be invasive rather than simply exotics that have become naturalized in Cumberland Plain Woodland [18]. Within both the native and exotic groups we included two C_3 and three C_4 species. Seeds for the 10 grass species were obtained from a commercial supplier (Nindethana Seed Service, Albany, WA, Australia). Information on the biology of each grass species is provided in table 1. Once collected, the seeds for each of the 10 grass species were germinated on moist paper towels within covered aluminium

trays. To spread the risk of germination failure, each grass species was germinated in a number of different aluminium trays.

Experimental design and treatments

The native and exotic grass species were grown together in mesocosms using a fully factorial experimental design with two factors: CO_2 concentration and drought treatment. The mesocosms consisted of 65 L tubs (60 cm long ×40 cm wide ×28 cm deep), each tub containing 55 L of soil mixture consisting of field-collected Cumberland Plain Woodland soil and coarse river sand in a ratio of 3:1. The Cumberland Plain Woodland soil was obtained from Mt Annan (34.07°S, 150.76°E) and Luddenham (33.88°S, 150.69°E) in western Sydney and was homogenized into a single batch. No specific approvals or permits were required for soil collection. The soil collection locations were not privately-owned or protected in any way and did not endanger any protected species. The river sand was obtained from a commercial supplier (Australian Native Landscapes, North Ryde, NSW, Australia). Seedlings were transplanted from the germination trays into the CO_2 treatment mesocosms at the stage of second true leaf emergence. All seedlings were planted within 24 hours of each other. The seedlings were planted in two rows of five with each species allocated a position randomly within each mesocosm. For each grass species multiple seedlings were transplanted into each mesocosm as insurance against seedling mortality. After six days, the excess seedlings were removed from the mesocosms, leaving one individual per species per mesocosm.

CO_2 treatments were set to two levels: ambient (380–420 ppm) and elevated (530–570 ppm) CO_2. These CO_2 concentration ranges were maintained and monitored daily by a CO_2 dosing and monitoring system (Canary Company Pty Ltd, Lane Cove, NSW, Australia). The lower concentrations of these ranges tended to occur at night-time while the higher concentrations occurred during the daytime. The elevated CO_2 treatment represents the predicted atmospheric CO_2 concentration by 2050 [19].

We defined a drought as the number of consecutive days with < 1 mm of rainfall. This definition is a part of the ETCCDI/CRD climate change indices [20]. We used the IPCC [2] definition of climatic extreme which is an extreme that occurs once every 20 years, on average. Gumbel I distributions were fitted to the annual drought extremes of the Cumberland Plain for each year from 1867–2010. Rainfall on the Cumberland Plain is not seasonal so

Table 1. Grass species used in the study.

Species	Origin	Seed mass (mg)	Photosynthetic pathway	Longevity
Chloris gayana Kunth	Exotic	0.4	C_4	Perennial
Eragostis curvula (Schrad.) Nees	Exotic	0.2	C_4	Perennial
Pennisetum clandestinum Hochst. Ex Chiov	Exotic	7.0	C_4	Perennial
Bromus catharticus Vahl	Exotic	10.7	C_3	Short lived perennial
Ehrharta erecta Lam.	Exotic	2.0	C_3	Perennial
Bothriochloa macra (Steud.) S.T.Blake	Native	1.2	C_4	Perennial
Chloris truncata R.Br.	Native	0.3	C_4	Perennial
Themeda australis (R.Br.) Stapf	Native	2.6	C_4	Perennial
Austrodanthonia racemosa (R.Br.) H.P.Linder	Native	0.7	C_3	Perennial
Microlaena stipoides (Labill.) R.Br.	Native	4.3	C_3	Perennial

Grass species (family: *Poaceae*) used in this study, with information on the origin, seed mass, photosynthetic pathway and longevity of each species. Average seed mass was obtained by oven drying 50 seeds from each species at 60°C for 48 hours and then weighing them. Data on taxonomy, physiology and longevity were obtained from PlantNET (www.plantnet.rbgsyd.nsw.gov.au).

the time of year that the annual drought extremes occurred differed from year to year. The data used were obtained from the Australian Bureau of Meteorology historical records of Brownlow Hill ($34.03°S$, $150.65°E$, 1867–1969), Kentlyn ($34.05°S$, $155.88°E$; 1970–1971), Camden Airport ($34.04°S$, $150.69°E$; 1972–1992, 1998–2001), Ruse ($34.06°S$, $150.85°E$; 1993–1997) and Mt Annan Botanical Gardens ($34.07°S$, $150.76°S$; 2002–2010). A one in 20 year extreme drought event for the Cumberland Plain was calculated to last for a period of 53 days. The extreme drought was simulated by turning off the watering system for the treatment period.

The extreme drought treatment mesocosms were replicated five times at each CO_2 level. These were called the 'drought treatment' mesocosms. In addition, five extra mesocosms were grown under each CO_2 level. These were called the 'before treatment' mesocosms. All mesocosms were mist watered for one minute twice daily which is representative of the average daily amount of rainfall (828 mm annually) on the Cumberland Plain. This daily rainfall average was based on the same Australian Bureau of Meteorology historical records from Camden airport (1943–2004) as described in the above paragraph. This design resulted in a total of 20 mesocosms each containing 10 grass species (i.e. [2 CO_2 treatments ×5 replicates]+10 extra mesocosms). The mesocosms within each CO_2 treatment were split between two glasshouses. The mesocosms were switched between glasshouses within each CO_2 treatment once during the growth period and once during the treatment period to reduce any glasshouse effect. The temperature of the glasshouses was set for a maximum of $24°C$ and a minimum of $16°C$. During the entire duration of the experiment the relative humidity within each glasshouse was measured every day at 9am and 3pm using a HOBO temperature/RH/2 external channel data logger (OneTemp, Parramatta, NSW, Australia).

The grasses were grown for 12 weeks at which stage they were mature and seeding. After the growth period the 'before treatment' mesocosms were harvested, washed free of soil and separated into their following components: leaf biomass, stem biomass and belowground biomass. The total leaf area of the leaf biomass was measured using a LI-3100C Area Meter (Li-Cor, Lincoln, NE, United States). The plant components were then oven-dried at $60°C$ for 72 hours and weighed using a Mettler Toledo B-S electronic balance.

The extreme drought treatment was then applied to the five 'drought treatment' mesocosms at each CO_2 level. After the 53 day treatment period the mesocosms were mist-watered twice daily as previously for four weeks to allow a recovery period. In field conditions on the Cumberland Plain four weeks would be ample time for the grasses to recover (i.e. resprout) from a dry period. For 30 days prior to and during the drought/recovery cycles the SWC of each mesocosm was measured at a depth of 15 cm every 10 days using a Hydrosense II Portable Soil Moisture System (Campbell Scientific Australia Pty Ltd, Garbutt, QLD, Australia). In addition, for the 30 days prior to the extreme drought treatment the stomatal conductance of each grass was measured every five days using a Porometer AP4 (Delta-T Devices, Burwell, CB, Uniting Kingdom). Measurements would begin at 8.30 am and would take approximately three and a half hours to complete. After the four week recovery period, the mortality of the grasses was recorded. Each grass was classified as dead or alive depending on if it showed signs of regeneration by the end of the recovery period. The grasses were then exposed to two more cycles of drought and recovery with mortality of the grasses recorded after each cycle. At the end of the recovery period

of the final cycle, every mesocosm was dug up to ensure that the grasses classified as dead had no living root material.

Data analysis

Leaf area index and biomass allocation analysis. We calculated LAI and root to shoot ratio (R:S) based on the harvested biomass data of the 'before treatment' mesocosms. LAI for each individual grass was calculated as total leaf area divided by ground area in the mesocosm (0.24 m^2). R:S for each individual grass was calculated as total root mass divided by total shoot mass. We used two-way ANOVAs to test for a CO_2 and species effect on LAI, R:S and total biomass across all grass species in the 'before treatment' mesocosms.

We used two-way ANOVAs to test for a CO_2 and plant type (physiology and origin) effect on LAI across all grass species in the 'before treatment' mesocosms.

CO2 level and plant type survival analysis. Kaplan-Meier survival curves were generated to determine the survival function across all the grass species in relation to (1) CO_2 level, (2) physiology, (3) origin and (4) CO_2 level combined with each plant type (origin and physiology). If there was a significant difference between the Kaplan-Meier survival curves within each CO_2 level/ plant type combination then the survival distributions within the combination were tested for a significant difference. All survival distributions were compared using log rank tests, with significance set at $P<0.05$.

Stomatal conductance, total canopy transpiration and soil water content analysis. We used General Estimating Equation (GEE) models with a Bonferroni adjustment to determine if there was a difference in the stomatal conductance between CO_2 level and each plant type (origin and physiology) combination during the 30 days prior to the extreme drought treatment. The same analyses were performed for SWC with the addition of testing for a difference between the CO_2 levels for the three drought/recovery cycles. For all GEE models we specified a Gamma with log-link model using an exchangeable correlation matrix for the continuous variables (stomatal conductance and SWC).

Total canopy transpiration for each species was calculated as the average stomatal conductance multiplied by the average LAI. We carried out species-pair comparisons using paired t-tests with a Bonferroni adjustment to test for a CO_2 effect on total canopy transpiration.

Relative humidity analysis. We used a paired t-test with a Bonferroni adjustment to determine if there was difference in the mean daily relative humidity between the ambient and elevated CO_2 glasshouses. The relative humidity data was then averaged within each CO_2 level and compared to the historical average relative humidity (9am-74%, 3pm-49%) of the Cumberland Plain using one-sample t-tests. This involved separate analyses for the relative humidity data obtained at 9am and 3pm. The historical average used was obtained from the Australian Bureau of Meteorology historical records of Camden Airport ($34.04°S$, $150.69°E$; 1943–2010).

All data analyses were performed using IBM SPSS statistical software, Version 21.0.0 (SPSS Inc., 2012, IBM, Illinois, United States, http://www.spss.com) with the significance level set at 0.05. Data were log_{10} transformed when necessary to fulfil the assumptions of ANOVA.

Results

Humidity analysis

There was no significant difference in the relative humidity between and ambient and elevated CO_2 glasshouses ($t_{1,38} = 1.00$, $p = 0.326$). However the relative humidity in the glasshouses at 9am ($t = 10.01$, $p<0.001$) and 3pm ($t = 17.78$, $p<0.001$) was significantly higher than the historical average for the same times of day.

Leaf area index and biomass allocation analysis

There was no significant interaction between CO_2 and species for LAI, R:S or total biomass but there was a significant difference between species for all traits. LAI ($F_{1,99} = 14.20$, $p<0.001$) and total biomass ($F_{1,99} = 19.97$, $p<0.001$) were significantly higher under elevated CO_2 compared to ambient CO_2 across all the grass species. R:S ($F_{1,99} = 0.26$, $p = 0.615$) did not significantly differ between ambient and elevated CO_2.

There was a significant interaction between CO_2 and physiology for LAI ($F_{1,99} = 5.15$, $p = 0.026$) with the C_4 grasses grown under ambient CO_2 having significantly lower LAI than all of the other $CO_2 \times$physiology combinations (figure 1). There was also a significant interaction between CO_2 and origin for LAI ($F_{1,99} = 4.21$, $p = 0.043$) with the native grasses grown under ambient CO_2 having significantly lower LAI than all of the other $CO_2 \times$physiology combinations (figure 1).

CO_2 level and plant type survival analysis

The average survival rate across all grass species was significantly higher under ambient compared to elevated CO_2 levels ($\chi^2 = 8.58$, df = 1, $p = 0.003$). The physiology ($\chi^2 = 3.24$, df = 1, $p = 0.072$) or origin ($\chi^2 = 0.03$, df = 1, $p = 0.855$) of the grasses did not significantly influence their survival rates.

There was a significant difference in survival between the different $CO_2 \times$physiology combinations ($\chi^2 = 13.29$, df = 3, $p = 0.004$; figure 2a). Both the C_3 ($\chi^2 = 10.01$, df = 1, $p = 0.002$) and C_4 ($\chi^2 = 7.91$, df = 1, $p = 0.005$) grasses grown under ambient CO_2 had significantly higher survival rates than the C_4 grasses grown under elevated CO_2.

There was a significant difference in survival between the different $CO_2 \times$origin combinations ($\chi^2 = 16.64$, df = 3, $p = 0.001$; figure 2b). The native grasses grown under ambient CO_2 had a significantly higher survival rate than the native ($\chi^2 = 17.45$, df = 1, $p<0.001$) and exotic ($\chi^2 = 6.52$, df = 1, $p = 0.011$) grasses grown under elevated CO_2. The exotic grasses grown under ambient CO_2 had a significantly higher survival rate than the native grasses grown under elevated CO_2 ($\chi^2 = 4.13$, df = 1, $p = 0.042$).

Stomatal conductance, total canopy transpiration and soil water content analysis

Stomatal conductance across all grass species was significantly lower under elevated CO_2 compared to ambient CO_2 (Wald $\chi^2 = 7.254$, df = 1, $p<0.007$).

There was no significant interaction between CO_2 and physiology for stomatal conductance (Wald $\chi^2 = 1.786$, df = 1, $p = 0.181$). The C_3 grasses had significantly higher stomatal conductance than the C_4 grasses (Wald $\chi^2 = 46.00$, df = 1, $p<0.001$; figure 3a). As described above, stomatal conductance was significantly lower under elevated CO_2 compared to ambient CO_2.

There was a significant interaction between CO_2 and origin for stomatal conductance (Wald $\chi^2 = 7.369$, df = 1, $p = 0.007$; figure 3b). The exotics grown under ambient CO_2 had significantly higher stomatal conductance than all of the other $CO_2 \times$origin combinations.

There was no significant difference in the total canopy transpiration between ambient and elevated CO_2 treatments ($t_8 = -0.93$, $p = 0.375$).

The SWC during the 30 days prior to the drought/recovery cycles was significantly higher under ambient CO_2 compared to elevated CO_2 (Wald $\chi^2 = 5.91$, df = 1, $p = 0.015$). The overall SWC of the mesocosms during the experiment drought/recovery cycles was significantly higher under ambient CO_2 compared to elevated CO_2 (Wald $\chi^2 = 11.48$, df = 1, $p = 0.001$; figure 4). However SWC differences varied between drought/recovery cycles. During the first drought/recovery cycle SWC was significantly higher in the ambient compared with elevated CO_2 treatment (Wald $\chi^2 = 8.98$, df = 1, $p = 0.003$). During the second and third drought/recovery cycles SWC did not significantly differ

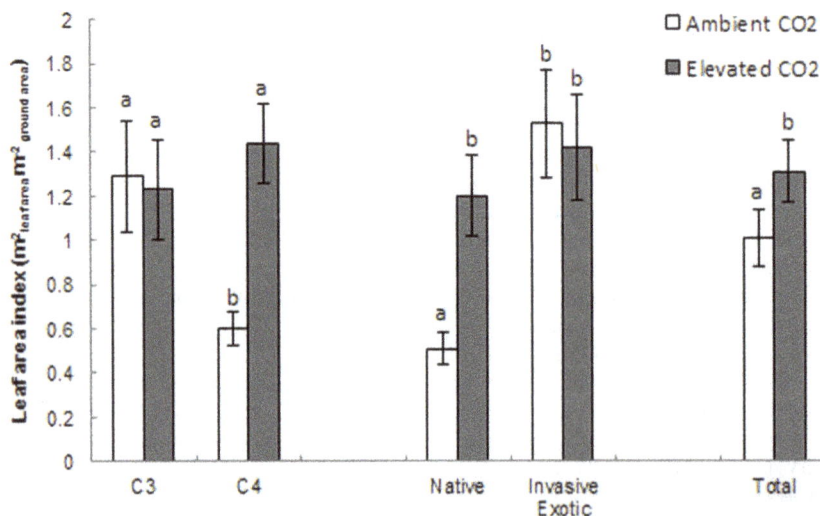

Figure 1. Leaf area index across all grass species for each $CO_2 \times$plant type combination. Mean leaf area index across all grass species for each $CO_2 \times$physiology and $CO_2 \times$origin combination. Error bars represent one standard error. Letters indicate significant differences at $p<0.05$.

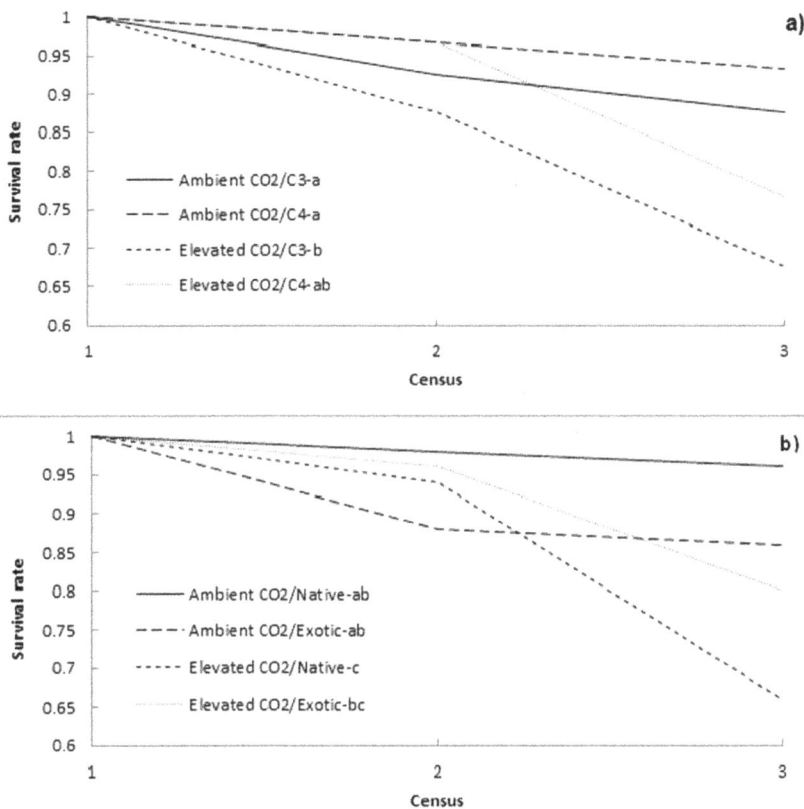

Figure 2. Survival rates across all grass species for each CO_2×plant type combination. The average survival rate across all grass species of each (a) CO_2×physiology and (b) CO_2×origin combination, at each survival census. A survival census was carried out after each extreme drought cycle which consisted of a 53 extreme drought period and 28 day recovery period. Letters indicate significant differences at p<0.05.

between CO_2 treatments (second cycle Wald $\chi^2 = 0.71$, df= 1, p = 0.400; third cycle Wald $\chi^2 = 2.504$, df= 1, p = 0.114).

Discussion

In this study we tested the vulnerability of an experimental grassland community to repeat extreme drought events under ambient and elevated CO_2 levels. Our results show that grasses grown under elevated CO_2 had significantly higher mortality in response to extreme drought than those grown under ambient CO_2. Our original hypothesis was that SWC of the mesocosms and thus vulnerability of the grasses to extreme drought would not differ between the CO_2 treatments because the reductions in stomatal conductance would be offset by increases in LAI. Our results did not support this hypothesis: soil water content was significantly lower in the elevated CO_2 treatment thus increasing extreme drought-related mortality in the grasses. The 32% reduction in stomatal conductance under elevated CO_2 was offset by a 30% increase in LAI. This offset is shown by the non-significant difference in total canopy transpiration of the grasses between the CO_2 treatments. These results contrast the results of previous studies that have found that the SWC in semi-arid grasslands increased under elevated CO_2 because the increases in total leaf area (related to LAI) were insufficient to offset the decreases in stomatal conductance [7], [21].

As the amount of water lost through total canopy transpiration did not differ between the CO_2 treatments, differences in SWC must have been due to the amount of water reaching the soil surface. Rain throughfall decreases with LAI because of increased interception and subsequent evaporation of rainfall from the

surfaces of leaves [10]. In this experiment the grasses had significantly higher LAI under elevated CO_2. This would have increased rainfall interception consequently causing differences in the SWC between the CO_2 treatments prior to the extreme drought treatments. As the experiment progressed through each drought/recovery cycle, the initial difference in SWC between the CO_2 treatments converged. This convergence of SWC between the CO_2 treatments coincided with the mortality of grasses in the elevated CO_2 mesocosms. We suggest that this is because there were fewer individuals (due to the extreme drought-related mortality) in the elevated CO_2 mesocosms which would have reduced the total canopy transpiration and rainfall interception in those mesocosms.

We hypothesised that decreases in stomatal conductance under elevated CO_2 would be greater for the C_4 and native grasses. Surprisingly we found that the exotics had the greatest reduction in stomatal conductance among the grasses under elevated CO_2. In contrast the C_4 and native grasses had greater increases in LAI under elevated CO_2 compared to the other grass species. It is often assumed that native plants are less influenced by water limited conditions (e.g. extreme drought) in comparison to exotic plants [22], [23]. Our findings suggest that the exotic grasses would be less influenced by extreme drought than C_4 and native grasses in a high CO_2 world. However it is difficult to suggest if this could significantly alter the species composition of grasslands. This is because changes in SWC can alter the structure and function of grasslands in aspects other than productivity that may influence species composition. SWC can contribute to species shifts within grasslands by changing seed production and seedling recruitment

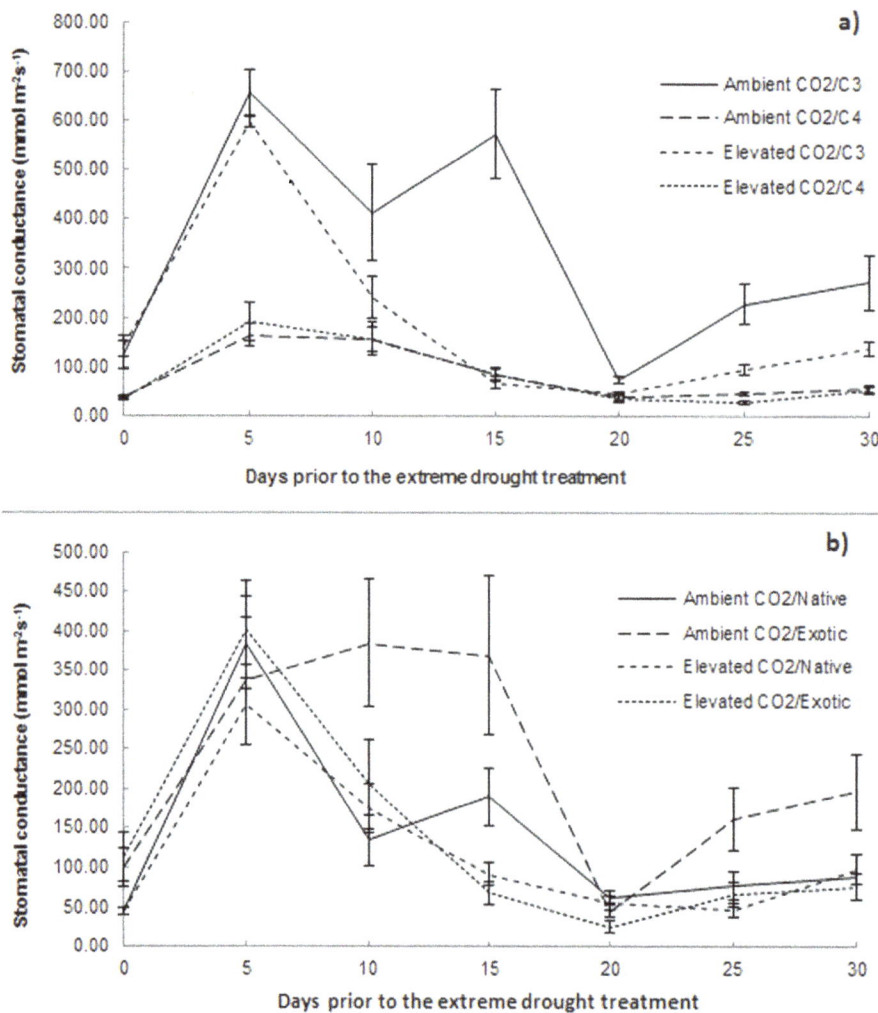

Figure 3. Stomatal conductance across all grass species for each CO$_2$×plant type combination. Mean stomatal conductance across all grass species for each (a) CO$_2$×physiology and (b) CO$_2$×origin combination, during the 30 days prior to the extreme drought treatment. Error bars represent one standard error.

among species [6], [24] and altering competitive interactions among established plants [25], [26], [27].

The findings from our study suggest that CO$_2$ concentration and soil water availability are important in mediating grassland productivity through grass dieback and mortality. It has been projected that a drastic shift in the annual global precipitation patterns may result in up to a 20% loss in soil water [28]. This is reinforced by global climate models which predict that large areas on every inhabitable continent will experience intensified droughts and widespread decreases in soil water [2]. However, net primary productivity in grasslands is projected to increase under elevated CO$_2$ conditions. Parton et al [29] modelled the effects of increased CO$_2$ for 31 temperate and tropical grassland sites, using the CENTURY model. They found that with the exception of cold desert steppe regions the net primary productivity of grasslands increased under elevated CO$_2$ conditions [29]. From our results we suggest that this projected global increase in the productivity of grasslands under elevated CO$_2$ levels may be negated (by grass dieback and mortality) due to the projected soil water constraints in the future, both as a direct consequence of changed precipitation and changes in plant-level traits such as stomatal conductance and LAI.

This study only examined the response of a single growth stage (i.e. mature) of the grasses to repeated extreme drought events under different CO$_2$ treatments. However, the vulnerability of plants to climatic extremes may be different in the early stages of their growth and development in comparison to mature plants. Therefore it is important to consider also the effect of climatic extremes on seed germination and seedling establishment as these stages are likely to strongly influence vegetation dynamics [30].

Predicting responses at local and regional scales to global change is often dependent on scaling up from plant-level mechanisms [10]. Although we attempted to make our glasshouse based mesocosm experiment as realistic a representation of the field conditions as possible, it is almost an impossible exercise to exactly recreate these conditions. For example, the soil depth in our mesocosms was 28 cm which is shallow compared to soil depth in the field. This may have caused higher mortality of the grasses in the mesocosms as they did not have access to deeper soil water as grasses in field conditions may have. Mesocosm studies such as this are important for understanding mechanisms and stimulating further research, rather than simply assessing outcomes [31]. It is important to note that we did not try to quantify exact mortality of grasses in response to extreme drought, but rather we

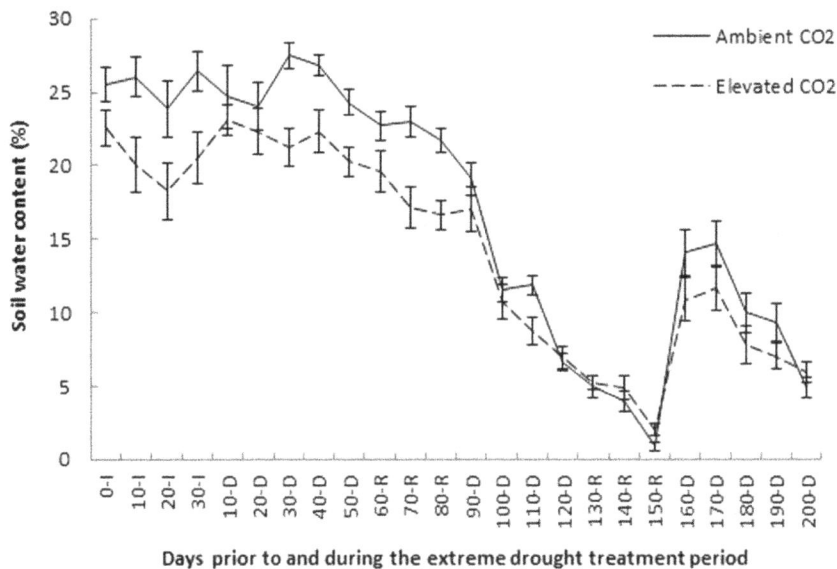

Figure 4. Soil water content across all mesocosms under ambient and elevated CO$_2$ levels. Mean soil water content across all mesocosms under ambient and elevated CO$_2$ levels over the 30 days prior to the extreme drought treatment and the drought/recovery cycles. The letters on the x-axis represent different phases of the experiment with I = the 30 day initial period prior to the extreme drought treatment, D = drought period and R = recovery period. Error bars represent one standard error.

aimed to test a potential mechanism that may be important in determining mortality which can then be scaled up to a local and regional scale.

This study has shown that vegetation response to climatic extremes is likely to be affected by CO$_2$ concentration, with extreme drought-related mortality of grasses increasing under elevated CO$_2$ levels. The suggested mechanism for this increased mortality is an increase in the LAI of the grasses which offset reductions in stomatal conductance and increased rainfall interception prior to extreme drought events. This highlights the importance of considering the interactions between climatic extremes and other aspects of climate change such as CO$_2$ concentration on plant communities. Our results also highlight the importance of better quantification of soil water in global climate

and vegetation models as this may be a key driver affecting global vegetation patterns and responses to climate change.

Acknowledgments

We gratefully acknowledge the Plant Invasion and Restoration Ecology Laboratory (PIREL) of Macquarie University for their input throughout the experiment and Muhammad Masood for assistance in the glasshouses. The experiments conducted complied with all current laws and regulations of Australia, where they were conducted.

Author Contributions

Conceived and designed the experiments: AM ML. Performed the experiments: AM. Analyzed the data: AM. Wrote the paper: AM ML.

References

1. Gutschick VP, BassiriRad H (2003) Extreme events as shaping physiology, ecology, and evolution of plants: Toward a unified definition and evaluation of their consequences. New Phytologist 160: 21–42.
2. IPCC (2011) Summary for policymakers. In Field CB, Barros V, Stocker TF, Qin D, Dokken D, Ebi KL, Mastrandrea MD, Mach KJ, Plattner G-K, Allen SK, Tignor M, Midgley PM, editors. Managing the risks of extreme events and disasters to advance climate change adaptation.
3. Easterling DR, Meehl GA, Parmesan C, Changnon SA, Karl TR, et al. (2000) Climate extremes: Observations, modelling, and impacts. Science 289: 2068–2074.
4. Asner GP, Elmore AJ, Olander LP, Martin RE, Harris AT (2004) Grazing systems, ecosystem responses and global change. Annual Review of Environment and Resources 29: 261–299.
5. Briggs JM, Knapp AK, Blair JM, Heisler JL, Hoch GA, et al. (2005) An ecosystem in transition: Causes and consequences of the conversion of mesic grassland to shrubland. BioScience 55: 243–254.
6. Morgan JA, Pataki DE, Korner C, Clark H, Del Grosso SJ, et al. (2004) Water relations in grassland and desert ecosystems exposed to elevated atmospheric CO$_2$. Oecologia 140: 11–25.
7. Morgan JA, LeCain DR, Pendall E, Blumenthal DM, Kimball BA, et al. (2011) C$_4$ grasses prosper as carbon dioxide eliminates desiccation in warmed semi-arid grassland. Nature 476: 202–205.
8. Fay PA, Kaufman DM, Nippert JB, Carlisle JD, Harper CW (2008) Changes in grassland ecosystem function due to extreme rainfall events: Implications for responses to climate change. Global Change Biology 14: 1600–1608.
9. Jentsch A, Kreyling J, Elmer M, Gellesch E, Glaser B, et al. (2011) Climate extremes initiate ecosystem-regulating functions while maintaining productivity. Journal of Ecology 99: 689–702.
10. Woodward FI (1990) Global change: Translating plant ecophysiological responses to ecosystems. Trends in Ecology and Evolution 5: 308–311.
11. Anderson LJ, Maherali H, Johnson HB, Polley HW, Jackson RB (2001) Gas exchange and photosynthetic acclimation over subambient to elevated CO$_2$ in a C$_3$–C$_4$ grassland. Global Change Biology 7: 693–707.
12. Morgan JA, LeCain DR, Mosier AR, Milchunas DG (2001) Elevated CO$_2$ enhances water relations and productivity and affects gas exchange in C$_3$ and C$_4$ grasses of the Colorado shortgrass steppe. Global Change Biology 7: 451–466.
13. Osborne CP, Sack L (2012) Evolution of C$_4$ plants: A new hypothesis for an interaction of CO$_2$ and water relations mediated by plant hydraulics. Philosophical Transactions of the Royal Society of London B: Biological Sciences 367: 583–600.
14. Cavaleri MA, Sack L (2010) Comparative water use of native and invasive plants at multiple scales: A global meta-analysis. Ecology 91: 2705–2715.
15. Newingham BA, Vanier CH, Charlet TN, Ogle K, Smith SD, et al. (2013) No cumulative effect of 10 years of elevated [CO$_2$] on perennial plant biomass components in the Mojave Desert. Global Change Biology 19: 2168–2181.
16. Wand SJE, Midgley GF, Jones MH, Curtis PS (1999) Responses of wild C$_4$ and C$_3$ grass (Poaceae) species to elevated atmospheric CO$_2$ concentration: A meta-analytic test of current theories and perceptions. Global Change Biology 5: 723–741.
17. Piao S, Friedlingstein P, Ciais P, de Noblet-Ducoudre N, Labat D, et al. (2007) Changes in climate and land use have a larger direct impact than rising CO$_2$ on

global river runoff trends. Proceedings of the National Academy of Sciences of the United States of America 104: 15242–15247.

18. Department of Infrastructure, Planning and Natural Resources (2003) Major weeds of the Cumberland Plain. In: Little D, editor. Bringing the bush back to western Sydney. pp. 47–55.

19. IPCC (2013) Climate change 2013: A physical science basis. In: Joussaume S, Penner J, Tangang F, editors. Working Group I Contribution to the IPCC 5th Assessment Report - Changes to the Underlying Scientific/Technical Assessment.

20. ETCCDI/CRD (2012) Climate change indices: Definitions of the 27 core indices. In: CC1/CLIVAR/JCOMM Expert Team, editors. Climate Change Detection and Indices.

21. LeCain DR, Morgan JA, Mosier AR, Nelson JA (2003) Soil and plant water relations determine photosynthetic responses of C_3 and C_4 grasses in a semi-arid ecosystem under elevated CO_2. Annals of Botany 92: 41–52.

22. Funk J, Zachary V (2010) Physiological responses to short-term water and light stress in native and invasive plant species in southern California. Biological Invasions 12: 1685–1694.

23. Diez JM, D'Antonio CM, Dukes JS, Grosholz ED, Olden JD, et al. (2012) Will extreme climatic events facilitate biological invasions? Frontiers in Ecology and the Environment 10: 249–257.

24. Niklaus PA, Leadley PW, Schmid B, Korner C (2001) A long-term field study on biodiversity x elevated CO_2 interactions in grassland. Ecological Monographs 71: 341–356.

25. Polley HW, Johnson HB, Derner JD (2003) Increasing CO_2 from subambient to superambient concentrations alters species composition and increases aboveground biomass in a C_3/C_4 grassland. New Phytologist 160: 319–327.

26. Dijkstra FA, Blumenthal D, Morgan JA, LeCain DR, Follett RF (2010) Elevated CO_2 effects on semi-arid grassland plants in relation to water availability and competition. Functional Ecology 24: 1152–1161.

27. Polley HW, Jin VL, Fay PA (2012) CO_2-caused change in plant species composition rivals the shift in vegetation between mid-grass and tallgrass prairies. Global Change Biology 18: 700–710.

28. Schiermeier Q (2008) Water: A long dry summer. Nature 452: 270–273.

29. Parton WJ, Scurlock JMO, Ojima DS, Schimel DS, Hall DO (1995) Impact of climate change on grassland production and soil carbon worldwide. Global Change Biology 1: 13–22.

30. Allen CD, Macalady AK, Chenchouni H, Bachelet D, McDowell N, et al. (2010) A global overview of drought and heat-induced tree mortality reveals emerging climate change risks for forests. Forest Ecology and Management 259: 660–684.

31. Benton TG, Solan M, Travis JMJ, Sait SM (2007) Microcosm experiments can inform global ecological problems. Trends in Ecology and Evolution 22: 516–521.

13

Can Timely Vector Control Interventions Triggered by Atypical Environmental Conditions Prevent Malaria Epidemics?

Peter Maes[1]*, Anthony D. Harries[2,3], Rafael Van den Bergh[4], Abdisalan Noor[5,6], Robert W. Snow[5,6], Katherine Tayler-Smith[4], Sven Gudmund Hinderaker[7], Rony Zachariah[4], Richard Allan[8]

1 Medical Department, Water, Hygiene and Sanitation Unit, Médecins Sans Frontières, Operational Center Brussels, Brussels, Belgium, 2 International Union Against Tuberculosis and Lung Disease (The Union), Paris, France, 3 London School of Hygiene and Tropical Medicine, London, United Kingdom, 4 Medical Department, Operational Research Unit (LuxOR), Operational Center Brussels, Médecins Sans Frontières -Luxembourg, Luxembourg, Luxembourg, 5 Malaria Public Health Department, KEMRI-University of Oxford-Wellcome Trust Collaborative Programme, Nairobi, Kenya, 6 Centre for Tropical Medicine, University of Oxford, Oxford, United Kingdom, 7 Centre for International Health, University of Bergen, Bergen, Norway, 8 The Mentor Initiative, Crawley, United Kingdom

Abstract

Background: Atypical environmental conditions with drought followed by heavy rainfall and flooding in arid areas in sub-Saharan Africa can lead to explosive epidemics of malaria, which might be prevented through timely vector-control interventions.

Objectives: Wajir County in Northeast Kenya is classified as having seasonal malaria transmission. The aim of this study was to describe in Wajir town the environmental conditions, the scope and timing of vector-control interventions and the associated resulting burden of malaria at two time periods (1996–1998 and 2005–2007).

Methods: This is a cross-sectional descriptive and ecological study using data collected for routine program monitoring and evaluation.

Results: In both time periods, there were atypical environmental conditions with drought and malnutrition followed by massive monthly rainfall resulting in flooding and animal/human Rift Valley Fever. In 1998, this was associated with a large and explosive malaria epidemic (weekly incidence rates peaking at 54/1,000 population/week) with vector-control interventions starting over six months after the massive rainfall and when the malaria epidemic was abating. In 2007, vector-control interventions started sooner within about three months after the massive rainfall and no malaria epidemic was recorded with weekly malaria incidence rates never exceeding 0.5 per 1,000 population per week.

Discussion and Conclusion: Did timely vector-control interventions in Wajir town prevent a malaria epidemic? In 2007, the neighboring county of Garissa experienced similar climatic events as Wajir, but vector-control interventions started six months after the heavy un-seasonal rainfall and large scale flooding resulted in a malaria epidemic with monthly incidence rates peaking at 40/1,000 population. In conclusion, this study suggests that atypical environmental conditions can herald a malaria outbreak in certain settings. In turn, this should alert responsible stakeholders about the need to act rapidly and preemptively with appropriate and wide-scale vector-control interventions to mitigate the risk.

Editor: Georges Snounou, Université Pierre et Marie Curie, France

Funding: Funding for the course was from the innumerable donors who collectively provide in 85% of the total MSF budget; was from the Department for International Development, UK; and was from Médecins Sans Frontières, Luxembourg. AMN is supported by the Wellcome Trust as an Intermediate Fellow (# 095127); RWS is supported by the Wellcome Trust as Principal Research Fellow (# 079080) and both acknowledge the support provided by the Wellcome Trust Major Overseas Programme grant to the KEMRI/Wellcome Trust Research Programme (#092654). The funders had no role in study design, data collection and analysis, decision to publish, or preparation of the manuscript.

Competing Interests: The authors have declared that no competing interests exist.

* E-mail: Peter.Maes@brussels.msf.org

Introduction

According to the World Health Organization (WHO) 2011 malaria report, there were 216 million cases of malaria globally in 2010, with over 80% occurring in sub-Saharan Africa [1]. There were 655,000 deaths, with 86% of the victims being children under the age of five years and 91% of malaria deaths occurring in Africa.

In Kenya, like many other African countries, malaria is a leading cause of morbidity and mortality, especially amongst children. Kenya has four malaria epidemiological zones: an endemic zone; seasonal malaria transmission zones; zones that are prone to malaria epidemics, and a zone of low malaria risk. There are limited data on effectiveness of interventions to prevent or control malaria outbreaks in zones where seasonal transmission or epidemics of malaria may occur. Control measures, when taken, are often implemented too late and with minimal coordination and expertise, and often under intense political pressure [2].

Wajir County in Northeast Kenya is usually dry and hot and is a zone classified as having seasonal malaria transmission. When certain environmental conditions occur, epidemics of malaria may ensue. Since 1932, there have been two recorded major malaria epidemics in the county, one in 1961 and one in late 1997. The epidemic in 1997–1998 was preceded by twelve months of high temperatures and drought that led to widespread malnutrition [3]. El Niño then caused virtually uninterrupted rainfall and the worst flooding in the county for over 50 years, resulting in large scale population displacement, and an outbreak of Rift Valley Fever (RVF) in the county. The remaining shallow flood waters covered an extensive area, and provided ideal breeding conditions for *Anopheles* mosquitoes (malaria vectors). This resulted in an exponential growth of the vector population and an explosive epidemic of malaria [4]. The poor nutritional status of the population, displacement from their homes, and lack of access to treatment services due to flooding and a national health staff strike, made individuals highly susceptible to malaria infection and to developing severe disease. Between February and May 1998, a total of 23,377 malaria cases (a malaria attack rate of 39% in the population) were reported [5]. The average crude mortality was approximately 9 per 10,000 per day and in the under-5 population this rose to approximately 28 per 10,000 per day.

Similar environmental conditions occurred in 2007 with protracted drought resulting in wide-spread under-nutrition, followed by very heavy rainfall and flooding across much of the province, causing mass population displacement and a RVF outbreak in the county. Prompted by these environmental conditions and before malaria cases had increased above normal baseline numbers, Médecins Sans Frontières (MSF) intervened with an emergency approach designed to mitigate the risk of a malaria epidemic occurring. The ongoing surveillance and case management activities in the county were promptly reinforced. Also, within ten days, malaria vector control interventions were launched with the aim of cutting malaria transmission and mitigating against the risk of a large scale malaria epidemic developing, as had occurred in 1997.

The aim of this study is to describe certain environmental conditions and the associated resulting burden of malaria in Wajir town at two time periods (1996–1998 and 2005–2007), each characterized by vector control interventions which started at different phases of the malaria transmission process.

Methods

Study Design, Data Variables and Sources
This was a cross-sectional descriptive and ecological study using previously collected data.

Data variables
The following data variables were collected:

i) for environmental conditions and their impact on the population: -daily temperature (°C); rainfall (mm per month); reported flooding in the county (yes/no); internationally reported cases of animal and/or human RVF in the county (yes/no); and malnutrition and severe acute malnutrition in the under-fives in the county and Wajir town as measured by two stage cluster surveys and defined by z-scores [6]: ii) for malaria vector control interventions:- number of houses that received indoor residual spraying; number of long-lasting impregnated nets (LLIN) distributed; and number of surface flood water bodies where

larvicides were applied: iii) for malaria:- number of admissions each month in Wajir hospital diagnosed with malaria based on clinical features; number of persons diagnosed with malaria each week in Wajir town per 1000 population (weekly malaria incidence). For the weekly malaria incidence figures, malaria was diagnosed based on clinical features of "any patient reporting a recent history of fever, in the absence of any other cause" confirmed by thin blood smears or malaria rapid diagnostic tests (Paracheck-Pf).

Data sources
Sources of data for the study included:- i) A Wellcome Trust electronic Excel 2003 database on rainfall, temperature and malaria from 1999 to 2008 [7]; ii) A Kenya Meteorological Department, Wajir station Excel 2003 data base for temperature and rainfall from 1998 to 2006; iii) MSF-Epicentre internal electronic reports on flooding, malnutrition and RVF cases in 1997/1998 and 2007; iv) MSF-Epicentre internal electronic reports on malaria vector control interventions in 1998 and 2007; v) MSF Mobile Clinic registers and Ministry of Health hospital and county reports on malaria cases in 1997/1998 and 2007; vi) The MENTOR Initiative internal electronic reports on a 2007 emergency Malaria Control intervention for the most vulnerable flood affected communities in Kenya; and vii) The Nutrition Information in Crisis Situation electronic database by the United Nations system Standing Committee on Nutrition for nutritional information..

Setting
General. Kenya is a large country in East Africa with a population of 42 million in 2011, according to the World Bank. It has eight provinces including the North Eastern Province, which was, at the time, divided into four districts, now counties, from north to south; Mandera, Wajir, Garissa and Ijara.

Wajir County. The study took place in Wajir County which is the second largest county in Kenya, covering an area of 56,501 km^2 [3]. Wajir County ranges from very arid in the north, to semi-arid in the south. In 2007, the county had 13 administrative divisions with a total of 74 locations and 88 sub-locations. Wajir County includes Wajir town, which consists of the centre of the town or Township surrounded by the five major villages of Jog Baru, Wagberi, Hodhan, Alimao and Barwaqo. The population was about 500,000 people with an annual growth of 2.5%, and the majority of the population consisted of nomadic pastoralists. Over 63% of the population depended solely on livestock for their livelihood, 57% lived in absolute poverty (<1 USD/day) and the overall literacy level was 12.5%. [8]. In 2007, the county reportedly had one district hospital in Wajir town, five sub-district hospitals, one health center and more than 30 dispensaries.

The county experiences the Sahel climate characterized by long dry spells and two rainy seasons each year from September to January and March to May. The mean annual precipitation between 1932 and 1998 was just over 300 mm [3]. These desert fringe conditions create seasonal malaria transmission during, and immediately following, the two rainy seasons. The intensity of the resulting malaria seasonal transmission varies considerably from year to year, depending largely on the climatic conditions. Temperatures are usually high and average rainfall creates surface water pools suitable for vector breeding sites. Extreme climatic conditions with unseasonal heavy rainfall can result in flooding in many parts of this county. Surface flood water, whilst initially washing away most existing mosquito larvae, creates unusually large vector breeding sites as the flood waters gradually recede and become shallow surface water in the following weeks. Mass vector breeding results in very high rates of malaria transmission, and epidemic outbreaks of malaria with high morbidity and mortality rates [9,10,11].

A RVF outbreak also requires heavy rainfall and extensive flooding of low lying grassland depressions associated with the rapid and mass emergence of *Aedes* mosquitoes reaching their maximum population about ten days after the flood event [12,13,14]. The breeding sites of the *Anopheles* mosquitoes that transmit malaria take much longer to recover from flooding compared with those of *Aedes*, and thus the *Anopheles* mosquito population reaches a maximum about one to two months after the flood event [15]. Thus, episodes of RVF often herald malaria epidemics in this sort of context.

Malaria control. In 1997/1998 MSF carried out water rehabilitation and cholera preparedness projects, and monitoring of RVF. MSF supported Wajir town hospital during this period, with out-patient and in-patient case management of malaria. Following atypical environmental conditions, an explosive malaria outbreak occurred in Wajir County between February and May 1998. MSF also started an Indoor Residual Spraying campaign in Wajir town about four to five weeks after the peak of the malaria outbreak in close collaboration with the District Public Health Office (DPHO). MERLIN, a UK non-governmental organization, worked in parallel with MSF, providing primary health care centre support, mobile clinics and insecticide treated bed-nets in flood affected areas in the county, but not in Wajir town. The Catholic mission, located in Wajir town provided additional clinical services, and the Swedish Rotarians had two volunteers providing laboratory and clinical support at the hospital. MSF left Wajir in 1998.

Heavy rains and flooding affected Wajir County and Greater Garissa from the end of September 2006, creating conditions very similar to those leading to the large malaria epidemic in 1997/1998. In response, MSF managed an outbreak of RVF in the neighbouring Garissa county and in January 2007 launched integrated malaria vector control activities in Wajir town.

Study participants

Study participants included all adult and pediatric (less than 15 years of age) inpatients diagnosed with malaria at Wajir district hospital and all outpatients diagnosed with malaria in Wajir town during the two time periods (1996–1998 and 2005–2007).

Analysis and statistics

Data were entered into an electronic Excel 2003 database and a simple descriptive analysis was carried out.

Ethics statement

This study met the Médecins Sans Frontières' Ethics Review Board (Geneva, Switzerland) -approved criteria for analysis of routinely-collected program data, and was also approved by the Ethics Advisory Group of the International Union Against Tuberculosis and Lung Disease, Paris, France. Analysis was done on previously collected program data only: patients were not contacted, and no identifying characteristics of patients were collected. As such, informed consent of involved patients was not indicated or sought.

Results

Environmental conditions and hospital malaria admissions: 1996–1998

The monthly rainfall, the presence of floods and RVF and the number of malaria admissions to Wajir hospital between January 1996 and December 1998 are shown in **Figure 1**. The seasonal periods of rainfall followed by drought are shown up to August 1997. From September 1997 to February 1998 there was very heavy monthly rainfall, at its peak reaching nearly 500 mm for the month of November. This resulted in extensive flooding between the months of October 1997 to February 1998 and RVF from November 1997 to February 1998. The number of malaria admissions from January 1996 to December 1997 was relatively stable ranging from 13 to 43 per month (with the exception of December when there was a hospital strike and no admissions). The increase in malaria admissions was first noted in January 1998 with 92 cases for the month and this peaked at 456 in February, before dropping to 427 in March and 69 in April.

During this three year period the daily temperature never dropped below 21°C. The prevalence of malnutrition and severe acute malnutrition in children under the age of five years in the county and the town during the three year period is shown in **Table 1**.

Environmental conditions and hospital malaria admissions: 2005–2007

The monthly rainfall, the presence of floods and RVF, and the number of malaria admissions in persons aged 15 years and below to Wajir hospital from January 2005 to December 2007 are shown in **Figure 2**. The seasonal periods of rainfall followed by drought are shown up to September 2006. At the end of September 2006, very heavy rains began, and in October the monthly rainfall reached almost 250 mm, declining to 95 mm in November and then back to the monthly average. This resulted in widespread flooding between the months of October and December 2006 and a RVF outbreak from November 2006 to January 2007. The number of malaria admissions in persons under the age of 15 years remained relatively stable during the whole three year period and there was no increase in cases during or after the time of rainfall and flooding.

During this three year period the daily recorded temperature never fell below 22°C. The prevalence of malnutrition and severe acute malnutrition in children under the age of five years in the county was 15.6% and 4.1% respectively in May 2006 and 23.0% and 2.8% in April 2007.

The scope of malaria vector control interventions: 1998 and 2007

The scope of malaria vector control interventions including the number of houses sprayed, LLINs distributed, modeled LLIN coverage based on the numbers distributed, and shallow pools treated with larvicides in Wajir town in 1998 and 2007 is shown in **Table 2**. Over 90% of houses were sprayed in the two periods, but in 1998 there were no LLINs distributed in the town, and no shallow pools were treated, in contrast to 2007.

Timing of malaria vector control interventions and malaria incidence: 1998 and 2007

In 1998, the incidence of malaria per 1000 population per week and the timing of indoor residual spraying are shown in **Figure 3**. The incidence of all malaria cases and malaria in children under the age of five years peaked at the end of March 1998 at 54 per

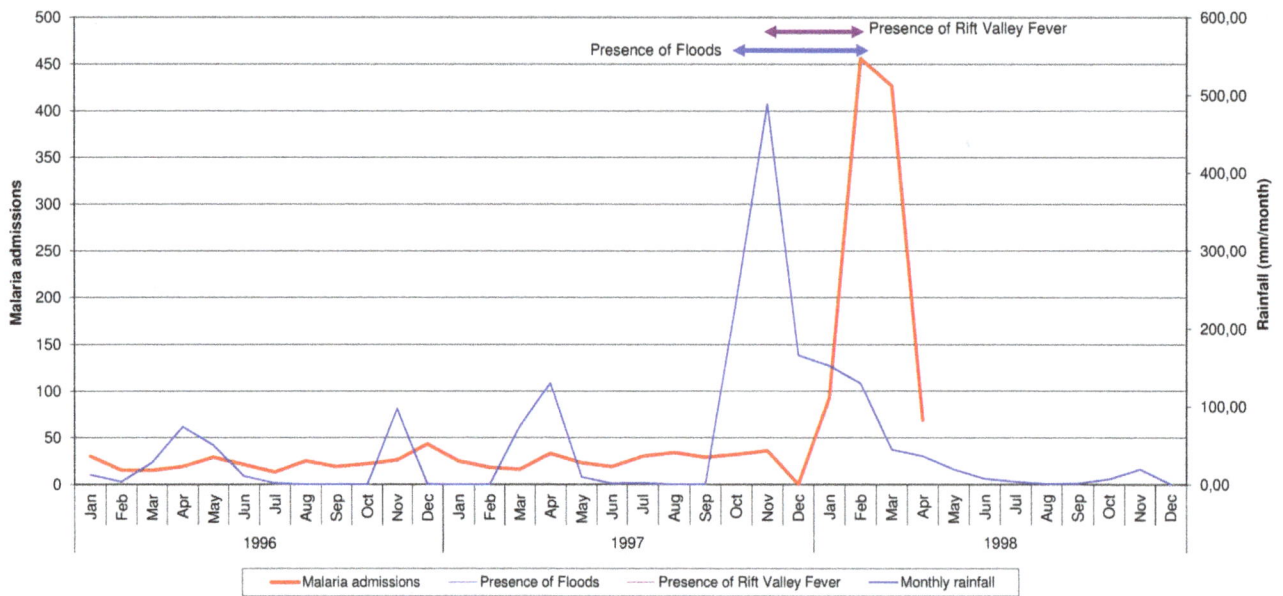

Figure 1. Monthly rainfall, presence of floods and Rift Valley Fever and malaria admissions to Wajir Hospital: January 1996–December 1998.

1000 population per week and 46 per 1000 per week respectively. Indoor residual spraying started in mid-April reaching full house coverage of Wajir Town after two weeks. However, the start of this intervention occurred at a time when malaria incidence was already decreasing.

In 2007, the incidence of malaria per 1000 population per week and the timing of indoor residual spraying and distribution of LLINs are shown in **Figure 4**. The two malaria vector control interventions illustrated in Figure 3 and the treatment of shallow pools with larvicides started in the third week of January before any increase in malaria incidence was observed. It took ten days between the decision to intervene and the start of activities. Seven weeks after starting activities, 93% of all houses in Wajir town were sprayed and there was 76% household coverage of LLINs meaning that 76% of Wajir town population had received one LLIN per 1.8 persons. The incidence of all malaria cases and malaria in children under the age of five years remained low at less than 0.5 per 1000 population per week between the months of January to March in Wajir Town. The malaria surveillance in the peripheral areas illustrated that the malaria incidence in Wajir

County did not exceed 1 per 1000 population per week between the months of January to March.

Discussion

This cross-sectional descriptive and ecological study showed that in 1997/1998 and 2006/2007 Wajir County in north-eastern Kenya experienced atypical environmental conditions with drought and malnutrition, followed by massive monthly rainfall resulting in flooding and animal/human RVF. In 1998, this was associated with a large and explosive epidemic of malaria resulting in large numbers of admissions to Wajir Hospital and a weekly malaria incidence of 40–55 cases per 1000 population per week in all persons and children. Vector control interventions at that time consisted solely of indoor residual spraying of houses, and these started over 6 months after the onset of heavy rains when the malaria epidemic was already abating. In 2007, as a result of similar environmental conditions, MSF instituted a package of vector control interventions that included indoor residual spraying, distribution of LLINs and applying larvicides to shallow pools. These interventions started over three months after the onset of heavy rains and no malaria epidemic was recorded with weekly

Table 1. Documented prevalence of malnutrition in Wajir county and Wajir town in children under the age of five years from 1996–1998.

Site	Year Month	Prevalence of Malnutrition %	Prevalence of severe acute malnutrition %	Source of data
Wajir County	Oct 1996	27.9	3.8	OXFAM
	Feb 1997	25.1	4.2	MERLIN
	Jul 1997	8.8	1.0	OXFAM/MOH
	Mar 1998	16.0	2.9	OXFAM
Wajir Town	Jul 1997	15.0	1.7	OXFAM/MOH[1]
	Mar 1998	25.3	3.7	MSF

[1]Ministry of Health.

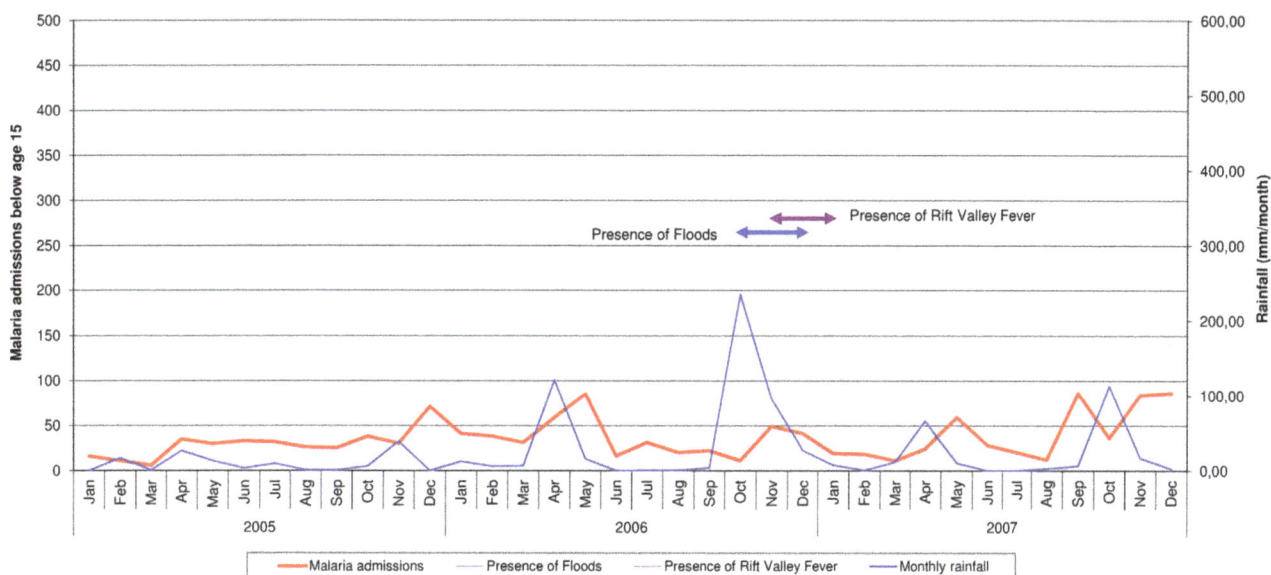

Figure 2. Monthly rainfall, presence of floods and Rift Valley Fever and malaria admissions in persons under 15 years of age to Wajir Hospital: January 2005–December 2007.

malaria incidence consistently below 0.5 per 1000 population per week (100 times less than in 1998).

The big question is whether these vector control interventions triggered by environmental cues, prevented a malaria epidemic? This is difficult to definitively answer, but the neighboring county of Greater Garissa County provides some sort of control arm. This county has similar characteristics to Wajir (semi arid) and experienced very similar atypical environmental conditions as Wajir County in 2006/2007 with massive rainfall followed by flooding, mass population displacement, and human/animal cases of RVF [16]. Malnutrition rates during the same time period were similar to those experienced in Wajir, reaching a mean of 24% global acute malnutrition in 2006 [17]. In this case, donors were slow to respond and large scale vector control interventions by the MENTOR initiative in Garissa did not start until March 2007, over six months after the start of heavy rains, which is similar to what happened in Wajir in 1997/1998. In Garissa County, there was almost a doubling of reported malaria cases from 7719 in 2006 to 13739 in 2007, and monthly malaria incidence peaked at 42 per 1000 population in January 2007 and continued at high levels of above 30 per 1000 per month for the next three to four months [16]. Thus, while impossible to formally compare these two regions due to differences in data collection systems, Garissa

can be said to have experienced similar climatic events as Wajir, malaria control interventions were started much later, and there was a significant increase in malaria cases [16].

There are several strengths to this study. Weekly malaria incidence figures at the town level in Wajir were based on clinical features observed first-hand by MSF combined with thin blood smears showing malaria parasites or positive malaria rapid diagnostic tests. There was also a robust multidisciplinary approach with several different and concurring sources of data on the measurement of environmental changes and the impact on the population such as rates of malnutrition and RVF.

Limitations relate to the operational nature of the study. First, malaria admission data from Wajir hospital during the two time periods were obtained from different sources, they did not describe the same populations and the diagnosis might have been made without parasite or serological confirmation. Second, the catchment area for Wajir Hospital might have been the county as well as the town, and it was principally the town that was targeted for vector control interventions by MSF. Third, we only have second-hand data over a limited period of time, from a control county such as Garissa, collected through separate channels precluding formal comparison - more data would have helped to strengthen the case for linking timely vector control interventions and malaria prevention in Wajir county in 2006/2007. Additionally, there was a report by Epicentre that in the neighboring county of Ijara, no malaria outbreak occurred between 22 January and 11 February 2007. While distribution of insecticide-treated bed nets reportedly also took place in this county, the scope and timing of such interventions could not be discerned, and it is therefore difficult to assess whether Ijara can be regarded as a control county for Wajir [18].

The rainfall was truly exceptional in 1996–1998. In 2006 there was massive rainfall, although not to the level or the duration experienced ten years previously. However, major outbreaks of RVF have occurred in this county only during the two study periods. This phenomenon requires heavy rainfall and extensive flooding of low lying grassland depressions associated with the rapid and mass emergence of *Aedes* mosquitoes, and such episodes

Table 2. Scope of malaria control interventions in 1998 and 2007 in Wajir Town.

Malaria control interventions	1998	2007
Number of houses	9468	28536
Number (%) houses sprayed	9372 (99%)	26544 (93%)
Number of LLINs needed for 100% coverage	No data	45708
Number (coverage %) LLINs distributed	0	34716 (76%)
Shallow pools treated with larvicides	0	5

LLIN = long-lasting insecticide treated nets.

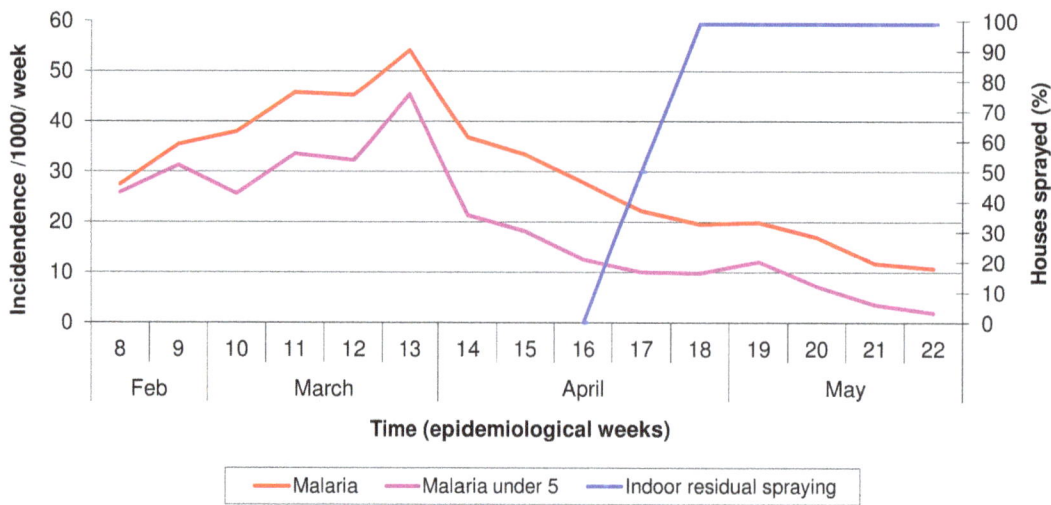

Figure 3. Incidence of malaria per 1000 population per week in Wajir town and timing of indoor residual spraying in 1998.

of RVF often herald malaria epidemics in this sort of context. Furthermore, in Wajir county the high temperatures which stayed above 22°C favor the development of the malaria parasite in the Anopheles mosquitoes [19].

Wajir County exemplifies the epidemiology of an unstable malaria zone: this is typified by limited overall rainfall, protracted periods of drought with few infective bites, limited transmission of *Plasmodium falciparum* and thus low collective immunity of the population for malaria which is then followed by heavy rainfall leading to conditions that are ripe for a true malaria epidemic. Other people living in moderate to high malaria transmission regions of the country experience many infective bites throughout the year, and the children that survive repeated malaria infections in their first five years of life develop partial protective immunity to severe *P. falciparum* malaria [5,19,20,21]. Periods of drought also cause widespread malnutrition and there is evidence that malaria prevalence is further increased in such circumstances [22].

Vector control interventions, such as the ones used in the study, are well-established methods for controlling malaria at the

community level [23,24,25]. However, in an area with seasonal malaria transmission it seems vital to implement these interventions in a timely way. For this, there needs to be sufficient information for Ministries of Health and Aid agencies to raise a red early warning flag to prevent a malaria epidemic from occurring [26]. This study suggests that very heavy rainfall in an area like Wajir County that is followed by flooding and reports of RVF should be enough to trigger a wide scale vector control intervention to prevent epidemic malaria. This concurs with previous suggestions [2,27] that area specific monitoring of rainfall is very important, and that this can provide key early epidemic warning indicators to inform responsible stakeholders of the need to mount rapid integrated epidemic prevention responses.

In conclusion, this study suggests that atypical environmental conditions with massive rainfall, flooding and RVF following a period of drought can herald a malaria outbreak in semi arid settings such as the North Eastern Province of Kenya, and that this could be prevented by timely, preemptive and rapid delivery of appropriate vector control interventions. A paradigm shift might

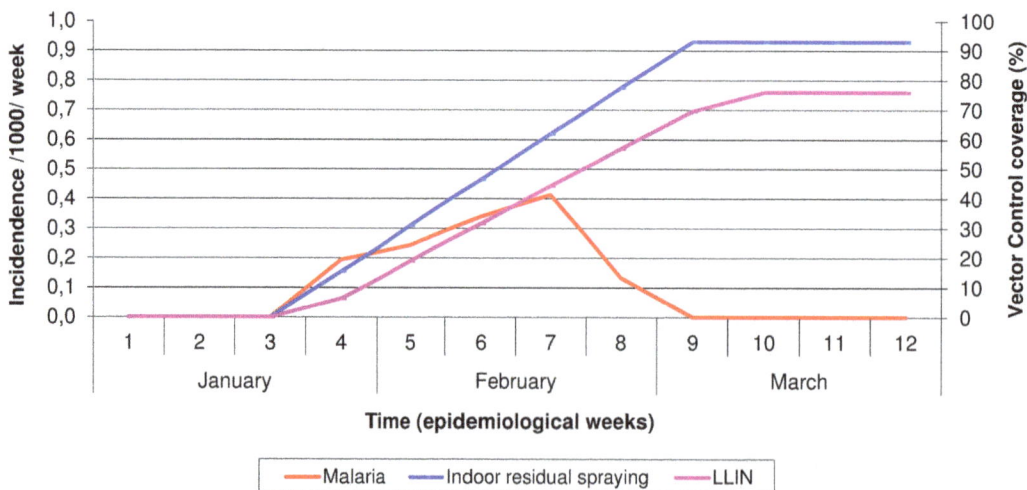

Figure 4. Incidence of malaria per 1000 population per week in Wajir town and timing of indoor residual spraying and modeled long-lasting insecticidal nets coverage in 2007.

be needed that allows resources to be made available to start rapid and widespread scale up of vector control interventions before any significant rise in malaria cases is declared.

Acknowledgments

This research was supported through an operational research course, which was jointly developed and run by the Operational Research Unit (LUXOR), Médecins Sans Frontières, Brussels-Luxembourg, The Centre for Operational Research, International Union Against Tuberculosis and Lung Disease, France, and The Union South-East Asia Regional Office. Additional support for running the course was provided by the Center for International Health, University of Bergen, Norway and the Institute of Tropical Medicine, Antwerp, Belgium. We are grateful to Dr Emelda Okiro for help with the data.

Author Contributions

Conceived and designed the experiments: PM ADH RVDB KTS SGH RZ RA. Performed the experiments: PM ADH RVDB AMN RWS RA. Analyzed the data: PM ADH RVDB KTS SGH. Contributed reagents/materials/analysis tools: PM ADH RVDB AMN RWS RZ SGH RA. Wrote the paper: PM ADH RVDB AMN RWS KTS SGH RZ RA.

References

1. World Health Organization (2011) WHO Global Malaria Report 2011. World Health Organization, Geneva, Switzerland.
2. WHO (2001) Malaria early warning systems. Concepts, indicators and partners. Roll Back Malaria, World Health Organization, Geneva, Switzerland. WHO/CDS/RBM/2001.32.
3. Snow RW, Ikoku A, Omumbo J, Ouma J (1999) The epidemiology, politics and control of malaria epidemics in Kenya:1900–1998. Report for the Roll Back Malaria. Resource network on epidemics, World Health Organization, Geneva, Switzerland.
4. Allan R, Nam S, Doull L (1998) MERLIN and malaria epidemic in north-east Kenya. Lancet 351: 1966–7.
5. Brown V, Issak MA, Rossi M, Barboza P, Paugam A (1998) Epidemic of malaria in North East Kenya. Lancet 352: 1356–1357.
6. WHO and UNICEF (2009) WHO child growth standards and the identification of sever acute malnutrition in infants and children. World Health Organization, Geneva, Switzerland. Available: http://www.who.int/nutrition/publications/severemalnutrition/9789241598163_eng.pdf (accessed 12 June 2013).
7. Okiro E A, Alegana V A, Noor A M, Mutheu J J, 1, Elizabeth Juma, et al (2009) Malaria paediatric hospitalization between 1999 and 2008 across Kenya. BMC Medicine 7:75 doi:10.1186/1741-7015-7-75.
8. Kenya National Bureau of Statistics (KNBS) and ICF Macro (2010) Kenya Demographic and Health Survey 2008–09. Calverton, Meryland: KNBS and ICF Macro.
9. Githeko AK, Lindsay SW, Confalonieri UE, Patz JA (2000) Climate change and vector-borne diseases – a regional analysis. Bull World Health Organ 78: 1136–1147.
10. Zhou G, Minakawa N, Githeko AK, Yan G (2004) Association between climate variability and malaria epidemics in the East African highlands. PNAS 101: 2375–2380
11. Peterson A (2009) Shifting suitability for malaria vectors across Africa with warming climates. BMC Infectious Diseases 9:59
12. Woods C, Karpati A, Grein T, McCarthy N, Gaturuku P, et al. (2002) An outbreak of Rift Valley Fever in Northeastern Kenya, 1997–98. Emerg Infect Dis 8: 138–144
13. Nguku PM, Sharif SK, Mutonga D, Amwayi S, Omolo J, et al. (2010) An investigation of a major outbreak of Rift Valley Fever in Kenya, 2006–2007. Am J Trop Med Hyg 83: 5–13.
14. Sang R, Kioko E, Lutomiah J, Warigia M, Ochieng C, et al. (2010) Rift Valley Fever Virus epidemic in Kenya, 2006/2007: the entomological investigations. Am J Trop Med Hyg 83: 28–37.
15. Linthicum KJ, Davies FG, Bailey CL (1983) Mosquito species succession in a dambo in an east African forest. Mosquito News 43: 464–470.
16. Allan R (2013) Basic data sheet. Extension and Expansion of Emergency Malaria Control for vulnerable communities in North Eastern and Coastal Province, Kenya. The MENTOR Initiative (contact person richard@mentor-intiative.net).
17. USAID (2006) Understanding nutrition data and the causes of malnutrition in Kenya. A special report by the Famine Early Warning Systems Network (FEWSNET).
18. Grandesso F, Wichman Ole. Investigation of a Rift valley Fever outbreak and implementation if a malaria surveillance system. Ijara County, Kenya 2007. Paris, EpiCentre Medicins sans Frontier: 2007. P. 3–4.
19. Craig MH, Snow RW, le Sueur (1999) A climate-based distribution model of malaria transmission in Sub-Saharan Africa. Parasitology Today 15: 105–111.
20. Hay S, Rogers DJ, Shanks GD, Myers MF, Snow RW (2001) Malaria early warning in Kenya. Trends Parasitol 17:95–99.
21. Reiter P, Thomas CJ, Atkinson PM, Hay SI, Randolph SE, et al. (2003). Global warming and malaria: a call for accuracy. The lancet Infectious diseases 4:323–324.
22. Khogali M, Zachariah R, Keiluhu A, Van den Brande K, Tayler-Smith K, et al. (2011) Detection of malaria in relation to fever and grade of malnutrition amongst malnourished children in Ethiopia. Public Health Action 1: 16–18.
23. Lengeler C (2009) Insecticide-treated bed nets and curtains for preventing malaria. The Cochrane Library, Issue 2.
24. Pluess B, Tanser FC, Lengeler C, Sharp BL (2010) Indoor residual spraying for preventing malaria. Cochrane Database Syst Rev 4: CD006657.
25. Alba S, Hetzel MW, Nathan R, Alexander M, Lengeler C (2011) Assessing the impact of malaria interventions on morbidity through a community-based surveillance system. Int J Epidemiol 40: 405–416.
26. Snow B, Hay S, Noor AM (2006) Rapid situation analysis of possible malaria epidemic in Wajir County, North Eastern Kenya and Somalia. Nairobi: Kemri-Wellcome Trust. Report.
27. Hay S, Renshaw M, Ochola S, Noor AM, Snow RW (2003) Performance of forecasting, warning and deterction of malaria epidemics in the highlands of western Kenya. Trends Parasitol 19: 394–399.

14

Plant Responses to Extreme Climatic Events: A Field Test of Resilience Capacity at the Southern Range Edge

Asier Herrero[1,2]*, Regino Zamora[1]

1 Department of Ecology, University of Granada, Granada, Andalusia, Spain, 2 Department of Life Sciences, University of Alcalá, Alcalá de Henares, Madrid, Spain

Abstract

The expected and already observed increment in frequency of extreme climatic events may result in severe vegetation shifts. However, stabilizing mechanisms promoting community resilience can buffer the lasting impact of extreme events. The present work analyzes the resilience of a Mediterranean mountain ecosystem after an extreme drought in 2005, examining shoot-growth and needle-length resistance and resilience of dominant tree and shrub species (*Pinus sylvestris* vs *Juniperus communis*, and *P. nigra* vs *J. oxycedrus*) in two contrasting altitudinal ranges. Recorded high vegetative-resilience values indicate great tolerance to extreme droughts for the dominant species of pine-juniper woodlands. Observed tolerance could act as a stabilizing mechanism in rear range edges, such as the Mediterranean basin, where extreme events are predicted to be more detrimental and recurrent. However, resistance and resilience components vary across species, sites, and ontogenetic states: adult *Pinus* showed higher growth resistance than did adult *Juniperus*; saplings displayed higher recovery rates than did conspecific adults; and *P. nigra* saplings displayed higher resilience than did *P. sylvestris* saplings where the two species coexist. *P. nigra* and *J. oxycedrus* saplings at high and low elevations, respectively, were the most resilient at all the locations studied. Under recurrent extreme droughts, these species-specific differences in resistance and resilience could promote changes in vegetation structure and composition, even in areas with high tolerance to dry conditions.

Editor: Edward Webb, National University of Singapore, Singapore

Funding: This study was supported by Ministerio de Ciencia e Innovación (Spanish Government) Projects CGL2008-04794 and CGL2011-29910 to R.Z., and by grant FPU-MEC (AP2005-1561) to A. H. The funders had no role in study design, data collection and analysis, decision to publish, or preparation of the manuscript.

Competing Interests: The authors have declared that no competing interests exist.

* E-mail: asier@ugr.es

Introduction

Extreme drought and warm events are closely related to growth reductions and mortality of woody species in forest ecosystems across the planet [1]. Recurrent and extreme droughts impact woody species performance differently through species-specific sensitivity, leading to changes in species composition [2,3,4,5,6,7]. In this respect, differences in drought sensitivity between functional types, such as trees and shrubs, can alter vegetation structure, shifting from a tree-dominated landscape to a shrub-dominated one [2,4,6]. However, stabilizing processes promoting community resilience can palliate and offset the aftermath of extreme events [8]. While resistance can be considered the force of an ecosystem, community or individual to oppose change exerted by an external disturbance [9], resilience is defined as the capacity to restore pre-disturbance structure and function (analogous to 'engineering resilience', see [10]). In this context, the analysis of woody species resistance and resilience is particularly crucial under the rising frequency of extreme events [11,12,13].

The study of ecosystem responses in terms of resistance and resilience to extreme events can help to forecast ecosystem changes, as future average conditions will be close to current extreme events [14]. At the community level, resistance and resilience after a single extreme event has been related to diversity [15] and resource availability [9]. However, assessments of the consequences of extreme climatic events at the individual and/or population level are limited by a lack of rigorous and testable methods that enable quantifications of plant responses to extreme events under field conditions [8].

The main objective of this study is to analyze the resistance and resilience of a Mediterranean mountain ecosystem to an extreme drought event in 2005, monitoring performance of dominant tree and shrubs species before, during, and afterwards. Boreo-alpine tree *Pinus sylvestris* L. subsp. *nevadensis* Christ and shrub *Juniperus communis* L. are the dominant species along the oromediterranean belt (1800–2000 m a.s.l.), while Mediterranean tree *Pinus nigra* Arnold and shrub *Juniperus oxycedrus* Sibth & Sm are the dominant ones in the supramediterranean belt (1400–1700 m). The species studied were situated close to their southernmost distribution limit, forming natural relict populations in the study area (particularly *P. sylvestris* and *J. communis*; [16]). The impact of extreme climatic events are expected to be more detrimental in populations living at the edge of the distribution range, as those populations are far from that species' optimum conditions. However, observed past persistence in relict populations at rear edges [17] suggest some degree of tolerance to extreme climatic events. Thus, alternatively, the examined rear-edge populations might show an acclimated response to the extreme drought thanks to different stabilizing processes, such as site-specific environmental conditions or stress tolerance capacity linked to local adaptation [8]. Analyses of plant resistance and resilience in rear-edge populations, as in the present work, will help to forecast future shifts in species distributions, as major range contractions are expected in southern ranges [18,19].

We compare resistance and resilience between species and environmental conditions, considering different ontogenetic states, in order to assess the tree and shrub dominant species response to an extreme drought event. Regarding life form (*Pinus* trees vs. *Juniperus* shrubs), lower resistance can be expected in trees due to stronger stomatal control during drought [20,7]. With respect to tree species comparison (*P. sylvestris* vs. *P. nigra*), we expect a lower resilience to an extreme drought for *P. sylvestris* due to its boreo-alpine biogeographical origin [21]. Concerning the environmental gradients, higher resilience can be expected for populations located at higher elevations and/or northern exposures than for those at lower elevations and/or southern exposures, because of wetter and cooler conditions in the former. Regarding ontogenetic stage, adults can show alternatively higher resilience to drought than saplings owing to deeper root system, or lower resilience due to higher vulnerability to xylem embolism, greater water use per unit of time [3], and/or slower shoot growth rates [22]. Such comprehensive analysis of tree and shrubs resistance and resilience allows the testing of community tolerance to extreme droughts and the identification of dynamics associated with predicted climatic changes.

In summary, the specific questions addressed in the present study are: 1) Do tree species show lower resilience and resistance than shrubs? 2) Does *P. sylvestris* show lower resilience and resistance than *P. nigra*? 3) Do pine species show lower resilience and resistance at a low elevation and/or southern exposition? 4) Do adults show lower resilience and resistance than saplings?

Materials and Methods

Study site and species

The study was conducted at Sierra de Baza Natural Park (SE Spain, 2°51′48″W, 37°22′57″N). All necessary permits for the field studies described herein (which did not involve endangered or protected species) were obtained thanks to Juan Romero, Director of Sierra de Baza Natural Park. The climate is Mediterranean, characterized by cold winters and hot summers, with pronounced summer drought (June-August). Precipitation is concentrated mainly in autumn and spring. The annual and summer rainfall is 495±33 mm and 31±9 mm, respectively (mean ± SE for period 1991–2006; Cortijo Narváez metereological station, 1360 m a.s.l.). The study species are dominant in their altitudinal belt, forming characteristic vegetation types. In the oromediterranean belt (1800–2000 m a.s.l.), while *P. sylvestris* subsp. *nevadensis* is the main tree species, *J. communis* is the main shrub covering the forest understory and open areas. On the other hand, in the supramediterranean belt (1400–1700), *P. nigra* and *J. oxycedrus* are the dominant tree and shrub species, respectively. In 2005 the most extreme drought in the last six decades occurred in Western Europe [23], with climate records in the study area (Cortijo Narváez meteorological station) registering the driest year since 1947.

Drought index

A drought index (DRI) was calculated for the study site to display the severity of the 2005 extreme drought. The DRI was calculated for the period 1947–2008 using the following formula:

$$DRI = P - PET$$

where P is equal to the sum of the precipitation from January to December, and PET equals the sum of estimated potential evapotranspiration for the same period as a function of monthly mean temperatures and geographical latitude (using Thornthwaite

formulation [24]). Monthly total precipitation data was recorded in Cortijo Narváez meteorological station (1360 m a.s.l.; at 900 m to the low altitude plots), very close to the study area. However, monthly mean temperature data was collected from the nearest meteorological station, at Baza village (2°46′24″W, 37°29′23″N), as there are no temperature records in Cortijo Narváez. Temperature data from Baza only cover the period 1990–2009, so data from the CRU TS 2.1 high-resolution gridded data set [25] was used to extend temperature data back to 1947. Linear regressions were performed between local temperature data and the CRU data set, being always significant at $P<0.05$, with R^2 ranging from 0.41 to 0.89 (approximately 60% of cases showed a R^2 higher or equal to 0.62). Thereafter, these linear regression equations were used to infer local temperature data from 1947 to 2008. More negative DRI values indicate more severe moisture deficits. DRI data are shown in Figure 1, with 2005 being the lowest value for the period 1947–2008. Thus, we consider 2005 an extreme drought year, since it presented a DRI value located at the lower end of the range of observed values for the studied period [26].

Sampling design

Different *P. sylvestris* and *P. nigra* populations were monitored in natural relict forests at Sierra de Baza. *P. sylvestris* populations were sampled on north- and south-facing slopes of the same valley (2000 m), while *P. nigra* populations were monitored following an altitudinal gradient: at high (2000 m), medium (1700 m), and low elevations (1500 m). South-facing *P. sylvestris* and high-elevation *P. nigra* populations coincide spatially, forming a mixed forest. *J. communis* and *J. oxycedrus* were sampled at the same north-facing locations of *P. sylvestris* and low-elevation *P. nigra* populations, respectively. For each location, two plots of 1–2 ha each were established, being at least 600 m away from each other. In each plot, large mature adults and non-reproductive saplings were sampled, avoiding individuals with significant herbivory or physical damages. See Table 1 for further information about monitored plots and adult and sapling sizes. All measurements of plant size (height, basal diameter, diameter at breast height, and cover area) were made in late autumn 2008.

Shoot- and needle-growth resistance and resilience were analyzed in the four dominant tree and shrubs species (*Pinus sylvestris* vs *Juniperus communis*, and *P. nigra* vs *J. oxycedrus*), which showed only one shoot- and needle-growth flush per year in the study area. The existing literature demonstrated that both shoot growth and needle length can be used as indicators of plant responses to water supply, providing a straightforward field sampling measure to analyze short term responses to extreme climatic events in an easy and testable way. For example, shoot growth has been used as an indicator of environmental favorability [27] as well as to measure the impact of drought conditions on plant growth [28,29,30,31]. On the other hand, needle length is also a good indicator of tree responses to water availability [32,33]. Due to the long retention time of needles in the species considered, branches bore multiple needle cohorts, enabling shoot- and needle-growth changes to be easily compared.

Trees. For tree species, 10 representative mature trees and 15 saplings of similar size were recorded haphazardly in each plot. Height and DBH (diameter at breast height) in adults, and height, basal diameter, and age in saplings were recorded (see Table 1). Adult height was measured using a Vertex IV hypsometer (Haglöf, Sweden). Sapling age was estimated by counting the number of annual bud scars or whorls [34,35,36] as the two pine species showed one flush per year in the study area. Longitudinal shoot growth in adults was measured in 10 branches per tree, five facing

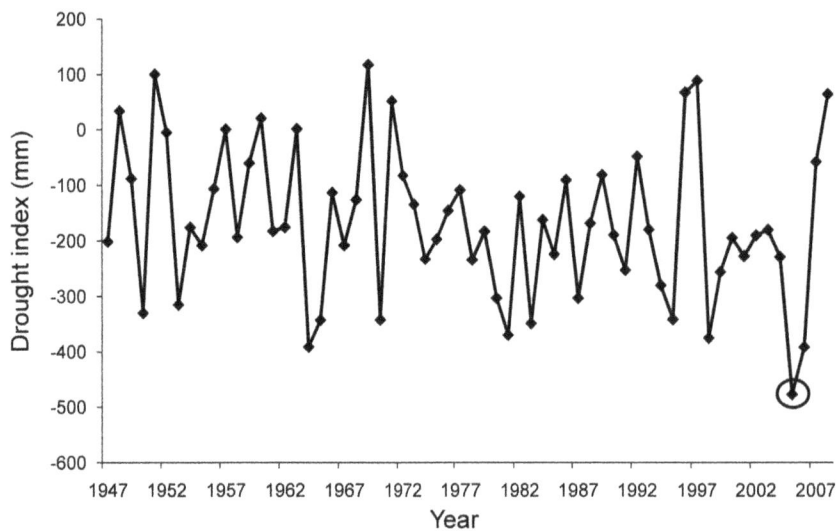

Figure 1. Drought index for 1947–2008 series. 2005, highlighted by a circle, was an extreme drought year.

north and five south. All the branches were tagged, measuring the same branches in the different samplings. Measured branches belonged to medium or low tree crown. Values from the ten branches were averaged to obtain a unique value per individual for each year. In saplings, shoot growth was measured in the leader shoot. Shoot growth of each year was identified using annual whorls and yearly bud scars from 2003 to 2008. Needle length was measured in three needles per shoot-growth cohort, which were randomly recorded. Shoot growth and needle length were measured from winter 2006 to late autumn 2008.

Shrubs. In each plot, 20 adults (10 males and 10 females) and 20 saplings of similar size were haphazardly recorded. All adults

were pooled due to the absence of significant differences between sexes in the recorded variables. Annual longitudinal shoot growth was measured in 10 and 5 branches for adults and saplings, respectively. Values from the measured branches were averaged to obtain a unique value per individual for each year. Measurements were made from the 2004 to 2008 cohort based on differences in color and diameter showed by the different cohorts. Needle length was also measured in three needles of each shoot-growth cohort. Shoot growth and needle length were measured from winter 2006 to late autumn 2008.

Table 1. Adults and saplings size in each sampled plot.

| Species | Altitude | Exposure | Plot | Adults | | | Saplings | | |
				Height (m)[a]	DBH (cm)	Cover area (m²)[c]	Height (cm)[a]	Basal diameter (cm)[b]	Cover area (m²)[c]
P. sylvestris	2065	N	1	9.55±0.69	44.86±2.84	-	112.2±5.85	4.44±0.28	-
P. sylvestris	2037	N	2	7.64±0.26	43.35±2.82	-	93.33±8.12	3.68±0.33	-
P. sylvestris	2008	S	1	8.89±0.61	43.42±3.92	-	92.63±8.74	5.09±0.64	-
P. sylvestris	2067	S	2	7.78±0.36	49.07±3.75	-	110.93±8.33	4.07±0.43	-
P. nigra	2008	S	1	9.74±0.67	46.9±4.04	-	111.29±6.55	4.11±0.33	-
P. nigra	2067	S	2	9.79±0.62	49.31±2.80	-	99.47±7.51	4.39±0.33	-
P. nigra	1753	NE	1	9.6±0.35	34.10±1.20		101.63±6.73	3.3±0.17	
P. nigra	1694	NW	2	8.74±0.52	35.86±2.51		103.87±6.10	4.52±0.25	
P. nigra	1525	NW	1	8.51±0.6	31.06±2.06	-	99.61±6.18	4.88±0.28	-
P. nigra	1544	NE	2	8.69±0.24	33.78±1.03	-	92.63±5.9	4.2±0.17	-
J. communis	2065	N	1	-	-	28.57±2.85	-	-	0.31±0.08
J. communis	2037	N	2	-	-	14.44±1.18	-	-	0.11±0.02
J. oxycedrus	1525	NW	1	1.96±0.08	-	4.29±0.48	0.34±0.02	-	0.06±0.01
J. oxycedrus	1544	NE	2	1.9±0.11	-	4.52±0.49	0.44±0.03	-	0.14±0.02

Cover area was calculated measuring maximum and minimum canopy diameters. Values are shown as mean ± standard error. DBH: diameter at breast height.
[a]Height was not recorded for J. communis, as it presents a prostrate growth form.
[b]Basal diameter was not quantified for the two Juniperus species due to measurement difficulties and to the common multi-trunk growth pattern.
[c]Maximum and minimum canopy diameters were measured to calculate the canopy cover area.

Resistance and resilience components

To analyze resistance and resilience to 2005 extreme drought in shoot growth and needle length of considered species, we calculated resistance, recovery, resilience, and relative resilience for both variables following the procedure of Lloret and others [37]. Resistance, the inverse of the performance reduction during the extreme drought, was calculated as the ratio between performance during and before drought. Recovery, the ability to recover relative to the performance reduction undergone during drought, was calculated as the ratio between performance after and during the extreme drought. Resilience, the capacity to return to pre-drought performance levels, was calculated as the ratio between the performance after and before drought. Relative resilience is the resilience weighted by the performance reduction during drought and was calculated using the following formula:

$$Relative\ resilience = (PostDr-Dr)/PreDr$$

where *PreDr*, *Dr* and *PostDr* indicate performance before, during, and after drought, respectively. Performance before and after drought were calculated as the average over a two-year period, and performance during drought as the values for the year 2005. However, we made some modifications taking into account the shoot-growth patterns of the species studied. For the shoot growth of pine species, 2003 and 2004 corresponded to pre-drought values, 2005 and 2006 to during-drought values, and 2007 and 2008 to post-drought values. In 2005, extreme drought affected 2005 and 2006 pine shoot cohorts, as the conditions during bud formation can affect the following year's shoot growth [38,39]. For shoot growth of shrub species, 2004 corresponded to pre-drought values, 2005 to during-drought values, and 2007 and 2008 to post-drought values. Pre-drought values included only 2004, as the identification of 2003 shoot cohort was not possible when the study began (winter 2006). In contrast to *Pinus*, *Juniperus* presented an indeterminate shoot growth, with only 2005 shoot cohort being affected by the extreme drought (see Fig. 1 and 2). Although only 2007 and 2008 were considered for post-drought values, the inclusion of 2006 values did not change the results. Finally, for needle growth, drought values include only the 2005 cohort for both pines and shrubs, as needle length appeared to respond to the dry conditions of the current season (Fig. S1 and S2; see also [38]).

Data analysis

Shoot-growth and needle-length resistance, recovery, resilience, and relative resilience were analyzed in a search for differences between species and locations (exposure and altitude) considering two different ontogenetic states (large adults/non-reproductive saplings). Our 'experimental unit' was the individual tree or shrub, for which shoot-growth and needle-length values were averaged. Afterwards, different resistance and resilience components were calculated as explained above. Three species comparisons were performed: 1) *P. sylvestris* vs. *J. communis* with a northern exposure at a high elevation; 2) *P. sylvestris* vs. *P. nigra* with a southern exposure at a high elevation; and 3) *J. oxycedrus* vs. *P. nigra* at a low elevation. For locations, two comparisons were made: 1) between northern and southern exposures for *P. sylvestris*; and 2) between high, medium, and low elevations for *P. nigra*. Differences between species and locations were analyzed using General Linear Mixed Models (GLMM), with species (or location), ontogenetic state and their interaction as fixed factors, and plot as a random factor. Shoot-growth or needle-length resistance, recovery, resilience or relative resilience was used as the dependent variable in each case. *Post hoc* comparisons between groups were performed using

Tukey's HSD test. All the analyses were performed using JMP 7.0 (SAS Institute Inc.). All results throughout this paper are given as mean ± standard error.

Results

Shoot growth

Figure 2 and 3 showed shoot growth for the period 2003–2008 (2004–2008 for *Juniperus*) for adults and saplings of the considered four species at sampled locations.

P. sylvestris vs. J. communis. *P. sylvestris* presented significantly higher resistance but lower relative resilience than did *J. communis* for adults as well as saplings (Fig. 4A, 4D; Table 2). Adult *P. sylvestris* showed slightly negative relative resilience, underlining the incomplete recovery in shoot growth for this case (Fig. 4D; Table 2). On the other hand, *J. communis* adults presented significantly higher recovery than did *P. sylvestris* adults (Fig. 4B; Table 2).

P. sylvestris vs. P. nigra. Saplings of both species displayed significantly higher recovery and relative resilience than conspecific adults, relative resilience also being significantly higher in *P. nigra* (Fig. 4B, 4D; Table 2). Finally, *P. nigra* saplings showed the greatest resilience values (Fig. 4C; Table 2).

P. nigra vs. J. oxycedrus. *J. oxycedrus* showed significantly higher recovery and relative resilience than *P. nigra*, values being significantly higher in saplings (Fig. 4B, 4D; Table 2). *P. nigra* adults showed significantly higher resistance than did conspecific saplings and *J. oxycedrus*, while *J. oxycedrus* saplings showed significantly higher resilience than did conspecific adults and *P. nigra* (Fig. 4A, 4C; Table 2).

P. sylvestris: Exposure. No significant differences were found between exposures in any resistance and resilience components (Fig. 5; Table 2). While adults showed significantly higher resistance than saplings, saplings showed significantly higher recovery (Fig. 5A, 5B; Table 2).

P. nigra: Altitude. Adults showed significantly lower recovery and relative resilience values than saplings (Fig. 5B, 5D; Table 2). Differences in altitude clearly appeared between saplings, with resistance and resilience being significantly stronger at the high altitude than at the low one (Fig. 5A, 5C; Table 2).

Needle length

Figure S1 and S2 showed needle length for the period 2003–2008 (2004–2008 for *Juniperus*) for adults and saplings of the considered four species at sampled locations.

P. sylvestris vs. J. communis. *J. communis* showed significantly higher resistance and resilience than *P. sylvestris* for both adults and saplings (Fig. S3A, S3C; Table 3). Furthermore, adults of *J. communis* displayed significantly higher recovery than did conspecific saplings and *P. sylvestris* (Fig. S3B; Table 3).

P. sylvestris vs. P. nigra. *P. nigra* displayed significantly higher recovery and relative resilience than *P. sylvestris*, values being significantly higher in saplings (Fig. S3B, S3D; Table 3). *P. nigra* also showed significantly higher resilience than did *P. sylvestris* (Fig. S3C; Table 3).

P. nigra vs. J. oxycedrus. *J. oxycedrus* showed higher resistance but lower recovery and relative resilience than did *P. nigra*, with saplings showing lower resistance but higher recovery and relative resilience (Fig. S3; Table 3). Finally, *J. oxycedrus* showed significantly higher resilience than did *P. nigra* (Fig. S3C; Table 3).

P. sylvestris: Exposure. *P. sylvestris* having a northern exposure showed significantly higher recovery and resilience than having a southern exposure, with saplings showing significant

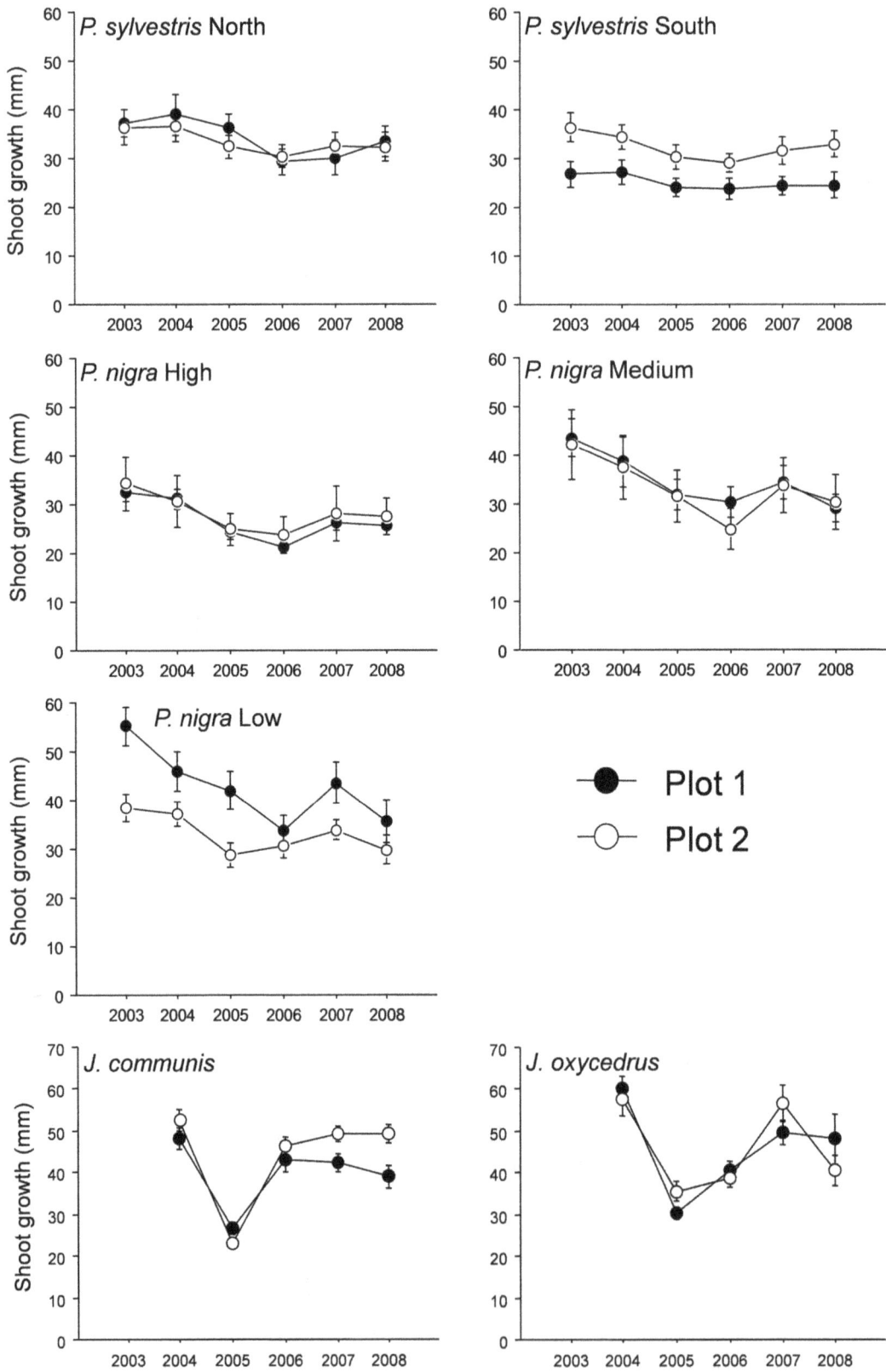

Figure 2. Shoot growth for the period 2003–2008 for adults of the four species at sampled locations. Data for a *Pinus sylvestris* with southern and northern exposure, for *P. nigra* at high (2000 m), medium (1700 m) and low elevation (1500 m), and for *Juniperus communis* and *J. oxycedrus* are shown.

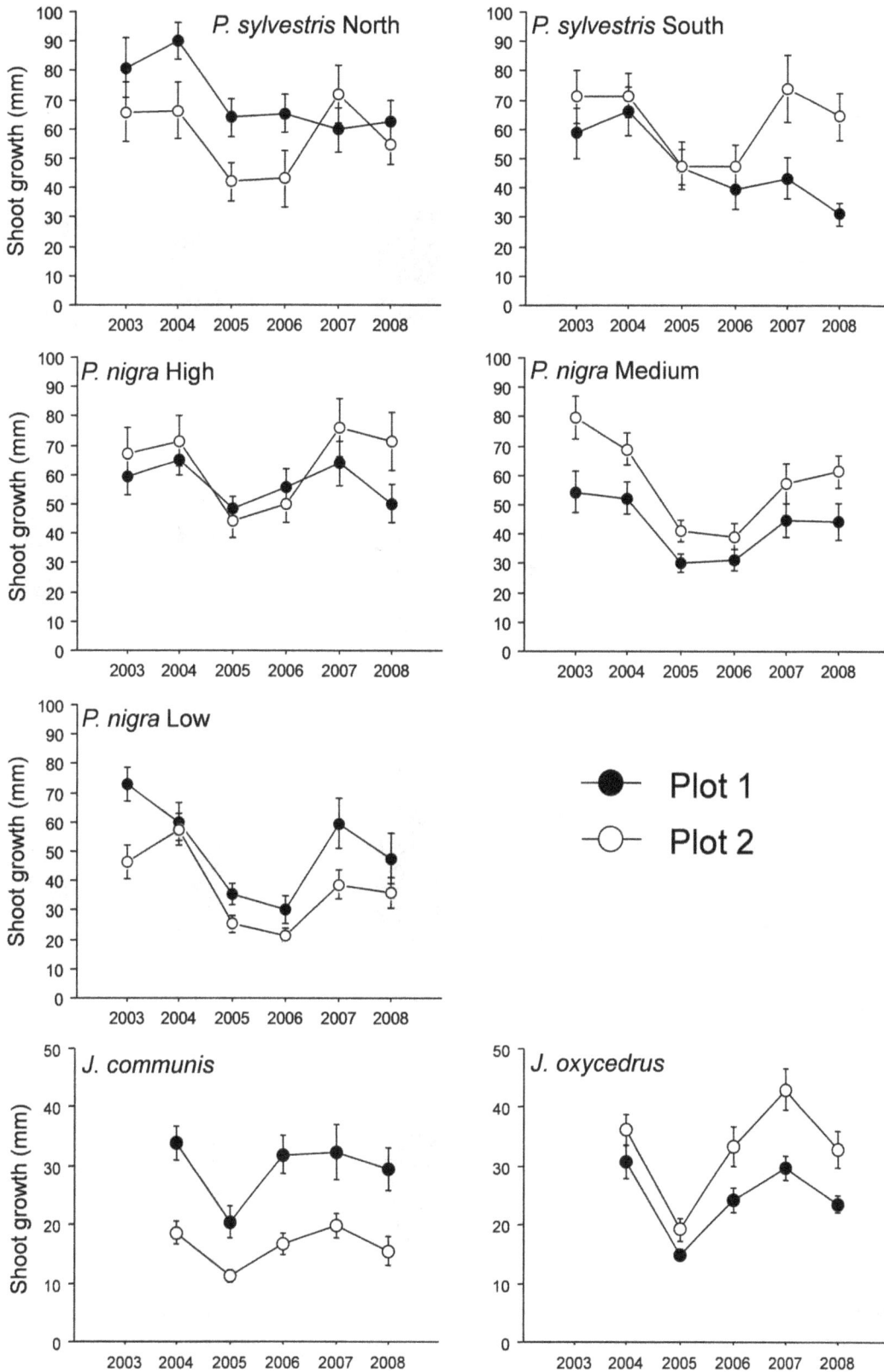

Figure 3. Shoot growth for the period 2003–2008 for saplings of the four species at sampled locations. Data for a *Pinus sylvestris* with southern and northern exposure, for *P. nigra* at high (2000 m), medium (1700 m) and low elevation (1500 m), and for *Juniperus communis* and *J. oxycedrus* are shown.

P. sylvestris vs J.communis

P. sylvestris vs P. nigra

P. nigra vs J. oxycedrus

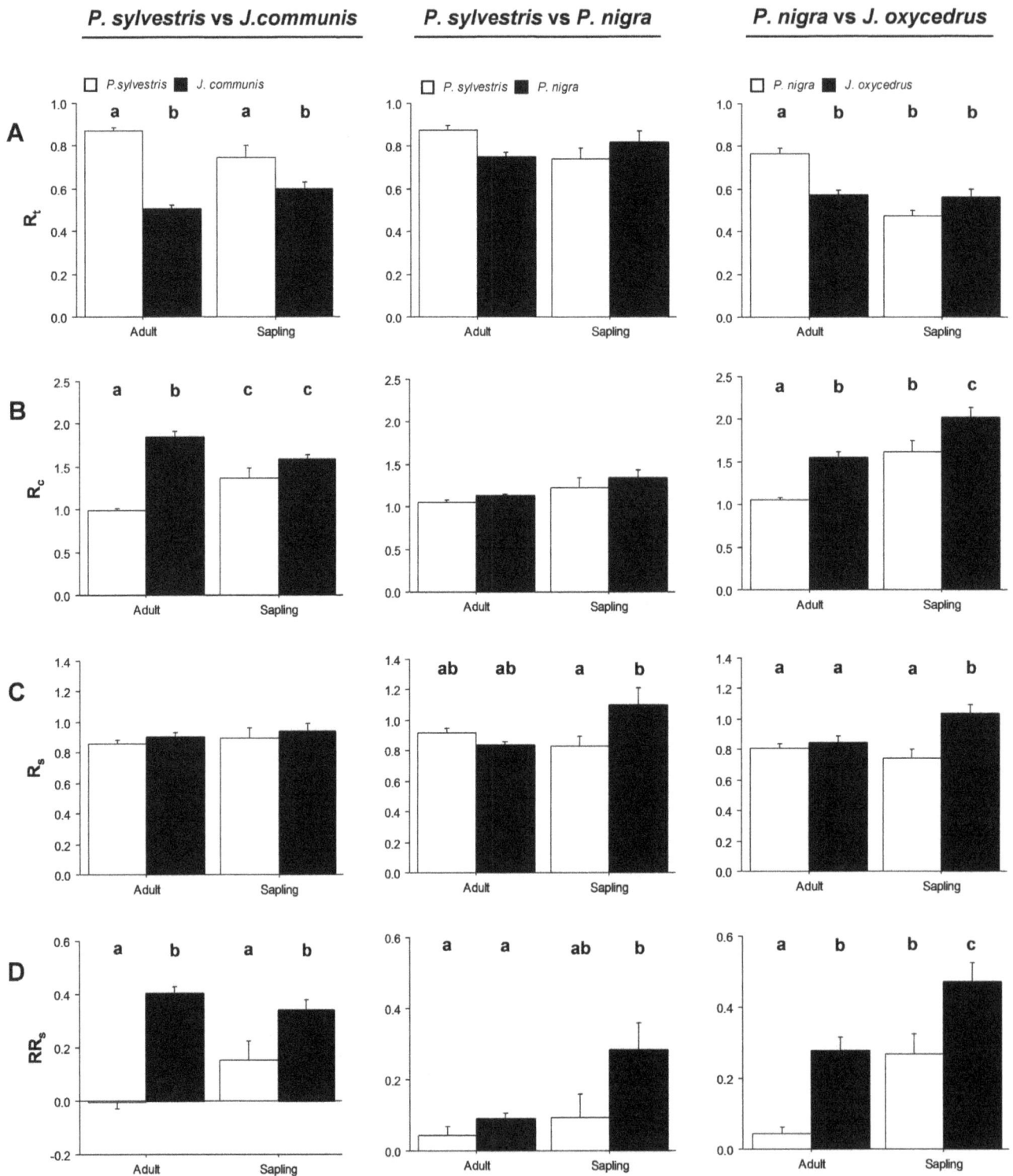

Figure 4. Differences in shoot-growth resistance (A), recovery (B), resilience (C) and relative resilience (D) between species and ontogenetic states (adults/saplings). Three comparisons are shown: *P. sylvestris* vs. *J. communis* with a northern exposure at high elevation; *P. sylvestris* vs. *P. nigra* with a southern exposure at high elevation; and *J. oxycedrus* vs. *P. nigra* at low elevation. Different letters above bars indicate significant *post hoc* differences between groups. Bars indicate the standard errors of calculated means.

higher recovery (Fig. S4B, S4C; Table 3). Saplings presented also significantly higher relative resilience than did adults (Fig. S4D, Table 3).

***P. nigra*: Altitude.** *P. nigra* trees showed significant differences in altitude for resistance, recovery, and relative resilience, especially for saplings (Fig. S4, Table 3). For resistance, the highest values were for high-elevation individuals and the lowest for low-

Table 2. Summary of GLMM analysis for shoot-growth resistance (R_t), recovery (R_c), resilience (R_s), and relative resilience (RR_s) for species and location comparisons.

	R_t		R_c		R_s		RR_s	
	F	P	F	P	F	P	F	P
P. sylvestris vs. *J. communis*								
Species	53.452	<0.0001	52.534	<0.0001	0.820	0.367	51.972	<0.0001
Ontogenetic state (Ont)	0.159	0.690	0.464	0.4968	0.593	0.443	1.486	0.2251
Species x Ont	10.132	**0.0018**	18.854	<0.0001	0.0001	0.991	7.289	**0.0079**
P. sylvestris vs. *P. nigra*								
Species	0.196	0.658	1.145	0.2874	1.637	0.2039	4.077	**0.0463**
Ontogenetic state (Ont)	0.501	0.481	4.491	**0.0367**	1.343	0.2494	4.299	**0.0408**
Species x Ont	4.783	**0.031**	0.069	0.7928	5.277	**0.0238**	1.556	0.2153
P. nigra vs. *J. oxycedrus*								
Species	2.915	0.0902	15.614	**0.0001**	8.719	**0.0038**	18.613	<0.0001
Ontogenetic state (Ont)	21.939	<0.0001	20.553	<0.0001	1.148	0.286	17.438	<0.0001
Sp x Ont	18.962	<0.0001	0.112	0.738	5.194	**0.0244**	0.098	0.754
P. sylvestris: Exposure								
Exposure	0.0001	0.992	0.206	0.651	0.014	0.905	0.013	0.907
Ontogenetic state (Ont)	6.796	**0.010**	7.623	**0.006**	0.201	0.655	3.809	0.053
Exposure x Ont	0.013	0.910	1.085	0.300	1.221	0.272	1.099	0.297
P. nigra: Altitude								
Altitude	10.216	<0.0001	0.633	0.532	4.258	**0.016**	0.244	0.784
Ontogenetic state (Ont)	16.649	<0.0001	27.785	<0.0001	2.359	0.127	25.598	<0.0001
Altitude x Ont	12.081	<0.0001	1.729	0.181	2.947	0.055	0.044	0.956

Species comparisons comprise *P. sylvestris* vs. *J. communis*, *P. sylvestris* vs. *P. nigra*, and *P. nigra* vs. *J. oxycedrus*. Location comparisons comprise exposure and altitude differences for *P. sylvestris* and *P. nigra*, respectively.

elevation ones, showing the opposite pattern in the case of recovery and relative resilience (Fig. S4).

Discussion

In this study, we empirically apply the concepts of resistance and resilience to patterns of tree and shrub growth, using shoot-length and needle-length as indicators of plant responses to an extreme drought event. The 2005 drought was the most extreme drought in the study area in the last six decades, even triggering pine sapling mortality in the nearby Sierra Nevada [40]. Our empirical results indicate that *Pinus* and *Juniperus* species at their southern distribution edge present great tolerance to an extreme drought event, as demonstrated by the high vegetative (shoot and needle growth) resilience values recorded across species, sites, and ontogenetic states. In fact, resilience values were in general higher than 0.8 which indicate that post-drought values were close to pre-drought ones ($R_t = 1$ indicate identical growth values before and after drought). Thus, the impact of the 2005 extreme drought after three years was rather low, supporting our hypothesis that dominant species of Mediterranean pine-juniper woodlands presents high tolerance and resilience to extreme droughts at their southern distribution edge. Although we cannot compare resilience capacity of southern populations with northern ones, which is beyond the scope of this study, our results are of special relevance under the climate change scenario, since strong distributional shifts and local extinctions are expected at the southern range edge associated with increasing aridity conditions [18,19].

Observed tolerance ability at the study area could be related to plant adaptation to Mediterranean dry conditions. In fact, high genetic differentiation of southern *P. sylvestris* and *P. nigra* populations [41,42] suggest high adaptation to the local environment. For instance, *P. sylvestris* population at the study area showed lower vulnerability to embolism than did other Northern European populations [43]. In addition, in an experimental study, Mediterranean *P. sylvestris* provenance showed higher emergence and survival than more northern provenance under different precipitation regimes [44]. Overall, the study species might present specific resilience component values above a hypothetic mortality threshold [37], as no die-back symptoms were detected. It is important to note that no mortality was observed in the study area associated with the 2005 extreme drought. Thus, dominance and maintenance of pine-juniper woodlands in Mediterranean mountains are fostered by the remarkable survival ability and longevity of mature individuals (persistence, *sensu* [45]) as well as high tolerance to extreme droughts of adults and saplings.

Despite that the overall high resilience, resistance and resilience components varied across species and ontogenetic states. Adults of both *Juniperus* species showed lower growth resistance (greater reduction of growth) than did *Pinus* adults. We expected the opposite pattern, as *Juniperus* present an anisohydric regulation, allowing higher stomatal conductance and thus higher photosynthetic uptake to be sustained under dry conditions than in isohydric *Pinus* [20,7]. However, the deeper root system of trees presumably provides them access to deeper groundwater, thereby boosting stomatal conductance during the 2005 extreme drought [46]. But *Juniperus* species displayed higher relative resilience than

P. sylvestris: Exposure

P. nigra: Altitude

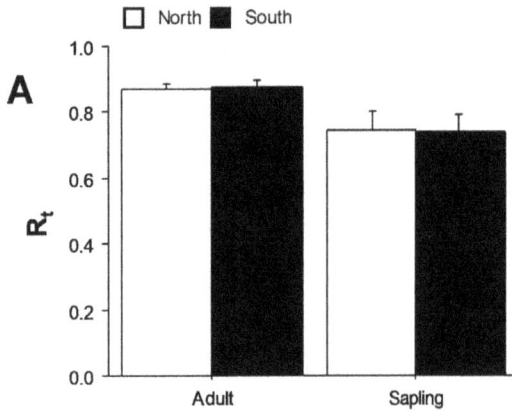

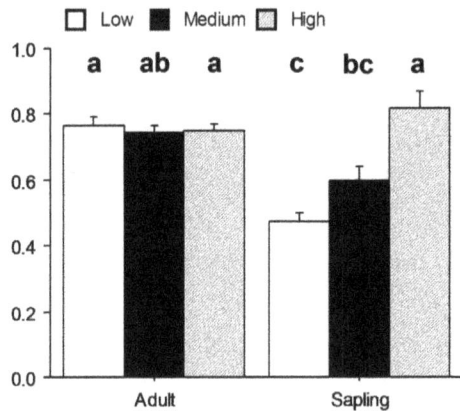

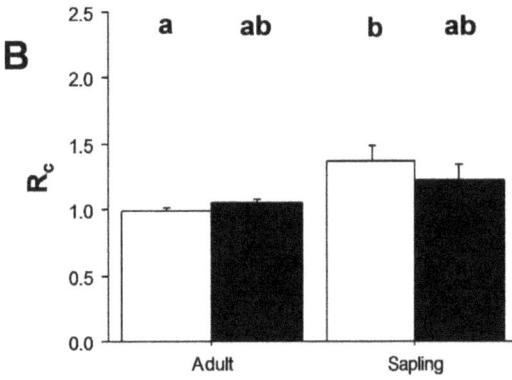

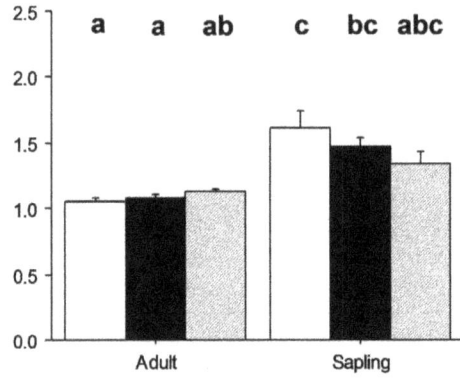

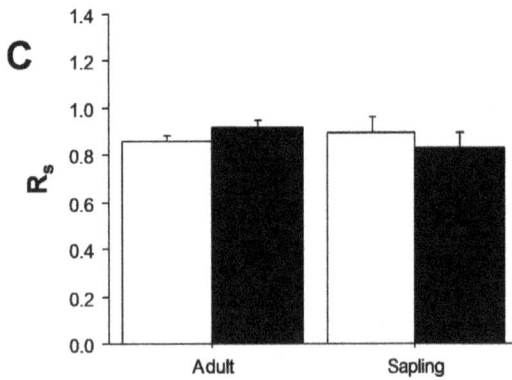

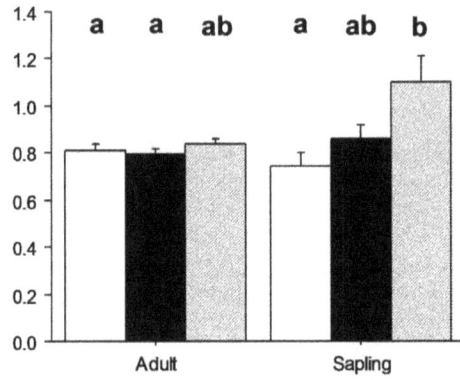

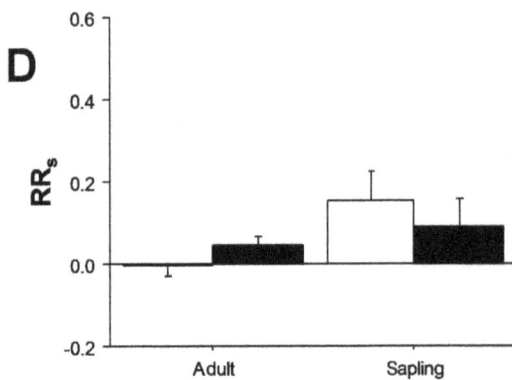

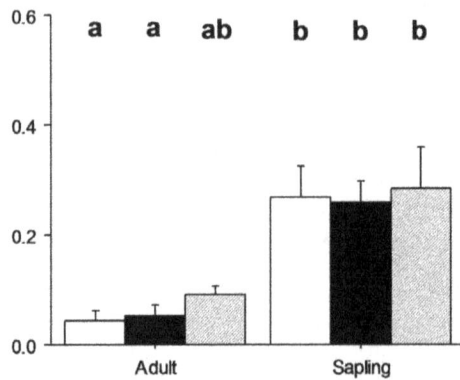

Figure 5. Differences in shoot-growth resistance (A), recovery (B), resilience (C), and relative resilience (D) between locations and ontogenetic states (adults/saplings). Two comparisons are shown: between northern and southern exposure for *P. sylvestris*; and between high (2000 m), medium (1700 m), and low (1500 m) elevations for *P. nigra*. Different letters above bars indicate significant *post hoc* differences between groups. Bars indicate the standard errors of calculated means.

did *Pinus* species, both for saplings and for adults, revealing the capacity of *Juniperus* to recover from heavier growth reductions than *Pinus* after an extreme drought event. Of special importance are the high resilience values registered by *J. oxycedrus* saplings at the low elevation, in comparison with coexisting *P. nigra* saplings. Higher drought-induced mortality for *Pinus* species in comparison with *J. monosperma* in the western USA [2,6,47] suggests less mortality risk for *Juniperus* species. Thus, differences in growth resilience between *P. nigra* and *J. oxycedrus* at sapling stage, as well as mortality risk for adults, could encourage a shift towards a shrub dominated forest at low elevations, as has been reported in other pine-juniper woodlands [2,6,47]. Overall, the extreme drought impact was stronger at the low altitude, as recorded in other studies [2,48,49].

Similarly, higher resilience of *P. nigra* saplings than *P. sylvestris* ones, may play an important role under a scenario of recurrent extreme droughts. Several studies indicate higher vulnerability to drought for *P. sylvestris* than for *P. nigra* over ontogeny in locations where the two species coexist [50,51,52,40]. In fact, in the last few years, drought-induced growth declines and mortality events have been recorded in many southern *P. sylvestris* populations [4,5,31,53,40]. Biotic factors, such pests or browsing, can exacerbate drought vulnerability, inflicting severe damage

[54,55]. In the study area, higher ungulate preference for *P. sylvestris* over *P. nigra* reinforced their climatic responses at the treeline, aggravating drought vulnerability of *P. sylvestris* [55]. Therefore, the higher resilience of *P. nigra* saplings, coupled with its lower vulnerability to drought and browsing, could favor a change in dominance toward this Mediterranean species at high elevations.

In general, both *Pinus* and *Juniper* saplings showed higher recovery than did adults for all the exposures and elevations considered. This recovery capacity might be due to the observed higher shoot-growth rate in saplings than in adults [22], promoting growth recovery after the extreme drought. In addition, *P. nigra* and *J. oxycedrus* saplings at high and low elevations, respectively, were the most resilient in terms of shoot growth. In fact, they were the only cases where shoot-growth resilience reached values higher than one, indicating greater growth values after drought than before drought.

Our study provided a new perspective on the analysis of vegetation responses to climatic events at the individual and population level. Differences in resistance and resilience between dominant tree and shrub species, as observed in this study, can heavily influence vegetation dynamics. Under recurrent extreme droughts, and progressively warmer and drier conditions, such

Table 3. Summary of GLMM analysis for needle-length resistance (R_t), recovery (R_c), resilience (R_s), and relative resilience (RR_s) for species and location comparisons.

	R_t		R_c		R_s		RR_s	
	F	P	F	P	F	P	F	P
P. sylvestris vs. *J. communis*								
Species	21.777	**<0.0001**	1.947	0.165	32.224	**<0.0001**	0.020	0.886
Ontogenetic state (Ont)	1.340	0.249	4.297	**0.0402**	0.159	0.690	1.527	0.219
Species x Ont	0.107	0.743	4.723	**0.0316**	4.497	**0.0359**	2.829	0.095
P. sylvestris vs. *P. nigra*								
Species	0.268	0.606	9.253	**0.0031**	8.606	**0.0043**	4.885	**0.0298**
Ontogenetic state (Ont)	2.888	0.093	10.187	**0.0020**	2.266	0.136	10.819	**0.0015**
Species x Ont	0.449	0.505	3.320	0.072	1.055	0.307	0.945	0.333
P. nigra vs. *J. oxycedrus*								
Species	63.698	**<0.0001**	74.103	**<0.0001**	4.1915	**0.0427**	57.762	**<0.0001**
Ontogenetic state (Ont)	23.592	**<0.0001**	49.452	**<0.0001**	0.075	0.7842	32.525	**<0.0001**
Sp x Ont	4.296	**0.0403**	25.462	**<0.0001**	0.097	0.7553	8.393	**0.0045**
P. sylvestris: Exposure								
Exposure	1.165	0.283	6.180	**0.0147**	4.334	**0.0401**	1.810	0.182
Ontogenetic state (Ont)	1.472	0.228	6.030	**0.0159**	1.490	0.225	5.816	**0.0179**
Exposure x Ont	0.007	0.932	0.831	0.364	0.287	0.593	0.231	0.632
P. nigra: Altitude								
Altitude	17.210	**<0.0001**	9.760	**0.0001**	1.894	0.154	4.525	**0.0125**
Ontogenetic state (Ont)	40.032	**<0.0001**	55.928	**<0.0001**	0.418	0.519	54.036	**<0.0001**
Altitude x Ont	2.996	0.0533	2.850	0.0613	0.935	0.395	0.067	0.934

Species comparisons comprise *P. sylvestris* vs. *J. communis*, *P. sylvestris* vs. *P. nigra* and *P. nigra* vs. *J. oxycedrus*. Location comparisons comprise exposure and altitude differences for *P. sylvestris* and *P. nigra*, respectively.

differences can promote changes in both structure and composition of vegetation along a gradient of environmental conditions, even in areas with high tolerance to dry conditions, such as the southern range edge. Our results are useful for forecasting plant responses and distributional shifts under a climate-change scenario, especially at species distribution limits such as the Mediterranean basin, where extreme events are predicted to be more detrimental and recurrent [56,57]. More interestingly, the great tolerance and/or higher recovery capacity to extreme droughts of both *Pinus* and *Juniper* species should be taken in account when species responses are modeled to future climatic conditions, as models predict sharp decreases in plant diversity and performance in Mediterranean mountains [19,57,58]. Our empirical results also indicated that, for an accurate evaluation of the resistance/resilience ability of current vegetation under a climate-change scenario, a realistic modeling approach requires empirical data to analyze plant responses to extreme events at the individual and population levels, considering both different environmental conditions and ontogenetic states (adults vs. saplings).

Supporting Information

Figure S1   Needle length for the period 2003–2008 for adults of the four species at sampled locations. Data for a *Pinus sylvestris* with southern and northern exposure, for *P. nigra* at high (2000 m), medium (1700 m) and low elevation (1500 m), and for *Juniperus communis* and *J. oxycedrus* are shown.

Figure S2   Needle length for the period 2003–2008 for saplings of the four species at sampled locations. Data for a *Pinus sylvestris* with southern and northern exposure, for *P. nigra* at high (2000 m), medium (1700 m) and low elevation (1500 m), and for *Juniperus communis* and *J. oxycedrus* are shown.

Figure S3   Differences in needle-length resistance (A), recovery (B), resilience (C) and relative resilience (D) between species and ontogenetic states (adults/saplings). Three comparisons are shown: *P. sylvestris* vs. *J. communis* with a northern exposure at high elevation; *P. sylvestris* vs. *P. nigra* with a southern exposure at high elevation; and *J. oxycedrus* vs. *P. nigra* at low elevation. Different letters above bars indicate significant *post hoc* differences between groups. Bars indicate the standard errors of calculated means.

Figure S4   Differences in needle-length resistance (A), recovery (B), resilience (C), and relative resilience (D) between locations and ontogenetic states (adults/saplings). Two comparisons are shown: between northern and southern exposure for *P. sylvestris*; and between high (2000 m), medium (1700 m), and low (1500 m) elevations for *P. nigra*. Different letters above bars indicate significant *post hoc* differences between groups. Bars indicate the standard errors of calculated means.

Acknowledgments

The authors acknowledge Edward Webb, Neil Cobb and three anonymous reviewers for insightful comments that improved the manuscript, and Francisco Lloret for his comments on an early version of the manuscript. We thank the Consejería de Medioambiente (Andalusian Government) and the direction of Sierra de Baza Natural Park for facilities to carry out the study. We also wish to thank Ignacio Villegas, Ramón Ruiz and Otilia Romera for field assistance. David Nesbitt checked the English of this paper.
   Data availability: The data are freely available upon request to the authors.

Author Contributions

Conceived and designed the experiments: RZ. Performed the experiments: AH. Analyzed the data: AH. Wrote the paper: AH RZ.

References

1. Allen CD, Macalady AK, Chenchouni H, Bachelet D, McDowell N, et al (2010) A global overview of drought and heat-induced tree mortality reveals emerging climate change risks for forests. For Ecol Manage 259: 660–684.
2. Allen CD, Breshears DD (1998) Drought-induced shift of a forest-woodland ecotone: Rapid landscape response to climate variation. Proc Natl Acad Sci USA 95: 14839–14842.
3. Slik JWF (2004) El Nino droughts and their effects on tree species composition and diversity in tropical rain forests. Oecologia 141: 114–120.
4. Mueller RC, Scudder CM, Porter ME, Trotter RT, Gehring CA, et al (2005) Differential tree mortality in response to severe drought: evidence for long-term vegetation shifts. J Ecol 93: 1085–1093.
5. Bigler C, Braker OU, Bugmann H, Dobbertin M, Rigling A (2006) Drought as an inciting mortality factor in Scots pine stands of the Valais, Switzerland. Ecosystems 9: 330–343.
6. Koepke DF, Kolb TE, Adams HD (2010) Variation in woody plant mortality and dieback from severe drought among soils, plant groups, and species within a northern Arizona ecotone. Oecologia 163: 1079–1090.
7. Zweifel R, Rigling A, Dobbertin M (2009) Species-specific stomatal response of trees to drought - a link to vegetation dynamics? J Veg Sci 20: 442–454.
8. Lloret F, Escudero A, Iriondo JM, Martínez-Vilalta J, Valladares F (2012) Extreme climatic events and vegetation: the role of stabilizing processes. Glob Change Biol 18: 797–805.
9. MacGillivray CW, Grime JP, Band SR, Booth RE, Campbell B, et al (1995) Testing Predictions of the Resistance and Resilience of Vegetation Subjected to Extreme Events. Funct Ecol 9: 640–649.
10. Holling CS (1996) Engineering resilience versus ecological resilience. In: Schulze P, editor. Engineering within ecological constraints. Washington: National Academy. pp. 31–44.
11. Della-Marta PM, Haylock MR, Luterbacher J, Wanner H (2007) Doubled length of western European summer heat waves since 1880. J Geophys Res-Atmos 112.
12. IPCC (2007) Climate Change 2007. The Physical Science Basis: Working Group I. Contribution to the Fourth Assessment Report of the IPCC. Cambridge: Cambridge University Press.
13. Briffa KR, van der Schrier G, Jones PD (2009) Wet and dry summers in Europe since 1750: evidence of increasing drought. Int J Climatol 29: 1894–1905.
14. Battisti DS, Naylor RL (2009) Historical Warnings of Future Food Insecurity with Unprecedented Seasonal Heat. Science 323: 240–244.
15. DeClerck FAJ, Barbour MG, Sawyer JO (2006) Species richness and stand stability in coniferous forest of the Sierra Nevada. Ecology 87: 2787–2799.
16. Blanco E, Casado MA, Costa M, Escribano R, García M, et al (1997) Los Bosques Ibéricos. Una Interpretación Geobotánica. Madrid: Planeta.
17. Hampe A, Petit RJ (2005) Conserving biodiversity under climate change: the rear edge matters. Ecol Lett 8: 461–467.
18. Thomas CD, Cameron A, Green RE, Bakkenes M, Beaumont LJ, et al (2004) Extinction risk from climate change. Nature 427:145–148
19. Thuiller W, Lavorel S, Araujo MB, Sykes MT, Prentice IC (2005) Climate change threats to plant diversity in Europe. Proc Natl Acad Sci USA 102: 8245–8250.
20. McDowell N, Pockman WT, Allen CD, Breshears DD, Cobb N, et al (2008) Mechanisms of plant survival and mortality during drought: why do some plants survive while others succumb to drought? New Phytologist 178: 719–739.
21. Castro J, Zamora R, Hódar JA, Gómez JM (2004a) Seedling establishment of a boreal tree species (*Pinus sylvestris*) at its southernmost distribution limit: consequences of being in a marginal Mediterranean habitat. J Ecol 92: 266–277.
22. Day ME, Greenwood MS (2011) Regulation of ontogeny in temperate conifers. In: Meinzer FC, Lachenbruch B, Dawson TE, editors. Size- and age-related changes in tree structure and function. New York: Springer.
23. García-Herrera R, Paredes D, Trigo RM, Trigo IF, Hernández E, et al (2007) The outstanding 2004/05 drought in the Iberian Peninsula: Associated atmospheric circulation. J Hydrometeorol 8: 483–498.
24. Thornthwaite CW (1948) An approach toward a Rational Classification of Climate. Geogr Rev 38:55–94
25. Mitchell TD, Jones PD (2005) An improved method of constructing a database of monthly climate observations and associated high-resolution grids. Int J Climatol 25: 693–712.
26. IPCC (2012) Managing the Risks of Extreme Events and Disasters to Advance Climate Change Adaptation. A Special Report of Working Groups I and II of

the Intergovernmental Panel on Climate Change. Field CB, Barros V, Stocker TF, Qin D, Dokken DJ, et al., editors. Cambridge University Press, Cambridge, UK, and New York, NY, USA.

27. Willms J, Rood SB, Willms W, Tyree M (1998) Branch growth of riparian cottonwoods: a hydrologically sensitive dendrochronological tool. Trees (Berl. West) 12:215–223.

28. Mutke S, Gordo J, Climent J, Gil L (2003) Shoot growth and phenology modeling of grafted stone pine (*Pinus pinea* L.) in Inner Spain. Ann For Sci 60: 527–537.

29. Peñuelas J, Gordon C, Llorens L, Nielsen T, Tietema A, et al (2004) Nonintrusive field experiments show different plant responses to warming and drought among sites, seasons, and species in a north-south European gradient. Ecosystems 7:598–612.

30. De Dato G, Pellizzaro G, Cesaraccio C, Sirca C, De Angelis P, et al (2008) Effects of warmer and drier climate conditions on plant composition and biomass production in a Mediterranean shrubland community. Ifor-Biogeosci For 1:39–48.

31. Thabeet A, Vennetier M, Gadbin-Henry C, Denelle N, Roux M, et al (2009). Response of *Pinus sylvestris* L. to recent climatic events in the French Mediterranean region. Trees (Berl. West) 23: 843–853.

32. Garret PW, Zahner R (1973). Fascicle density and needle growth responses of pine to water supply over two seasons. Ecology 54: 1328–1334.

33. Royce EB, Barbour MG (2001) Mediterranean climate effects. II. Conifer growth phenology across a Sierra Nevada ecotone. Am J Bot 88: 919–932.

34. Edenius L, Danell K, Nyquist H (1995) Effects of Simulated Moose Browsing on Growth, Mortality, and Fecundity in Scots Pine: Relations to Plant Productivity. Can J For Res 25:529–535.

35. Zamora R, Gómez JM, Hódar JA, Castro J, García D (2001) Effect of browsing by ungulates on sapling growth of Scots pine in a Mediterranean environment: consequences for forest regeneration. For Ecol Manage 144: 33–42.

36. Debain S, Chadoeuf J, Curt T, Kunstler G, Lepart J (2007) Comparing effective dispersal in expanding population of *Pinus sylvestris* and *Pinus nigra* in calcareous grassland. Can J For Res 37:705–718.

37. Lloret F, Keeling EG, Sala A (2011) Components of tree resilience: effects of successive low-growth episodes in old ponderosa pine forests. Oikos 120: 1909–1920.

38. Dobbertin M, Eilmann B, Bleuler P, Giuggiola A, Pannatier EG, et al (2010) Effect of irrigation on needle morphology, shoot and stem growth in a drought-exposed *Pinus sylvestris* forest. Tree Physiol 30:346–360.

39. Isik K (1990) Seasonal course of height and needle growth in *Pinus nigra* grown in summer-dry Central Anatolia. For Ecol Manage 35:261–270.

40. Herrero A, Castro J, Zamora R, Delgado-Huertas A, Querejeta JI (2013) Growth and stable isotope signals associated with drought-related mortality in saplings of two coexisting pine species. Oecologia 173:1613–1624.

41. Prus-Glowacki W, Stephan BR (1994) Genetic-Variation of *Pinus sylvestris* from Spain in Relation to Other European Populations. Silvae Genet 43: 7–14.

42. Afzal-Rafii Z, Dodd RS (2007) Chloroplast DNA supports a hypothesis of glacial refugia over postglacial recolonization in disjunct populations of black pine (*Pinus nigra*) in western Europe. Mol Ecol 16:723–736.

43. Martínez-Vilalta J, Cochard H, Mencuccini M, Sterck F, Herrero A, et al (2009) Hydraulic adjustment of Scots pine across Europe. New Phytologist 184:353–364.

44. Richter S, Kipfer T, Wohlgemuth T, Calderón Guerrero C, Ghazoul J, et al (2012) Phenotypic plasticity facilitates resistance to climate change in a highly variable environment. Oecologia 169: 269–279.

45. García D, Zamora R (2003) Persistence, multiple demographic strategies and conservation in long-lived Mediterranean plants. J Veg Sci 14: 921–926.

46. Lloret F, Siscart D, Dalmases C (2004) Canopy recovery after drought dieback in holm-oak Mediterranean forests of Catalonia (NE Spain). Glob Change Biol 10: 2092–2099.

47. Floyd ML, Clifford M, Cobb NS, Hanna D, Delph R, et al (2009) Relationship of stand characteristics to drought-induced mortality in three Southwestern pinon-juniper woodlands. Ecol Appl 19:1223–1230

48. Adams HD, Kolb TE (2004) Drought responses of conifers in ecotone forests of northern Arizona: tree ring growth and leaf sigma C-13. Oecologia 140: 217–225.

49. Linares JC, Tíscar PA (2010) Climate change impacts and vulnerability of the southern populations of *Pinus nigra* subsp *salzmannii*. Tree Physiol 30: 795–806.

50. Martínez-Vilalta J, Piñol J (2002) Drought-induced mortality and hydraulic architecture in pine populations of the NE Iberian Peninsula. For Ecol Manage 161: 247–256.

51. Castro J, Zamora R, Hódar JA, Gómez JM, Gómez-Aparicio L (2004b) Benefits of using shrubs as nurse plants for reforestation in Mediterranean mountains: A 4-year study. Restor Ecol 12: 352–358.

52. Boulant N, Kunstler G, Rambal S, Lepart J (2008) Seed supply, drought, and grazing determine spatio-temporal patterns of recruitment for native and introduced invasive pines in grasslands. Divers Distrib 14: 862–874.

53. Galiano L, Martínez-Vilalta J, Lloret F (2010) Drought-Induced Multifactor Decline of Scots Pine in the Pyrenees and Potential Vegetation Change by the Expansion of Co-occurring Oak Species. Ecosystems 13:978–991.

54. Hódar JA, Castro J, Zamora R (2003) Pine processionary caterpillar *Thaumetopoea pityocampa* as a new threat for relict Mediterranean Scots pine forests under climatic warming. Biological Conservation 110: 123–129.

55. Herrero A, Zamora R, Castro J, Hódar JA (2012) Limits of pine forest distribution at the treeline: herbivory matters. Plant Ecol 213: 459–469.

56. Beniston M, Stephenson DB, Christensen OB, Ferro CAT, Frei C, et al (2007) Future extreme events in European climate: an exploration of regional climate model projections. Clim Change 81: 71–95.

57. Lindner M, Maroschek M, Netherer S, Kremer A, Barbati A, et al (2010) Climate change impacts, adaptive capacity, and vulnerability of European forest ecosystems. For Ecol Manage 259: 698–709.

58. Reich PB, Oleksyn J (2008) Climate warming will reduce growth and survival of Scots pine except in the far north. Ecol Lett 11: 588–597.

15

Variability of Carbon and Water Fluxes Following Climate Extremes over a Tropical Forest in Southwestern Amazonia

Marcelo Zeri[1]*, Leonardo D. A. Sá[2], Antônio O. Manzi[3], Alessandro C. Araújo[4], Renata G. Aguiar[5], Celso von Randow[1], Gilvan Sampaio[1], Fernando L. Cardoso[6], Carlos A. Nobre[7]

1 Centro de Ciência do Sistema Terrestre, Instituto Nacional de Pesquisas Espaciais, Cachoeira Paulista, SP, Brazil, 2 Centro Regional da Amazônia, Instituto Nacional de Pesquisas Espaciais, Belém, PA, Brazil, 3 Instituto Nacional de Pesquisas da Amazônia (INPA), Manaus, Amazonas, Brazil, 4 Embrapa Amazônia Oriental, Belém, Pará, Brazil, 5 Universidade Federal de Rondônia, Porto Velho, Rondônia, Brazil, 6 Universidade Federal de Rondônia, Ji-Paraná, Rondônia, Brazil, 7 Secretaria de Políticas e Programas de Pesquisa e Desenvolvimento, Ministério da Ciência, Tecnologia e Inovação, Brasília, DF, Brazil

Abstract

The carbon and water cycles for a southwestern Amazonian forest site were investigated using the longest time series of fluxes of CO_2 and water vapor ever reported for this site. The period from 2004 to 2010 included two severe droughts (2005 and 2010) and a flooding year (2009). The effects of such climate extremes were detected in annual sums of fluxes as well as in other components of the carbon and water cycles, such as gross primary production and water use efficiency. Gap-filling and flux-partitioning were applied in order to fill gaps due to missing data, and errors analysis made it possible to infer the uncertainty on the carbon balance. Overall, the site was found to have a net carbon uptake of ≈ 5 t C ha^{-1} year^{-1}, but the effects of the drought of 2005 were still noticed in 2006, when the climate disturbance caused the site to become a net source of carbon to the atmosphere. Different regions of the Amazon forest might respond differently to climate extremes due to differences in dry season length, annual precipitation, species compositions, albedo and soil type. Longer time series of fluxes measured over several locations are required to better characterize the effects of climate anomalies on the carbon and water balances for the whole Amazon region. Such valuable datasets can also be used to calibrate biogeochemical models and infer on future scenarios of the Amazon forest carbon balance under the influence of climate change.

Editor: Dafeng Hui, Tennessee State University, United States of America

Funding: M. Zeri is grateful to São Paulo Research Foundation (FAPESP) — grant 2011/04101-0 — for the support during the preparation of this manuscript. Leonardo Sá is particularly grateful to CNPq — Conselho Nacional de Pesquisa e Desenvolvimento Tecnológico — for his research grant (process 303.728/2010-8). The funders had no role in study design, data collection and analysis, decision to publish, or preparation of the manuscript.

Competing Interests: The authors have declared that no competing interests exist.

* E-mail: marcelo.zeri@inpe.br

Introduction

The intra-annual variability of carbon and water fluxes over forest and pasture sites in the Amazon region have been reported in many studies in the last several decades. The area covered by the world's largest tropical forest includes sites with evergreen species, semi-deciduous and transitions to *Cerrado*, among other classifications [1]. Sites with distinct vegetation types or topographies – and subjected to different sums of rainfall – are also different regarding the annual trends of fluxes of carbon, evapotranspiration and sensible heat flux. Southern sites (between latitudes $10°$ and $20°$ S) tend to have longer dry seasons while northern locations (between the Equator and $10°$ S) receive more rainfall due to the proximity to the Intertropical Convergence Zone (ITCZ), a migrating band of clouds and precipitation over the Equator, and the proximity of the Atlantic Ocean where the incoming air provides the moisture that forms precipitation over the Amazon [2]. Different rainfall patterns and annual cycles of air temperature, vapor pressure deficit and incoming solar radiation contribute to different trends in the fluxes of carbon, water and heat between the surface and the atmosphere [3,4].

Previous works on water and heat fluxes for a group of forest and savanna sites across the Amazon region revealed that evapotranspiration increases during the dry season in sites with higher annual precipitation and shorter dry seasons [5]. This apparent contradiction seems to be explained by the higher availability of incoming radiation and the hypothesized ability of trees for reaching water deep into the soil [1,6–9]. On the other hand, savanna and pasture sites were reported to have decreased evapotranspiration during the dry period [5,10]. For the carbon fluxes, while an increase in net ecosystem exchange (NEE) during the dry season is reported in some studies [11,12], others report a decrease of carbon assimilation during the same period [10].

The effects of droughts in 2005 and 2010 on the carbon balance of Amazonian forests have been extensively reported in recent years. A green-up effect following the drought of 2005 [13] was hypothesized to be related to the ability of trees to extract water using deep roots [14]. However, this effect was not observed in another study, which concluded that only 11–12% of Amazonian forests subjected to the drought exhibited greening during the dry season of 2005 [15]. The drought of 2010 was associated with low precipitation in 40% of the vegetated area and low water levels in several rivers of the Amazon basin, such as Rio Negro, near Manaus [16,17]. As a consequence, a decline of 7% in net primary production (NPP) between July and September of 2010 were reported in a remote sensing study which found that 0.5 Pg C were not sequestered in that year due to the dry conditions [18]. Climate change and droughts in the last decade were related to a decline in global NPP, specifically related to increases in air temperature over the Amazon, which increased autotrophic respiration [19]. Finally, deforestation was also found to play a role in drought events, caused by disturbances in evapotranspiration which affect other regions via regional circulation patterns [20].

In this study, fluxes of carbon dioxide and evapotranspiration were investigated using a dataset composed of seven years of measurements in a forest site within the Jaru Biological Reserve, in Brazil. The time series of fluxes reported here are the longest ever reported for a tropical forest in the southwestern region of the Amazon, enabling the investigation of the impacts on fluxes of extreme climatic events that affected the region, such as the droughts of 2005 and 2010 and the rainy year of 2009 [21–25]. The intra-annual variability of carbon flux and evapotranspiration was described using monthly averages and compared to common drivers of the carbon and water cycles, such as air temperature, vapor pressure deficit and incoming solar radiation (direct and diffuse).

The tower is located $\approx$ 14 km from the site described in other studies of this area [5,10,26], making it possible to test the spatial homogeneity of the forest-atmosphere exchanges of carbon and water by the similarity of some results, such as mean daily cycles of fluxes. The partitioning of net ecosystem productivity (NEP) in gross primary production (GPP) and ecosystem respiration (R_e) allowed the investigation of the intra-annual trends of those components of the carbon cycle as well as different metrics of water use efficiency [27–29].

Site and Data

Measurements were carried out at the Jaru Biological Reserve, near the city of Ji-Paraná, Rondônia, Brazil (Figure 1). Authorization for field studies in this area was provided by IBAMA (National Institute of Environment and Renewable Resources). The tower was mounted in 2004 at the location marked with the red star in Figure 1 (10° 11' 21.2712" S, 61° 52' 15.1674" W, at 145 m above sea level), which is approximately 14 km south-southeast of the old location (orange star, zoomed in detail) up until November of 2002. The old tower, which was disassembled in 2002, was used in several experiments and studies [10,26,30–39] of the LBA (Large-Scale Biosphere-Atmosphere Experiment in Amazonia) project [30,34,40]. The forest was previously characterized as an open tropical rain forest with leaf area index ranging from 4 to 6 m^2 m^{-2} [30,41]. Trees are 35 m high, on average, but some reached up to 45 m. Soil depth at the old site ranged from 1 – 2 m and its texture was classified as sandy loam [30].

The new tower was equipped with an eddy covariance (EC) system and micrometeorological sensors. The EC system was installed at 63.4 m and included a 3D sonic anemometer (model Solent 1012R2, Gill Instruments, UK) and an open-path infra-red gas analyzer (IRGA) model LICOR 7500 (LICOR Inc., Nebraska, USA), both operating at 10.4 Hz. A wind vane placed at 62 m (model W200P, Vector Instruments, UK) was used for measurements of wind direction while a barometer (model PTB100A, Vaisala, Helsinki, Finland) was used for recording air pressure at 40 m. Wind speed was measured at several heights above and below the canopy using cup anemometers (model A100R, Vector

Instruments, UK) placed at 30, 41, 50.5 and 62.4 m. A vertical profile of thermohygrometers (model Temp107, Campbell, Logan, USA), for measurements of air temperature and relative humidity, was set up at heights 1.6, 11.2, 21.2, 33 and 49 m, while a model HMP45D (Vaisala) was placed at 61.5 m. Incoming and reflected shortwave solar radiation were measured with a pyranometer model CM21, from Kipp&Zonen (The Netherlands), while the incoming and emitted longwave components were measured with a pirgeometer model CG1 (Kipp&Zonen) mounted over arms extending from the tower. Both radiation sensors were installed at 57.5 m. The photosynthetically active component of solar radiation (PAR) was measured with a sensor model SKE 510 (Skye Instruments, UK), also mounted at 57.5 m. Lastly, a vertical profile of intakes for measurements of CO_2 concentration was setup in 2008 at the following heights: 62, 50, 34, 22, 12 and 2 m.

Methodology

Fluxes were calculated using the eddy covariance technique [42–45], which is implemented in the software Alteddy (Jan Elbers, Alterra Group, Wageningen University, The Netherlands). The software was set up to apply the planar fit rotation [46,47] to the coordinate system in order to make the vertical velocity zero for different sectors of wind direction. The effects of humidity on the temperature measured by the sonic anemometer were corrected [48] while the influence of air density on the measurements from the infra-red gas analyzer were adjusted using the WPL correction [49–51]. In addition, known algorithms were applied to compensate for: a) losses in the high frequency end of the spectra due to spatial separation between sonic anemometer and IRGA [52–54]; b) the effects of heating of lenses in the open-path IRGA [55]. Finally, quality control of time series was based on the level of stationarity, i.e., the variability of statistical moments over time [56–58]. Stationarity was calculated [59] and summarized in flags from 1 to 9, which are proportional to the level of non-stationarity. For example, fluxes with flags ranging from 1 to 3 may have up to 50% of variability in their statistical moments during the period of 30 min used for averages. Data with flags 1–5 were accepted based on the good energy balance closure of this range, which was previously reported for the old Jaru forest site [26].

Recent developments in the dynamics of air flow past a sonic anemometer's body led to new findings about the errors of vertical velocity measurements. As a result, different configurations of an anemometer's transducers (orthogonal or non-orthogonal) may have an impact on fluxes [60,61]. Those errors can be on the order of $\approx$10% for sensible heat flux and can propagate to other fluxes through direct measurements or corrections. Due to their recent nature and uncertain impact on fluxes, such corrections were not included in the calculations. However, we expect that the errors calculated in this work (random and gap-filling errors, described next) will account for most of the uncertainty on annual sums of evapotranspiration and carbon fluxes.

The balance of carbon in an ecosystem – the net ecosystem production, NEP – can be calculated as follows:

$$NEP = \overline{w'c'} + \int_0^{zi} \frac{\partial \overline{c}}{\partial t} dz \qquad (1)$$

where w' and c' are the departures from the mean of vertical velocity and concentration of CO_2, respectively, zi is the measurement height and z is the vertical coordinate. Positive

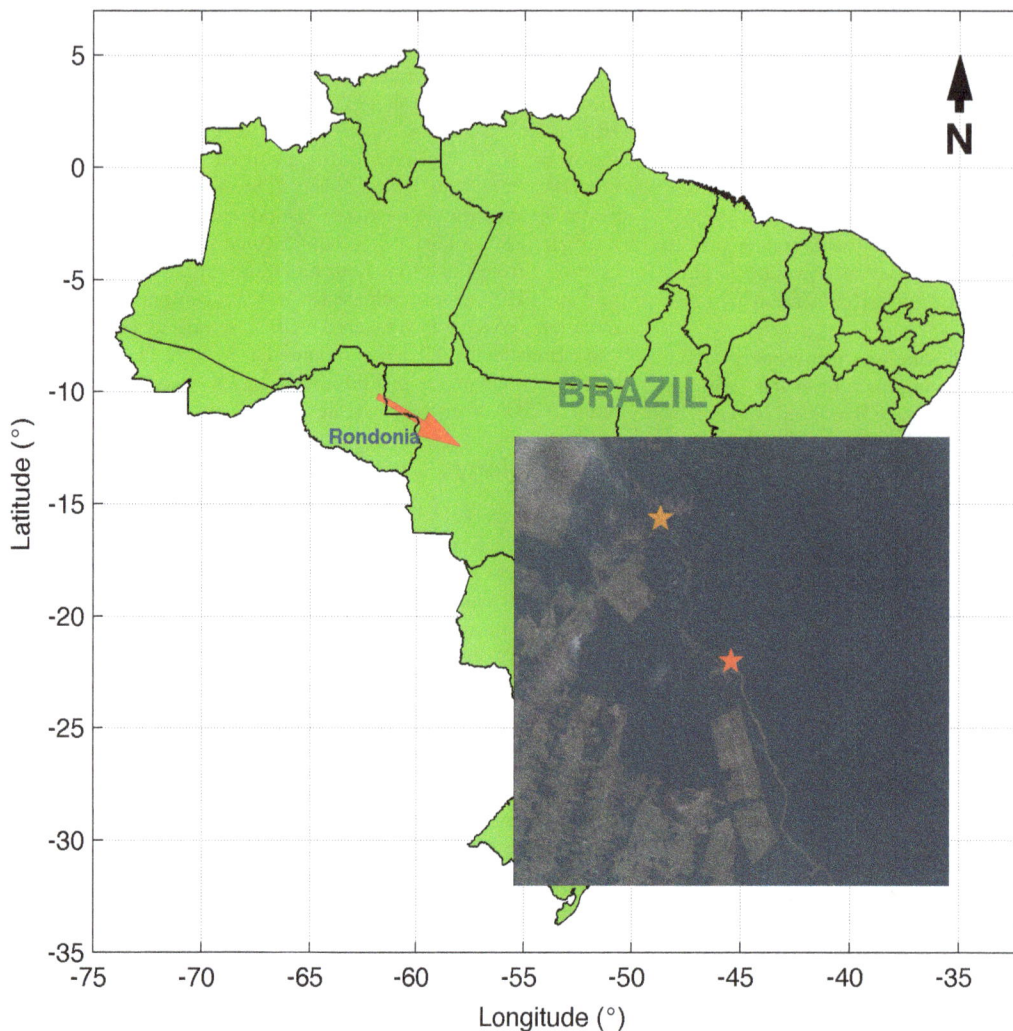

Figure 1. Location of the tower marked with the red star in the detail. The old tower was located at approximately 14 km northwest of the current location (orange star). Satellite picture recorded by Landsat 7 on October 1st 2002.

values of NEP denote accumulation of carbon by the ecosystem, according to the biological convention. The first term in Equation 1 is the eddy covariance flux, which accounts for the exchanges of carbon by the fast turbulent motions. The second term accounts for the storage of carbon below the measurement point at the top of the tower during conditions of low turbulent motions. The storage of carbon is usually calculated from a vertical profile of CO_2 concentrations above and below the canopy. The changes in concentration from one half-hour to the next are integrated vertically and contribute to a large fraction of NEP around sunrise due to the stratification of cold – and CO_2-enriched – air below the canopy during calm nighttime conditions.

The vertical profiles of CO_2 concentration were not available during the first four years of data, from 2004 to 2007. For this reason, an artificial time series of storage was calculated based on the average values of 2008 to 2010. First, the mean daily cycle of storage was calculated for each month from 2008 to 2010. Next, the daily cycles were grouped by month and averaged over the years. The resulting twelve diurnal cycles were then replicated to fill the respective month, creating a series of 30 minutes averages from January 1st to December 31st. The artificial series represented well the real data when comparing the annual impact on the

carbon balance: while the average annual sum of storage from 2008 to 2010 was an uptake of ≈ 1.7 t C ha^{-1} year^{-1}, the annual sum resulting from the artificial storage was of ≈ 1.9 t C ha^{-1} year^{-1}. The contribution of the artificial storage is likely to be small to the carbon balance of the first years since its magnitude is close to the average uncertainty derived from the error analysis and gap-filling, which was estimated to be $\approx \pm 1.7$ t C ha^{-1} year^{-1}.

The validity of the eddy covariance method relies on the sufficient intensity of wind speed and turbulence in the surface layer, so that vertical exchanges can be averaged over several vortices passing by the tower [62]. The level of turbulence can be inferred by the value of u∗, the friction velocity, which is calculated as $u_* = \left[(\overline{w'u'})^2 \right]^{1/4}$, where $\overline{w'v'}$ is the longitudinal vertical flux of momentum. The transversal component of the vertical flux $\overline{w'v'}$ is ignored after rotating the coordinate system to follow the average wind direction [47]. Nighttime conditions usually have light winds and low levels of turbulence, resulting in underestimated fluxes and high values CO_2 storage below the measuring height (Figure 2). The curves in Figure 2 are used to estimate the u∗-threshold, used to filter out nighttime or daytime fluxes which are later replaced by modeled values in the gap-filling analysis [63,64].

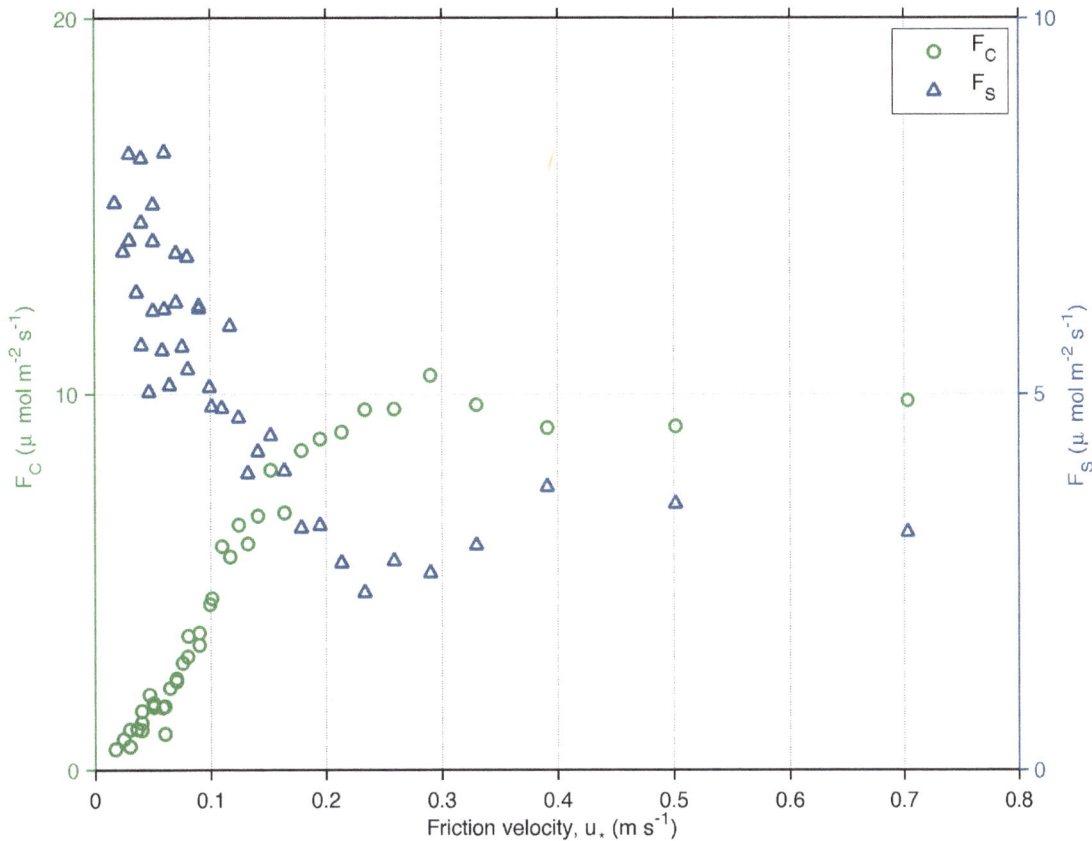

Figure 2. CO_2-flux (F_C) and storage (F_S) plotted versus classes of friction velocity. Atmospheric convention used to denote nocturnal respiration as positive. Only data from 2009 was used for the averages since this was the year with longer data availability of vertical profiles of CO_2.

However, the choice of the threshold can be subjective and may change the carbon balance depending on the fraction of data replaced by models [26,65]. Here, we chose the threshold as 0.1 m s^{-1}, a value that separates the top 60% of the storage values. A test of sensitivity to this choice was made and the results are presented in the section Results and discussion.

In recent years, two additional terms were proposed to equation 1 to account for horizontal and vertical advection, which are contributions to the flux caused by slopes in the terrain around the tower [66–71]. An inclined terrain can lead to horizontal and vertical transports of air that are usually small in flat areas and hence not considered in traditional eddy covariance applications. The effects of advection in Amazonian sites were already investigated over sites with complex topography [72,73]. An

indication of the importance of advection to a site is the curve of CO_2 storage versus friction velocity, as shown in Figure 2. If the storage is high for low values of $u*$, then cold air is not being flushed out of the ecosystem by horizontal transports and thus advection is not occurring. Since this is the case in Figure 2, it is clear that the effects of advection can be ignored for this site.

Finally, the diffuse component of PAR – caused by scattering of light by clouds, aerosols, smoke or other particles [37,74,75] – was calculated using total PAR and the clearness index, an indicator of cloudiness used in solar radiation research. This index is defined as the ratio of global solar radiation at the surface over the radiation received from the Sun at the top of the atmosphere [76]. In addition, the calculations make use of characteristics of the site

Table 1. Accumulated rainfall and annual sums of total water use (TWU), gross primary production (GPP) and net ecosystem production (NEP) for each seasonal year (integration from September to August).

Year	Rainfall (mm)	TWU (mm)	GPP (t C ha-1 yr-1)	NEP (t C ha-1 yr-1)
2004/2005	1552.8	1095.6±39.8	22.1±0.6	1.7±0.7
2005/2006	1683.8	1000.4±119.1	20.0±1.9	−0.7±1.9
2006/2007	2114.4	1224.7±101.5	22.0±1.1	3.0±1.0
2007/2008	1975	1231.9±88.7	22.0±1.8	4.3±2.6
2008/2009	1964.8	1378.6±61.7	22.7±0.8	6.3±1.3
2009/2010	1861.4	1321.7±41.3	22.7±1.6	4.8±2.5

Table 2. Similar to Table 1, but using the calendar year, which uses data integrated from January to December.

Year	Rainfall (mm)	TWU (mm)	GPP (t C ha^{-1} yr^{-1})	NEP (t C ha^{-1} yr^{-1})
2004	2181.8	968.8±20.0	29.9±1.2	1.9±1.6
2005	1315.2	1172.8±9.1	33.5±0.6	2.2±1.0
2006	2075.1	1153.4±48.9	29.7±1.9	−1.2±2.1
2007	1942	804.3±22.4	35.7±1.1	4.8±1.1
2008	1782.2	1195.0±24.7	36.0±1.7	6.0±2.2
2009	2258.4	1498.7±13.4	35.4±0.9	10.4±1.2
2010	1551.2	961.7±27.2	38.7±1.8	7.4±2.5

such as solar elevation angle, latitude and declination of the Sun [76].

Gap-filling and flux partitioning

Gaps in the time series of fluxes (CO_2, water vapor) and meteorological variables (air temperature, relative humidity, etc.)

need to be filled if annual sums of those variables are calculated. Short gaps – up to 1 hour – are filled by linear interpolation. Longer gaps are filled either by using the average daily cycle for each month, for meteorological variables, or by using other algorithms such as look-up tables, for fluxes. Initially, gaps in air temperature and other meteorological variables are filled since they are required in the gap-filling of fluxes. Next, it is applied an algorithm [26,64,77] to fill the fluxes of CO_2 and evapotranspiration. The algorithm is based on a "look-up table" approach, which tries to fill each gap with an average of good records taken on similar environmental conditions of net radiation, air temperature and vapor pressure deficit. The search starts within ±5 days from the gap and extends up to ±100 days, until it finds at least 5 records to average and fill the gap. In years with fewer long gaps, such as 2005 or 2009, the fraction of gaps due to the low quality flag or low u* was ≈40% before the filling process and 10% afterwards. For years with longer gaps due to instrument malfunction or maintenance, the fraction of missing data could reach up to 70% but the filling algorithm was still able to leave 10–20% of gaps. Those remaining gaps were filled with monthly mean daily cycles, for evapotranspiration, or with modeled fluxes for night and day, for CO_2 fluxes.

Daytime gaps were filled by light response curves of NEP versus PAR for two classes of air temperature: below and above 25°C.

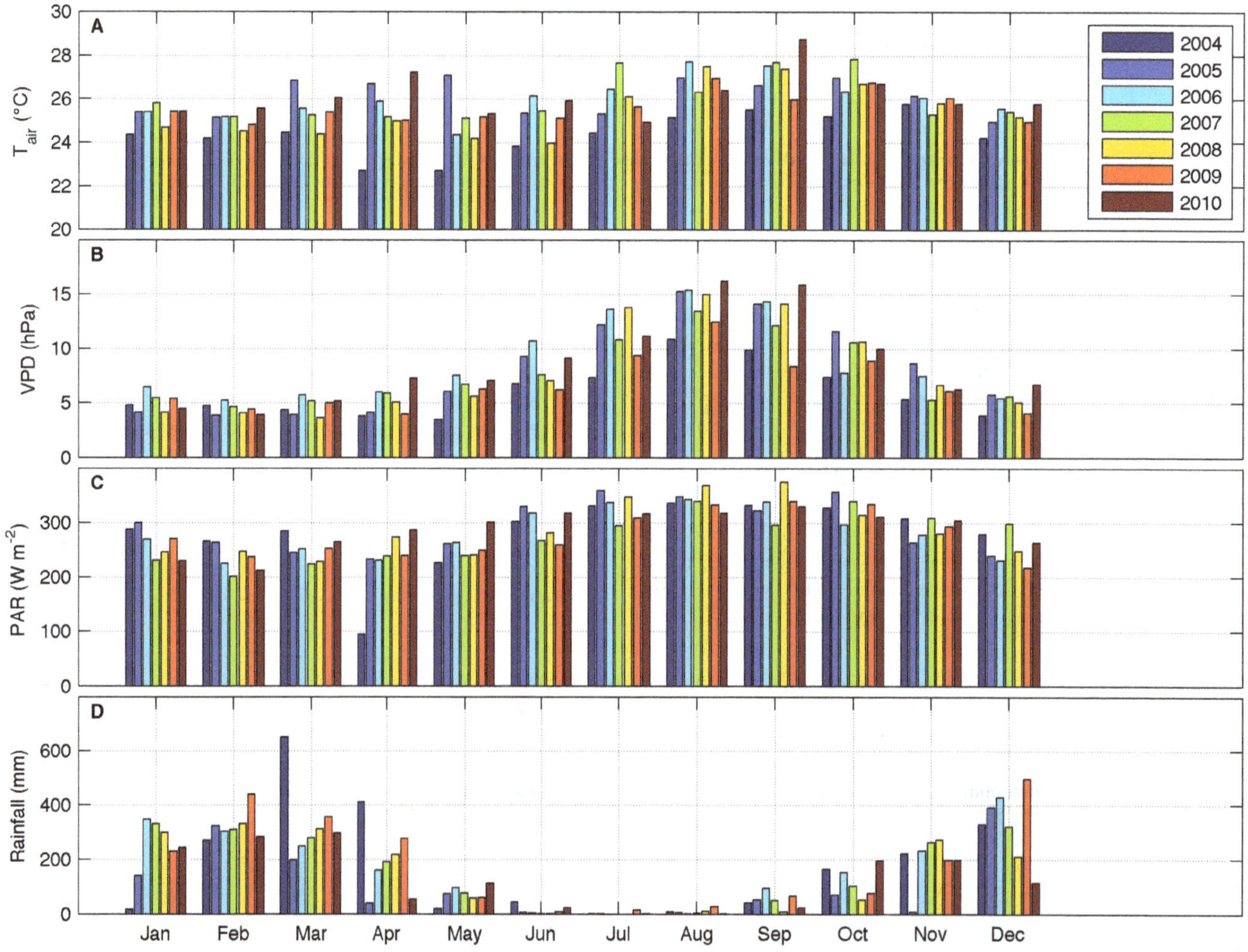

Figure 3. Monthly averages (median) of meteorological drivers. A: air temperature; B: vapor pressure deficit; C: maximum value of incoming PAR at noon; D: accumulated rainfall.

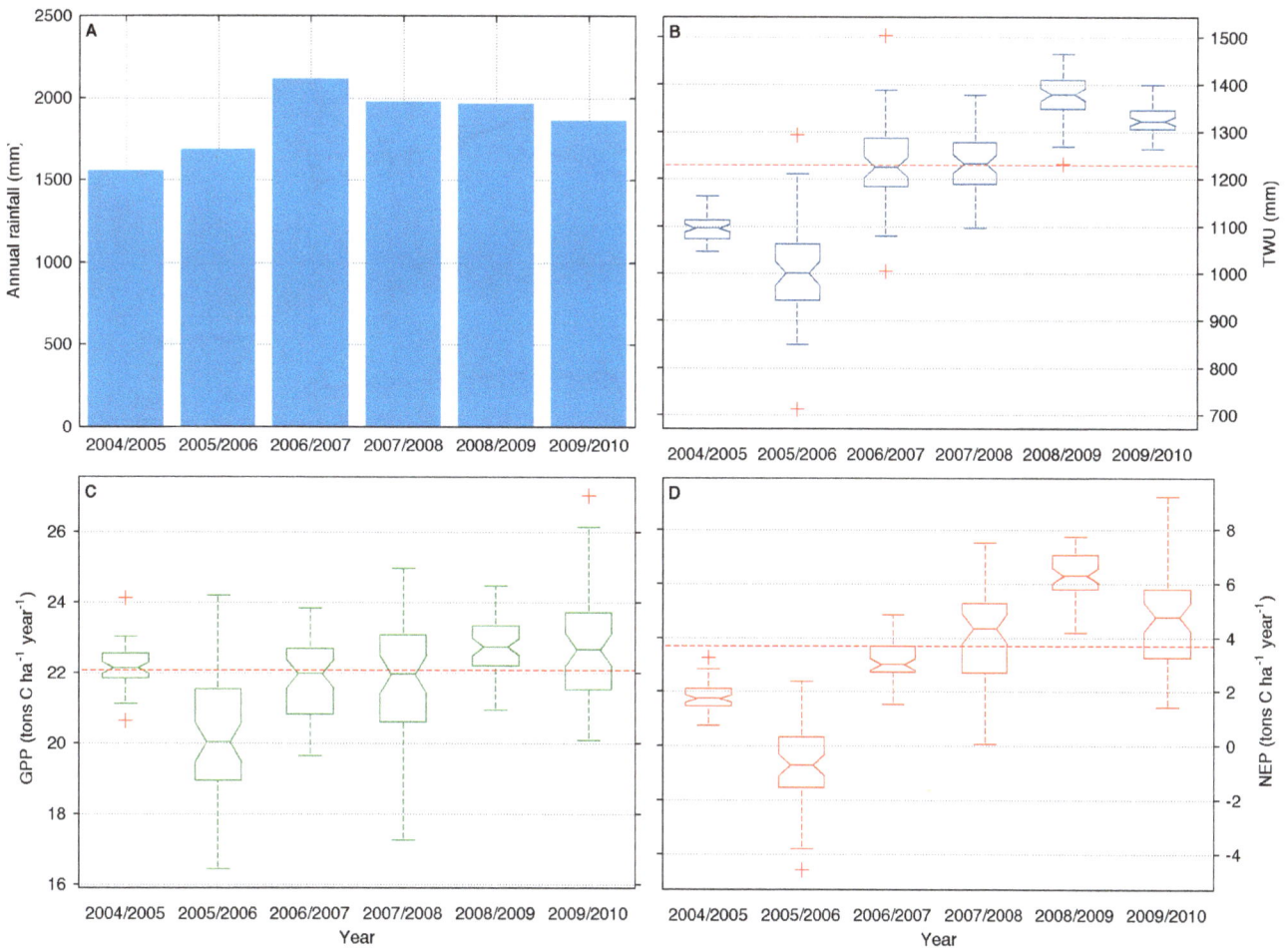

Figure 4. Interannual variability of annual sums for precipitation (A), evapotranspiration, or total water used (B), gross primary production (C), and net ecosystem production (D). Horizontal lines in panels B, C and D denote overall median. Labels in x-axis indicate seasonal year starting in September and ending in August (e.g. from Sep-2004 to Aug-2005).

Nighttime fluxes, or ecosystem respiration R_e, was modeled using an Arrhenius function [78]:

$$Re = R_{ref} \exp \left[E_0 \left(\frac{1}{T_{ref} - T_0} - \frac{1}{T - T_0} \right) \right] \quad (2)$$

where R_{ref} is the respiration at the reference temperature T_{ref}, which was set to 293.15 K, E_0 is the activation energy and T_0 is a constant. The constant T_0 was set to 227.13 K, as previously reported [78]. The independent variable T is referred to air temperature since no measurements of soil temperature were available at the site. The constants E_0 and R_{ref} were calculated by using a non-linear least-squares regression method [64]. The ecosystem respiration calculated in Equation 2 in combination with the gap-filled NEP made it possible to calculate gross primary production (GPP) following the relation NEP = GPP − R_e. GPP was used in the analysis of intra-annual variability of fluxes and meteorological drivers, as well as in the calculation of monthly water use efficiency. For the analysis of annual sums (Table 1), the fluxes were integrated from September to August so that one cycle included full wet and dry seasons. This approach will be referred as the seasonal year to distinguish the periods in the discussions

that use the regular calendar year. Values computed using the regular calendar year are shown in Table 2.

Two metrics of water use efficiency were calculated using NEP, GPP and the amount of water used by the ecosystem in one year, i.e., total water use (TWU). TWU was calculated as the cumulative sum of evapotranspiration, which is also referred as the latent heat flux in atmospheric sciences. The first metric of water use efficiency was GWUE = GPP/TWU and the second was EWUE = NEP/TWU [27–29,79]. While GWUE measures the water use efficiency of the vegetation exclusively, EWUE measures the efficiency of the whole ecosystem, taking into account inputs and outputs of carbon.

The uncertainty in the annual fluxes was estimated by calculating the random (σ_{rand}) and gap-filling (σ_{gap}) errors. The random error was calculated from the variability of fluxes measured in successive days and under similar conditions [80]. Environmental variables such as net radiation, air temperature and vapor pressure deficit were used to select fluxes subjected to similar physical drivers. The random error σ was estimated as the standard deviation of all differences, normalized by $\sqrt{2}$ [81]. Then, it was then averaged in several classes and a linear relationship with the magnitude of the CO_2 flux was found. Next, random noise from a normal distribution with mean 0 and standard deviation σ was created and added to each class of flux,

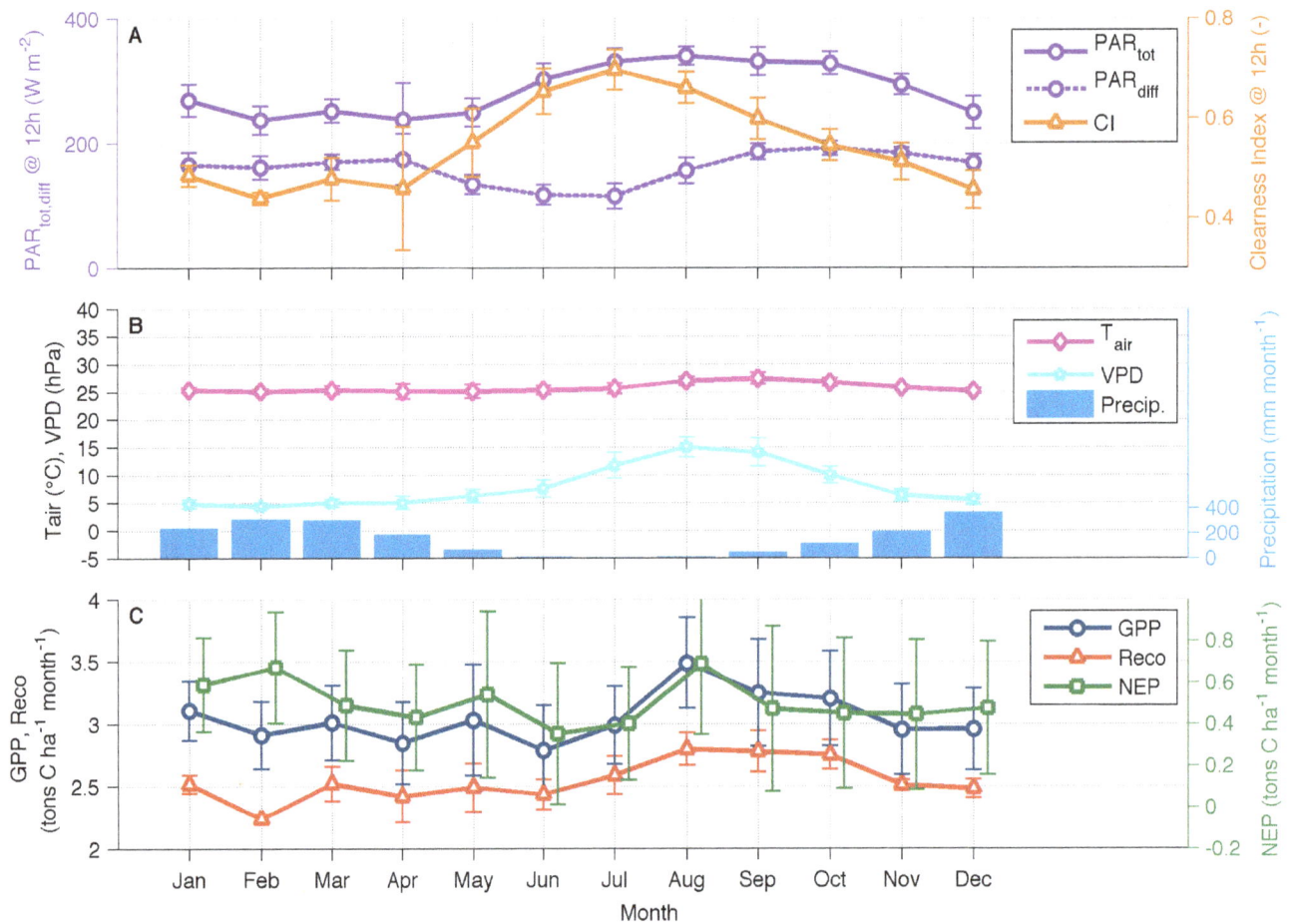

Figure 5. Median intra-annual cycles of PAR (total and diffuse) at noon and clearness index (A); air temperature, vapor pressure deficit and precipitation (B); and gross primary production, net ecosystem production and ecosystem respiration (D).

in order to have a synthetic flux with errors. This artificial flux was then gap-filled and its annual sum stored. Those steps were repeated 50 times, generating 50 versions of the original noisy flux, and the uncertainty due to the random error (σ_{rand}) was calculated as the standard deviation of all cumulative fluxes. The gap-filling error was calculated by inserting new gaps in the filled flux in the same proportion of the missing data found for day and night after filtering for high quality and turbulent conditions. First, the new random gaps were filled and the annual sum of the flux of CO_2 was calculated. Then, after 50 iterations, the gap-filling error σ_{gap} was calculated as the standard deviation of all 50 annual sums. The final error for the CO_2 flux was calculated as $\sqrt{\sigma_{rand}^2 + \sigma_{gap}^2}$.

Results and Discussion

The most important meteorological drivers of fluxes (air temperature, vapor pressure deficit, radiation and precipitation) for the period of 2004 to 2010 are shown in Figure 3. On average, air temperature (Figure 3A) was $\approx 25°C$, from January to March and November to December, and $\approx 27°C$, from July to September. Vapor pressure deficit (Figure 3B) was highest, i.e. drier conditions, from June to October, the same period of the year when incoming PAR was maximal (Figure 3C). The high values of VPD and PAR are in synchronicity with the dry season at this site, as can be noticed by reduced precipitation in the period from May to October (Figure 3D). The remainder of the year is

characterized by abundant rainfall, contributing to lower values of air temperature, VPD and PAR, the latter caused by overcast skies.

Abnormal values for some variables are evident in some months when compared to the average of all seven years of data. The period of March to April of 2004 registered two of the three highest values of monthly accumulated precipitation in the dataset. It caused a minimum in PAR for April of 2004 most likely due to cloudy skies. For the following year, 2005, most of the wet season months (Jan, Mar, Apr, Sep, Oct, Nov) had the lowest values of rainfall, in accordance to the extensive drought that affected the Amazon region in this year [23]. Air temperature in March and May of 2005 was the highest among all years. On the other hand, the months of February, March, April and December of 2009 presented high values of accumulated precipitation.

The cumulative sums of rainfall, TWU, NEP and GPP (Figure 4, Table 1) help to explain the impacts of dry and wet years in the water and carbon cycles. The seasonal year was used in this figure, i.e., integration from September of 2004 to August of 2005. The median value of annual rainfall was 1913 mm, while the minimum was 1552.8 (2004/2005) and the maximum was 2114.4 (2006/2007). The minimum of rainfall was followed by another dry period in 2005/2006, which caused sharp drops in median values of TWU, GPP and NEP. As a result of this long drought, this forest was a net source of carbon for the period of 2005/2006, when NEP was -0.7 ± 1.9 t C ha^{-1} year^{-1}. The years that

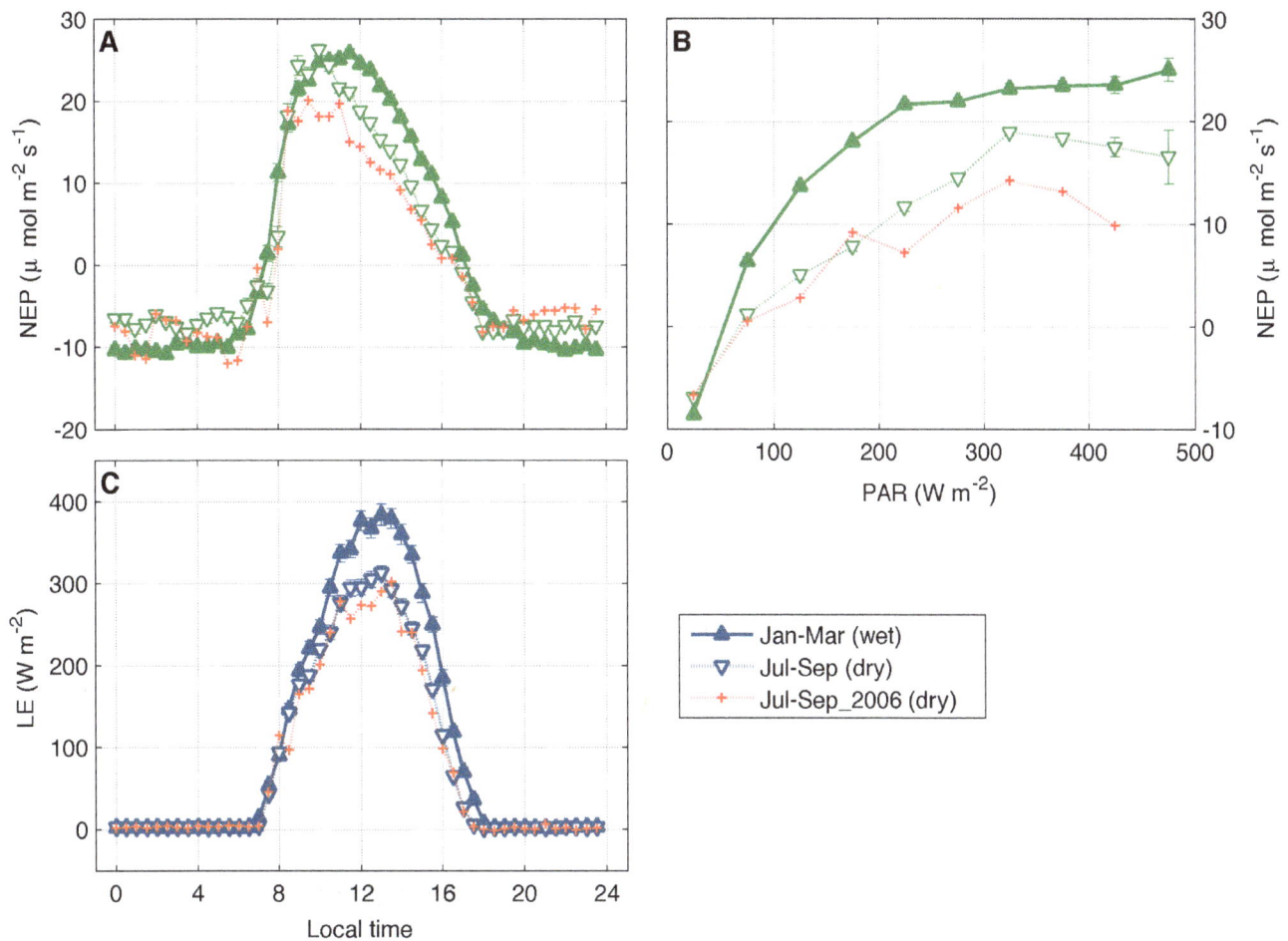

Figure 6. Median daily cycles of net ecosystem production (A), NEP versus incoming PAR (B), and average daily cycles of latent heat flux (C). Measured half-hourly values were averaged for hours of the day (A, C) and for classes of PAR.

followed this drought had above average precipitation, TWU, GPP and NEP. The peaks in TWU and NEP in 2008/2009 were likely influenced by the high availability of soil water and stand regeneration following three seasonal cycles with high precipitation [82]. The next cycle was characterized by a reduction in the annual rainfall caused by the drought of 2010 in the Amazon [21], which caused drops in TWU and NEP. However, those drops were smaller than the reduction in NEP and TWU immediately after the drought in 2004/2005. It is likely that a further reduction in NEP has occurred in 2011 (data not available), similar to the reduction in 2006.

The annual balances presented in Figure 4 are sensitive to the filtering used to remove nighttime conditions with low levels of turbulence. In general such filtering is based on a threshold of the friction velocity u*, below which values are replaced by modeled data in the gap-filling analysis. The results in Figure 4 are based on a u* threshold of 0.1 m s^{-1}. To test the sensitivity of this choice, the threshold was changed to 0.15 m s^{-1} and the resulting carbon balance was calculated. The new threshold changed the annual carbon balance within 41% and 63% of the uncertainty generated by the error and gap-filling analysis, for 2004 and 2005, respectively. Hence, it is unlikely that a higher threshold would change the carbon balance beyond the uncertainty already determined.

The average intra-annual variability of some meteorological drivers helped to explain the variability of the carbon and water cycles (Figure 5). Total PAR, represented by the maximum at noon in Figure 5A, is highest in August, near the end of the dry season. The month of August is also the time of the year when the trend of clear-sky conditions reverses, as evident by the decrease in the clearness index and consequent increase of diffuse radiation. Despite the high values of T_{air} and VPD in August, which forces leaves to close stomata and reduce photosynthesis, GPP and NEP present a temporary maximum at this time of the year. The increased ecosystem production and net uptake of carbon is most likely due to the combination of several factors, such as: the maximum in total PAR; the increase in diffuse PAR – which is known to be highly effective for NEP since light is able to reach leaves not directly exposed to the Sun [74,75,83]; new leaves being produced at the end of the dry season; and the availability of water via deep roots. It has been reported before that trees can reach water deep into the soil [7,14,84], maintaining high levels of productivity even during the dry season provided that the soil is recharged by rainfall each year during the wet season. Nonetheless, the peak in NEP and GPP was not sustained in September, on average, most likely due to the reduction in the availability of soil water at the end of the dry season.

Despite the temporary increase in NEP and GPP at the end of the dry season, this ecosystem has lower net uptake of carbon

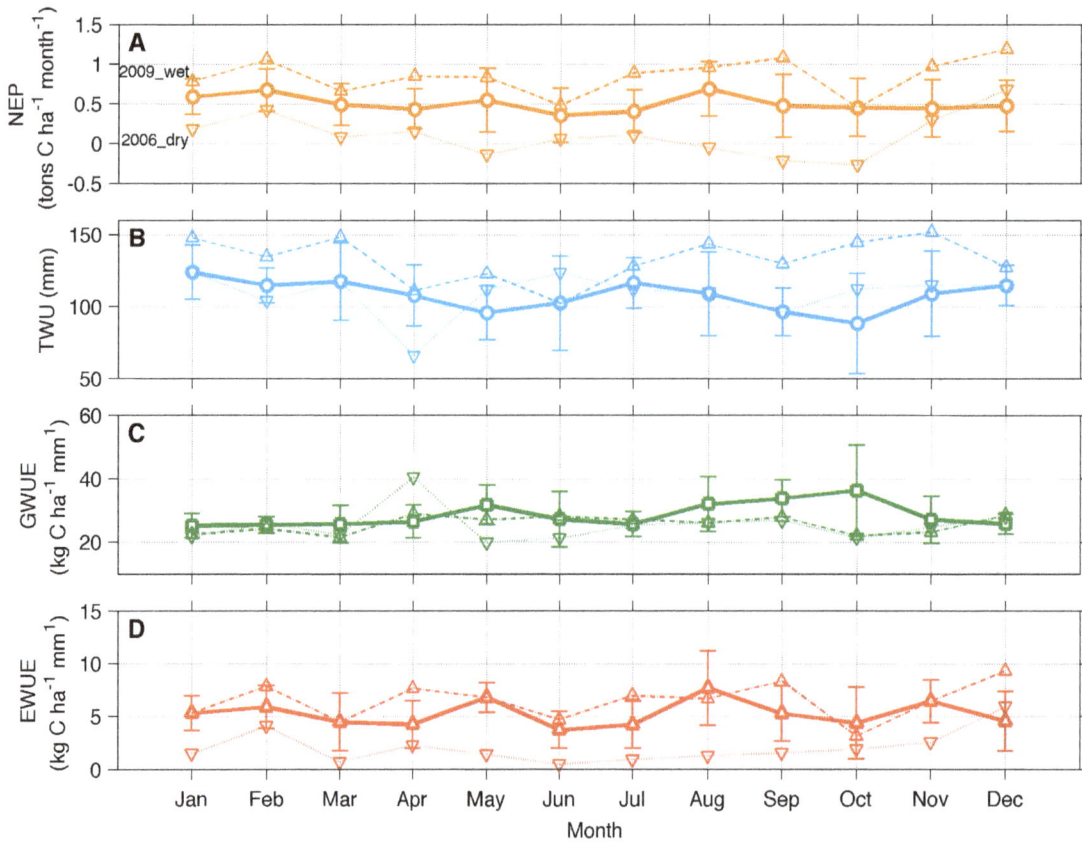

Figure 7. Annual cycles of NEP (A), TWU (B), GWUE (C), and EWUE (D), for average conditions and for years under anomalous climate conditions (dry 2006 and wet 2009). Vertical bars denote the interquartile range.

during this period, as can be seen in Figure 6, after averaging the daily cycle of NEP and evapotranspiration over a wet (Jan-Mar) and a dry period (Jul-Sep). Those periods were the same used in a previous accounting of the carbon balance for the old tower at this forest [10]. Daytime NEP is lowest from late morning to late afternoon, indicating lower net uptake of carbon (Figure 6A). The evolution of NEP versus PAR (Figure 6B) confirms the reduction in NEP during the dry season due to a combination of higher temperature, higher VPD and drier soils. In fact, the lower daytime evapotranspiration during the dry season confirms the lower vertical transport of water [85]. Nocturnal values of NEP during the dry season are less negative, indicating lower CO_2 emissions through soil respiration. This is caused by the drier soil, imposing limitations to microbial activity, which is responsible for CO_2 emissions from soils. Such lower emissions during nighttime are not enough to offset the lower uptake during daytime, causing the cumulative average NEP during the dry season (1.5 g C m^{-2} day^{-1}) to be 21% lower than the cumulative for the wet season (1.9 g C m^{-2} day^{-1}).

The dry season of 2006 presented a higher reduction on carbon and water fluxes compared to the usual reduction observed during a typical dry season (Figure 7). The impact of the drought of 2005 was strongest at this ecosystem during the dry season of 2006. Water limitations from September to November of 2005 likely reduced the recharge of soils to the next year. This impact is evident in the average daily cycle of NEP, its relation to PAR, and the daily cycle of latent heat flux (LE) in Figure 6. To further explore the effects of this climate extreme, the monthly variability of NEP and TWU for 2006 was compared with the average cycles

and with the wet year of 2009 (Figure 7A, B). In addition, the annual cycles of two water use efficiency ratios were investigated for both extreme years.

Monthly values of NEP were below average for most of 2006, a yearlong effect on fluxes likely caused by the water limitation experienced by this site in 2005 and during the first months of 2006 (February, March and April). This drought reduced net uptake of carbon most likely due to tree mortality and/or decomposition of fallen branches or trees. TWU was below or close to the average for the most part of the first semester of the year. The water use efficiencies (panels 7C,D) were also affected by the dry and wet years. For the dry year of 2006, most of the values of GWUE were below the average due to the decrease of GPP in that year (Figure 4C). For 2009, GPP did not strongly respond to the higher water availability and, when combined with the higher evapotranspiration (Figure 4B), resulted in a GWUE which was also below the average for most of the year. On the other hand, EWUE benefited from the wet year of 2009 with higher values above the monthly means, whereas the drought of 2006 contributed to lower values of EWUE. In conclusion, the drought that affected this forest in 2005 and 2006 strongly influenced the variability of carbon and water fluxes since it altered efficiencies of the ecosystem when using water and accumulating carbon.

Conclusions

Previous works about the carbon balance of Amazon forest sites revealed that the annual uptake could vary from 1 to ≈6 t C ha^{-1} year^{-1} [10,86], while climate disturbances, such as droughts, may cause net release of carbon even during the wet season of the year [11]. The determination of the carbon balance of different forest ecosystems in the Amazon requires the long-term studies of fluxes in order to capture the typical conditions as well as transient influences such as droughts or floods. The time series of annual NEP analyzed in this work is the longest ever published for this ecosystem, making it possible to notice the influence of such extreme climate events.

According to results from the first year of measurements at this site [87] – 2004 – evapotranspiration during the dry season (from July to September) decreased by 20%, which is similar to the decrease of $\approx 22\%$ found in this study when comparing the maximum values at midday for wet and dry seasons (Figure 6). In addition, the integration of carbon fluxes in that study resulted in an annual uptake of carbon of ≈ 5 t C ha^{-1} year^{-1}, a value close to the range reported here (4.7 ± 1.7 t C ha^{-1} year^{-1}, Table 1).

The carbon flux at this site has a peculiar annual cycle with an increase in net uptake of carbon at the end of the dry season. This lack of synchronicity with the monthly-accumulated rainfall disrupts the positive correlation of the carbon cycle with precipitation. Instead, the effects of incoming solar radiation, new leaves with increased assimilation rates, and probable water use through deep roots [14,84], enable this ecosystem to increase its net carbon uptake at the end of the dry season. Long-term measurements of soil moisture and soil respiration would surely add valuable information to the carbon and water balances of this ecosystem.

The response of this ecosystem to the climate anomalies of the last decade is in agreement with the results found in many studies of the larger-scale impacts of the droughts of 2005 and 2010 [17–20]. The annual carbon balance of this forest (NEP) ranged from a net source, in 2006, to an increasing sink afterwards. A decline in NEP was observed in 2009/2010, probably caused by the drought

in 2010. In spite of the typical increase in NEP at the end of the dry season, such mechanism was not enough to revert the long-term rainfall deficit experienced in 2005. In fact, the dry conditions affected the following seasonal year (2005/2006), contributing to turn the ecosystem into a carbon source. Therefore, it is unlikely that a green-up effect could have occurred over this forest [15,17].

Different parts of the Amazon have distinct annual cycles of meteorological drivers, longer or shorter dry seasons, and consequently different responses to changes in the region's atmospheric conditions [5,6,88,89]. Based on the variability in the results for NEP, the annual carbon balance is close to 5 t C ha^{-1} year^{-1}, with significant changes from year to year depending on climatological conditions. Continuous monitoring of the biogeochemical cycles at this site as well as other ecosystems in the Amazon would help to identify the typical carbon balance, when the ecosystem is not under the influence of climate extremes. Moreover, long-term series of fluxes and meteorological drivers are crucial for the calibration of biogeochemical models that simulate the exchanges of energy and scalars between the biosphere and the atmosphere.

Acknowledgments

The authors are grateful to Daniele S. Nogueira for improving the readability of the text.

Author Contributions

Conceived and designed the experiments: LDAS AOM ACA CAN. Performed the experiments: MZ RGA CvR FLC. Analyzed the data: MZ LDAS RGA. Contributed reagents/materials/analysis tools: AOM ACA RGA FLC GS. Wrote the paper: MZ LDAS AOM.

References

1. Betts AK, Dias MAFS (2010) Progress in Understanding Land-Surface-Atmosphere Coupling from LBA Research. J Adv Model Earth Sy 2. DOI: 10.3894/James.2010.2.6.

2. Costa MH, Foley JA (1999) Trends in the hydrologic cycle of the Amazon Basin. J Geophys Res Atmos 104: 14189–14198. DOI: 10.1029/1998JD200126.

3. Betts AK, Fisch G, von Randow C, Silva Dias MAF, Cohen JCP, et al. (2009) The Amazonian boundary layer and mesoscale circulations. In: Keller M, editor. Amazonia and Global Change. Washington, D.C.: AGU.

4. Betts AK (2007) Coupling of water vapor convergence, clouds, precipitation, and land-surface processes. J Geophys Res Atmos 112. DOI: 10.1029/2006jd008191.

5. da Rocha HR, Manzi AO, Cabral OM, Miller SD, Goulden ML, et al. (2009) Patterns of water and heat flux across a biome gradient from tropical forest to savanna in Brazil. J Geophys Res Biogeosci 114: G00B12. DOI: 10.1029/2007jg000640.

6. da Rocha HR, Goulden ML, Miller SD, Menton MC, Pinto LDVO, et al. (2004) Seasonality of Water and Heat Fluxes over a Tropical Forest in Eastern Amazonia. Ecol Appl 14: 22–32. DOI: 10.1890/02-6001.

7. Bruno RD, da Rocha HR, de Freitas HC, Goulden ML, Miller SD (2006) Soil moisture dynamics in an eastern Amazonian tropical forest. Hydrological Processes 20: 2477–2489. DOI: 10.1002/hyp.6211.

8. Costa MH, Biajoli MC, Sanches L, Malhado ACM, Hutyra LR, et al. (2010) Atmospheric versus vegetation controls of Amazonian tropical rain forest evapotranspiration: Are the wet and seasonally dry rain forests any different? J Geophys Res Biogeosci 115: G04021. DOI: 10.1029/2009jg001179.

9. Hutyra LR, Munger JW, Saleska SR, Gottlieb E, Daube BC, et al. (2007) Seasonal controls on the exchange of carbon and water in an Amazonian rain forest. J Geophys Res Biogeosci 112. DOI: 10.1029/2006JG000365.

10. von Randow C, Manzi AO, Kruijt B, de Oliveira PJ, Zanchi FB, et al. (2004) Comparative measurements and seasonal variations in energy and carbon exchange over forest and pasture in South West Amazonia. Theor Appl Clim 78: 5–26.

11. Saleska SR, Miller SD, Matross DM, Goulden ML, Wofsy SC, et al. (2003) Carbon in Amazon forests: unexpected seasonal fluxes and disturbance-induced losses. Science 302: 1554–1557. DOI: 10.1126/science.1091165.

12. Goulden ML, Miller SD, da Rocha HR, Menton MC, de Freitas HC, et al. (2004) Diel and seasonal patterns of tropical forest CO2 exchange. Ecol Appl 14: S42–S54.

13. Saleska SR, Didan K, Huete AR, da Rocha HR (2007) Amazon Forests Green-Up During 2005 Drought. Science 318: 612–612.

14. Nepstad DC, de Carvalho CR, Davidson EA, Jipp PH, Lefebvre PA, et al. (1994) The role of deep roots in the hydrological and carbon cycles of Amazonian forests and pastures. Nature 372: 666–669.

15. Samanta A, Ganguly S, Hashimoto H, Devadiga S, Vermote E, et al. (2010) Amazon forests did not green-up during the 2005 drought. Geophys Res Lett 37.

16. Lewis SL, Brando PM, Phillips OL, van der Heijden GMF, Nepstad D (2011) The 2010 Amazon Drought. Science 331: 554–554.

17. Xu L, Samanta A, Costa MH, Ganguly S, Nemani RR, et al. (2011) Widespread decline in greenness of Amazonian vegetation due to the 2010 drought. Geophys Res Lett 38: L07402. DOI: 10.1029/2011GL046824.

18. Potter C, Klooster S, Hiatt C, Genovese V, Castilla-Rubio JC (2011) Changes in the carbon cycle of Amazon ecosystems during the 2010 drought. Environmental Research Letters 6: 034024.

19. Zhao M, Running SW (2010) Drought-Induced Reduction in Global Terrestrial Net Primary Production from 2000 Through 2009. Science 329: 940–943.

20. Bagley JE, Desai AR, Harding KJ, Snyder PK, Foley JA (2013) Drought and Deforestation: Has land cover change influenced recent precipitation extremes in the Amazon? J Clim. DOI: 10.1175/JCLI-D-12-00369.1.

21. Marengo JA, Tomasella J, Alves LM, Soares WR, Rodriguez DA (2011) The drought of 2010 in the context of historical droughts in the Amazon region. Geophys Res Lett 38. DOI: 10.1029/2011gl047436.

22. Zeng N, Yoon JH, Marengo JA, Subramaniam A, Nobre CA, et al. (2008) Causes and impacts of the 2005 Amazon drought. Environmental Research Letters 3. DOI: 10.1088/1748-9326/3/1/014002.

23. Marengo JA, Nobre CA, Tomasella J, Oyama MD, Oliveira GS, et al. (2008) The drought of Amazonia in 2005. J Clim 21: 495–516. DOI: 10.1175/2007JCLI1600.1.

24. Marengo JA, Nobre CA, Tomasella J, Cardoso MF, Oyama MD (2008) Hydro-climate and ecological behaviour of the drought of Amazonia in 2005. Philos Trans R Soc Lond B Biol Sci 363: 1773–1778. DOI: 10.1098/rstb.2007.0015.

25. Marengo J, Tomasella J, Soares WR, Alves LM, Nobre CA (2012) Extreme climatic events in the Amazon basin. Theor Appl Clim 107: 73–85.

26. Zeri M, Sá LDA (2010) The impact of data gaps and quality control filtering on the balances of energy and carbon for a Southwest Amazon forest. Agric For Meteorol 150: 1543–1552. DOI: 10.1016/j.agrformet.2010.08.004.

27. Suyker AE, Verma SB (2009) Evapotranspiration of irrigated and rainfed maize-soybean cropping systems. Agric For Meteorol 149: 443–452. DOI: 10.1016/J.Agrformet.2008.09.010.

28. Zeri M, Hussain MZ, Anderson-Teixeira KJ, DeLucia EH, Bernacchi CJ (2013) Water use efficiency of perennial and annual bioenergy crops in central Illinois. J Geophys Res Biogeosci 118: 1–9. DOI: 10.1002/jgrg.20052.

29. Keenan TF, Hollinger DY, Bohrer G, Dragoni D, Munger JW, et al. (2013) Increase in forest water-use efficiency as atmospheric carbon dioxide concentrations rise. Nature. DOI: 10.1038/nature12291.

30. Andreae MO, Artaxo P, Brandao C, Carswell FE, Ciccioli P, et al. (2002) Biogeochemical cycling of carbon, water, energy, trace gases, and aerosols in Amazonia: The LBA-EUSTACH experiments. J Geophys Res Atmos 107: LBA 33-31–LBA 33-25. DOI: 10.1029/2001jd000524.

31. Bolzan MJA, Vieira PC (2006) Wavelet analysis of the wind velocity and temperature variability in the Amazon forest. Braz J Phys 36: 1217–1222. DOI: 10.1590/S0103-97332006000700018.

32. Bolzan MJA, Ramos FM, Sa LDA, Neto CR, Rosa RR (2002) Analysis of fine-scale canopy turbulence within and above an Amazon forest using Tsallis' generalized thermostatistics. J Geophys Res Atmos 107. DOI: 10.1029/2001JD000378.

33. Campanharo AS, Ramos FM, Macau EE, Rosa RR, Bolzan MJ, et al. (2008) Searching chaos and coherent structures in the atmospheric turbulence above the Amazon forest. Philos Transact A Math Phys Eng Sci 366: 579–589. DOI: 10.1098/rsta.2007.2118.

34. Andreae MO, Rosenfeld D, Artaxo P, Costa AA, Frank GP, et al. (2004) Smoking Rain Clouds over the Amazon. Science 303: 1337–1342.

35. von Randow C, Sa LDA, Gannabathula PSSD, Manzi AO, Arlino PRA, et al. (2002) Scale variability of atmospheric surface layer fluxes of energy and carbon over a tropical rain forest in southwest Amazonia - 1. Diurnal conditions. J Geophys Res Atmos 107: 1–12.

36. Kruijt B, Elbers JA, von Randow C, Araujo AC, Oliveira PJ, et al. (2004) The robustness of eddy correlation fluxes for Amazon rain forest conditions. Ecol Appl 14: S113.

37. Yamasoe MA, von Randow C, Manzi AO, Schafer JS, Eck TF, et al. (2006) Effect of smoke and clouds on the transmissivity of photosynthetically active radiation inside the canopy. Atmos Chem Phys 6: 1645–1656. DOI: 10.5194/acp-6-1645-2006.

38. Fisch G, Tota J, Machado LAT, Dias MAFS, Lyra RFD, et al. (2004) The convective boundary layer over pasture and forest in Amazonia. Theor Appl Clim 78: 47–59. DOI: 10.1007/S00704-004-0043-X.

39. Zeri M, Sá LDA (2011) Horizontal and Vertical Turbulent Fluxes Forced by a Gravity Wave Event in the Nocturnal Atmospheric Surface Layer Over the Amazon Forest. Boundary-Layer Meteorol 138: 413–431. DOI: 10.1007/s10546-010-9563-3.

40. Dias MAFS, Rutledge S, Kabat P, Dias PLS, Nobre C, et al. (2002) Cloud and rain processes in a biosphere-atmosphere interaction context in the Amazon Region. J Geophys Res Atmos 107. DOI: 10.1029/2001JD000335.

41. Kruijt B, Malhi Y, Lloyd J, Norbre AD, Miranda AC, et al. (2000) Turbulence statistics above and within two Amazon rain forest canopies. Boundary-Layer Meteorol 94: 297–331. DOI: 10.1023/A:1002401829007.

42. Montgomery RB (1948) Vertical Eddy Flux Of Heat In The Atmosphere. J Meteorol 5: 265–274. DOI: 10.1175/1520-0469(1948)005<0265:vefohi>2.0.co;2.

43. Swinbank WC (1951) The Measurement Of Vertical Transfer Of Heat And Water Vapor By Eddies In The Lower Atmosphere. J Meteorol 8: 135–145. DOI: 10.1175/1520-0469(1951)008<0135:tmovto>2.0.co;2.

44. Goulden ML, Munger JW, Fan SM, Daube BC, Wofsy SC (1996) Measurements of carbon sequestration by long-term eddy covariance: Methods and a critical evaluation of accuracy. Global Change Biol 2: 169–182. DOI: 10.1111/J.1365-2486.1996.Tb00070.X.

45. Baldocchi DD, Hicks BB, Meyers TP (1988) Measuring Biosphere-Atmosphere Exchanges of Biologically Related Gases with Micrometeorological Methods. Ecology 69: 1331–1340. DOI: 10.2307/1941631.

46. Wilczak JM, Oncley SP, Stage SA (2001) Sonic anemometer tilt correction algorithms. Boundary-Layer Meteorol 99: 127–150.

47. Kaimal JC, Finnigan JJ (1994) Atmospheric boundary layer flows: their structure and measurement. New York: Oxford University Press. 289 p.

48. Schotanus P, Nieuwstadt FTM, Debruin HAR (1983) Temperature-Measurement with a Sonic Anemometer and Its Application to Heat and Moisture Fluxes. Boundary-Layer Meteorol 26: 81–93. DOI: 10.1007/Bf00164332.

49. Webb EK, Pearman GI, Leuning R (1980) Correction of Flux Measurements for Density Effects Due to Heat and Water-Vapor Transfer. Q J Roy Meteorol Soc 106: 85–100.

50. Leuning R (2006) The correct form of the Webb, Pearman and Leuning equation for eddy fluxes of trace gases in steady and non-steady state, horizontally homogeneous flows. Boundary-Layer Meteorol 123: 263–267. DOI: 10.1007/s10546-006-9138-5.

51. Gu L, Massman WJ, Leuning R, Pallardy SG, Meyers T, et al. (2012) The fundamental equation of eddy covariance and its application in flux measurements. Agric For Meteorol 152: 135–148. DOI: 10.1016/j.agrformet.2011.09.014.

52. Philip JR (1963) The Damping of a Fluctuating Concentration by Continuous Sampling through a Tube. Aust J Phys 16.

53. Moore CJ (1986) Frequency response corrections for eddy correlation systems. Boundary-Layer Meteorol 37: 17–35.

54. Leuning R, King KM (1992) Comparison of eddy-covariance measurements of CO2 fluxes by open- and closed-path CO2 analysers. Boundary-Layer Meteorol 59: 297–311. DOI: 10.1007/bf00119818.

55. Burba GG, McDermitt DK, Grelle A, Anderson DJ, Xu L (2008) Addressing the influence of instrument surface heat exchange on the measurements of CO2 flux from open-path gas analyzers. Global Change Biol 14: 1854–1876. DOI: 10.1111/j.1365-2486.2008.01606.x.

56. Wilks DS (1995) Statistical Methods In The Atmospheric Sciences: An Introduction. San Diego: Academic Press. 467 p.

57. Foken T, Wichura B (1996) Tools for quality assessment of surface-based flux measurements. Agric For Meteorol 78: 83–105. DOI: 10.1016/0168-1923(95)02248-1.

58. Vickers D, Mahrt L (1997) Quality control and flux sampling problems for tower and aircraft data. J Atmos Oceanic Tech 14: 512–526.

59. Foken T, Goeckede M, Mauder M, Mahrt L, Amiro BD, et al. (2004) Post-field data quality control. In: Lee X, Massman WJ, Law B, editors. Handbook of micrometeorology: a guide for surface flux measurement and analysis. Dordrecht, The Netherlands: Kluwer Academic Publishers. pp. 181–208.

60. Kochendorfer J, Meyers T, Frank J, Massman W, Heuer M (2012) How Well Can We Measure the Vertical Wind Speed? Implications for Fluxes of Energy and Mass. Boundary-Layer Meteorol 145: 383–398. DOI: 10.1007/s10546-012-9738-1.

61. Frank JM, Massman WJ, Ewers BE (2013) Underestimates of sensible heat flux due to vertical velocity measurement errors in non-orthogonal sonic anemometers. Agric For Meteorol 171–172: 72–81. DOI: 10.1016/j.agrformet.2012.11.005.

62. Stull RB (1988) An Introduction to Boundary Layer Meteorology. Dordrecht, Boston, London: Kluwer Academic Publishers. 666 p.

63. Falge E, Baldocchi D, Olson R, Anthoni P, Aubinet M, et al. (2001) Gap filling strategies for defensible annual sums of net ecosystem exchange. Agric For Meteorol 107: 43–69.

64. Reichstein M, Falge E, Baldocchi D, Papale D, Aubinet M, et al. (2005) On the separation of net ecosystem exchange into assimilation and ecosystem respiration: review and improved algorithm. Global Change Biol 11: 1424–1439.

65. Anthoni PM, Freibauer A, Kolle O, Schulze ED (2004) Winter wheat carbon exchange in Thuringia, Germany. Agric For Meteorol 121: 55–67. DOI: 10.1016/S0168-1923(03)00162-X.

66. Finnigan J (1999) A comment on the paper by Lee (1998): "On micrometeorological observations of surface-air exchange over tall vegetation". Agric For Meteorol 97: 55–64. DOI: 10.1016/S0168-1923(99)00049-0.

67. Lee XH (1998) On micrometeorological observations of surface-air exchange over tall vegetation. Agric For Meteorol 91: 39–49. DOI: 10.1016/S0168-1923(98)00071-9.

68. Lee XH, Hu XZ (2002) Forest-air fluxes of carbon, water and energy over non-flat terrain. Boundary-Layer Meteorol 103: 277–301. DOI: 10.1023/A:1014508928693.

69. Aubinet M, Heinesch B, Yernaux M (2003) Horizontal and vertical CO2 advection in a sloping forest. Boundary-Layer Meteorol 108: 397–417. DOI: 10.1023/A:1024168428135.

70. Feigenwinter C, Bernhofer C, Vogt R (2004) The influence of advection on the short term CO2-budget in and above a forest canopy. Boundary-Layer Meteorol 113: 201–224. DOI: 10.1023/B:Boun.0000039372.86053.Ff.

71. Feigenwinter C, Bernhofer C, Eichelmann U, Heinesch B, Hertel M, et al. (2008) Comparison of horizontal and vertical advective CO2 fluxes at three forest sites. Agric For Meteorol 148: 12–24. DOI: 10.1016/J.Agrformet.2007.08.013.

72. de Araújo AC, Dolman AJ, Waterloo MJ, Gash JHC, Kruijt B, et al. (2010) The spatial variability of CO2 storage and the interpretation of eddy covariance fluxes in central Amazonia. Agric For Meteorol 150: 226–237. DOI: 10.1016/j.agrformet.2009.11.005.

73. Tóta J, Fitzjarrald DR, Staebler RM, Sakai RK, Moraes OMM, et al. (2008) Amazon rain forest subcanopy flow and the carbon budget: Santarém LBA-ECO site. J Geophys Res Biogeosci 113. DOI: 10.1029/2007jg000597.

74. Doughty CE, Flanner MG, Goulden ML (2010) Effect of smoke on subcanopy shaded light, canopy temperature, and carbon dioxide uptake in an Amazon rainforest. Global Biogeochem Cycles 24. DOI: 10.1029/2009gb003670.

75. Bai YF, Wang J, Zhang BC, Zhang ZH, Liang J (2012) Comparing the impact of cloudiness on carbon dioxide exchange in a grassland and a maize cropland in northwestern China. Ecol Res 27: 615–623. DOI: 10.1007/S11284-012-0930-Z.

76. Gu L, Fuentes JD, Shugart HH, Staebler RM, Black TA (1999) Responses of net ecosystem exchanges of carbon dioxide to changes in cloudiness: Results from two North American deciduous forests. J Geophys Res Atmos 104: 31421–31434. DOI: 10.1029/1999jd901068.

77. Zeri M, Anderson-Teixeira K, Hickman G, Masters M, DeLucia E, et al. (2011) Carbon exchange by establishing biofuel crops in Central Illinois. Agric Ecosyst Environ 144: 319–329. DOI: 10.1016/j.agee.2011.09.006.

78. Lloyd J, Taylor JA (1994) On the Temperature-Dependence of Soil Respiration. Funct Ecol 8: 315–323.

79. VanLoocke A, Twine TE, Zeri M, Bernacchi CJ (2012) A regional comparison of water use efficiency for miscanthus, switchgrass and maize. Agric For Meteorol 164: 82–95. DOI: 10.1016/j.agrformet.2012.05.016.

80. Richardson AD, Hollinger DY (2007) A method to estimate the additional uncertainty in gap-filled NEE resulting from long gaps in the CO2 flux record. Agric For Meteorol 147: 199–208.

81. Richardson AD, Hollinger DY, Burba GG, Davis KJ, Flanagan LB, et al. (2006) A multi-site analysis of random error in tower-based measurements of carbon and energy fluxes. Agric For Meteorol 136: 1–18.

82. Malhi Y, Pegoraro E, Nobre AD, Pereira MGP, Grace J, et al. (2002) Energy and water dynamics of a central Amazonian rain forest. J Geophys Res Atmos 107. DOI: 10.1029/2001JD000623.

83. Butt N, New M, Malhi Y, da Costa ACL, Oliveira P, et al. (2010) Diffuse radiation and cloud fraction relationships in two contrasting Amazonian rainforest sites. Agric For Meteorol 150: 361–368. DOI: 10.1016/J.Agrformet.2009.12.004.

84. Wright IR, Nobre CA, Tomasella J, da Rocha HR, Roberts JM, et al. (1996) Towards a GCM surface parameterization of Amazonia. In: Gash JHC, editor. Amazonian Deforestation and Climate. New York: John Wiley. pp. 473–504.

85. Betts AK (2004) Understanding hydrometeorology using global models. Bull Am Meteorol Soc 85: 1673–1688. DOI: 10.1175/Bams-85-11-1673.

86. Grace J, Lloyd J, McIntyre J, Miranda AC, Meir P, et al. (1995) Carbon Dioxide Uptake by an Undisturbed Tropical Rain Forest in Southwest Amazonia, 1992 to 1993. Science 270: 778–780.

87. Aguiar RG (2005) Fluxos de massa e energia em uma floresta tropical no sudoeste da Amazônia [Masters Thesis]. Cuiabá, MT: Universidade Federal de Mato Grosso. 78 p.

88. Baker IT, Prihodko L, Denning AS, Goulden M, Miller S, et al. (2008) Seasonal drought stress in the Amazon: Reconciling models and observations. J Geophys Res Biogeosci 113. DOI: 10.1029/2007jg000644.

89. Fisher JB, Malhi Y, Bonal D, Da Rocha HR, De Araujo AC, et al. (2009) The land-atmosphere water flux in the tropics. Global Change Biol 15: 2694–2714. DOI: 10.1111/J.1365-2486.2008.01813.X.

16

A Mixed Method to Evaluate Burden of Malaria Due to Flooding and Waterlogging in Mengcheng County, China

Guoyong Ding[1,2,3], Lu Gao[1,3], Xuewen Li[3,4], Maigeng Zhou[5], Qiyong Liu[3,6], Hongyan Ren[7], Baofa Jiang[1,3]*

1 Department of Epidemiology and Health Statistics, School of Public Health, Shandong University, Jinan City, Shandong Province, P.R.China, 2 Department of Occupational and Environmental Health, School of Public Health, Taishan Medical College, Taian City, Shandong Province, P.R.China, 3 Shandong University Climate Change and Health Center, Jinan City, Shandong Province, P.R.China, 4 Department of Environment and Health, School of Public Health, Shandong University, Jinan City, Shandong Province, P.R.China, 5 National Center for Chronic and Noncommunicable Disease Control and Prevention, China CDC, Beijing City, P.R.China, 6 State Key Laboratory for Infectious Diseases Prevention and Control, National Institute for Communicable Disease Control and Prevention, China CDC, Beijing City, P.R.China, 7 State Key Laboratory of Resources and Environmental Information System, Institute of Geographic Sciences and Natural Resources Research, Chinese Academy of Sciences, Beijing City, P.R.China

Abstract

Background: Malaria is a highly climate-sensitive vector-borne infectious disease that still represents a significant public health problem in Huaihe River Basin. However, little comprehensive information about the burden of malaria caused by flooding and waterlogging is available from this region. This study aims to quantitatively assess the impact of flooding and waterlogging on the burden of malaria in a county of Anhui Province, China.

Methods: A mixed method evaluation was conducted. A case-crossover study was firstly performed to evaluate the relationship between daily number of cases of malaria and flooding and waterlogging from May to October 2007 in Mengcheng County, China. Stratified Cox models were used to examine the lagged time and hazard ratios (HRs) of the risk of flooding and waterlogging on malaria. Years lived with disability (YLDs) of malaria attributable to flooding and waterlogging were then estimated based on the WHO framework of calculating potential impact fraction in the Global Burden of Disease study.

Results: A total of 3683 malaria were notified during the study period. The strongest effect was shown with a 25-day lag for flooding and a 7-day lag for waterlogging. Multivariable analysis showed that an increased risk of malaria was significantly associated with flooding alone [adjusted hazard ratio (AHR) = 1.467, 95% CI = 1.257, 1.713], waterlogging alone (AHR = 1.879, 95% CI = 1.696, 2.121), and flooding and waterlogging together (AHR = 2.926, 95% CI = 2.576, 3.325). YLDs per 1000 of malaria attributable to flooding alone, waterlogging alone and flooding and waterlogging together were 0.009 per day, 0.019 per day and 0.022 per day, respectively.

Conclusion: Flooding and waterlogging can lead to higher burden of malaria in the study area. Public health action should be taken to avoid and control a potential risk of malaria epidemics after these two weather disasters.

Editor: Rick Edward Paul, Institut Pasteur, France

Funding: This work was supported by the National Basic Research Program of China (973 Program) (Grant No. 2012CB955502). The funders had no role in study design, data collection and analysis, decision to publish, or preparation of the manuscript.

Competing Interests: The authors have declared that no competing interests exist.

* E-mail: bjiang@sdu.edu.cn

Introduction

Climate change is a current global concern and has an influence on the epidemiology of vector-borne diseases [1,2]. In particular, climatic disasters play a major role in affecting the emergence and prevalence of vector-borne diseases [3]. Heavy rainfall may cause flooding or waterlogging. Flooding is an overflow of surface runoff that submerges towns and farmland, which is often caused by long-lasting heavy storms. Waterlogging is one of the most hazardous natural disasters, which can also be called as submergence, wet damage, moisture damage, and is often caused by long lasting rainfall without a heavy precipitation intensity [4].

On average, floods and other hydrological events accounted for over 50% of the disasters between 2001–2010 in the world [5]. In late June and July 2007, the persistent and heavy rainfall caused several floods in the Huaihe River Basin, China. The floods in 2007 forced an evacuation of thousands of people from homelands, with at least 89 counties and over 15.1 million people affected in Anhui Province [6]. It is important to study the impact of floods on human health for forecasting and informing the population, in order to help minimize negative consequences.

The health effects of flooding or waterlogging are complex and far-reaching, which may include increased mortality and morbidity from Malaria. Malaria, a highly climate-sensitive vector-borne infectious disease, is a major public health problem in most developing countries. At the global level, malaria is considered the world's most important vector-borne disease. According to the World Malaria Report 2012, there were an estimated 219 million cases of malaria and 660,000 deaths in 2010 [7]. Historically, a higher incidence of malaria was observed in the Huang-Huai River region of central China and the total number of malaria cases accounted for 91.2% of the total reported cases in the country in the 1970s [8]. At present, this disease still represents a significant public health problem in this region, with dramatic re-emergence since 2001 [8]. A total 27,307 malaria cases in Anhui Province were notified in the annual case reporting system with accounting for 58.5% of the total number of reported cases in China in 2007 [9]. The incidence of malaria in the northern areas of Anhui Province was higher than that in the middle and southern Anhui Province since 2000 [10]. Mengcheng, one of the northern counties of Anhui, has one of the highest burden of malaria with a peak 3,803 malaria cases reported in 2007, particularly in July and August (Figure 1) [9]. While, the rainfall in Mengcheng County brought about a severe flooding and a waterlogging before the peak incidence of cases. And it was the largest floods since the 1954 Huaihe River floods in this region [6].

Few studies have been conducted about the impact of flooding on malaria. Flooding may wash away existing mosquito-breeding sites, standing water caused by heavy rainfall or overflow of rivers can create new breeding site. This situation can result in an increase in the vector population and the potential for malaria transmission. These studies have described the disease status of malaria during post-disaster periods [11–14], but there was no quantitative examination on the relationship between malaria and flooding. To our knowledge, relevant studies on the association between malaria and waterlogging have not been reported. The association between these two weather events and malaria is far from clear. In addition, given little research has been conducted in China, effects of the 2007 flooding and waterlogging on malaria remain unknown. In order to know the epidemiological information on the malaria situation caused by the 2007 Huaihe River floods and to provide reliable data for the control programs in the county of Mengcheng, this study was conducted to quantify the impact of flooding and waterlogging on malaria in 2007.

Methods

Ethical statement

Disease surveillance data used in this study were permitted by Chinese Center for Disease Control and Prevention. All data are unidentified. The study was approved by the research institutional review board of Public Health of Shandong University.

Study site

Mengcheng County is located in Northern Anhui Province, between 32°56′ and 33°29′ of latitude north and between 116°15′ and 116°49′ of longitude east (Figure 2). Mengcheng has an area of 2091 km^2 and a population of 1.2 million. The county is generally characterized by a sub-humid warm temperate continental monsoon with mild climate and plentiful rainfall, with an annual average temperature of 14.8°C, an annual average precipitation of 843 mm, and an annual average relative humidity of 70.4%. The geographic landscape and climate situation, such as suitable temperature and humidity, abundant rainfall, and existence of water bodies, provided favorable breeding sites for *Anopheles*. Studies conducted in the areas along the Huang and Huaihe River show that *An. sinensis* plays an important role in the *P. vivax* malaria transmission [15–17]. The main crops of the

county are wheat, soybean, corn, and a small amount of rice. During summers, most of local residents tend to sleep outdoors.

Data collection

Disease surveillance data. Malaria data were collected for the period of 2005–2010 from the National Notifiable Disease Surveillance System (NDSS). All malaria cases were defined based on the diagnostic criteria and principles of management for malaria (GB 15989-1995) issued by Ministry of Health of the People's Republic of China. Only the cases confirmed clinically and by laboratory test, including thick and thin blood smear, were included in our study. Information of cases included age, gender, type of disease, date of onset, and date of death. In China, malaria is a statutory notifiable category B infectious disease. Therefore, physicians in hospitals must report every case of malaria to the local health authority within 24 hours. Therefore, it is believed that the degree of compliance in disease notification over the study period was consistent. Demographic data were obtained from the Center for Public Health Science Data in China (http://www.phsciencedata.cn/).

Data on flooding and waterlogging. Meteorological events data were collected from the Yearbook of Meteorological Disasters in China and Chinese Agro-meteorological Disasters Information Data (http://cdc.cma.gov.cn/choiceStation.do). Exceptionally heavy rains occurred during the main flood season in 2007 caused serious disasters in Huaihe River Basin [6]. Mengcheng County was one of the worst affected areas. From 3 July to 9 July, the county had experienced a severe flooding which had a duration of 7 days and hit 43 thousand hectares of crops [18]. A continuous rain process during 15 July to 26 July led to a waterlogging disaster which had a duration of 12 days and hit more than 67 thousand hectares of crops [18].

Meteorological data. After receiving permission from National Meteorological Information Center of China, daily meteorological data were obtained from the China Meteorological Data Sharing Service System (http://cdc.cma.gov.cn/). The meteorological variables included daily average temperature, daily average relative humidity, daily rainfall, and daily sunshine duration. Because the effect of meteorological variables on the incidence of malaria is not linear [19–21], average temperature, average relative humidity, rainfall and sunshine duration were transformed to categorical variables. Average temperature was grouped into three levels: <20°C, 20–30°C, and >30°C. Average relative humidity was grouped into <60%, 60–80%, and >80%. Such categorization of temperature and relative humidity was based on the reports of the climatic conditions thought to be suitable for transmission malaria by *Anopheles* [21–25]. According to the scale of precipitation in China, rainfall was grouped into four levels in our study: light rain (0.1–9.9 mm per day), moderate rain (10–24.9 mm per day), heavy rain (25–49.9 mm per day) and rainstorm (>50 mm per day) [26]. Sunshine duration was grouped into two levels based on median of variables: <6 h and ≥6 h.

Study design and statistical analysis

A method combining a case-crossover study and attributable burden of disease was adopted to carry out the risk assessment on malaria caused by flooding and waterlogging. Firstly, a descriptive analysis was performed to describe characteristics of cases of malaria and the distribution of meteorological factors. Secondly, a case-crossover study, which was proposed for the study of transient outcomes that were impacted by short-term events or exposures [27,28], was conducted to examine whether flooding and waterlogging were related to the number of malaria cases. The case-crossover design, which is a special case-control design where

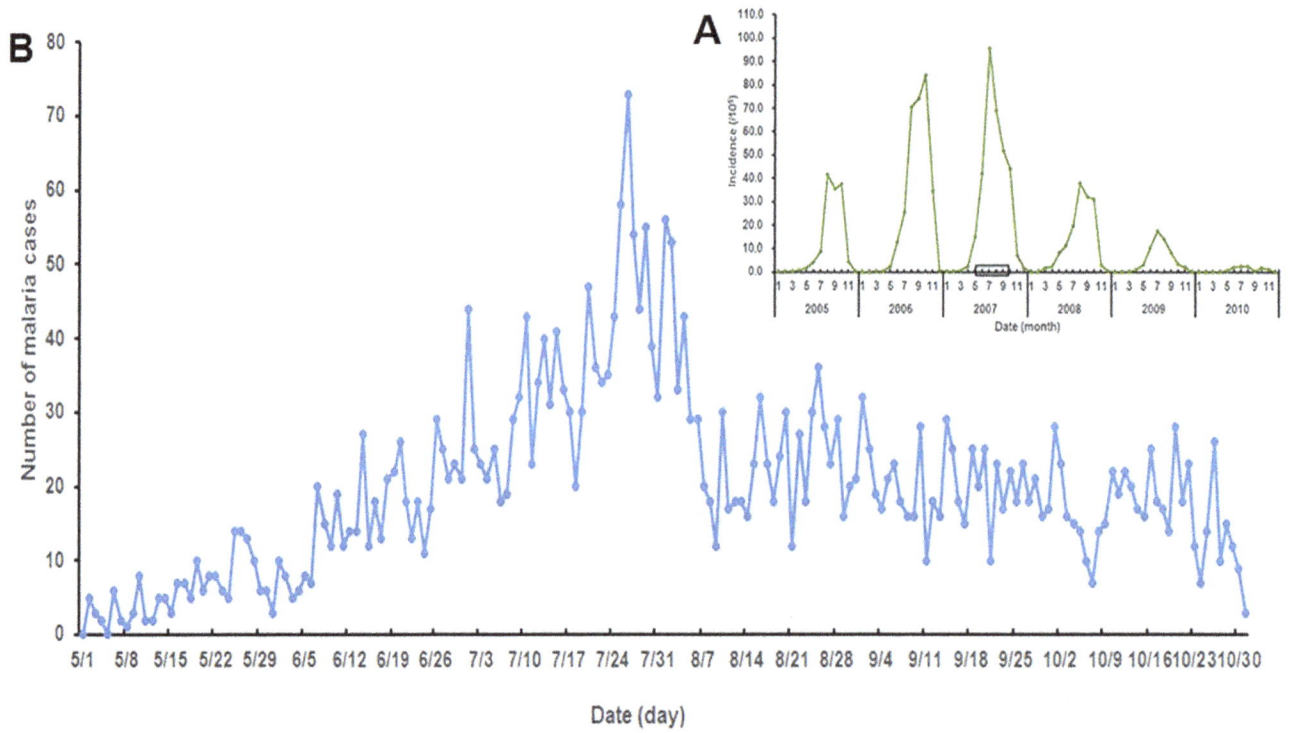

Figure 1. Monthly incidence of malaria from 2005–2010 (A) and daily cases of malaria during the study period (B) in Mengcheng County.

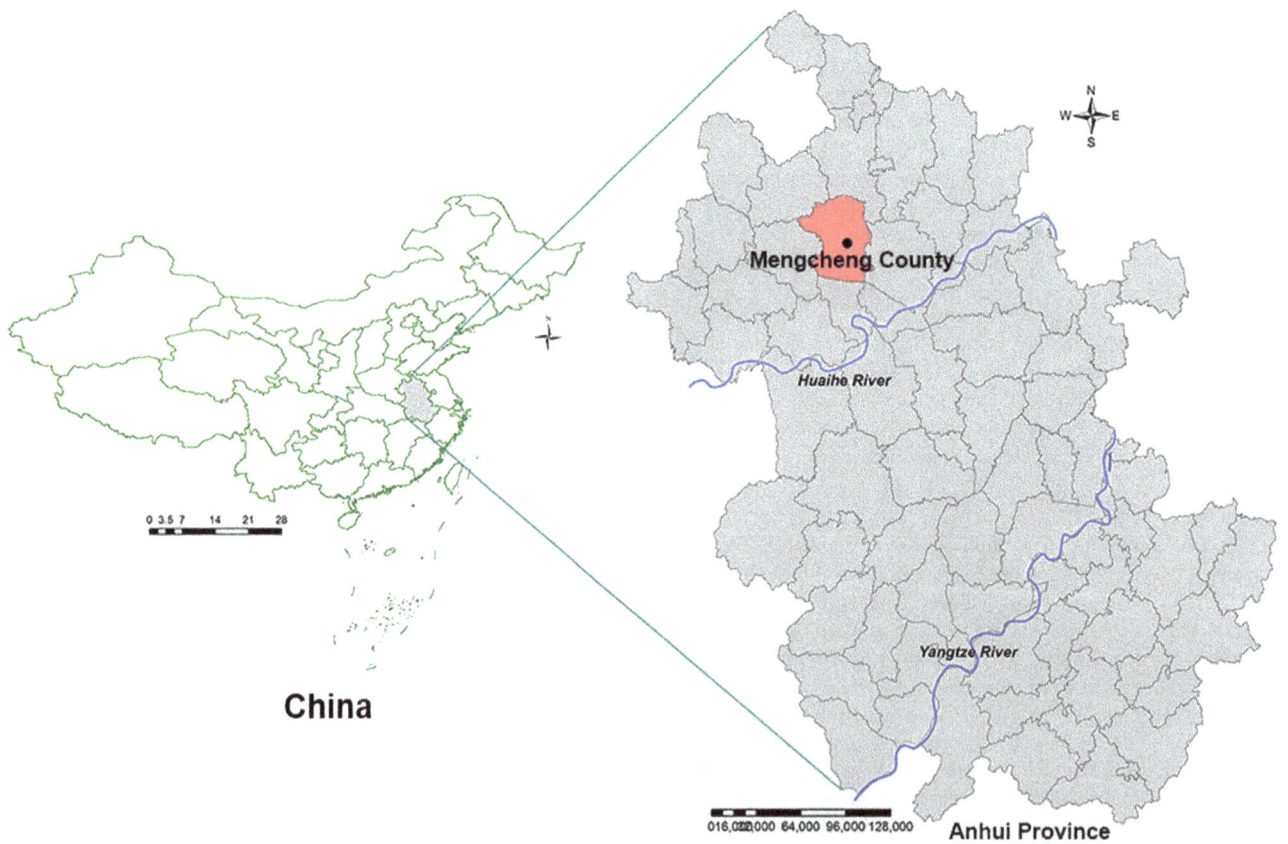

Figure 2. Location of the study area in Anhui Province, China.

every case serves as its own control, offers the ability to control confounders by design rather than by modeling [29,30]. Therefore, potential confounding due to age, sex, personality, and other fixed characteristics is eliminated [31]. The period of May-October 2007 was selected as our study period. The 1:3 symmetric bidirectional design was applied for selecting references to overcome the time trend of exposure and confounding, because referents were within the same season and on the same day of the week as the index time [32]. Six referents 7, 14 and 21 days before and after the event day were selected as control days. For example, when the event day was 9 July 2007, we considered 18 June, 25 June, 2 July, 16 July, 23 July, and 30 July 2007 as reference days.

The association between the number of cases of malaria and flooding and waterlogging was estimated by using the hazard ratios (HRs) and their 95% confidence intervals (CI) on the basis of stratified Cox models. We used fitting a stratified Cox model with the "Breslow" option for handing tied failure times [33]. The effects of exposure to flooding and waterlogging were explored for the duration of the effect-period using univariate stratified Cox models. The lagged 0 day was labeled "L0", and the lagged 1 day was "L1". The lagged 2 days was "L2", and so forth. Meteorological events may indirectly lead to an increase in vector-borne diseases through the expansion in the number and range of vector habitats [34]. Under suitable conditions, the duration from egg development to adult mosquitoes is about 9-15 days. The incubation period of *P. vivax* malaria ranges from 6–21 days. Thus, the lag effect up to 60 days was assessed by the stratified Cox regression analysis. The maximum lag time was selected based on the maximum HRs (at this point, the best estimate of duration had minimal nondifferential misclassification [27]). After adjusting for average temperature, average relative humidity, rainfall, and sunshine duration in multivariate stratified Cox models, HRs and 95% CI of malaria due to the exposure to flooding and waterlogging were calculated in each model. All statistical analyses were performed using SAS 9.1.3 (SAS Institute Inc., USA).

Thirdly, years lived with disability (YLDs) were calculated to estimate the burden of disease due to malaria during exposure effect-period of flooding and waterlogging. Since there was no death of malaria notified during the study period, we adopted YLDs to estimate the burden of disease with the consideration of lagged effects. The method of estimating YLDs as recommended by the World Health Organization (WHO) was used to calculate burden of malaria during exposure effect-period of flooding and waterlogging [35]. Calculations of YLDs and YLD per 1000 were made using DisMod II (WHO, 2001) and Microsoft Office Excel 2003 (Microsoft Corp., USA).

Lastly, the potential impact fraction (PIF) and attributable YLDs were estimated for the percentage of burden of disease due to malaria that was attributed to flooding and waterlogging. The following formula for PIF was $PIF = \frac{(\sum P_i RR_i - 1)}{\sum P_i RR_i}$ [36].

Where: P_i = Proportion of the population in exposure category i. RR_i = Relative risk at exposure category i compared to the reference level (If the rare disease assumption holds, HR is a good approximation to RR [37]).

The YLDs for the population were multiplied by PIF to calculate the fraction of malaria attributable to flooding and waterlogging for the study population, as shown in the following equation. *Attributable YLDs* = *PIF* × *YLDs* [36].

Results

Descriptive analysis for the disease and meteorological data

From 2005 to 2010, a total of 11491 malaria were reported in the study area with a mean monthly incidence rate of $13.76/10^5$. The monthly incidence peaked in July 2007, reaching $95.78/10^5$. A seasonal distribution of incidence was observed with most cases occurred in summer and autumn (Figure 1-A). There were 3683 (2146 males and 1537 females) notified malaria cases during the study period, accounted for 96.2% of the total reported cases in this county in 2007. By age groups: 22.7% aged below 14 years, 55.4% aged between 15 and 59 years, and 21.9% aged over 60 years. There was a distinct difference in the daily number of malaria cases among different months with a peak from late July to early August, and the maximum number of malaria cases reached 73 on 26 July (Figure 1-B). Figure 3 shows the distribution of daily meteorological factors during the study period. The mean daily average temperature over the study period was 23.9°C (Figure 3-A) and mean daily average relative humidity was 74.3% (Figure 3-B). The total precipitation in Mengcheng was 1177 mm from May to October with maximum rainfall occurred on 3 July (Figure 3-C). Besides, the mean daily sunshine duration was 5.3 h (Figure 3-D).

Analysis for lagged effects

Results from the lag time analysis are summarized in Figure 4. Flooding significantly increased the number of cases from L5 to L9 and from L20 to L29 (HRs>1, P<0.05), and waterlogging was from L0 to L14 (HRs>1, P<0.05). The strongest effects were observed at L25 days (HR = 1.695, 95% CI = 1.505, 1.910) for flooding, and L7 days (HR = 1.838, 95% CI = 1.654, 2.042) for waterlogging, respectively. After adjusting the durations of exposure and lagged effects, exposure effect-period for flooding was from 28 July to 3 August, and effect-period for waterlogging was from 22 July to 2 August (Figure 5). Common exposure effect-period for both flooding and waterlogging was from 28 July to 2 August (Figure 5).

Multivariate analysis

HRs of flooding and waterlogging on the risk of malaria are presented in Table 1. Flooding alone was significantly associated with an increased risk of malaria with adjustment for meteorological factors (AHR = 1.467, 95% CI = 1.257, 1.713). The AHR of malaria for waterlogging alone was 1.879 (95% CI = 1.696, 2.121) in the multivariate model. The risk for malaria was significantly associated with flooding and waterlogging together (AHR = 2.926, 95% CI = 2.576, 3.325).

Analysis for attributable YLDs of malaria

Tables 2, 3 and 4 display the burden of disease due to malaria caused by flooding and waterlogging. As shown in Table 2, the incidence rate and YLD per 1000 of malaria caused by flooding alone between 2 August and 3 August were $4.569/10^5$ and 0.028, respectively. The YLD per 1000 of male at this stage was higher than that of female (0.032 vs. 0.024). The YLD per 1000 of malaria was highest in old people above 80 years of age (0.090), followed by the 60–69 years old age group (0.050). Table 3 shows that the incidence rate and YLD per 1000 of malaria caused by waterlogging alone between 22 July and 28 July were $25.604/10^5$ and 0.242, respectively. The YLD per 1000 of male was also higher than that of female (0.273 vs. 0.209). The age of 5–14 years had the highest YLD per 1000 (0.706), followed by the age of 60–69 years 0.496). The burden of disease due to malaria caused by

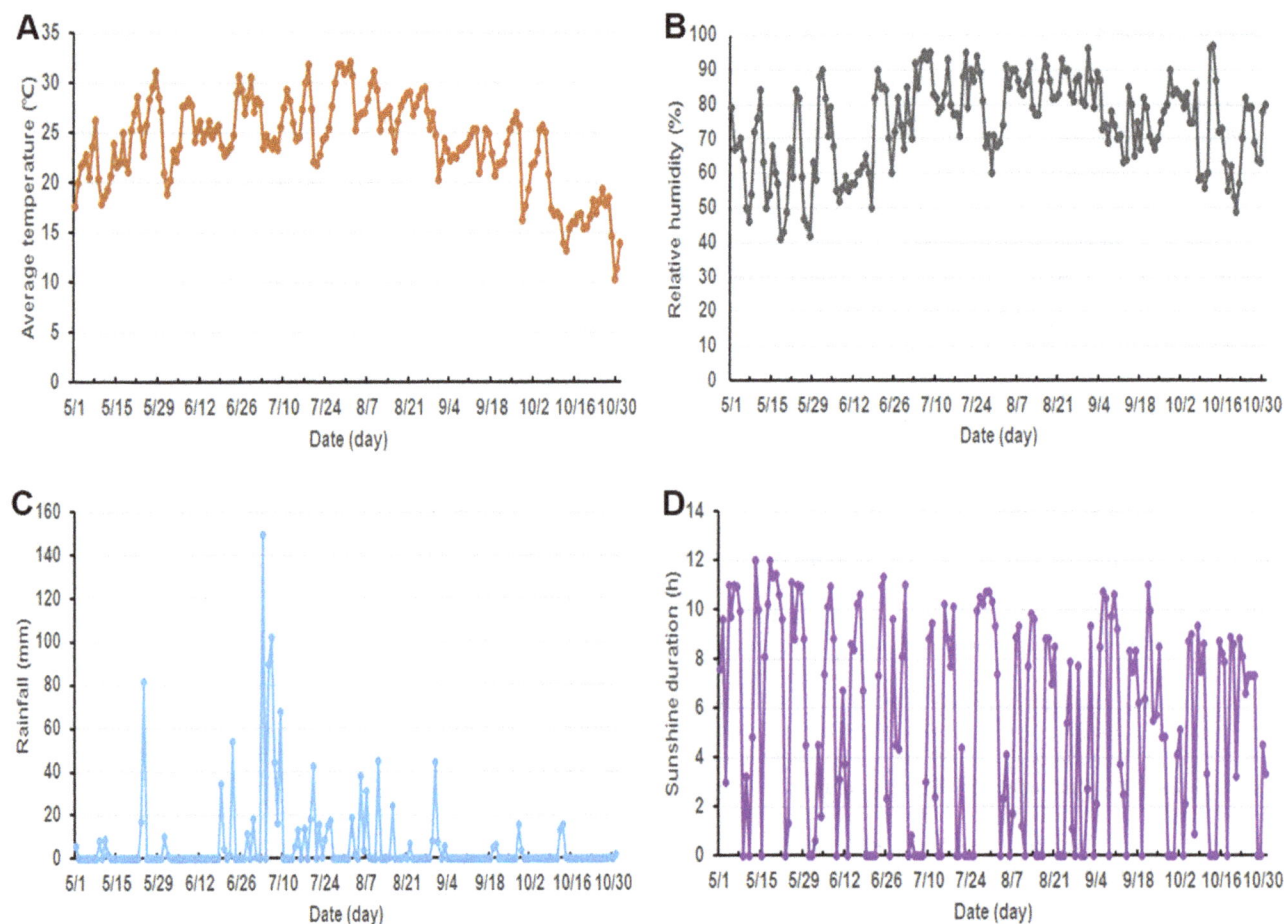

Figure 3. The distribution of daily meteorological factors during the study period. (A) Daily average temperature in Mengcheng County; (B) daily average relative humidity in Mengcheng County; (C) daily rainfall in Mengcheng County; (D) daily sunshine duration in Mengcheng County.

flooding and waterlogging together from 28 July to 2 August is given in Table 4. The incidence rate and YLD per 1000 of malaria were $19.483/10^5$ and 0.168, respectively. The YLD per 1000 for malaria in male (0.182) was higher than that in female (0.153). The highest YLD per 1000 of malaria was in old people above 80 years of age (0.351), and the second was in people aged between 70 and 79 years (0.305).

We assumed proportion of the study population in exposing these two disasters at 100 percent (i.e. $P_i = 1$). Based on the estimates of HRs and the formula of PIF above, PIFs of the study population exposed to flooding alone, waterlogging alone, and flooding and waterlogging together were 0.318, 0.473, and 0.658, respectively. The PIFs were considered in the further calculation of attributable YLDs. Figure 6 shows YLD per 1000 and attributable YLD per 1000 of the study population during different exposure effect-period of disasters. The attributable YLD per 1000 during common exposure effect-period for both flooding and waterlogging (0.111/5 = 0.022 per day) was higher than that exposed to flooding alone (0.009/1 = 0.009 per day) and waterlogging alone (0.114/6 = 0.019 per day).

Discussion

Our results indicate that flooding and waterlogging play an important role in the epidemic of malaria during the flood season. This was first time that the study quantified the association

between malaria and flooding or waterlogging using a mixed method in Mengcheng County of Anhui Province, China. The study confirms that exposure to flooding and waterlogging will affect burden of malaria. Although the study is based on only one area in Anhui Province, the real burden of malaria due to flooding and waterlogging will be much higher than the estimates from this study, given the larger population at risk in China. Determining the effect of these two events on burden of disease due to malaria would be beneficial for malaria risk assessment and thus providing a basis for the policy making for malaria control technologies.

Increased numbers of cases of malaria have been noted after floods in some countries. In Africa, an epidemiological study found that the incidence of endemic malaria increased four to fivefold following the 2000 flooding in Mozambique [38]. Another study reported that malaria was one main impact of flooding on human health in Gambella region [39]. After flooding, there was an increased risk of malaria epidemics in Khartoum [40–42]. WHO found that the flooding in the Dominican Republic in 2004 led to malaria outbreaks [34]. For other countries, flooding has also been associated with changes in habitat that were beneficial for breeding and preceded an extreme rise in malaria cases [43–46]. Additionally, periodic flooding linked to El Niño has been associated with malaria epidemics in Peru, Bolivia and the USA [47–49]. Similar findings have been reported in our study. Our study shows that the risk for malaria epidemics following flooding is very high. Malaria is sensitive to environmental change.

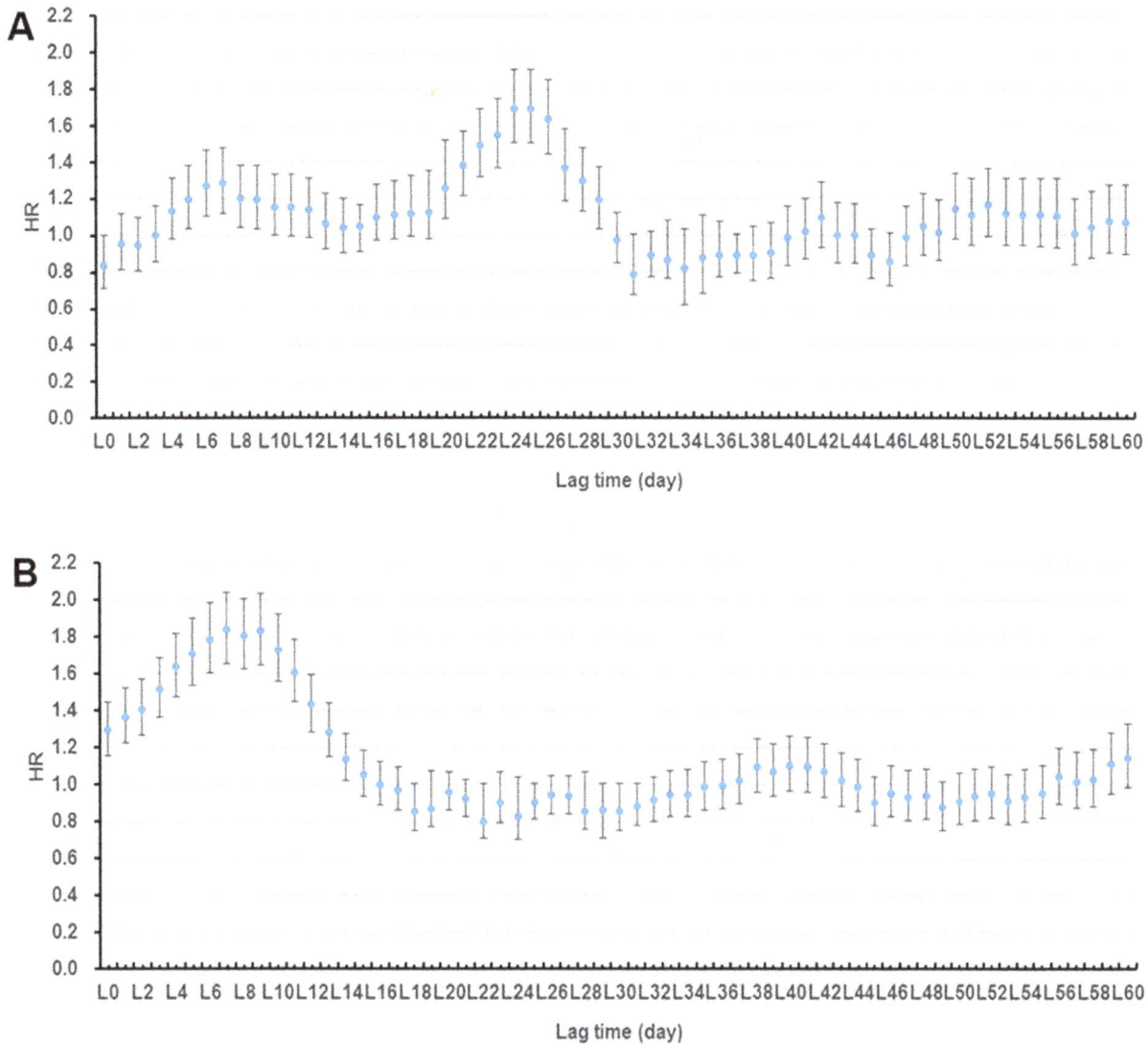

Figure 4. HR estimates of flooding (A) and waterlogging (B) on the risk of malaria in different lagged days.

Climatic variables have been established as important environmental drivers of malaria transmission [22], because climatic factors can impact on the growth and reproduction rates of mosquitoes, the temporal activity pattern of the population as well as the life cycle of *Plasmodium* [50,51]. After fitting meteorological factors, flooding and waterlogging were significantly associated

with an increased risk of malaria in our study, which was not just due to the seasonal fluctuation. Results indicate that flooding and waterlogging played an important role in the peak of malaria incidence from late July to early August in 2007.

Evidence of malaria associated with extreme dry weather is mixed. In South America, one study found that malaria mortality

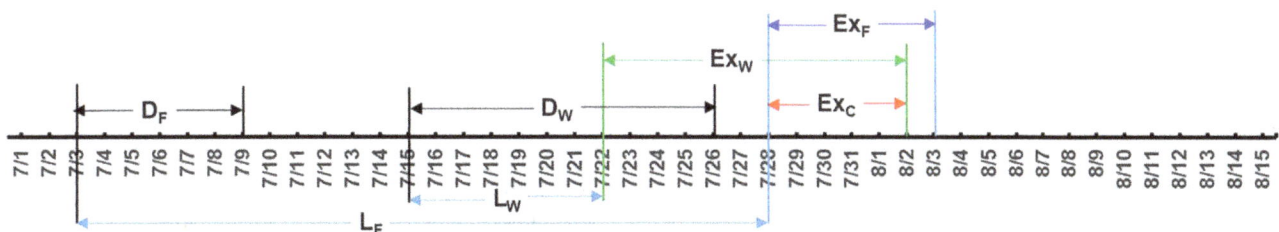

Figure 5. A timeline of exploring for duration of the effect-period. D_F: duration of flooding; D_W: duration of waterlogging; L_F: lag time of flooding; L_W: lag time of waterlogging; Ex_F: the exposure effect-period for flooding alone; Ex_W: the exposure effect-period for waterlogging alone; Ex_C: the common exposure effect-period for both flooding and waterlogging.

Table 1. AHRs of flooding and waterlogging on the risk of malaria in multivariate stratified Cox models.

Model	AHR (95% CI) Flooding alone	Waterlogging alone	Both flooding and waterlogging
Model 1[a]	1.687 (1.498–1.901)	1.837 (1.653–2.041)	2.642 (2.335–2.988)
Model 2[b]	1.695 (1.505–1.910)	1.818 (1.635–2.020)	2.905 (2.568–3.286)
Model 3[c]	1.515 (1.297–1.768)	1.919 (1.717–2.146)	2.366 (2.116–2.647)
Model 4[d]	1.687 (1.487–1.914)	1.837 (1.651–2.042)	2.395 (2.131–2.691)
Model 5[e]	1.467 (1.257–1.713)	1.897 (1.696–2.121)	2.631 (2.341–2.956)

AHR: adjusted hazard ratio; CI: confidence intervals.
[a]adjusted for average temperature; [b]adjusted for average relative humidity; [c]adjusted for rainfall; [d]adjusted for sunshine duration; [e]adjusted for average temperature, average relative humidity, rainfall, and sunshine duration.

is strongly related to drought in the year before outbreaks [52]. Another study showed that droughts favor the development of malaria epidemics in Colombia and Guyana, and epidemics lag a drought by 1 year in Venezuela [53]. However, there are some studies showing decreases in risk of malaria associated with droughts. Studies from the Sahel revealed that the decreases in malaria prevalence and incidence are likely due to the disappearance of the *An. funestus* as a result of severe droughts [54,55]. Sultan's study showed that malaria cases are rare in the dry season and during drought [56]. Another study of *Plasmodium falciparum* transmission by *An. arabiensis* and *An. funestus* during a period of drought (2004–2005) in Zambia reported reduced mosquito activity and reduced numbers of malaria cases during the period of drought [57]. Our study area has a sub-humid warm temperate continental monsoon with mild climate and plentiful rainfall. No extreme dry weather occurred during the study period.

This study has identified a longer lagged effect of flooding on malaria than that of waterlogging. Standing water caused by heavy rainfall or overflow of rivers in the flooding-period or waterlogging-period can create new breeding sites. This situation can result (with typically some weeks' delay) in an increase of the vector population and potential for disease transmission [58]. In this study, the strongest lagged effect of the flooding was observed at 25 lagged days. During the early flooding, floodwaters may wash away breeding sites and, hence, no increasing mosquito-borne transmission [59]. But mosquito breeding comes back when the waters recede. The flooding in 2007 may indirectly lead to an

increase in malaria through the expansion in the number and range of *An. sinensis*. Considering the incubation periods of the parasite in the mosquito and the human, the lagged time of 7 days between waterlogging and increased malaria transmission is not biologically feasible. Thus, we assume that the waterlogging in 2007 may indirectly affect malaria through providing proper environmental conditions for adult mosquitoes' activity, because activity of adult *An. sinensis* had a certain bearing on rainfall, humidity and air temperature [60].

The hazard ratio and attributable YLDs of flooding and waterlogging together were higher than those of flooding alone and waterlogging alone, which suggests that burden of disease due to malaria caused by flooding and waterlogging together is more severe than their individual independent burden alone. The environment during exposure effect-period of the flooding and waterlogging together may indirectly lead to an increase in malaria through the expansion in the number, range and activity of vector habitats [34]. Standing water caused by heavy rainfall or overflow of rivers can act as breeding sites for *An. sinensis*, and therefore enhance the potential for exposure of the disaster-affected population and emergency workers to increasing risk of malaria [34]. Suitable temperature and rainfall during waterlogging period forced wild *An. sinensis* into indoor residential spaces and increased the chance to be bitten by mosquitoes. This indicates that flooding and waterlogging together can make large-scale ecological changes for creating an environment favorable for the more *An. sinensis* and increasing the survival and longevity of the adult *An. sinensis*.

Table 2. The epidemiological burden of malaria caused by flooding alone during exposure effect-period for flooding.

Age (years)	Case	Incidence (1/10⁵)	YLD per 1000 Males	Females	Persons
0–4	4	4.810	0.019	0.054	0.036
5–14	9	5.554	0.020	0.040	0.028
15–29	6	1.829	0.010	0.021	0.015
30–44	15	4.773	0.043	0.025	0.034
45–59	7	4.665	0.047	0.000	0.024
60–69	6	9.257	0.064	0.033	0.050
70–79	4	8.983	0.059	0.018	0.038
80+	2	15.319	0.211	0.020	0.090
Total	53	4.569	0.032	0.024	0.028

YLD: year lived with disability.

Table 3. The epidemiological burden of malaria caused by waterlogging alone during exposure effect-period for waterlogging.

Age (years)	Case	Incidence (1/10⁵)	YLD per 1000		
			Males	Females	Persons
0–4	16	19.240	0.246	0.137	0.194
5–14	50	30.856	0.803	0.573	0.706
15–29	44	13.412	0.151	0.138	0.145
30–44	60	19.091	0.097	0.075	0.086
45–59	51	33.989	0.168	0.175	0.171
60–69	44	67.888	0.460	0.537	0.496
70–79	24	53.896	0.389	0.281	0.333
80+	8	61.275	0.072	0.289	0.209
Total	297	25.604	0.273	0.209	0.242

YLD: year lived with disability.

This study has also indicated that burden of malaria caused by waterlogging alone is more severe than that by flooding alone. It is biologically plausible that moderate rainfall and moist environment during the waterlogging-day increase adult mosquitoes' activity and susceptible people become sick easily after being bitten by adult *An. sinensis* [10,60]. While excessive floodwater during the flooding-day may partly destroy breeding sites and flush out the mosquitoes larvae [61]. This effect partially detracted burden of malaria caused by flooding. Additionally, we found that burden of disease due to malaria among males was more than females, and people who are older and children were vulnerable groups of malaria. This may be because that males participated in more relief work and engaged more frequently in emergency than females did, leading to a higher exposure to adverse environment among males [62]. In addition, there is a custom that Chinese men remove their shirts when it is hot and suffocating weather during summer in China. Children have immature immune system, and older people may have weak immune systems to in responding to malaria. Hence, males, older people and children are the population groups that are most vulnerable for malaria after flooding and waterlogging.

There are some strengths in applying a mixed approach. Firstly, based on our approach, the attributable burden of disease caused by meteorological conditions could be estimated explicitly, which could be borrowed and validated by other studies in this field. Secondly, the symmetric bidirectional case-crossover design can avoid bias resulting from time trend in the exposure series [63], and can quantitatively assess the risk of the spread of infections caused by environmental factors. Thirdly, we have controlled other meteorological factors in the multivariate models with consideration of lagged effects of flooding and waterlogging.

There are some limitations in our study. Firstly, not all environmental factors were taken into account for analysis the risk of malaria. As with other vector-borne diseases, malaria typically was driven by climatic, ecological and human factors. We have only analyzed the effect of flooding and waterlogging on malaria after adjusting climatic factors. Other factors, e.g. human activities, mosquitoes' activity, availability of health services, could not be included in this analysis. Secondly, the malaria data were from the NDSS and under-reporting bias is inevitable. Some people with mild clinical symptoms and self-treated cases might not seek medical help. This could lead to an underestimation of attributable YLD due to malaria. Thirdly, only two meteorological

Table 4. The epidemiological burden of malaria caused by flooding and waterlogging together during exposure common effect-period for both flooding and waterlogging.

Age (years)	Case	Incidence (1/10⁵)	YLD per 1000		
			Males	Female	Persons
0–4	9	10.822	0.071	0.042	0.057
5–14	37	22.834	0.251	0.306	0.274
15–29	34	10.364	0.122	0.068	0.097
30–44	46	14.637	0.076	0.077	0.077
45–59	45	29.991	0.241	0.214	0.228
60-69	26	40.115	0.503	0.064	0.298
70–79	20	44.914	0.207	0.396	0.305
80+	9	68.934	0.091	0.503	0.351
Total	226	19.483	0.182	0.153	0.168

YLD: year lived with disability.

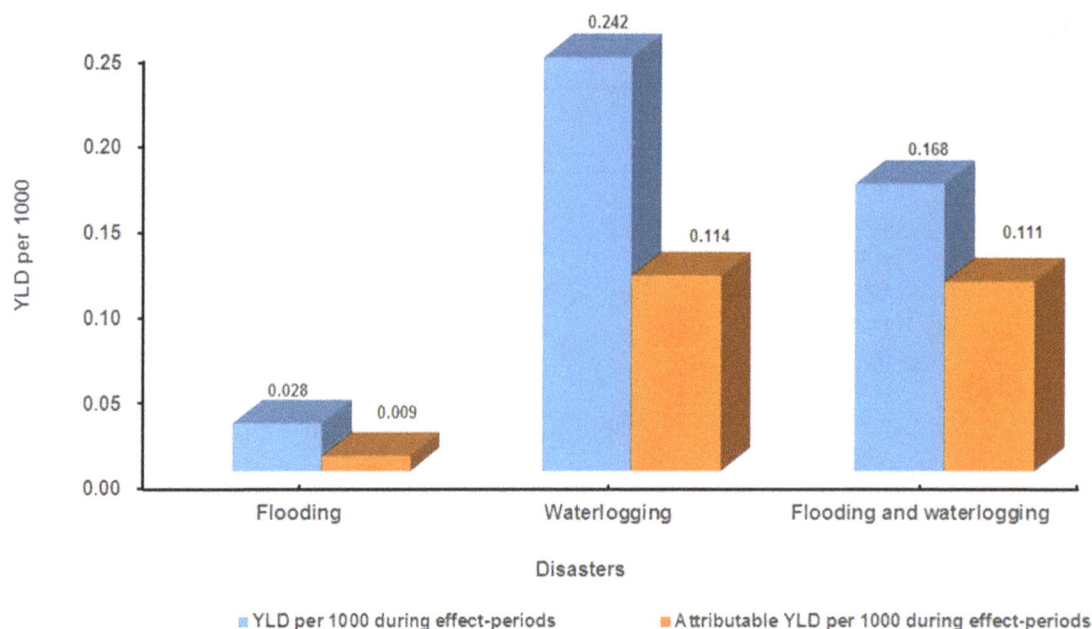

Figure 6. YLD per 1000 and attributable YLD per 1000 of malaria caused by flooding and waterlogging during exposure effect-period.

events and one study area in Anhui are selected in the analysis. Moreover, the transmission of malaria is very complicated, and more studies in other floods affected regions in China with different climatic, ecological and human conditions are still needed to assess the risk from ecology.

Conclusions

A key conclusion of this study is that flooding and waterlogging contribute to unusually high incidence of malaria in the study region. In addition, risk of malaria caused by both flooding and waterlogging is greater than their individual risk alone. Therefore, effective preventive and treatment interventions should be developed to avoid and control a potential risk of malaria epidemics after flooding and waterlogging. Particular vulnerable groups, including males, older people and children, should be paid

more attention in developing strategies to prevent and reduce the health impact of flooding and waterlogging.

Acknowledgments

We thank Chinese Center for Disease Control and Prevention, National Meteorological Information Center of China, and Data center for Institute of Geographic Sciences and Natural Resources Research of China sharing with us the data needed for this study. We would like to thank Dr. Ying Zhang from the University of Sydney for the earnest assistance in revising and editing this manuscript.

Author Contributions

Conceived and designed the experiments: GYD BFJ XWL. Performed the experiments: GYD LG. Analyzed the data: LG XWL. Contributed reagents/materials/analysis tools: MGZ QYL HYR. Wrote the paper: GYD.

References

1. Githeko AK, Lindsay SW, Confalonieri UE, Patz JA (2000) Climate change and vector-borne diseases: a regional analysis. Bull World Health Organ 78: 1136–1147.

2. Bezirtzoglou C, Dekas K, Charvalos E (2011) Climate changes, environment and infection: facts, scenarios and growing awareness from the public health community within Europe. Anaerobe 17: 337–340.

3. Kouadio IK, Aljunid S, Kamigaki T, Hammad K, Oshitani H (2012) Infectious diseases following natural disasters: prevention and control measures. Expert Rev Anti Infect Ther 10: 95–104.

4. Ahmed F, Rafii MY, Ismail MR, Juraimi AS, Rahim HA, et al. (2013) Waterlogging tolerance of crops: breeding, mechanism of tolerance, molecular approaches, and future prospects. Biomed Res Int 2013: 963525. Available: http://dx.doi.org/10.1155/2013/963525 Accessed 2013 April 28.

5. Guha-Sapir D, Vos F, Below R, Ponserre S (2011) Annual Disaster Statistical Review 2010: The Numbers and Trends. Brussels: CRED. Available: http://www.cred.be/sites/default/files/ADSR_2010.pdf. Accessed 2013 January 17.

6. Xiao ZN (2008) Yearbook of meterorological disasters in China, 1st edition [in Chinese]. Beijing: China Meteorological Press. 233 p.

7. WHO (2012) World malaria report: 2012. WHO Global Malaria Programme. Geneva, Switzerland. Available: http://www.who.int/entity/malaria/publications/world_malaria_report_2012/wmr2012_full_report.pdf. Accessed 2013 March 25.

8. Zhou SS, Huang F, Wang JJ, Zhang SS, Su YP, et al. (2010) Geographical, meteorological and vectorial factors related to malaria re-emergence in Huang-Huai River of central China. Malar J 9: 337. Available: http://www.malariajournal.com/content/9/1/337. Accessed 2013 March 26.

9. Data center for Public Health. National notifiable infectious disease database [in Chinese]. Available: http://www.phsciencedata.cn/Share/ky_sjml.jsp. Accessed 2013 March 25.

10. Gao HW, Wang LP, Liang S, Liu YX, Tong SL, et al. (2012) Change in rainfall drives malaria re-emergence in Anhui Province, China. PLoS One 7: e43686. Available: http://www.plosone.org/article/info%3Adoi%2F10.1371%2Fjournal.pone.0043686. Accessed 2013 March 26.

11. Pawar AB, Bansal RK, Kumar M, Jain NC, Vaishnav KG (2008) A rapid assessment of mosquito breeding, vector control measures and treatment seeking behaviour in selected slums of Surat, Gujarat, India, during post-flood period. J Vector Borne Dis 45: 325–327.

12. Majambere S, Pinder M, Fillinger U, Ameh D, Conway DJ, et al. (2010) Is mosquito larval source management appropriate for reducing malaria in areas of extensive flooding in The Gambia? A cross-over intervention trial. Am J Trop Med Hyg 82: 176–184.

13. Harrison BA, Whitt PB, Roberts LF, Lehman JA, Lindsey NP, et al. (2009) Rapid assessment of mosquitoes and arbovirus activity after floods in southeastern Kansas, 2007. J Am Mosq Control Assoc 25: 265–271.

14. Hashizume M, Kondo H, Murakami T, Kodama M, Nakahara S, et al. (2006) Use of rapid diagnostic tests for malaria in an emergency situation after the flood disaster in Mozambique. Public Health 120: 444–447.
15. Wang M, Tang LH, Gu ZC, Jiang WK, Zhu JM, et al. (2007) Stuy on threshold density of An.sinensis for transmission of malaria in the Northern Anhui Province [in Chinese]. Journal of Tropical Medicine 7: 597–599.
16. Shen YZ (2006) Investigation on transmission factors of malaria in Anopheles sinensis areas in Anhui Province [in Chinese]. Journal of Pathogen Biology 1: 301–303.
17. Liu XB, Liu QY, Guo YH, Jiang JY, Ren DS, et al. (2011) The abundance and host-seeking behavior of culicine species (Diptera: Culicidae) and Anopheles sinensis in Yongcheng city, People's Republic of China. Parasit Vectors 4: 221. Available: http://www.parasitesandvectors.com/content/4/1/221. Accessed 2013 March 29.
18. China Meteorological Data Sharing Service System. Chinese Agro-meteorological Disasters Information Data [in Chinese]. Available: http://cdc.cma.gov.cn/choiceStation.do. Accessed 2013 March 21.
19. Huang F, Zhou S, Zhang S, Wang H, Tang L (2011) Temporal correlation analysis between malaria and meteorological factors in Motuo County, Tibet. Malar J 10: 54. Available: http://www.malariajournal.com/content/10/1/54. Accessed 2013 March 26.
20. Alemu A, Abebe G, Tsegaye W, Golassa L (2011) Climatic variables and malaria transmission dynamics in Jimma town, South West Ethiopia. Parasit Vectors 4: 30. Available: http://www.parasitesandvectors.com/content/4/1/30. Accessed 2013 May 2.
21. Bi P, Tong S, Donald K, Parton KA, Ni J (2003) Climatic variables and transmission of malaria: a 12-year data analysis in Shuchen County, China. Public Health Rep 118: 65–71.
22. Ye Y, Louis VR, Simboro S, Sauerborn R (2007) Effect of meteorological factors on clinical malaria risk among children: an assessment using village-based meteorological stations and community-based parasitological survey. BMC Public Health 7: 101. Available: http://www.biomedcentral.com/1471-2458/7/101. Accessed 2013 May 5.
23. Garg A, Dhiman RC, Bhattacharya S, Shukla PR (2009) Development, malaria and adaptation to climate change: a case study from India. Environ Manage 43: 779–789.
24. Yang GJ, Gao Q, Zhou SS, Malone JB, McCarroll JC, et al. (2010) Mapping and predicting malaria transmission in the People's Republic of China, using integrated biology-driven and statistical models. Geospat Health 5: 11–22.
25. Barati M, Keshavarz-valian H, Habibi-nokhandan M, Raeisi A, Faraji L, et al. (2012) Spatial outline of malaria transmission in Iran. Asian Pacific Journal of Tropical Medicine 5: 789–795.
26. Wang Z, Shen S, Liu R (2011) Impact analysis of precipitation in different classes on Anuual Precipitation Change in recent 40 years in China [in Chinese]. Meteorological and Environmental Sciences 34: 7–13.
27. Maclure M (1991) The case-crossover design: a method for studying transient effects on the risk of acute events. Am J Epidemiol 133: 144–153.
28. Maclure M, Mittleman MA (2000) Should we use a case-crossover design? Annu Rev Public Health 21: 193–221.
29. Wang S, Linkletter C, Maclure M, Dore D, Mor V, et al. (2011) Future cases as present controls to adjust for exposure trend bias in case-only studies. Epidemiology 22: 568–574.
30. Turin TC, Kita Y, Rumana N, Nakamura Y, Ueda K, et al. (2012) Short-term exposure to air pollution and incidence of stroke and acute myocardial infarction in a Japanese population. Neuroepidemiology 38: 84–92.
31. Lee JT, Schwartz J (1999) Reanalysis of the effects of air pollution on daily mortality in Seoul, Korea: A case-crossover design. Environ Health Perspect 107: 633–636.
32. Bateson TF, Schwartz J (1999) Control for seasonal variation and time trend in case-crossover studies of acute effects of environmental exposures. Epidemiology 10: 539–544.
33. Wang SV, Coull BA, Schwartz J, Mittleman MA, Wellenius GA (2011) Potential for bias in case-crossover studies with shared exposures analyzed using SAS. Am J Epidemiol 174: 118–124.
34. WHO. Flooding and communicable diseases fact sheet. Risk assessment and preventive measures. Available: http://www.who.int/hac/techguidance/ems/flood_cds/en/index.html. Accessed 2013 April 2.
35. Mathers CD, Vos T, Lopez AD, Salomon J, Ezzati M (ed.) (2001) National Burden of Disease Studies: A Practical Guide. Edition 2.0. Global Program on Evidence for Health Policy. Geneva: World Health Organization. Available: http://www.who.int/entity/healthinfo/nationalburdenofdiseasemanual.pdf. Accessed 2011 October 23.
36. Prüss-Üstün A, Mathers C, Corvalán C, Woodward A (2003) Introduction and methods: assessing the environmental burden of disease at national and local levels. Geneva: World Health Organization. (WHO Environmental Burden of Disease Series, No. 1). Available: http://www.who.int/quantifying_ehimpacts/publications/en/9241546204.pdf. Accessed 2011 December 12.
37. Viera AJ (2008) Odds ratios and risk ratios: what's the difference and why does it matter? South Med J 101: 730–734.
38. Kondo H, Seo N, Yasuda T, Hasizume M, Koido Y, et al. (2002) Post-flood—infectious diseases in Mozambique. Prehosp Disaster Med 17: 126–133.
39. Wakuma Abaya S, Mandere N, Ewald G (2009) Floods and health in Gambella region, Ethiopia: a qualitative assessment of the strengths and weaknesses of coping mechanisms. Glob Health Action 2. Available: http://www.ncbi.nlm.nih.gov/pmc/articles/PMC2792158/pdf/GHA-2-2019.pdf. Accessed 2013 June 1.
40. El Sayed BB, Arnot DE, Mukhtar MM, Baraka OZ, Dafalla AA, et al. (2000) A study of the urban malaria transmission problem in Khartoum. Acta Trop 75: 163–171.
41. McCarthy MC, Haberberger RL, Salib AW, Soliman BA, El-Tigani A, et al. (1996) Evaluation of arthropod-borne viruses and other infectious disease pathogens as the causes of febrile illnesses in the Khartoum Province of Sudan. J Med Virol 48: 141–146.
42. Woodruff BA, Toole MJ, Rodrigue DC, Brink EW, Mangoub ES, et al. (1990) Disease Surveillance and Control After a Flood: Khartoum, Sudan, 1988. Disasters 14: 151–163.
43. Saenz R, Bissell RA, Paniagua F (1995) Post-disaster malaria in Costa Rica. Prehosp Disaster Med 10: 154–160.
44. Moreira Cedeño JE (1986) Rainfall and flooding in the Guayas river basin and its effects on the incidence of malaria 1982–1985. Disasters 10: 107–111.
45. Russac PA (1986) Epidemiological surveillance: Malaria epidemic following the Niño phenomenon. Disasters 10: 112–117.
46. Mathur KK, Harpalani G, Kalra NL, Murthy GG, Narasimham MV (1992) Epidemic of malaria in Barmer district (Thar desert) of Rajasthan during 1990. Indian J Malariol 29: 1–10.
47. Gagnon AS, Smoyer-Tomic KE, Bush AB (2002) The El Niño southern oscillation and malaria epidemics in South America. Int J Biometeorol 46: 81–89.
48. Valencia Telleria A (1986) Health consequences of the floods in Bolivia in 1982. Disasters 10: 88–106.
49. Gueri M, González C, Morin V (1986) The effect of the floods caused by "El Niño" on health. Disasters 10: 118–124.
50. Craig MH, Kleinschmidt I, Nawn JB, Le Sueur D, Sharp BL (2004) Exploring 30 years of malaria case data in KwaZulu-Natal, South Africa: part I. The impact of climatic factors. Trop Med Int Health 9: 1247–1257.
51. Hoshen MB, Morse AP (2004) A weather-driven model of malaria transmission. Malar J 3: 32. Available: http://www.malariajournal.com/content/3/1/32. Accessed 2013 May 12.
52. Bouma MJ, Dye C (1997) Cycles of malaria associated with El Niño in Venezuela. JAMA 278: 1772–1774.
53. Gagnon AS, Smoyer-Tomic KE, Bush AB (2002) The El Niño southern oscillation and malaria epidemics in South America. Int J Biometeorol 46: 81–89.
54. Mouchet J, Faye O, Juivez J, Manguin S (1996) Drought and malaria retreat in the Sahel, west Africa. Lancet 348: 1735–1736.
55. Mouchet J, Manguin S, Sircoulon J, Laventure S, Faye O, et al. (1998) Evolution of malaria in Africa for the past 40 years: impact of climatic and human factors. J Am Mosq Control Assoc 14: 121–130.
56. Theander TG (1998) Unstable malaria in Sudan: the influence of the dry season. Malaria in areas of unstable and seasonal transmission. Lessons from Daraweesh. Trans R Soc Trop Med Hyg 92: 589–592.
57. Kent RJ, Thuma PE, Mharakurwa S, Norris DE (2007) Seasonality, blood feeding behavior, and transmission of Plasmodium falciparum by Anopheles arabiensis after an extended drought in southern Zambia. Am J Trop Med Hyg 76: 267–274.
58. Watson JT, Gayer M, Connolly MA (2007) Epidemics after natural disasters. Emerg Infect Dis 13: 1–5.
59. Sidley P (2000) Malaria epidemic expected in Mozambique. BMJ 320: 669.
60. Duo-quan W, Lin-hua T, Heng-hui L, Zhen-cheng G, Xiang Z (2013) Application of structural equation models for elucidating the ecological drivers of Anopheles sinensis in the three gorges reservoir. PLoS One 8: e68766. Available: http://www.plosone.org/article/info%3Adoi%2F10.1371%2Fjournal.pone.0068766#pone-0068766-g003. Accessed 2013 May 25.
61. McMichael AJ, Martens P (1995) The health impacts of global climate change: grasping with scenarios, predictive models, and multiple uncertainties. Ecosystem Health 1: 23–33.
62. Li X, Tan H, Li S, Zhou J, Liu A, et al. (2007) Years of potential life lost in residents affected by floods in Hunan, China. Trans R Soc Trop Med Hyg 101: 299–304.
63. Janes H, Sheppard L, Lumley T (2005) Case-crossover analyses of air pollution exposure data: referent selection strategies and their implications for bias. Epidemiology 16: 717–726.

17

Land Cover and Rainfall Interact to Shape Waterbird Community Composition

Colin E. Studds[1]*[¤], William V. DeLuca[2], Matthew E. Baker[3], Ryan S. King[4], Peter P. Marra[1]

1 Smithsonian Conservation Biology Institute, Migratory Bird Center, National Zoological Park, Washington, D. C., United States of America, **2** Department of Environmental Conservation, University of Massachusetts, Amherst, Massachusetts, United States of America, **3** Department of Geography and Environmental Systems, University of Maryland, Baltimore, Maryland, United States of America, **4** Center for Reservoir and Aquatic Systems, Department of Biology, Baylor University, Waco, Texas, United States of America

Abstract

Human land cover can degrade estuaries directly through habitat loss and fragmentation or indirectly through nutrient inputs that reduce water quality. Strong precipitation events are occurring more frequently, causing greater hydrological connectivity between watersheds and estuaries. Nutrient enrichment and dissolved oxygen depletion that occur following these events are known to limit populations of benthic macroinvertebrates and commercially harvested species, but the consequences for top consumers such as birds remain largely unknown. We used non-metric multidimensional scaling (MDS) and structural equation modeling (SEM) to understand how land cover and annual variation in rainfall interact to shape waterbird community composition in Chesapeake Bay, USA. The MDS ordination indicated that urban subestuaries shifted from a mixed generalist-specialist community in 2002, a year of severe drought, to generalist-dominated community in 2003, of year of high rainfall. The SEM revealed that this change was concurrent with a sixfold increase in nitrate-N concentration in subestuaries. In the drought year of 2002, waterbird community composition depended only on the direct effect of urban development in watersheds. In the wet year of 2003, community composition depended both on this direct effect and on indirect effects associated with high nitrate-N inputs to northern parts of the Bay, particularly in urban subestuaries. Our findings suggest that increased runoff during periods of high rainfall can depress water quality enough to alter the composition of estuarine waterbird communities, and that this effect is compounded in subestuaries dominated by urban development. Estuarine restoration programs often chart progress by monitoring stressors and indicators, but rarely assess multivariate relationships among them. Estuarine management planning could be improved by tracking the structure of relationships among land cover, water quality, and waterbirds. Unraveling these complex relationships may help managers identify and mitigate ecological thresholds that occur with increasing human land cover.

Editor: Simon Thrush, National Institute of Water & Atmospheric Research, New Zealand

Funding: This study was funded by a grant from the United States Environmental Protection Agency's (USEPA) Science to Achieve Results (STAR) Estuarine and Great Lakes (EaGLe) program to the Atlantic Slope Consortium, USEPA Agreement #R-82868401. Although the research described in this article was funded by the USEPA, it was not subjected to the Agency's peer and policy review and therefore does not necessarily reflect the views of the Agency and no official endorsement should be inferred. The funders has no role in study design, data collection and analysis, decision to publish, or preparation of the manuscript.

Competing Interests: The authors have declared that no competing interests exist.

* E-mail: c.studds@uq.edu.au

¤ Current address: School of Biological Sciences, University of Queensland, Brisbane, Queensland, Australia

Introduction

Conversion of natural habitats to human dominated landscapes has led to worldwide deterioration of estuarine ecosystems [1]. In many regions, the continued spread of human land cover has been accompanied by greater variation in climatic conditions, such as strong cycles of drought and precipitation [2]. Together, human land cover and rainfall shape estuarine condition directly by reducing coastal habitat quality or indirectly by lowering water quality. Both agriculture and urban development directly impair estuaries through degradation or loss of coastal wetlands, modification of other natural shoreline areas, and habitat fragmentation and isolation [3–5]. These land cover types also indirectly degrade estuaries by carrying nutrients and contaminants from terrestrial watersheds into coastal water bodies. Rainfall events amplify the hydrological connectivity between watersheds and estuaries, potentially leading to severe eutrophi-

cation [6,7]. With watersheds in many coastal areas undergoing dynamic changes in land cover and climate, management and restoration programs could benefit from understanding how rainfall interacts with expanding human development to shape estuarine condition.

The greatest risk posed by eutrophication of coastal waters is the decrease in dissolved oxygen (DO) when blooms of aquatic algae die and are consumed through microbial respiration [8]. Hypoxia occurs when DO falls below 2 mg/L and has been shown to cause mortality of sessile benthic invertebrates and trigger emigration of commercially harvested crab and fish species to more oxygen-rich waters [9,10]. Despite these well-documented responses of lower trophic level organisms, the consequences for higher order consumers are essentially unknown. Because conservation of upper trophic level wildlife is mandated by state and federal management agencies, there is a need to understand how these species respond to eutrophication and other disturbances associated with increasing human land cover.

Bird communities can be robust indicators of biological condition because they often occupy the highest trophic level in an ecosystem, causing them to integrate the effects of abiotic stressors acting on species at lower trophic levels [11,12]. DeLuca et al. [13] developed an index of waterbird community integrity in Chesapeake Bay, USA and evaluated its sensitivity to anthropogenic disturbance. They found that even low levels of urban development, particularly when located close to the estuarine shoreline, severely impaired the waterbird community. This finding strongly suggests that waterbird communities are directly degraded by urban development, but, because this study did not assess potential indirect pathways that compromise water quality, the underlying causes remain unclear. Waterbird abundance depends on both habitat quality and food availability [14]. Therefore, waterbird communities should also be sensitive to changes in water quality because these species primarily consume fish and invertebrates, some of whose local abundances decline when eutrophic conditions prevail [15].

In 2002 and 2003, we monitored the waterbird community, nitrate-N concentrations, dissolved oxygen levels, and salinity in 27 subestuaries of Chesapeake Bay. In 2002, a severe region-wide drought limited freshwater flow and nutrient inputs to the Bay to near historic lows [16,17]. In contrast, the Bay received more than twice the amount of freshwater in 2003 and the third highest amount since 1937, much of which was concentrated in the spring and summer months. This massive influx nearly tripled nitrogen export from the previous year, resulting in sustained hypoxia throughout the estuary [16]. Here, we explore how this temporal variation in rainfall interacted with spatial variation in land cover to shape waterbird community composition. We used non-metric multidimensional scaling and structural equation modeling to evaluate direct and indirect effects of land cover on waterbird community composition and to assess change in their strength between years. We predicted waterbird community composition would be limited directly by urban development in both years, and indirectly by lower water quality due to greater hydrological connectivity between watersheds and subestuaries in 2003.

Methods

Fieldwork was conducted from 2002–2003 in 27 subestuaries of Chesapeake Bay, USA (39° 23′ N–36° 48′ N, 76° 45′ W–75° 44′ W). Each subestuary had a distinct embayment that separated it from major tributaries of the Bay and a watershed drained by a third to fifth order stream [18]. We quantified land cover within each watershed by using ArcGIS software (ESRI, Redlands, CA, USA) and the 1992 National Land Cover Database derived from 30-m resolution Landsat Thematic mapper images [19]. We calculated the percentage watershed area covered by agriculture and urban development and the percentage area of emergent marsh within 500 m of the shoreline. We summarized the percentage of urban development in each watershed weighted by its squared inverse distance (IDW) to the shoreline because of past evidence that ecological indicators in Chesapeake Bay are sensitive to this land cover in close proximity to subestuaries [13,18]. We sampled 17 subestuaries in 2002, 20 in 2003, and a subset of nine in both years (Fig. 1). This design permitted us to balance spatial coverage and temporal replication of data collection.

We surveyed the waterbird community along three, 1-km transects in each subestuary. Transects were positioned in the upper, middle, and lower thirds of subestuaries and were oriented

parallel to and 100 m from the shoreline (Fig. 1 inset). Adjacent transects within subestuaries were separated by >500 m to reduce the probability of counting an individual more than once. We surveyed waterbirds with the double observer approach from a boat traveling at three knots along transects [20]. For this study, we defined waterbirds as all non-passerine bird species that forage exclusively or opportunistically on aquatic estuarine organisms (e.g., gulls, terns, waders, raptors, kingfishers, and waterfowl). Observers counted all individuals on the water, in the air, or perched along the shoreline within 100 m of transects. We surveyed each transect three times from 15 May–15 August between 0600 h and 1300 h and used program DOBSERV to calculate abundance estimates corrected for imperfect detection probabilities [20]. Deluca et al. [13] incorporated these corrected abundance estimates along with information about foraging and nesting niche breadth, migratory behavior, and regional rarity into an index of waterbird community integrity for each subestuary. We employed this index in the present study to characterize spatial and temporal variation in waterbird community composition (WCC) because subestuaries with high scores supported high abundances of specialist species with high conservation value and those with lower scores harbored high abundances of generalist species with lower conservation value (see Text S1 for index development).

Six water quality sampling stations were distributed via a random sampling approach nearby to waterbird survey transects in each subestuary. Salinity, dissolved oxygen (DO), and nitrate-N were measured approximately 50 m from shore at two locations at every sampling station. Salinity (ppt) and DO (mg/L and percent saturation) were measured using an YSI 556 multiparameter instrument (YSI Inc., Yellow Springs, OH, USA) at 10 cm below the water surface and 10 cm above the bottom. The difference between percent saturation of DO at the surface and bottom (DO difference) was used as a metric of potential benthic hypoxia. This index also helped alleviate differences in DO due to diel fluctuations among sampling stations because measurements were not collected at the same time of day across all stations. Nitrate-N (μg/L) samples were collected near the water surface in acid-washed polyethylene bottles, stored on ice, and returned to the laboratory for analysis [21]. No specific permits were required for the described field studies.

We used non-metric multidimensional scaling (MDS) to describe spatial and temporal variation in the abundance of generalist and specialist species in each subestuary following steps outlined by [22]. We used corrected abundance estimates of each species to compute Bray-Curtis distances among subestuaries for each year separately. The number of MDS axes was chosen by minimizing stress, a measure of the mismatch between distance among species indicated by the Bray-Curtis matrix and distance in the ordination. Species centroids were mapped in ordination space by weighted averaging. We used rotational vector fitting to relate land cover and water quality indices to the ordination [23]. Significance of vectors was estimated using 1000 random permutations. Ordinations and vector fitting were performed in program R 2.11 using the vegan package [24,25].

We used structural equation modeling (SEM) to investigate a set of hypothesized causal relationships among land cover, water quality, and WCC. These relationships can be visualized through a path model in which arrows indicate the proposed effect of one variable on another. The path diagram was drawn based on past research from Chesapeake Bay and on established linkages between land cover and estuarine condition [13,21,26]. Direct paths were expected to act on WCC as a function of the amount of emergent marsh habitat and urban development in each

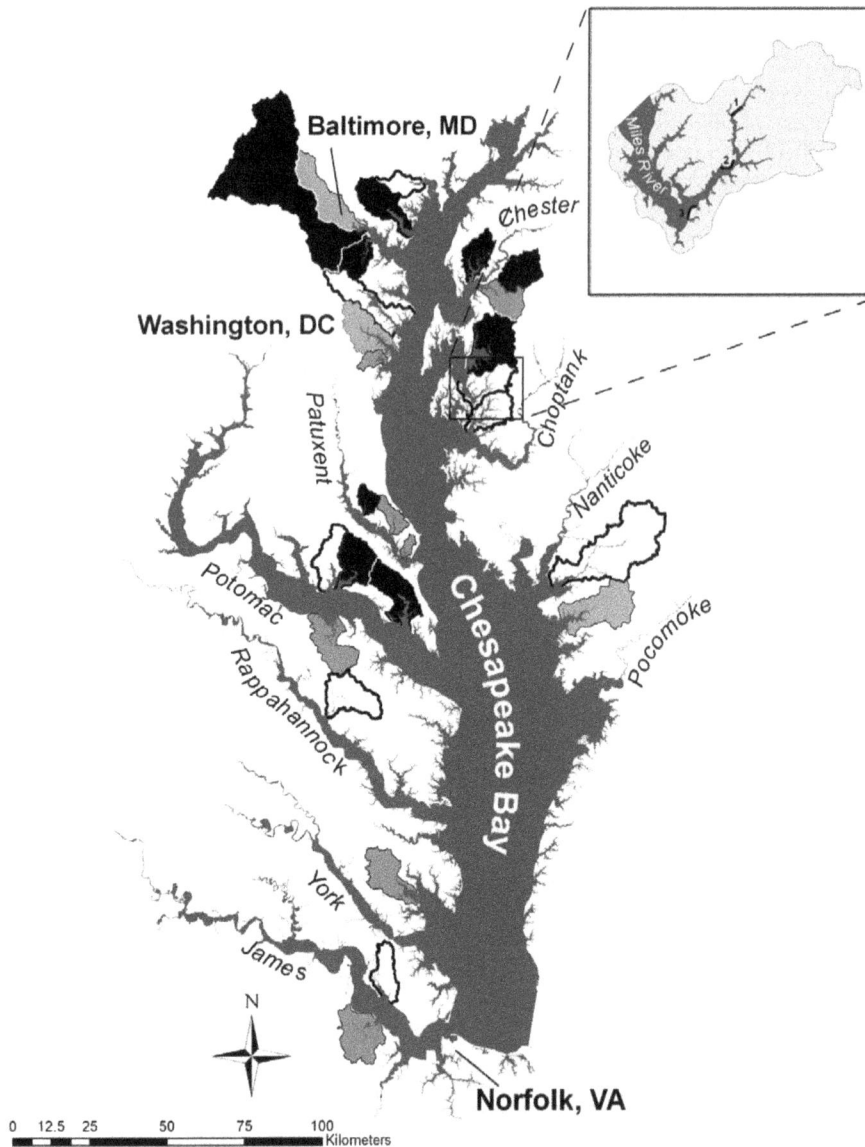

Figure 1. Distribution of 27 watersheds and associated subestuaries in Chesapeake Bay, USA where relationships between land cover, water quality, and waterbird communities were studied in the dry year of 2002 (white), the wet year of 2003 (gray), and in both years (black). The inset shows an example distribution of the three waterbird sampling transects in the lower, middle, and upper reaches of the subestuary.

subestuary [13]. Subestuaries with a high percentage of emergent marsh were predicted to have high WCC scores because many waterbirds forage in this habitat, and it serves as a nursery for many fish species consumed as prey. Subestuaries adjacent to watersheds with a high percentage of urban land cover were hypothesized to have low WCC scores due to degradation of marsh and shoreline habitat. Such an effect could occur through habitat fragmentation and isolation, habitat alteration from shoreline hardening, or prevalence of invasive vegetation [3,18,27].

Indirect pathways proceeding from urban development and agriculture were hypothesized to act on WCC through their effect on water quality. Agriculture and urban development are important determinants of nitrate-N and DO levels in estuaries [28]. Subestuaries with a high percentage of these land cover types in nearby watersheds were predicted to have low WCC scores

owing to high nitrate-N concentrations and large DO differences. The model posited that these two indices of water quality influence a conceptual and unmeasured latent variable labeled as "Eutrophication". We included a measure of surface salinity in the model to index the spatial position of subestuaries on the landscape. Freshwater inputs are higher in northern compared to southern reaches of Chesapeake Bay, and such gradients can lead to spatial variation in nutrient concentration in estuaries [29,30]. At the northern end of the Bay, the Susquehanna River is a major source of freshwater, and agricultural discharges to this river account for a substantial fraction of nitrogen loads to the Bay [31]. Relationships among land cover and salinity were drawn using double-headed arrows because these variables often covary with one another, particularly in this landscape [32].

We used multi-group sampling to identify the path model that best fit annual variation in the data that could have occurred due to differences in rainfall. Multi-group sampling allows parameters of interest to be constrained to equality for subsets of data, thus permitting evaluation of a priori hypotheses [33]. In the first model, we allowed all paths to vary between years to test the hypothesis that annual variation in the data occurred due to differences in both direct and indirect effects. The second model tested the hypothesis that annual variation in the data was driven by differences in degradation of marsh and coastal habitat by requiring only indirect paths to be equal. In the third model, we fixed only the direct paths to be equal to test the hypothesis that data varied between years due to changes in water quality. The final model imposed equality constraints on all paths to test the hypothesis that there was no discernable annual variation in the data.

The path coefficients in each model are values that maximize the likelihood of the covariance structure in the data given the covariance structure proposed by the path model. We assessed the fit of each model with a chi-squared goodness-of-fit test, where a significant chi-squared indicates that the model does not fit the data. This statistic is an asymptotic approximation of a chi-squared distribution when the data follow a multivariate normal distribution. We also assessed model fit with the comparative fit index (CFI) and root mean square error (RMSEA), two commonly used indices that provide an approximate measure of model fit. Models with a CFI >0.95 and a RMSEA of <0.05 indicate a good fit to the data [34]. Because our sample size was smaller than desired for structural equation models estimated through maximum likelihood, we performed Monte Carlo permutation tests with 1000 replicates to evaluate the robustness of each model fit. Significance of path coefficients for the best model was judged based on an empirical distribution generated from 1000 bootstrap replicates of the data. All SEM analyses were done with AMOS 19.0 [35].

We used least squares regression to evaluate temporal change in the relationship among land cover, water quality, and WCC for the subset of subestuaries studied in both years. For this analysis, we considered only predictors judged important in the best-fit model from the SEM multi-group analysis. We expected statistical inference from this analysis might be somewhat limited given that only nine subestuaries were studied in both years. However, our primary goal in this analysis was to assess concordance between regression coefficients from this model and path coefficients from the best-fit multi-group model that were fit with data from all subestuaries. Regression analyses were done with program R 2.11 [24].

Results

The MDS ordination of WCC revealed distinct inter-annual patterns in the distribution of generalist and specialist species across the 27 subestuaries. Two-dimensional solutions were chosen because stress was relatively low for both years (2002: stress = 0.176; 2003: stress = 0.183). Both generalists and specialists were widely distributed among subestuaries in the drought year of 2002 and demonstrated no clear association with land cover or water quality indices (Fig. 2A, see Table S1 for species names and WCC scores). Conversely, when nutrient flow to the Bay reached a near record high in 2003, generalist species exhibited a pronounced shift toward subestuaries in developed watersheds with high nitrate-N concentrations and large differences between surface and bottom DO, whereas specialists exhibited a weaker but opposite trend (Fig. 2B).

The multi-group SEM indicated strong annual variation in the paths affecting WCC. The model in which all paths were unconstrained and thus required to vary between years provided

the best fit to the data ($\chi^2 = 11.49$, $df = 12$, $P = 0.487$; RMSEA $< 10^{-3}$; CFI = 1.00). The model that fixed direct paths to be equal between years provided a reasonable fit to the data based on the goodness of fit test, but CFI and RMSEA values indicated somewhat poor fit ($\chi^2 = 17.40$, $df = 14$, $P = 0.236$; RMSEA = 0.09; CFI = 0.96). Models that required either indirect paths or all paths in the model to be equal between years fit the data poorly ($\chi^2 = 27.72$, $df = 19$, $P = 0.089$; RMSEA = 0.12; CFI = 0.90 and $\chi^2 = 35.33$, $df = 21$, $P = 0.026$; RMSEA = 0.14; CFI = 0.83, respectively). Monte Carlo permutation tests indicated that these estimates of model fit were robust (all models P<0.05). Examination of multivariate kurtosis values revealed moderate departure from multivariate normality (2002: kurtosis = 6.48, critical ratio = 1.12; 2003: kurtosis = 8.11, critical ratio = 1.62), but a Bollen-Stine bootstrap indicated that model fit was not compromised ($P = 0.878$). Based on these results, we interpreted the model in which all paths varied independently between 2002 and 2003. Pearson correlation coefficients, variances, and covariances among variables in these models are presented in Table S2.

Observed correlations among predictors showed strong concordance with those implied by the model (Table 1). These relationships suggested that differences between observed correlations and total effects resulted from non-causal or spurious correlations among predictors, most likely due to unanalyzed relationships or residual spatial autocorrelation that could not be accounted for by covariance relationships specified in the path model. Still, several predictors exhibited differences between their total effects and observed correlations with WCC. In particular, surface salinity (2002) and percent cropland (2002 and 2003) had moderate positive correlations with WCC, but had total effects that changed sign or were notably lower (Table 1). Conditioning of predictors by SEM, however, indicated that these correlations were enhanced partially by strong negative associations with urban development (Table 1). Discrepancies in how land cover acted on DO difference were also apparent. Observed correlations, implied correlations, and total effects between percentage cropland and DO difference were consistently negative, whereas those between percentage development and DO difference were uniformly positive, an outcome most likely attributable to the strong negative correlation between these land cover types and differences in their pollutant export dynamics (Table 1, Table S2).

Fitted path coefficients indicated substantial annual variation in how land cover and water quality shaped WCC, but also revealed a consistent negative effect of urban development. In the dry year of 2002, only the direct effect of urban development was a strong predictor of WCC (Fig. 3A). WCC scores were lower in subestuaries with a high percentage of urban development ($\lambda = -0.74$, $P = 0.046$). The difference in percent saturation of DO between the surface and bottom of the water column was greater in subestuaries with more urban development ($\beta = 0.67$, $P = 0.009$), but did not markedly increase the estimated eutrophication variable ($\beta = -0.34$, $P = 0.596$). Nitrate-N levels were higher in subestuaries with lower salinity, reflecting elevated concentrations in relatively fresh northern compared to brackish southern subestuaries ($\lambda = -0.44$, $P = 0.048$). This trend led to greater predicted values of the latent eutrophication variable ($\beta = 1.00$, $P = 0.002$), but it was not substantial enough to alter WCC ($\beta = 0.18$, $P = 0.666$).

In the wet year of 2003, WCC again depended on the direct effect of urban development but also on the indirect effects of urban development and salinity on water quality. WCC scores remained lower in subestuaries with a high percentage of urban development ($\lambda = -0.52$, $P = 0.047$; Fig. 3B). Nitrate-N concen-

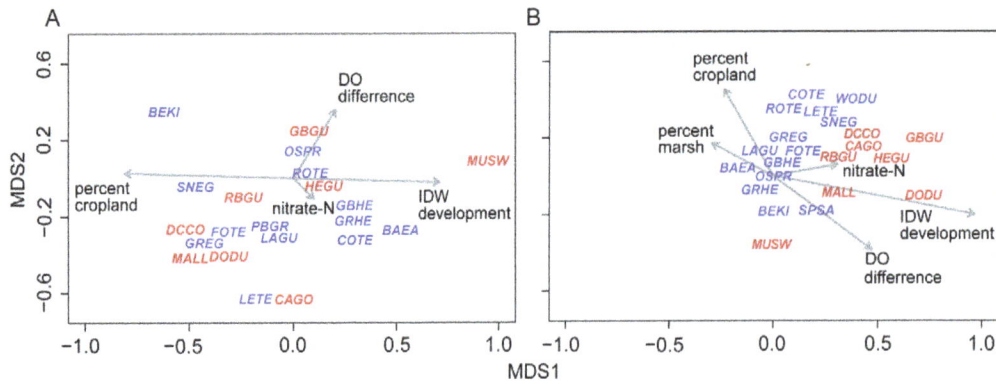

Figure 2. Non-metric multidimensional scaling (MDS) of Chesapeake Bay, USA waterbird communities in (A) 17 subestuaries in the drought year of 2002 (stress = 0.176) and (B) 20 subestuaries in the near record wet year of 2003 (stress = 0.183). Species centroids were mapped in 2-dimensional ordination space and rotational vector fitting was used to relate land cover, water quality, and waterbird community composition (WCC) scores to the ordination. WCC scores are a measure of waterbird community composition, where subestuaries with high abundances of specialist species (blue) received high scores and those with high abundances of generalists (red) received lower scores (DeLuca et al. 2008). Species codes and WCC scores are shown in Table S1. IDW development is the percent urban development in each watershed weighted by the square of its inverse distance to the shoreline. The vector for percentage marsh in 2002 was short and was not plotted for clarity.

Table 1. Standardized total effects, model correlations, and observed correlations from the best-fit structural equation model (SEM) in which all paths were free to vary between the drought year of 2002 and the wet year of 2003.

Year	Predictor	Analysis	Response Nitrate-N	DO difference	WCC
2002	Salinity	Total effects	**−0.437**	–	−0.086
		Model r	**−0.617**	−.258	0.230
		Observed r	**−0.617**	−.320	**0.360**
	Cropland	Total effects	0.092	−0.061	0.022
		Model r	−0.127	**−0.418**	**0.435**
		Observed r	−0.127	**−0.418**	**0.393**
	Development	Total effects	0.447	**0.665**	**−0.735**
		Model r	0.570	**0.698**	**−0.721**
		Observed r	0.570	**0.698**	**−0.715**
	Marsh	Total effects	–	–	0.332
		Model r	−0.127	−0.119	0.438
		Observed r	−0.041	−0.246	0.458
		Multiple r^2 for prediction	**0.513**	**0.490**	**0.652**
2003	Salinity	Total effects	**−0.538**	–	**0.149**
		Model r	**−0.709**	−0.012	**0.466**
		Observed r	**−0.709**	−0.197	**0.563**
	Cropland	Total effects	0.220	**−0.371**	−0.009
		Model r	0.188	**−0.533**	0.290
		Observed r	0.188	**−0.533**	**0.393**
	Development	Total effects	**0.364**	0.345	**−0.667**
		Model r	**0.429**	0.519	**−0.769**
		Observed r	**0.429**	0.519	**−0.768**
	Marsh	Total effects	–	–	0.257
		Model r	−0.285	−0.116	0.438
		Observed r	−0.105	**−0.319**	0.458
		Multiple r^2 for prediction	**0.579**	**0.377**	**0.652**

Differences between observed and model correlations give specific estimates of residual error in model fit. Total effects are the sum of all direct and indirect effects between predictor and response variables, where indirect effects are the product of all direct effects along a hypothesized casual pathway. Coefficients in bold differed from 0 based on 1000 bootstrap replicates.

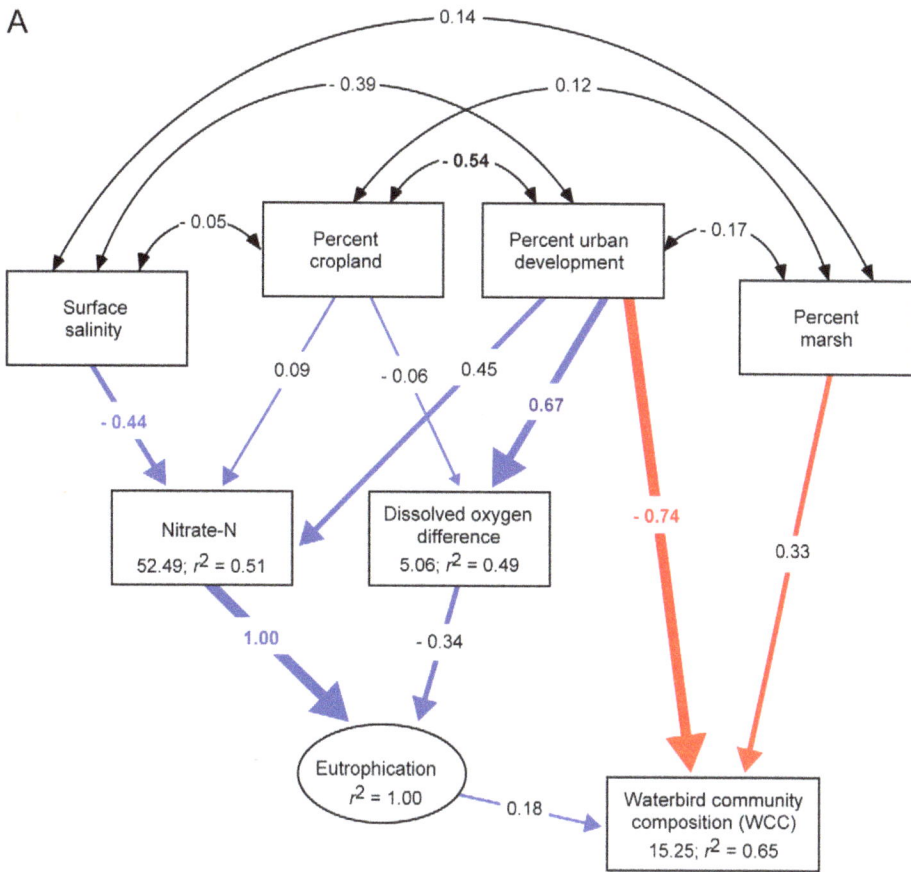

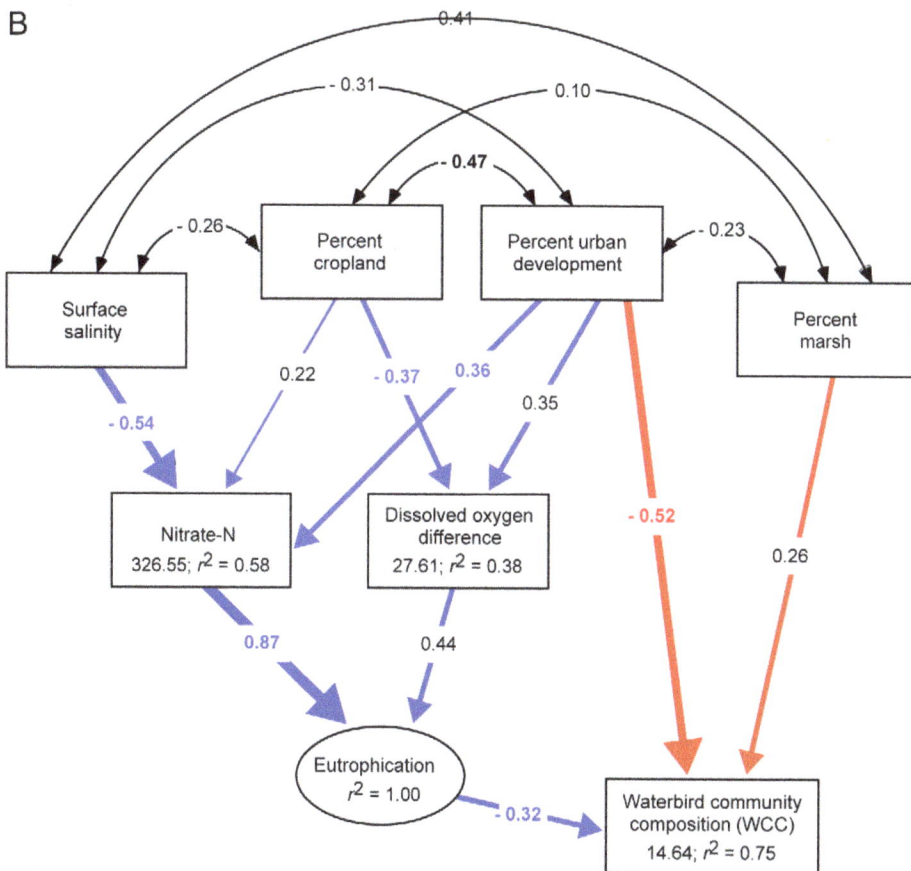

Figure 3. Structural equation models (SEM) testing hypothesized causal effects of land cover and rainfall on waterbird community composition (WCC) in Chesapeake Bay, USA. The best-fit model indicated variation in the strength of direct (red) and indirect paths (blue) between (A) 2002, a year of severe drought, and (B) 2003, a year of near record rainfall ($\chi^2 = 11.49$, $df = 12$, $P = 0.487$; RMSEA $<10^{-3}$; CFI = 1.00). Values between variables are standardized path coefficients. Arrow widths are proportional to the size of path coefficients, but only values in color differed from zero based on a parametric bootstrap of the data with 1000 replicates. Estimated intercepts and the proportion of variance explained by that part of the model appear below variable names. IDW development is the percent urban development in each watershed weighted by the square of its inverse distance to the shoreline.

trations were greater in subestuaries surrounded by a high percentage of urban development ($\lambda = 0.36$, $P = 0.043$), and in those with lower salinity ($\lambda = -0.54$, $P = 0.021$), indicating elevated nutrient levels in northern compared to southern subestuaries. Together, these effects were associated with a mean nitrate-N concentration in subestuaries over six times greater than in the previous year and a marked rise in predicted values of the latent eutrophication variable ($\beta = 0.87$, $P = 0.002$; Table 2). DO difference was greater in subestuaries adjacent to watersheds with a lower percentage of cropland ($\lambda = -0.37$, $P = 0.049$), however this negative pattern also reflected a large DO difference in subestuaries with a large amounts of urban development (Table 1). This latter relationship is more consistent with predicted responses of DO to land cover and with the observed deterioration in mean water quality from 2002 to 2003 (Table 2, Table 3). In contrast to 2002, predicted eutrophication in 2003 was strong enough to alter WCC ($\beta = -0.32$, $P = 0.047$), a change implied by the MDS ordination to include a decrease in the abundance of specialist species and an increase in generalists.

Annual variation in WCC associated with elevated nitrate-N concentration in subestuaries was also evident when we considered this relationship in only the subset of nine subestuaries that were studied in both years. During the dry year of 2002, WCC did not vary in relation to nitrate-N levels (Fig. 4; $r2 = 0.12$. $P = 0.368$). In contrast, in the wet year of 2003, WCC scores in the same subestuaries decreased with increasing nitrate-N concentration ($r2 = 0.51$. $P = 0.032$). These data suggest that higher nitrate-N concentrations observed in 2003 were unlikely due to spatial variation resulting from sampling a group of different subestuaries in 2002 and 2003.

Discussion

Our results support the hypothesis that increased hydrological connectivity between the terrestrial landscape and the aquatic environment can lower water quality enough to alter the composition of estuarine waterbird communities. The MDS

ordination indicated that urban watersheds shifted from a mixed generalist-specialist community in 2002 to generalist-dominated community in 2003. The multi-group SEM revealed that this change in WCC was associated with higher overall trophic status of subestuaries as indicated by elevated nitrate-N concentrations. High nitrate-N concentrations in estuarine grab samples can indicate that nitrogen demand has been temporarily satisfied, that primary producers have yet to respond to continued enrichment, or both [28]. Even though our models did not incorporate rainfall or nitrogen loads explicitly, concurrent research, including long-term monitoring of Chesapeake Bay water quality, provides independent evidence that higher rainfall in 2003 facilitated a strong increase in nitrogen loading compared to 2002 [16,17,36]. Together, these data imply that continued expansion of urban development and strong rainfall events that flush accumulated pollutants from the landscape may interact to promote estuarine waterbird communities increasingly dominated by generalist species.

High predicted eutrophication in the wet year of 2003 likely represents elevated nitrate-N discharge from at least three distinct terrestrial sources that were differentiated from one another only after accounting for spatial autocorrelation among land cover types. Agricultural lands, and cropland in particular, typically supply the majority of nutrients to estuaries [21,37,38]. The Susquehanna River, situated at the northern end of the Bay, drains a region of extensive agriculture and accounts for a large fraction of annual nitrogen loads [31]. The increase in nitrate-N concentrations in lower salinity northern subestuaries probably reflects lagged nutrient discharges from the Susquehanna River

Table 2. Water quality indices (mean ± SE) from six sampling stations in each of 27 subestuaries of Chesapeake Bay, USA.

Water quality	Year	
	2002	2003
Salinity (ppt)	10.58±1.17	5.48±0.59
Nitrate-N (µg/L)	27.41±13.46	169.53±54.72
Bottom DO (% saturation)	65.69±4.51	52.01±4.61
Minimum bottom DO (% saturation)	41.07±4.04	28.46±4.54
Hypoxia frequency (prop. of stations)*	0.08	0.18

*Hypoxia frequency is the proportion of stations across all subestuaries where bottom DO concentration was <2 mg/L.

Table 3. Dissolved oxygen indices (mean ± SE) from six sampling stations in each of a subset of 19 subestuaries where percent cropland or IDW development covered >15 percent of the surrounding watershed in Chesapeake Bay, USA.

Year	Dissolved oxygen	Land cover	
		Cropland	IDW development*
2002	Surface DO (% saturation)	73.61±3.64	91.26±12.15
	Bottom DO (% saturation)	66.00±3.53	63.48±11.87
	DO difference (% saturation)	7.61±2.62	27.78±10.51
2002	Surface DO (% saturation)	73.66±5.28	92.24±10.53
	Bottom DO (% saturation)	54.70±7.11	49.56±8.29
	DO difference (% saturation)	18.96±4.84	42.67±7.55

*IDW development is the percent urban development in the watershed weighted by its squared inverse distance from the shoreline.

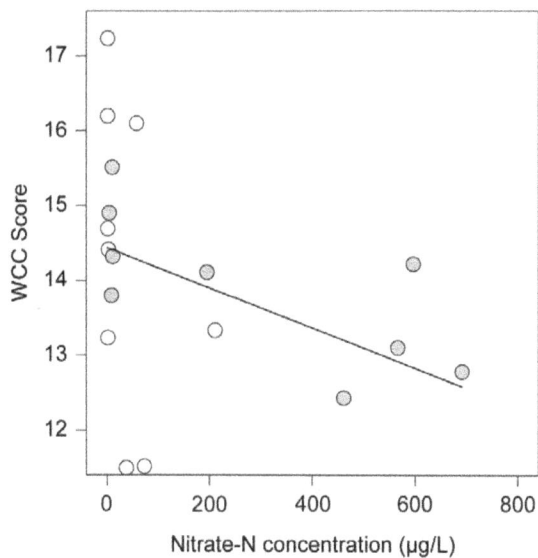

Figure 4. Relationship between nitrate-N concentration and waterbird community composition (WCC) in the nine subestuaries of Chesapeake Bay, USA that were sampled in both the drought year of 2002 and the wet year of 2003. Nitrate-N levels did not limit waterbird community composition during the drought year of 2002 (white; $r^2 = 0.12$. $P = 0.368$), but were associated with degraded waterbird communities during the wet year of 2003 (grey; $r^2 = 0.51$. $P = 0.032$).

that originated from distant, upstream cropland, potentially as far north as New York. Conversely, local cropland in watersheds surrounding subestuaries appeared to have little effect on measured nitrate-N concentrations. One explanation for this pattern is that nitrogen export from cropland peaks in early spring before green vegetation emerges and evapotranspiration reduces flow rates of smaller rivers [16]. Nitrate-N arriving to subestuaries in spring would have been converted to algal biomass and organic detritus by the time water our quality sampling occurred. Nitrate-N enrichment from local cropland therefore was detectible primarily through its effect on bottom DO, whereas DO difference remained largely unresponsive to increasing cropland (Table 3). Unlike those dominated by agriculture, urban watersheds have comparatively lower evapotranspiration and appeared to load similar amounts of nitrate-N throughout the summer, leading to greater observed concentrations, supersaturation in surface DO, pronounced DO difference, and high predicted values of the latent eutrophication variable (Table 3). Therefore, the negative correlation between DO difference and cropland likely was driven by DO trends in urban watersheds and the negative correlation between cropland and development that exists in this region [32].

We did not measure the response of lower trophic level organisms to eutrophication, but it is likely that the increase in generalist waterbird species in 2003 reflected disturbance propagating from the base to the top of the estuarine food web [39,40]. All of the specialists recorded in this study prey on fish species that concentrate near estuarine wetlands during the summer breeding season, whereas half of the generalists observed (mute swan [*Cygnus olor*], mallard [*Anas platyrhynchos*], domestic duck [*Anas* and *Cairina spp.*], and Canada goose [(*Branta canadensis*)] incorporate aquatic or terrestrial plants in their diet. Eutrophic, oxygen depleted conditions, such as those present in 2003, have been linked to reduced abundance and mortality of fish species consumed as prey by piscivorous waterbirds in Chesapeake Bay and other shallow

estuaries [41–43]. Because some waterbirds travel long distances between breeding and foraging areas, their distribution while foraging should reflect the availability of prey and the overall quality of estuarine habitat [44,45]. Thus, it is possible that the sensitivity of WCC to the annual change in water quality was due to piscivorous species tracking food resources to areas minimally affected by eutrophication and hypoxia. Low salinity in estuaries may also reduce fish abundance through temporary emigration to more saline waters [46]. Lower salinity in 2003 could therefore have facilitated the increase in generalist waterbird species, particularly in urban subestuaries at northern reaches of the Bay (Table 2). Nonetheless, the total effect of salinity on WCC was less than half that of nitrate-N in 2003, suggesting that nutrient loading was the primary aspect of water quality affecting waterbirds (Table 1; Fig. 3B).

Two lines of evidence suggest that we were successful at capturing variation in water quality across years and that the observed effects were not due to spatial variation caused by sampling different subestuaries in each year. First, the intercept estimated for nitrate-N concentration in the best-fit model was more than six times greater in 2003 compared to 2002 (Table 2). This difference is in line with annual variation in nitrate-N concentration in small streams and major rivers entering Chesapeake Bay found by other studies during the same time period [16,36]. Second, annual variation in the relationship between nitrate-N and WCC in the subset of nine subestuaries studied in both 2002 and 2003 closely paralleled the trend found in the full data set, which included sites surveyed in only one of two years.

Even though elevated nitrate-N concentrations in 2003 were linked to a strong change in WCC, urban development directly limited the waterbird community in both years, an outcome likely due in part to fragmentation and loss of both near-shore terrestrial and wetland habitat. Fragmentation of terrestrial shoreline habitat by human development can reduce habitat suitability for piscivorous species that use shoreline foraging or nesting perches (e.g., belted kingfisher [*Ceryle alcyon*] and osprey [*Pandion haliaetus*]) or prefer unbroken stretches of natural shoreline (e.g., great egret [*Ardea alba*] and green heron [*Butorides virescens*]). In addition, shoreline modification can facilitate colonization of wetlands by terrestrial and invasive vegetation, especially *Phragmites australis* [18]. This species can fragment native vegetation and render even large wetlands inhospitable to foraging waterbirds. Moreover, fragmentation and isolation of natural vegetation by urban development may facilitate increases in mammalian predators and limit nesting habitat suitability, potentially forcing waterbirds to commute longer distances to foraging areas.

Some unknown portion of the variation we have attributed to direct pathways could also have occurred through unmeasured aspects of indirect pathways, including phosphorus export, sewage overflow, and the presence of toxins [47–49]. Nitrate-N concentrations in subestuaries increased with the percentage developed land in watersheds not only in the wet year 2003, but also in the absence of substantial rainfall in 2002 (Fig. S1). This pattern is consistent with a greater contribution of distinct point sources of nitrate-N with high amounts of urban development in the dry year of 2002. Because urban areas can have point sources that do not require strong rainfall events to deliver nutrients to estuarine environments (e.g., water treatment facilities, sewage leaks), it is probable that they were responsible some of the negative effect of urban development on WCC, especially in the drought year of 2002. Urban environments also contain sources of nitrogen that may be continually replenished throughout the growing season (e.g., lawn fertilizers, atmospheric deposition). Increasing development has been shown to shift the mode of nitrogen export from base flows toward event flows, so flushing of these sources likely increases with more intense precipitation events [36,50].

The expansion of human development in the Chesapeake Bay watershed is likely to result in greater nutrient inputs and enhanced conditions for foraging generalists, a situation that could be exacerbated by predicted increases in severe storms [2]. These changes may not only reduce foraging habitat quality for piscivorous specialists, but could also limit their populations through increased competition and predation by generalists. Agonistic interactions with double-crested cormorants (*Phalacrocorax auritus*) can reduce reproductive success of other colonial nesting species, and have been implicated in displacement of heron and egret rookeries [51,52]. Cormorants can also severely deplete fish populations, potentially limiting prey availability for other piscivorous birds [53]. Greater black-blacked (*Larus marinus*) and herring gulls (*Larus argentatus*) are important nest predators of common tern (*Sterna hirundo*) and least tern (*Sterna antillarum*) nests, and growing populations of these species are thought to have contributed to the decline and abandonment of tern colonies in Chesapeake Bay [52]. In addition, herbivores such as Canada goose and mute swan may overgraze wetland vegetation and deteriorate nesting habitat for marsh-breeding birds. Moreover, guano from these species is rich in nitrogen and phosphorus, and large aggregations have been linked to changes in water chemistry that promote local eutrophication, although only in small water bodies [54].

A critical challenge faced by restoration programs is to decide on criteria for management targets [55]. The ability of our model to describe spatial and short-term temporal variation in estuarine condition suggests a strategy for defining such targets using existing monitoring data. Many of the data analyzed in our research are also collected in a number of estuarine monitoring programs, yet statistical analyses typically are done using traditional ordination or regression approaches that may not be conducive to tracking long-term change in covariance relationships among ecological variables and a large set of potential stressors. Where long time series data are available, SEM or other casual models could be used in conjunction with techniques such as changepoint analysis to identify critical historical thresholds in estuarine condition that could then be used as targets for restoration or mitigation.

Our findings also may contribute to adaptive management of estuarine waterbird communities. As in many coupled natural-human systems, management planning in Chesapeake Bay and other estuaries requires iterated decision making in the face of uncertainty about future changes in land cover and climate [56]. Restoring and maintaining fish and wildlife populations is mandated by state and federal management agencies, but current monitoring in Chesapeake Bay assigns a separate grade to each

part of the system and therefore is not entirely consistent with the decision context [57]. Our results provide a snapshot of the chief pathways through which land cover and water quality interact to limit waterbird communities under two extremes of annual rainfall. By themselves, these findings may not cover a sufficient time scale to immediately facilitate management decisions; however, the strong linkages we illustrate among waterbirds, water quality, and land cover indicate the need for a long-term monitoring program to explicitly track the structure of these relationships. Such an approach could help align monitoring efforts with management decision-making and potentially facilitate the knowledge feedback that is central to adaptive management.

Supporting Information

Figure S1 Results of a generalized additive model showing the relationship between percent urban development in the watershed and the partial residuals of nitrate-N concentration (adjusted for percent cropland) during (A) the drought year of 2002 ($r2 = 0.61$, $P = 0.001$) and (B) the wet year of 2003 ($r2 = 0.62$, $P = 0.001$). Dashed lines depict 95% confidence intervals.

Text S1 Methods of WCC index development.

Table S1 List of bird species detected on transect surveys in Chesapeake Bay, USA subestuaries in the dry year of 2002 and the wet year of 2003 and their waterbird community composition (WCC) scores.

Table S2 Pearson's correlations (below diagonal), variances (diagonal; in bold), and covariances (above diagonal) for variables used in the best-fit structural equation model (SEM) in which all paths were free to vary between the drought year of 2002 and the wet year of 2003.

Acknowledgments

We thank Anne Balogh, Suzanne Conrad, Sacha Mkheidze, Dan Mummert, Ryan Peters, and Beth Wright for their assistance with fieldwork. We also thank Donald Weller for his contribution to the GIS components of the study and two anonymous reviewers for their comments.

Author Contributions

Conceived and designed the experiments: CES WVD RSK PPM. Performed the experiments: CES WVD RSK. Analyzed the data: CES WVD MEB RSK. Wrote the paper: CES WVD PPM.

References

1. Lotze HK, Lenihan HS, Bourque BJ, Bradbury RH, Cooke RG, et al. (2006) Depletion, degradation, and recovery potential of estuaries and coastal seas. Science 312: 1806–1809.
2. Easterling DR, Meehl GA, Parmesan C, Changnon SA, Karl TR, et al. (2000) Climate extremes: Observations, modeling, and impacts. Science 289: 2068–2074.
3. Paracuellos M, Telleria JL (2004) Factors affecting the distribution of a waterbird community: The role of habitat configuration and bird abundance. Waterbirds 27: 446–453.
4. Thrush SF, Halliday J, Hewitt JE, Lohrer AM (2008) The effects of habitat loss, fragmentation, and community homogenization on resilience in estuaries. Ecological Applications 18: 12–21.
5. Traut AH, Hostetler ME (2004) Urban lakes and waterbirds: effects of shoreline development on avian distribution. Landscape and Urban Planning 69: 69–85.
6. Howarth RW, Swaney DP, Boyer EW, Marino R, Jaworski N, et al. (2006) The influence of climate on average nitrogen export from large watersheds in the Northeastern United States. Biogeochemistry 79: 163–186.
7. Rabalais NN, Turner RE, Diaz RJ, Justic D (2009) Global change and eutrophication of coastal waters. Ices Journal of Marine Science 66: 1528–1537.
8. Diaz RJ, Rosenberg R (2008) Spreading dead zones and consequences for marine ecosystems. Science 321: 926–929.
9. Essington TE, Paulsen CE (2010) Quantifying Hypoxia Impacts on an Estuarine Demersal Community Using a Hierarchical Ensemble Approach. Ecosystems 13: 1035–1048.
10. Vaquer-Sunyer R, Duarte CM (2008) Thresholds of hypoxia for marine biodiversity. Proceedings of the National Academy of Sciences of the United States of America 105: 15452–15457.
11. O'Connell TJ, Jackson LE, Brooks RP (2000) Bird guilds as indicators of ecological condition in the central Appalachians. Ecological Applications 10: 1706–1721.
12. DeLuca WV, Studds CE, Rockwood LL, Marra PP (2004) Influence of land use on the integrity of marsh bird communities of Chesapeake Bay, USA. Wetlands 24: 837–847.

13. DeLuca WV, Studds CE, King RS, Marra PP (2008) Coastal urbanization and the integrity of estuarine waterbird communities: Threshold responses and the importance of scale. Biological Conservation 141: 2669–2678.

14. Erwin RM (1996) Dependence of waterbirds and shorebirds on shallow-water habitats in the mid-Atlantic coastal region: An ecological profile and management recommendations. Estuaries 19: 213–219.

15. Kennish MJ (2002) Environmental threats and environmental future of estuaries. Environmental Conservation 29: 78–107.

16. Langland MJ, Phillips SW, Raffensperger JP, Moyer DL (2004) Changes in streamflow and water quality in selected nontidal sites in the Chesapeake Bay Basin, 1985–2003. Scientific Investigations Report 2004–5259: U.S. Geological Survey, Reston, VA, USA.

17. Acker JG, Harding LW, Leptoukh G, Zhu T, Shen SH (2005) Remotely-sensed chl a at the Chesapeake Bay mouth is correlated with annual freshwater flow to Chesapeake Bay. Geophysical Research Letters 32.

18. King RS, Deluca WV, Whigham DF, Marra PP (2007) Threshold effects of coastal urbanization on Phragmites australis (common reed) abundance and foliar nitrogen in Chesapeake Bay. Estuaries and Coasts 30: 469–481.

19. United States Enivronmental Protection Agency (USEPA) (2000) Multi-Resolution Land Characteristics Consortium (MRLC) Database. Available: http://www.mrlc.gov/nlcd1992.php. Accessed: 10 June 2009.

20. Nichols JD, Hines JE, Sauer JR, Fallon FW, Fallon JE, et al. (2000) A double-observer approach for estimating detection probability and abundance from point counts. Auk 117: 393–408.

21. Jordan TE, Weller DE, Correll DL (2003) Sources of nutrient inputs to the Patuxent River estuary. Estuaries 26: 226–243.

22. McCune B, Grace JB (2002) Analysis of ecological communities. Gleneden Beach, OR, USA: MJM Software Design. 304 p.

23. Faith DP, Norris RH (1989) Correlation of environmental variables with pattersn of distribution and abundance of common and rare freshwater macroinvertebrates Biological Conservation 50: 77–98.

24. R Development Core Team (2011) R: A language and environment for statistical computing R Foundation for Statistical Computing, Vienna, Austria. ISBN 3-900051-07-0. Available: http://www.R-project.org/. Accessed 3 March 2011.

25. Oksanen J, Blanchet FG, Kindt R, Legendre P, Minchin PR, et al. (2011) vegan: Community Ecology Package. R package version 2.0-2. Available: http://CRAN.R-project.org/package=vegan. Accessed: 3 March 2011.

26. Mitsch WJ, Gosselink JG (2000) Wetlands. New York, NY, USA: Wiley. 920 p.

27. Bulleri F, Chapman MG (2010) The introduction of coastal infrastructure as a driver of change in marine environments. Journal of Applied Ecology 47: 26–35.

28. Paerl HW (2009) Controlling Eutrophication along the Freshwater-Marine Continuum: Dual Nutrient (N and P) Reductions are Essential. Estuaries and Coasts 32: 593–601.

29. Yu K, DeLaune RD, Seo DC (2008) Influence of salinity level on sediment denitrification in a Louisiana estuary receiving diverted Mississippi River water. Archives of Agronomy and Soil Science 54: 249–257.

30. Connell DW, Miller GJ (1984) Chemistry and ecotoxicology of pollution. New York, NY, USA: John Wiley & Sons.

31. Preston SD, Brakebill JW (1999) Application of spatially referenced regression modeling for the evaluation of total nitrogen loading in the Chesapeake Bay watershed. Water Resources Investigations Report 99–4054: U.S. Geological Survey, Baltimore, Maryland, USA.

32. King RS, Baker ME, Whigham DF, Weller DE, Jordan TE, et al. (2005) Spatial considerations for linking watershed land cover to ecological indicators in streams. Ecological Applications 15: 137–153.

33. Shipley B (2000) Cause and Correlation in Biology: a user's guide to path analysis, structural equations, and causal inference. Cambridge, UK: Cambridge University Press. 317 p.

34. Bollen KA, Long JS, eds (1993) Testing Structural Equation Models. Newbury Park, CA, USA: Sage.

35. Arbuckle JL (2010) Amos (version 19). Chicago, IL, USA: IBM SPSS.

36. Kaushal SS, Groffman PM, Band LE, Shields CA, Morgan RP, et al. (2008) Interaction between urbanization and climate variability amplifies watershed nitrate export in Maryland. Environmental Science & Technology 42: 5872–5878.

37. Howarth RW, Sharpley A, Walker D (2002) Sources of nutrient pollution to coastal waters in the United States: Implications for achieving coastal water quality goals. Estuaries 25: 656–676.

38. Boyer EW, Goodale CL, Jaworsk NA, Howarth RW (2002) Anthropogenic nitrogen sources and relationships to riverine nitrogen export in the northeastern USA. Biogeochemistry 57: 137–169.

39. Bundy MH, Breitburg DL, Sellner KG (2003) The responses of Patuxent River upper trophic levels to nutrient and trace element induced changes in the lower food web. Estuaries 26: 365–384.

40. Wazniak CE, Hall MR, Carruthers TJB, Sturgis B, Dennison WC, et al. (2007) Linking water quality to living resources in a mid-Atlantic lagoon system, USA. Ecological Applications 17: S64–S78.

41. Chesney EJ, Houde ED (1989) Laboratory studies on the effect of hypoxic waters on the survival of eggs and yolk-sac larvae of the bay anchovy, Anchoa mitchilli, 184–191. In Chesney EJ, Houde ED, Newberger TA, Vasquez AV, Zastrow CE, Morin LG, Harvey HR, Gooch JW (eds.) Population Biology of Bay Anchovy in Mid-Chesapeake Bay. Final Report to Maryland Sea Grant Ref. No. (UM-CEES) CBL 89–141. Solomons, Maryland, USA.

42. Miller DS, Poucher SL, Coiro L (2002) Determination of lethal dissolved oxygen levels for selected marine and estuarine fishes, crustaceans, and a bivalve. Marine Biology 140: 287–296.

43. Esssington TE, Paulson CE (2010) Quantifying hypoxia imapcts on an estuarine demersal community using a hierarchical ensemble approach. Ecosystems 13: 1035–1048.

44. Haas K, Kohler U, Diehl S, Kohler P, Dietrich S, et al. (2007) Influence of fish on habitat choice of water birds: A whole system experiment. Ecology 88: 2915–2925.

45. Kloskowski J, Nieoczym M, Polak M, Pitucha P (2010) Habitat selection by breeding waterbirds at ponds with size-structured fish populations. Naturwissenschaften 97: 673–682.

46. Jung S, Houde ED (2003) Spatial and temporal variabilities of pelagic fish community structure and distribution in Chesapeake Bay, USA. Estuarine and Coastal Shelf Science 58: 335–351.

47. Fear JM, Paerl HW, Braddy JS (2007) Importance of submarine groundwater discharge as a source of nutrients for the Neuse River Estuary, North Carolina. Estuaries and Coasts 30: 1027–1033.

48. Lawrie RA, Stretch DD, Perissinotto R (2010) The effects of wastewater discharges on the functioning of a small temporarily open/closed estuary. Estuarine Coastal and Shelf Science 87: 237–245.

49. King RS, Beaman JR, Whigham DF, Hines AH, Baker ME, et al. (2004) Watershed land use is strongly linked to PCBs in white perch in Chesapeake Bay subestuaries. Environmental Science & Technology 38: 6546–6552.

50. Shields CA, Band LE, Law N, Groffman PM, Kaushal SS, et al. (2008) Streamflow distribution of non-point source nitrogen export from urban-rural catchments in the Chesapeake Bay watershed. Water Resources Research 44.

51. Somers CM, Lozer MN, Quinn JS (2007) Interactions between double-crested cormorants and herring gulls at a shared breeding site. Waterbirds 30: 241–250.

52. Brinker DF, McCann JM, Williams B, Watts BD (2007) Colonial-nesting seabirds in the Chesapeake Bay region: Where have we been and where are we going? Waterbirds 30: 93–104.

53. Rudstam LG, VanDeValk AJ, Adams CM, Coleman JTH, Forney JL, et al. (2004) Cormorant predation and the population dynamics of walleye and yellow perch in Oneida lake. Ecological Applications 14: 149–163.

54. Ronicke H, Doerffer R, Siewers H, Buttner O, Lindenschmidt KE, et al. (2008) Phosphorus input by nordic geese to the eutrophic Lake Arendsee, Germany. Fundamental and Applied Limnology 172: 111–119.

55. Thorpe AS, Stanley AG (2011) Determining appropriate goals for restoration of imperilled communities and species. Journal of Applied Ecology 48: 275–279.

56. Lyons JE, Runge MC, Laskowski HP, Kendall WL (2008) Monitoring in the Context of Structured Decision-Making and Adaptive Management. Journal of Wildlife Management 72: 1683–1692.

57. Chesapeake Bay Foundation (2010) 2010 State of the Bay. Annapolis, MD, USA: Chesapeake Bay Foundation. Available: Available: http://www.cbf.org. Accessed 1 April 2012.

18

Metabolite Adjustments in Drought Tolerant and Sensitive Soybean Genotypes in Response to Water Stress

Sonia Silvente[1]*, Anatoly P. Sobolev[2], Miguel Lara[1]

1 Centro de Ciencias Genómicas, Universidad Nacional Autónoma de México, Cuernavaca, Morelos, México, 2 Istituto di Metodologie Chimiche, Laboratorio di Risonanza Magnetica "Annalaura Segre", CNR, Monterotondo, Rome, Italy

Abstract

Soybean (*Glycine max* L.) is an important source of protein for human and animal nutrition, as well as a major source of vegetable oil. The soybean crop requires adequate water all through its growth period to attain its yield potential, and the lack of soil moisture at critical stages of growth profoundly impacts the productivity. In this study, utilizing ^{1}H NMR-based metabolite analysis combined with the physiological studies we assessed the effects of short-term water stress on overall growth, nitrogen fixation, ureide and proline dynamics, as well as metabolic changes in drought tolerant (NA5009RG) and sensitive (DM50048) genotypes of soybean in order to elucidate metabolite adjustments in relation to the physiological responses in the nitrogen-fixing plants towards water limitation. The results of our analysis demonstrated critical differences in physiological responses between these two genotypes, and identified the metabolic pathways that are affected by short-term water limitation in soybean plants. Metabolic changes in response to drought conditions highlighted pools of metabolites that play a role in the adjustment of metabolism and physiology of the soybean varieties to meet drought effects.

Editor: Haibing Yang, Purdue University, United States of America

Funding: This work was supported by Dirección General de Asuntos del Personal Académico/UNAM, México (grant no. PAPIIT: IN212110). The funders had no role in study design, data collection and analysis, decision to publish, or preparation of the manuscript.

Competing Interests: The authors have declared that no competing interests exist.

* E-mail: silvente@ccg.unam.mx

Introduction

Soybean (*Glycine max*) is one of the most important grain legumes. It represents not only an essential source of protein, oil and micronutrients in human and animal diets, but is also an attractive crop for the production of biodiesel [1]. Soybean growth is affected by unfavorable environmental factors such as extreme temperatures, drought, nutrient deficiency and soil acidity, which form major constraints for soybean crop production.

Soybean plants form root nodule symbiosis with nitrogen-fixing bradyrhizobia, thus rendering the plant independent of N fertilizers. Nodulation and symbiotic nitrogen fixation have long been recognized as being sensitive to environmental stresses, particularly drought [2]. Water stress reduces nitrogen fixation as a result of a decrease in photosynthate supply [3] or a reduction in the O_2 flux into the nodule as well as through overloading nodules with nitrogenous compounds [4–7].

Some of the most important responses of a plant against drought stress are associated with the accumulation of minerals [8] and the enhanced synthesis of osmoprotectants, or compatible solutes, which are part of normal metabolism. The accumulation of these compounds helps the stressed cells in water retention [9] and in the maintenance of the structural integrity of the cell membranes [10].

The types of osmoprotectant metabolites and their relative contribution in lowering the osmotic potential differ greatly among plant species. Osmotic adjustment has been reported in legumes

with a high tolerance to water stress [11,12]. Metabolic adjustments in response to the adverse environmental conditions may highlight pools of metabolites that play important roles in metabolism and physiology and may indicate which pathways have been perturbed by the stress.

Nuclear magnetic resonance (NMR) spectroscopy can be used to monitor and quantify the degree of metabolic impact induced by drought or other environmental disturbances [13,14], since NMR can bring "high-throughput" spectroscopic/structural information on a wide range of metabolites simultaneously with high analytical precision. One of its main advantages is that it avoids biases against various classes of compounds. Molecular identification is easy and straightforward as it can be deduced from the NMR spectrum of the mixture itself by means of 1D and 2D experiments, standard additions and by comparison with database of standard compounds.

In the present investigation, ^{1}H NMR-based metabolic profiling combined with the physiological studies were conducted in two genotypes of soybean differing in their tolerance to drought in order to elucidate metabolite adjustments in relation to the physiological responses in the nitrogen-fixing plants towards water stress. To our knowledge this is the first report on metabolite profiling in soybean under drought stress. NMR based metabolic profiling approach [15,16] adopted in this study enabled the identification and quantification of a number of metabolites

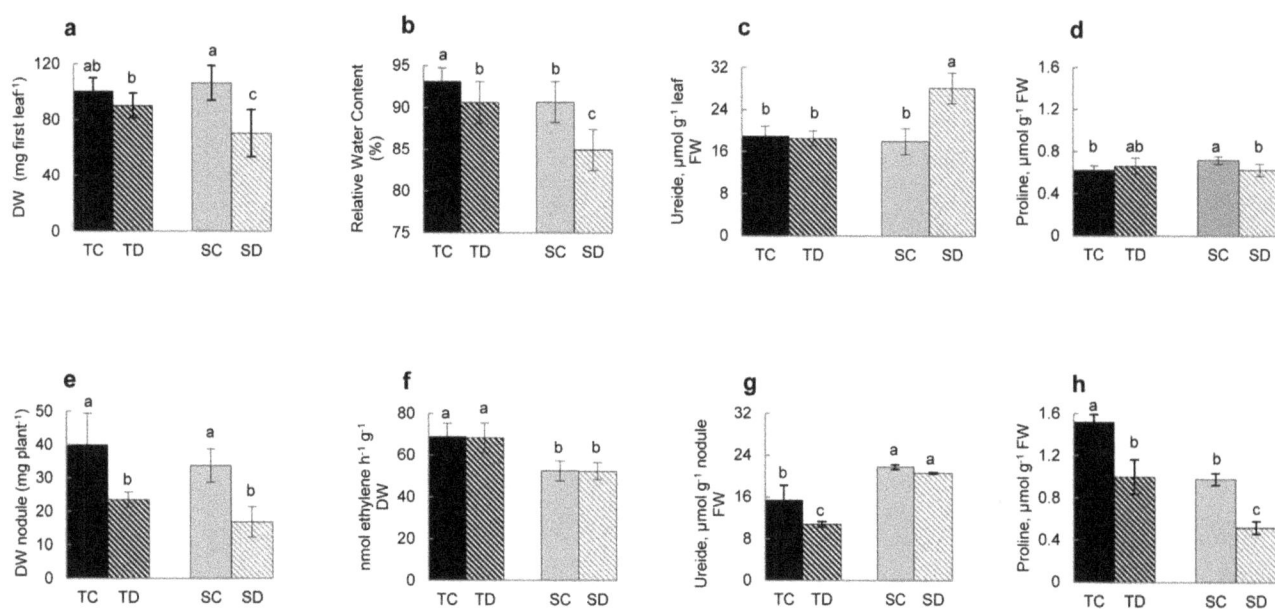

Figure 1. Responses of drought tolerant and sensitive varieties of soybean towards short-term water stress. (a–d) trifoliate leaves and (e–h) nodules. (a, e) dry weight (DW); (b) relative water content (RWC), (f) nitrogenase activity, (c, g) ureide levels and (d, h) proline levels in 17-day-old plants subjected to 10 days water sufficient and deficient conditions. TC: Tolerant genotype Control; TD: Tolerant subjected to Drought, SC: Sensitive genotype Control; SD: Sensitive subjected to Drought. Data represent the mean ± standard deviation (SD) of five to 15 replicates. Different letters at the top of each bar indicate significant differences at $P<0.01$.

belonging to various classes of compounds from the crude extracts, without involving any separation step.

Results

The two soybean cultivars used in the present study were categorized at CIAP-INTA (Centro de Investigaciones Agropecuarias – Instituto Nacional de Tecnología Agropecuaria, Argentina) as tolerant (NA5009RG) and sensitive (DM50048) to drought stress based on their ability to maintain relative water content (RWC) and growth, and withstand oxidative stress through the modulation of cellular malondialdehide (MDA) levels. The drought tolerant line maintained higher RWC and showed greater ability to withstand oxidative damage owing to lower production of MDA, and exhibited sustained growth at reduced soil moisture conditions (Dr. Celina Luna, personal communication). In the present study, a comparison of these two soybean drought tolerant and sensitive genotypes was undertaken to determine the differences in their metabolic profiles/responses during water stress. Drought was imposed on nodulated soybean plants 17 days after inoculation by withholding water during 10 days, and physiological characteristics such as dry weight (DW), relative water content (RWC), chlorophyll and nitrogenase activity (ARA), and metabolite profiles were analyzed in order to establish the effect of water stress on these plant parameters (Figure 1). The results of the study showed that the water stress produced varied effects on leaf and nodule metabolism in both drought tolerant and sensitive soybean varieties. Under the drought condition imposed, both RWC and DW of the leaves showed a reduction, more remarkably in the sensitive variety (a decrease of about 10% in RWC and 42% in DW in the sensitive genotype as compared to 9% and 15%, respectively, in the tolerant one; Figure 1a, b). In addition, water stress affected proline and ureide contents in leaves of the sensitive variety but not in the tolerant genotype (Figure 1c, d). On the other hand, no effect of drought stress was observed on

chlorophyll content as well as chlorophyll a/b ratio in both the genotypes (data not shown).

In case of nodules, in both genotypes, water stress conditions even for 10 days resulted in a dramatic reduction in DW, even though the nitrogenase activity per unit DW remained unaffected (Figure 1e, f). Similar to DW, drought conditions also caused fall in both proline and ureide contents in the tolerant variety (Figure 1g, h). In contrast, in the sensitive genotype, only proline declined while ureide content remained unchanged in the nodules as compared to that in well watered plants.

Comparison of the metabolite profiles of leaf and nodule tissues of tolerant and sensitive varieties subjected to drought

Comparison of the ^{1}H NMR spectra revealed no major qualitative differences in the metabolites between leaf and nodule tissues except for minor aromatic compounds (Figure 2 and Table 1). As an example, a typical ^{1}H NMR spectra of leaf and nodule extracts of well watered tolerant soybean plants together with the assignment of the most abundant metabolites (amino acids, sugars, organic acids) are shown in the Figures 2a and 2b, respectively. Among the resonances in the 9.5 – 7 ppm region, characteristic for aromatic and heteroaromatic compounds, only trigonelline was assigned in leaves extracts. On the other hand, asparagine was detected exclusively in nodules.

Basically, regardless of the genotype (tolerant or sensitive plants) or watering conditions, all the ^{1}H NMR spectra of extracts from the same type of tissues (leaves or nodules) share the same signals, although their relative intensity is variable. The intensity of selected signals (Table 1 and Materials and Methods section) was used to calculate the relative molecular abundance of about 15 assigned metabolites. On the other hand, the assignment of minor components was hindered by the scarcity of data on these metabolic compounds in literature. Although the number of compounds identified by NMR is limited, the NMR spectra

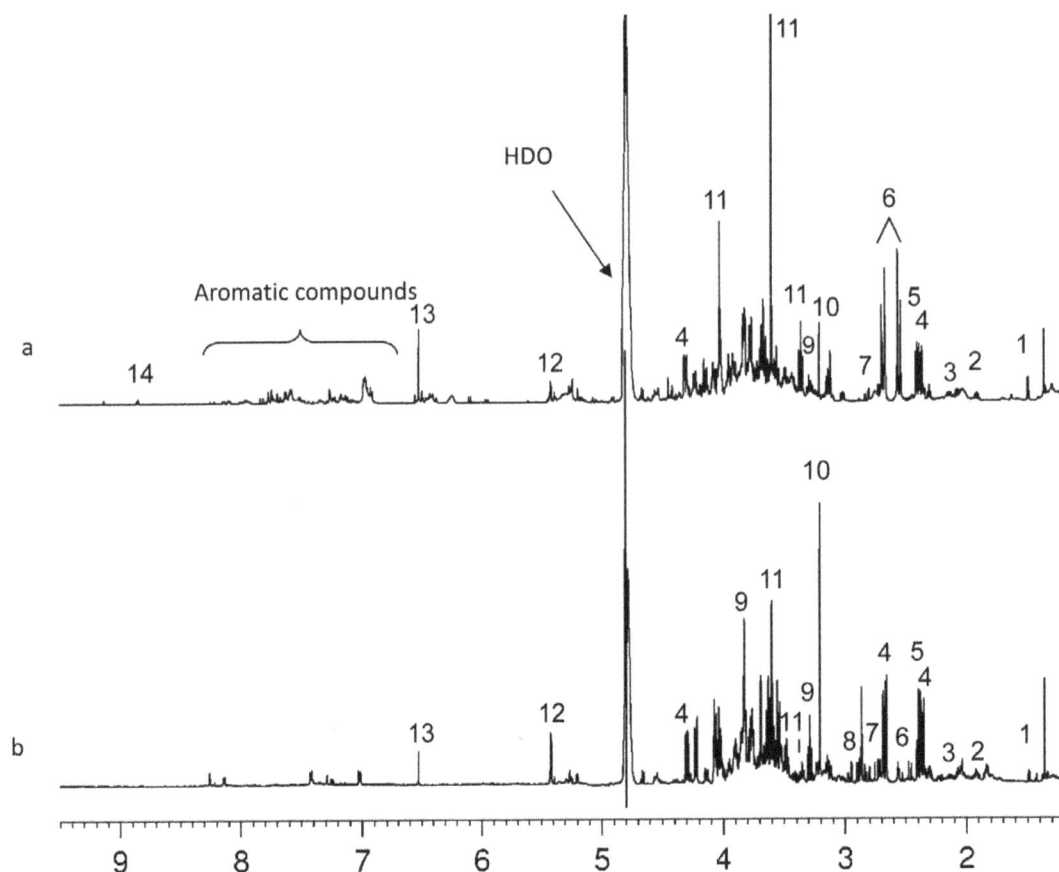

Figure 2. ¹H NMR spectra of (a) leaves and (b) nodules of water-soluble extracts from well-watered tolerant soybean plants. Assignments: 1, alanine; 2, GABA; 3, glutamine; 4, malic acid; 5, succinic acid; 6, citric acid; 7, aspartate; 8, asparagine; 9, myo-inositol; 10, choline; 11, pinitol; 12, sucrose; 13, fumaric acid; 14, trigonelline. HDO: deuterated water.

Table 1. List of variables used in statistical analysis for leaves and nodules samples.

Compound	Abbreviation	¹H Chem. shift, ppm
Ala		1.49
GABA		1.91
Gln		2.14
Malic A	MA	2.38
Succinic A	SU	2.41
2-Oxoglutaric A	Ox	2.45
Citric A	CI	2.53
Asp		2.83
Asn (*)		2.98
Choline	CH	3.21
Myo-inositol	MI	3.29
Pinitol	PI	3.36
Allantoin	AL	5.39
Sucrose	SUCR	5.42
Fumaric A	FU	6.52
Trigonelline (**)	Trig	8.85

Note: (*) nodules, (**) leaves.

indeed gave a good picture of what really is present in the plant extracts examined. Results of the study demonstrated that water stress induces several changes in various metabolic pathways in both genotypes; the effect being more pronounced in the leaves than in nodules (Figure 3 and Tables 2,3,4).

Under drought conditions, levels of the individual sugars varied considerably among the genotypes: for example, sucrose and myo-inositol levels in the leaves decreased drastically in the sensitive genotype, but no significant changes were observed in the tolerant variety. In contrast, in the leaves of the sensitive genotype, pinitol levels increased under drought while it decreased in the tolerant one. In nodules, however, sucrose content decreased in drought in both varieties while pinitol levels increased. Myo-inositol content, on the other hand, did not alter in the nodules of the both varieties when the water stress was imposed.

Individual organic acids that mainly contributed to the differences in total organic acids under drought were 2-oxoglutaric acid, succinic acid and malic acid. Of these three, only succinic acid levels rose while malic acid content decreased in the leaves of drought stressed plants as compared to well-watered plants in both the genotypes, with no significant changes in nodules. 2-oxoglutaric acid, on the other hand, showed downward trend only in the leaves of sensitive variety.

With regard to the free amino acids in the leaves under drought, the contents of alanine and glutamine decreased in both the genotypes. On the other hand, GABA declined only in the tolerant one, whereas aspartate levels increased in the sensitive genotype.

Pinitol ← ← Myo-Inositol ← ← Glc-6P ← Glucose   Fructose

Sucrose

P-hydroxyruvate ← 3--PGA   PEP

P-serine   Asparagine   Pyruvate →   Alanine

P-ethanolamine   Acetyl-CoA

Aspartate   Oxalacetic A   Citric A

Phosphatidylcholine   Malic A   Isocitric A   Proline

Choline   Fumaric A   2-Oxoglutaric A   Glutamate

Succinic A   Succinyl-Coa   Glutamine

GABA ←   Putrescine ← Ornithine   Urea Cycle   Arginine

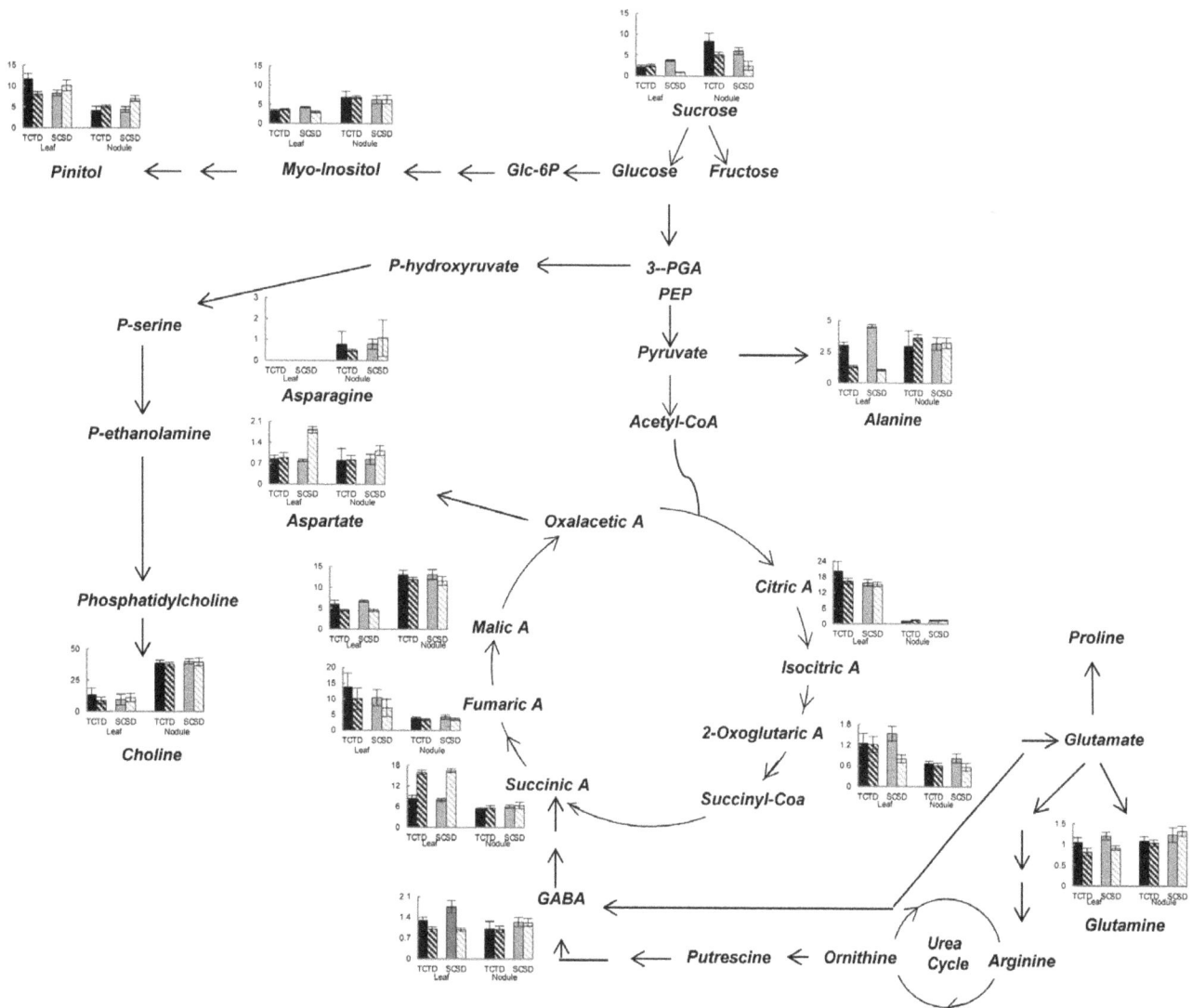

Figure 3. Schematic representation of the selected metabolic pathways affected by drought in two soybean genotypes contrasting in sensitivity/tolerance to water stress. Histograms represent relative changes in the level of the metabolites (arbitrary units) in trifoliate leaves and nodules in the plants subjected to water stress. Values are presented as the mean ± standard deviation (SD) of nine independent biological determinations. TC: Tolerant control; TD: Tolerant Drought, SC: Sensitive Control; SD: Sensitive Drought.

In contrast, no significant differences were observed in the amino acid contents in the nodules of both cultivars under control and stress conditions (Figure 3 and Table 4).

Principal component analysis (PCA) is one of the most popular explorative methods used to reduce multivariate data complexity. This is a method of choice for identifying patterns, and expressing data in ways that highlight similarities and differences between samples [17]. In our study, PCA was applied to ^{1}H NMR spectral data of control and stressed leaves and nodules derived from two soybeans genotypes with varied tolerance to drought, in order to authenticate the differences between the metabolic profiles of the control and stressed tissues statistically and to identify the main metabolites responsible for the differences.

A scores scatter plot of the first two PCs obtained considering all ^{1}H NMR data derived from the leaves shows a good separation of all four groups (LTC: leaf tolerant control, LTD: leaf tolerant drought, LSC: leaf sensitive control and LSD: leaf sensitive drought) along PC1 axis (Figure 4a). It seems that this separation is due to the treatment (control vs stressed) with further separation between stressed sensitive and stressed tolerant genotypes. The greatest separation along PC1 is between LSC and LSD groups whereas the separation between LTC and LTD is less apparent along PC1 axis, but noticeable along PC2 (see arrows on Figure 4a). This behaviour of the data evidences a markedly more profound metabolic impact of drought stress on sensitive plants with respect to tolerant ones. The separation between leaf samples of well watered and stressed plants along PC1 axis seemed to be mainly attributable to aspartate, succinic acid, sucrose, malic acid, alanine, GABA, myo-inositol and 2-oxoglutaric acid as shown in the complementary PCA loading plot (Figure 4b). In case of the tolerant genotype, leaf samples of control and stressed plants are well separated along PC2 due to the metabolites pinitol, citric acid, choline, and allantoin.

A comparison of metabolite mean levels between LSC and LSD samples (sensitive plants) and between LTC and LTD ones (tolerant plants) was performed using ANOVA (Table 3). ANOVA

Table 2. Two factors ANOVA with a 2×2 between groups design (drought tolerant vs. sensitive plants, well-watered vs. drought stressed plants), leaves samples.

Metabolites	Control vs Stressed		Tolerant vs Sensitive		Interaction	
	F	p-level	F	p-level	F	p-level
Ala	2207.38	0.00E+00	121.02	2.09E-12	274.79	2.89E-17
Gln	71.66	1.13E-09	17.84	1.86E-04	0.96	3.36E-01
Asp	199.14	2.73E-15	134.92	5.15E-13	171.96	2.04E-14
GABA	159.3	5.73E-14	27.94	8.64E-06	34.04	1.75E-06
MA	112.1	5.50E-12	3.99	5.45E-02	3.93	5.59E-02
CI	9.21	4.75E-03	14.66	5.65E-04	4.64	3.88E-02
SU	1588.19	7.58E-29	0.24	6.27E-01	5.45	2.59E-02
OX	27.56	9.60E-06	0.92	3.45E-01	23.12	3.46E-05
FU	9.38	4.43E-03	7.62	9.47E-03	0.03	8.54E-01
MI	31.17	3.64E-06	1.56	2.20E-01	94.60	4.45E-11
PI	6.77	1.39E-02	4.21	4.84E-02	62.40	5.15E-09
SUCR	181.79	9.59E-15	0.03	8.72E-01	295.46	1.02E-17
CH	1.48	2.32E-01	0.33	5.71E-01	5.20	2.93E-02
AL	3.07	8.91E-02	49.18	6.00E-08	121.49	1.99E-12
Trig	11.57	1.82E-03	9.06	5.06E-03	0.01	9.07E-01

results confirm the observations obtained with PCA. In fact, the levels of 11 out of 15 metabolites was significantly changed in sensitive plants upon the application of drought stress, whereas lesser number of metabolites (7 out of 15) were influenced by the stress in tolerant plants. Considering drought treatment and genotype as two independant factors and possible interaction between them, ANOVA has been applied using a 2×2 between

group design (Table 2). This approach was aimed to give a statistical measure of significance for each factor and interaction between them for each variable. The criterion of statistically significant difference between the mean values was p-level less than 0.01. For 8 variables (alanine, aspartate, GABA, 2-oxoglutaric acid, myo-inositol, pinitol, sucrose, allantoin) out of 15 the interaction between two factors was found to be statistically

Table 3. ANOVA on single groups, two types of grouping (control vs stressed, and tollerant vs sensitive) leave samples.

	Control vs Stressed				Tolerant vs Sensitive			
	Tolerant		Sensitive		Control		Stressed	
Metabolite	LTC vs LTD		LSC vs LSD		LTC vs LSC		LTD vs LSD	
	F	p-level	F	p-level	F	p-level	F	p-level
Ala	291.72	1.10E-11	4862.95	2.60E-21	220.19	9.00E-11	56.94	1.20E-06
Gln	20.73	0.00033	68.84	3.40E-07	10.03	0.00598	8.08	0.01175
Asp	0.34	0.56602	675.71	1.60E-14	2.07	0.16953	209.71	*1.30E-10
GABA	36.97	1.60E-05	123.69	6.10E-09	36.95	1.60E-05	0.46	0.50727
MA	22.61	0.00022	217.86	9.70E-11	4.69	0.04581	0.00	0.98729
CI	7.57	0.01422	1.76	0.20349	10.02	0.00599	6.56	0.02094
SU	477.46	2.40E-13	1691.62	1.20E-17	1.50	0.23907	4.63	0.04702
Ox	0.07	0.79541	84.41	8.80E-08	5.65	0.03023	24.13	0.00016
FU	3.82	0.06824	6.67	0.02006	3.82	0.06824	3.83	0.06796
MI	7.55	0.01432	135.87	3.10E-09	67.38	4.00E-07	32.48	3.30E-05
PI	56.65	1.20E-06	13.67	0.00195	46.31	4.20E-06	18.37	0.00057
SUCR	3.84	0.06779	2236.63	1.30E-18	108.64	1.50E-08	226.12	7.40E-11
CH	5.85	0.02785	0.59	0.45256	2.98	0.1037	2.31	0.14815
AL	66.89	4.20E-07	55.08	1.40E-06	8.09	0.01171	161.52	8.90E-10
TRIG	4.41	0.05201	7.97	0.01223	3.87	0.06674	5.32	0.03473

Table 4. Two factors ANOVA with a 2×2 between groups design (drought tolerant vs. sensitive plants, well-watered vs. drought stressed plants), nodules samples.

Metabolite	Control vs Stressed		Tolerant vs Sensitive		Interaction	
	F	p-level	F	p-level	F	p-level
Ala	2.70	1.12E-01	0.27	6.05E-01	2.02	1.67E-01
Gln	0.37	5.51E-01	18.12	2.38E-04	1.79	1.92E-01
Asp	3.81	6.16E-02	4.38	4.63E-02	2.74	1.10E-01
GABA	0.03	8.74E-01	15.28	5.92E-04	0.02	8.80E-01
MA	12.52	1.54E-03	0.25	6.22E-01	0.39	5.38E-01
CI	22.78	6.13E-05	1.72	2.01E-01	4.62	4.11E-02
SU	3.86	6.04E-02	7.80	9.69E-03	0.05	8.27E-01
OX	16.31	4.22E-04	2.63	1.17E-01	8.04	8.74E-03
FU	10.30	3.52E-03	2.39	1.34E-01	0.84	3.67E-01
MI	0.03	8.57E-01	2.07	1.62E-01	0.00	9.65E-01
PI	52.88	1.01E-07	20.71	1.10E-04	10.04	3.90E-03
SUCR	52.81	1.02E-07	27.54	1.75E-05	0.00	9.49E-01
CH	0.44	5.14E-01	3.71	6.51E-02	0.12	7.37E-01
AL	4.32	4.77E-02	23.31	5.29E-05	1.92	1.78E-01
Asn	0.01	9.42E-01	2.27	1.44E-01	2.12	1.57E-01

significant, evidencing different responses of sensitive and tolerant plants to the drought stress on the molecular level. In fact, for aspartate, 2-oxoglutaric acid, myo-inositol, pinitol, sucrose and allantoin the trends of changes upon the application of drought stress is opposite in tolerant and sensitive plants. For example, we can see that in tolerant plant samples, the level of pinitol is higher in control than in water stressed plants, while this is reversed when the sensitive plants were subjected to water stress. It is seems that the idea of considering treatment and genotype as independant factors is not adequate, as the stress produces different results in tolerant and sensitive samples.

In nodules, the PCA analysis (Figure 5a) showed that the first two PCs represented 48.2% of the initial variability contained in the original data. The scores plot exhibited separation between all four groups (NSC: Nodule Sensitive Control, NSD: Nodule Sensitive Drought, NTC: Nodule Tolerant Control and NTD: Nodule Tolerant Drought) when PC2 and PC1 were used as variables. It seems that with a few exceptions, the samples of tolerant plants are separated from sensitive ones along PC1, while control samples are separated from stressed ones along PC2. Plot of loadings (Figure 5b) show the variables responsible for this separation. The metabolites sucrose, aspartate, glutamine, GABA, allantoin, and succinic acid play a crucial role in the separation of tolerant from sensitive samples. On the other hand, the separation of controls from the stressed samples is due to the variations in the levels of malic acid, 2-oxoglutaric acid, fumaric acid, and sucrose.

The ANOVA analysis (Table 4) confirmed the same variables (that were identified by PCA) as statistically significant for the separation of groups. In addition to this, ANOVA revealed that the level of pinitol and citric acid is significantly different in control and stressed nodule tissues. It is notewothy, that only in the case of two metabolites (pinitol and 2-oxoglutaric acid) the interaction between genotype and drought treatment was significant.

Discussion

In plants, the level of tolerance or sensitivity to water stress depends on the species and genotype, length and severity of water loss, as well as on the developmental stage. In *Aeluropus lagopoide*, Mohsenzadeh S, *et al* [18] found significant correlation between leaf relative water content (RWC) and relative growth rate, net photosynthesis rate, chlorophyll and proline contents. The leaf RWC directly reflects the water status of plants and it can be used to identify the genotypes tolerant to stress [19]. In our study, the highest level of RWC was found in the cv. NA5009RG (Figure 1b), and according to Rampino *et al.* [19] this cultivar could be defined as a genotype tolerant to water deficit. In addition, even under water limitation, this variety showed only a marginal reduction in leaf RWC as compared to the sensitive genotype (DM50048) where RWC was significantly affected. Moreover, our results showed a notable reduction in leaf DW only in the sensitive genotype (Figure 1b-a). The decrease in DW under drought in the sensitive soybean variety might be related to the depletion of sucrose in the leaves of this genotype (Figure 3). These results are in agreement with the findings of Reddy *et al.* [20] who reported that water stress inhibited dry mater production due to limitation of photosynthesis. Differences in the reduction in leaf DW under water stress has also been reported for different legume species: 78% in mungbean, 60 % in cowpea and 37% in peanut compared with the unstressed plants [21].

Despite of the reduction in the RWC and dry weight of the leaves of the sensitive genotype under drought, no parallel decrease in chlorophyll content was observed (data not shown). This result is in agreement with Ashraf and Iram [12] who also observed a lack of effect of drought on chlorophyll content, and suggested that it could be due to the mild moisture stress to which the experimental plants were exposed.

Enhanced tolerance of plants to low water availability is attributed to the accumulation of soluble sugars in water-stressed tissues [22], acting as osmoprotectants [23,24]. In contrast to this, the results of the present study with the twenty seven-day old plants showed no enhanced accumulation of soluble sugars such as sucrose and myo-inositol in the leaves of both genotypes (Figure 3; indeed sugar content decreased in the sensitive variety), indicating that these sugars do not play an osmoprotectant role at least at the early stages of the plant growth. Similar findings have been reported for other legume species exposed to osmotic stress [25].

In legumes, pinitol is a common sugar alcohol and it has been described as a common osmoprotectant [11,26]. Our results showed that the tolerant genotype has higher amounts of pinitol even under normal conditions as compared to the sensitive variety. On the other hand, pinitol synthesis was found to be enhanced in the sensitive genotype under water stress. Future work can only establish the exact role of pinitol in osmoprotection of soybean plants.

Accumulation of amino acids was suggested to aid stress tolerance in plants, through osmotic adjustment, detoxification of reactive oxygen species and by intracellular pH regulation [27,28]. An equivalent role for most of the amino acids, detected in our experiments, seems unlikely because their content, with the exception of aspartic acid, tended to decrease or remain constant even under drought in both genotypes (Figure 3). These results are in agreement with the findings in *Phaseolus vulgaris* [29].

Proline, an imino acid, is widely regarded as a main osmoprotectant in water stress tolerance in plants. In the present study with the plants at the vegetative stage, we found that water stress does not trigger enhanced proline synthesis (Figure 1d, h). In contrast, when the drought treatment was imposed during the

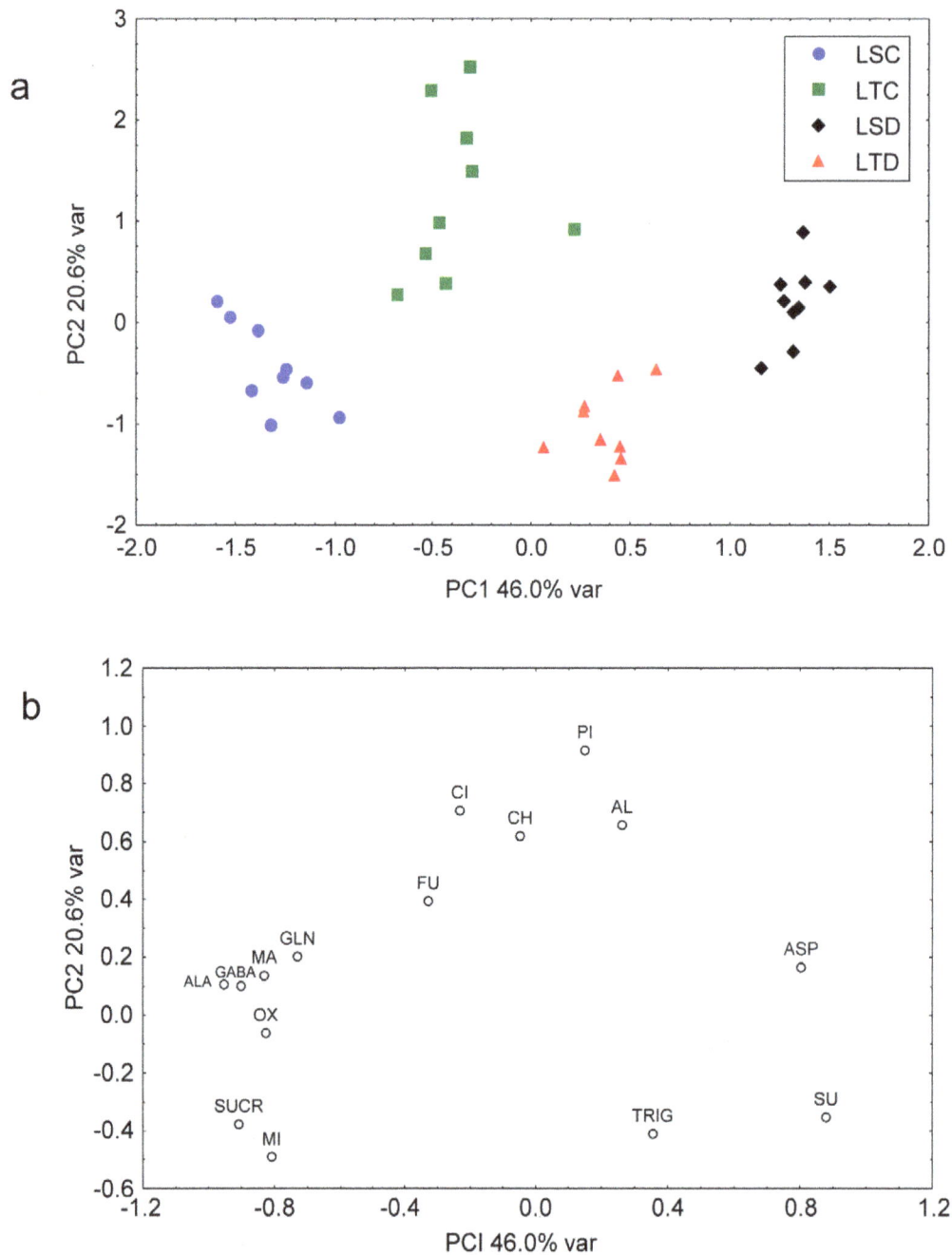

Figure 4. Principal component analysis (PCA) of 15 metabolites in the leaves from the plants grown under water sufficient and deficient conditions. Score (a) and loading plot (b) of soybean leaf samples. LTC: Leaf Tolerant Control, LTD: Leaf Tolerant Drought, LSC: Leaf Sensitive Control, LSD: Leaf Sensitive Drought.

flowering stage, there was a considerable increase in proline levels in the leaves and nodules of both genotypes (data not shown). Increases in proline level were also observed in others varieties of soybean, when drought was imposed at the reproductive stage and the RWC was lower than the values observed in the present work [30,31]. Fukutoku and Yamato [32] reported that in intact soybean leaves remarkable proline accumulation occurred only when water stress became severe and protein metabolism was disturbed. These results suggest that even under water stress, the stage of the plant and RWC of the leaves seem to be critical in promoting proline synthesis. Role of proline in stress tolerance

remains controversial as some authors have reported high proline levels in the susceptible cultivars subjected to stress conditions [33,34], while the others have observed the opposite trend [35]. It has been suggested that proline functions as an indicator of plant water status but not a measure of level of tolerance [36].

In both soybean cultivars, drought did not trigger the accumulation of organic acids excepting succinate as its concentration was doubled in the leaves of both genotypes after the imposition of drought. Sassi *et al.* [29] reported a decrease in the total amount of organic acids in the leaves of a sensitive line of bean plants subjected to osmotic stress.

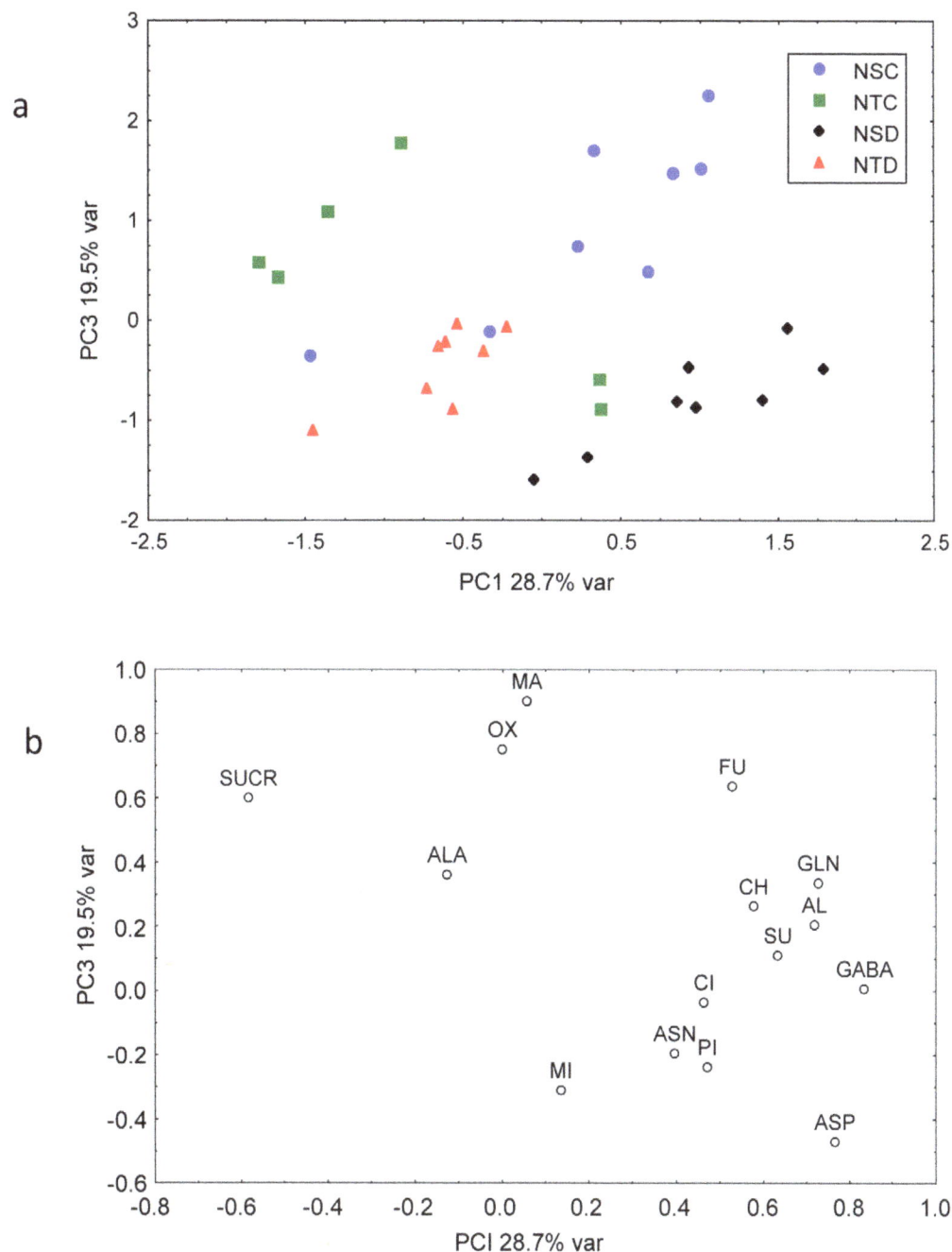

Figure 5. Principal component analysis (PCA) of 15 metabolites in nodules from the plants subjected to water sufficient and deficient conditions. Score (a) and loading plot (b) of soybean nodule samples. NTC: Nodule Tolerant Control, NTD: Nodule Tolerant Drought, NSC: Nodule Sensitive Control, NSD: Nodule Sensitive Drought.

Present study demonstrated that water stress has varied effects on the metabolic processes of leaves and nodules in the soybean cultivars tested; Nodule DW was more affected than the leaf, and in the nodules of both genotypes the decrease in DW mirrored the drought-induced decline in sucrose. However, drought didn't affect the nitrogenase activity. The lack of response of nitrogenase activity to drought in these soybeans cultivars is in contrast to earlier reports [37,38,7], and it perhaps reflects the level of intensity of the stress imposed in various studies. In several soybean cultivars, inhibition of nitrogen fixation under drought stress has been attributed to ureide accumulation in leaves [39]. In

the present investigation, ureide accumulation in response to drought stress was observed only in the leaves of the sensitive genotype, but not in the tolerant variety (Figure 1). Since the nitrogenase activity was unaffected under drought stress in both the genotypes, it seems that ureide accumulation in the leaves does not have a feedback inhibitory effect on nitrogen fixation. Alamillo *et al.* [40] suggested that ureide accumulation and nitrogen fixation follow different kinetics and are probably not causally related.

In nodules, malic acid is the most abundant organic acid and is the main carbon substrate for bacteroid respiration and nitrogen fixation activity. It had been suggested that a decrease in nodule malic acid content under certain environmental conditions may lead to the inhibition of nitrogen fixation [3]. Also in the present work, genotype differences in malic acid content or interactions between genotypes and drought were found to be absent (Figure 3), a situation that is consistent with the results of ARA activity under stress condition (Figure 1).

The analysis of metabolites contributes to the understanding of stress biology of plants through the identification of the compounds and the part they play in acclimation or tolerance response. In the present study, metabolite fingerprinting and profiling based on ^{1}H NMR spectra were used to analyze the similarities and differences among leaf and nodule samples obtained from two soybean genotypes with the aim of identifying markers useful for pinpointing water stress response. In this context the results of our study point to six metabolites in leaves (aspartate, 2-oxoglutaric acid, myo-inositol, pinitol, sucrose, allantoin) and two in nodule (2-oxoglutaric acid and pinitol) that were affected differentially in the genotypes when drought was imposed at the vegetative stage in the nodulated soybean plants. These data provide information that may, with further experimentation, allow elucidation of biochemical pathway underlying stress tolerance in soybean.

The results of the study demonstrated that a combination of ^{1}H NMR and multivariate analyses allows comparisons of overall metabolite fingerprints and that this technique can be applied to conclusively identify differences that are due to stress or genotype. The differences under stress conditions between the two genotypes discussed above are reflected in the PCA models of metabolite content as well. PCA of the present study clearly demonstrated that the major variability in metabolites levels (associated with PC1 in PCA) is due to treatment (control vs stressed) in the case of leaves, while in the case of nodules the major variability is due to genetic makeup (tolerant vs sensitive). The phenomena observed in the case of nodules likely depends on the cumulative effects of plant-bacterial genotypes, specificity of interaction and the resultant symbiosis which in turn alters the metabolism outcome.

Materials and Methods

Plant materials and growth conditions

Seeds of the soybean genotypes (*Glycine max* L. Merr.), namely NA5009RG (drought tolerant) and DM50048 (drought sensitive) were obtained from CIAP-INTA (Centro de Investigación Agropecuaria-Instituto Nacional de Tecnología Agropecuaria), Argentina. Seeds were surface sterilized with 20% (v/v) commercial bleach, washed extensively with sterile-distilled water, and germinated on a sterile moist filter paper at 28°C, in the dark, for 3 days. Subsequently, the seedlings were transferred to vermiculite in pots and inoculated with 1 cm^3 of *Bradyrhizobium japonicum* USDA110, and grown in a greenhouse at 26°C/19°C (day/night temperatures). Plants received nitrogen-free Summerfield nutrient solution [41] twice a week until the stress treatments were imposed.

Drought stress was imposed on 17-day old plants at the vegetative stage by withholding water supply for 10 days until soil water content reached 23% (0.230 g H$_2$O g^{-1} dry soil). A set of well-watered plants served as a control. Measurements of dry weight, relative water content and nitrogen fixation, as well as sampling of the plant tissues (nodules and first trifoliate leaves) for chlorophyll, proline, ureide and metabolome analysis were performed at the end of the stress period.

The results presented are the means with standard deviations of five to 15 replicates. All data obtained was subjected to one-way analysis of variance (ANOVA), and the mean differences were compared by lowest standard deviations (LSD) test using the STATISTICA package for Windows (version 5.1, 1997). Comparisons with P values<0.01 were considered significantly different.

Leaf relative water content (RWC)

RWC was measured according to Barrs and Weatherly [42]. Briefly, immediately after sampling, the leaves were weighted and then soaked overnight in distilled water at 4°C. After the cold incubation, the leaves were blotted dry and weighed prior to oven-drying at 80°C for 48 h. Subsequently, dry weight of the plant samples was determined. The leaf relative water content was calculated using the following formula: RWC = (FW − DW)/ (TW-DW)) ×100, where FW is fresh weight, DW is dry weight, and TW is turgid weight (weight after the leaf was kept in distilled water for overnight).

Nitrogen fixation

Nitrogenase activity was determined by acetylene reduction assay (ARA), [43].

Estimation of ureide content

Concentration of ureides present in leaf and nodule cell-free extracts was measured using the colorimetric detection method of Vogels and Van Der Drift [44]. Allantoic acid dissolved at a concentration of 10 mM in water served as a standard for ureide estimation.

Determination of proline

Samples of fresh plant tissues (0.5 g) were homogenized in 5 ml of 3% aqueous sulphosalicyclic acid and supernatant was collected after centrifugation. Two mL extract was reacted with 2 mL acid-nihydrin and 2 mL glacial acetic acid, and incubated for 1 h in a boiling water bath. The reaction was terminated by placing the test tubes on ice after which the reaction mixture was vigorously mixed with 2 ml of toluene. After warming to 25°C, proline present in the upper toluene layer was measured at 520 nm [45].

Metabolite extraction and NMR analysis

Comparative metabolite profiling was performed in leaves and nodules derived from the drought tolerant and sensitive soybean genotypes subjected to water stress. First trifoliate expanded leaves and nodules were collected after 10 days of drought treatment, and immediately frozen in liquid nitrogen, lyophilized and stored at −80°C till they were subjected to NMR analysis.

Water-soluble extracts were derived from 20 mg of lyophilized tissues mixed with 0.9 mL of CH$_3$CN:H$_2$O (1:1 v/v); Extracts were clarified by centrifugation at 10,000 rpm (5,600 g) for 7 min, and the supernatant obtained was filtered and lyophilized. The dry residue was dissolved in 0.7 mL of 400 mM D$_2$O phosphate buffer (pD = 6.5) containing 1.0 mM of 3-(trimethylsilyl) propionic−2,2,3,3-d4 acid sodium salt (TSPA) and transferred into a standard 5 mm NMR tube. NMR spectra of the extracts were recorded at 300 K on a Bruker AVANCE AQS600 spectrometer operating at the proton frequency of 600.13 MHz and equipped with a Bruker multinuclear z-gradient inverse probe-head capable of producing gradients in the z-direction with the strength of 55.4 G/cm.

Proton spectra were referenced to the signals of TSPA methyl group at δ = 0.00 ppm in D$_2$O phosphate buffer. The ^{1}H spectra of the aqueous extracts were acquired by co-adding 512 transients with a recycle delay of 2.5 s and 32 K data points (acquisition time 40 min). The residual HDO signal was suppressed using a presaturation during the relaxation delay with a long single soft pulse. To avoid possible saturation effects, the experiment was carried out by using a 45° flip angle pulse of 8.0 μs 2D NMR experiments, namely ^{1}H–^{1}H TOCSY and ^{1}H–^{13}C HSQC, were performed using the same experimental conditions as previously reported by Sobolev et al [46].

Fifteen metabolites in leaves and nodules extracts were identified and used for statistical analysis, see Tables 1 and 2. Metabolites were assigned and identified using 2D experiments ^{1}H–^{1}H TOCSY and ^{1}H–^{13}C HSQC and by comparison with the literature data [47].

The signal heights of selected ^{1}H resonances of water-soluble metabolites (Table 1) were measured with respect to the height of TSPA signal used as internal standard. The height of TSPA signal was normalized to 100. The obtained values (relative molecular abundances of selected metabolites) were used in statistical analysis.

Statistical Analysis of NMR data

The statistical treatment of the NMR data was performed using the STATISTICA package for Windows (version 5.1, 1997). Two

factors ANOVA has been performed with a 2×2 between groups design (drought tolerant vs. sensitive plants, well-watered vs. drought stressed plants). Principal component analysis (PCA) was performed using all 15 variables for leaves and nodules. Before the PCA analysis the variables were mean-centered and each variable was divided by its standard deviation (autoscaling). The effects and interactions represented in bold in the Tables 2,3,4 were statistically significant within the 99% confidence interval.

Acknowledgments

We would like to thank PM Reddy for critically reading the manuscript, Nestor Mariano for his valuable help in statistical analysis and Celina Luna (CIAP-INTA (Centro de Investigaciones Agropecuarias – Instituto Nacional de Tecnología Agropecuaria)) for his kind gift of soybean seeds.

Author Contributions

Conceived and designed the experiments: SS APS. Performed the experiments: SS APS. Analyzed the data: SS APS. Contributed reagents/materials/analysis tools: SS APS. Wrote the paper: SS APS ML.

References

1. Pimentel D, Patzek TW (2008) Ethanol production using corn, switch grass and wood; biodiesel production using soybean. In Pimentel, D. (ed.), Biofuels, Solar and Wind as Renewable Energy Systems: Benefits and Risks. Springer, Dordrecht, The Netherlands, 375–396.

2. Sprent JI (1972) The effects of water stress on nitrogen fixing root nodules. IV Effects on whole plants of Vicia faba and Glicine max. New Phytol 71: 603–611.

3. Arrese-Igor C, Gonzalez EM, Gordon AJ, Minchin FR, Galvez L, et al. (1999) Sucrose synthatse and nodule nitrogen fixation under drought and other environmental stresses. Symbiosis 27: 189–212.

4. Neo HH, Layzell DB (1997) Phloem glutamine and the regulation of O_2 diffusion in legume nodules. Plant Physiol 113: 259–267.

5. Serraj R, Vadez V, Sinclair TR (2001) Feedback regulation of symbiotic N_2 fixation under drought stress. Agronomie 21: 621–626.

6. King CA, Purcell LC (2005) Inhibition of N_2 fixation in soybean is associated with elevated ureides and amino acids. Plant Physiol 137: 1389–1396.

7. Marino D, Frendo P, Ladrera R, Zabalza A, Puppo A, et al. (2007) Nitrogen fixation control under drought stress. Localized or systemic? Plant Physiol 143: 1968–1974.

8. Samarah N, Mullen R, Cianzio S (2004) Size distribution and mineral nutrients of soybean seeds in response to drought stress. J Plant Nutr 275: 815–835.

9. Hare PD, Cress WA, Van Staden J (1998) Dissecting the roles of osmolyte accumulation during stress. Plant Cell Environ 21: 535–553.

10. Conroy MJ, Goldsberry JR, Hines JE, Stotts DB (1988) Evaluation of aerial transects surveys for wintering American black ducks. Journal of Wildlife Management 52: 694–703.

11. Ford CW (1984) Accumulation of low molecular weight solutes in water stressed tropical legumes. Phytochem 23: 1007–1015.

12. Ashraf MY, Iram A (2005) Drought stress induced changes in some organic substances in nodules and other plant parts of two potential legumes differing in salt tolerance. Flora 200: 535–546.

13. Bligny R, Douce R (2001) NMR and plant metabolism. Curr Opin Plant Biol 4: 191–196.

14. Charlton AJ, Donarski JA, Harrison M, Jones SA, Godward J, et al. (2008) Responses of the pea (Pisum sativum L.) leaf metabolome to drought stress assessed by nuclear magnetic resonance spectroscopy. Metabolomics 4: 312–327.

15. Fiehn O (2002) Metabolomics – the link between genotypes and phenotypes. Plant Mol Biol 48: 155–171.

16. Colquhoun IJ (2007) Use of NMR for metabolic profiling in plant systems. J Pestic Sci 32: 200–212.

17. Goodacre R, Shann B, Gilbert RJ, Timmins EM, McGovern AC, et al. (2000) Detection of the dipicolinic acid biomarker in Bacillus spores using Curie-point pyrolysis mass spectrometry and Fourier transform infrared spectroscopy. Anal Chem 72: 119–127.

18. Mohsenzadeh S, Malboobi MA, Razavi K, Farrahi-Aschtiani S Razavi K (2006) Physiological and molecular responses of Aeluropus lagopoides (Poaceae) to water deficit. Environ Exp Bot 56: 314–322.

19. Rampino P, Pataleo S, Gerardi C, Perotta C (2006) Drought stress responses in wheat: Physiological and molecular analysis of resistant and sensitive genotypes. Plant Cell Envir 29: 2143–2152.

20. Reddy AR, Chaitanya KV, Vivekanandan M (2004) Drought-induced responses of photosynthesis and antioxidant metabolism in higher plants. J Plant Physiol 161: 1189–1202.

21. Pandey RK, Herrera WAT, Villepas AN, Pendelton JW (1984) Drought response grain legumes under irrigation gradient. III. Plant Growth Agron J 76: 557–560.

22. McManus MT, Bieleski RL, Caradus JR, Barker DJ (2000) Pinitol accumulation in mature leaves of white clover in response to a water deficit. Env Exp Botany 43: 11–18.

23. Ingram J, Bartels D (1996) The molecular basis of dehydration tolerance in plants. Annu Rev Plant Physiol Plant Mol Biol 47: 377–403.

24. Sanchez FJ, Manzanares M, Andres EF, Tenorio JL, Ayerbe L (1998) Turgor maintenance, osmotic adjustment and soluble sugar and proline accumulation in 49 pea cultivars in response to water stress. Field Crops Res 59: 225–235.

25. Pinheiro C, Passarinho JA, Ricardo CP (2004) Effect of drought and rewatering on the metabolism of Lupinus albus organs. J Plant Physiol 161: 1203–1210.

26. Keller R, Ludlow M (1993) Carbohydrate metabolism in drought-stressed leaves of pigeonpea (Cajanus cajan). J Exp Bot 44: 1351–1359.

27. De Ronde JA, Van Der Mescht A, Steyn HSF (2000) Proline accumulation in response to drought and heat stress in cotton. Afr Crop Sci J 8: 85–92.

28. Alia MP, Matysik J (2001) Effect of proline on the production of singlet oxygen. Amino Acids 21: 195–200.

29. Sassi S, Aydi S, Hessini K, Gonzalez EM, Arrese-Igor C, et al. (2010) Long-term mannitol-induced osmotic stress leads to stomatal closure, carbohydrate accumulation and changes in leaf elasticity in Phaseolus vulgaris leaves. Afr J Biotechnol 9: 6061–6069.

30. Angra S, Kaur S, Singh K, Pathania D, Kaur N, et al. (2010) Water-deficit stress during seed filling in contrasting soybean genotypes: Association of stress sensitivity with profiles of osmolytes and antioxidants. Int Agric Res 5: 328–345.

31. Masoumi H, Darvish F, Daneshian J, Nourmohammadi G, Habibi D (2011) Chemical and biochemical responses of soybean (Glycine max L.) cultivars to water deficit stress. Aust J Crop Sci 5: 544–553.

32. Fukutoku Y, Yamada Y (1981) Sources of Proline-nitrogen in Water-stressed Soybean (Glycine max L.) I. Protein Metabolism and Proline Accumulation. Plant Cell Physiol 22: 1387–1404.

33. Premachandra GS, Hahn DT, Rhodes D, Joly RJ (1995) Leaf water relations and solute accumulations in two grain sorghum lines exhibiting contrasting drought tolerance. J Exp Bot 46: 1833–1841.

34. Sundaresan S, Sudhakaran PR (1995) Water stress-induced alternations in the proline metabolism of drought-susceptible and tolerant cassava (Manihot esculenta) cultivars. Physiol Plant 94: 635–642.

35. Hien DT, Jacobs M, Angenon G, Hermans C, Thu TT, et al. (2003) Proline accumulation and Δ1-pyrroline-5-carboxylase syhthetase gene properties in three rice cultivars differing in salinity and drought tolerance. Plant Sci 165: 1059–1068.

36. Lazcano-Ferrat I, Lovatt CJ (1999) Relationship between relative water content, nitrogen pools, and growth of Phaseolus vulgaris L. and P. acutifolius A. Gray during water deficit. Crop Sci 39: 467–475.

37. González EM, Aparicio-Tejo PM, Gordon AJ, Minchin FR, Royuela M, et al. (1998) Water-deficit effects on carbon and nitrogen metabolism of pea nodules. J Exp Bot 49: 1705–1714.

38. Gálvez L, Gónzalez EM, Arrese-Igor C (2005) Evidence for carbon flux shortage and strong carbon/nitrogen interactions in pea nodules at early stages of water stress. J Exp Bot 56: 2551–2561.

39. Serraj R, Sinclair TR, Purcell LC (1999) Symbiotic N_2 fixation response to drought. J Exp Bot 50: 143–155.

40. Alamillo JM, Díaz-Leal JL, Sánchez-Moran MV, Pineda M (2010) Molecular analysis of ureide accumulation under drought stress in *Phaseolus vulgaris* L. Plant Cell Environ 33: 1828–1837.

41. Summerfield RJ, Huxley PA, Minchin F (1977) Plant husbandry and management techniques for growing grain legumes under simulated tropical conditions in controlled environments. Exp Agr 13: 81–92.

42. Barrs HD, Weatherley PE (1962) A re-examination of the relative turgidity technique for estimating water deficits in leaves. Aust J Biol Sci 15: 413–428.

43. Hardy RWF, Holsten RD, Jackson EK, Burns RC (1968) The acetylene-ethylene assay for N_2 fixation: Laboratory and field evaluation. Plant Physiol 43: 1185–1207.

44. Vogels GD, Van Der Drift C (1970) Differential analyses of glyoxylate derivatives. Anal Biochem 33: 143–157.

45. Bates LS, Waldren RP, Teare ID (1973) Rapid determination of free proline for water-stress studies. Plant Soil 39: 205–207.

46. Sobolev AP, Segre A, Lamanna R (2003) Proton high-field NMR study of tomato juice. Magn Reson Chem 41: 237–245.

47. Fan TWM (1996) Metabolite profiling by one- and two-dimensional NMR analysis of complex mixtures. Prog Nucl Magn Reson Spectrosc 28: 161–219.

19

Plant Functional Group Composition Modifies the Effects of Precipitation Change on Grassland Ecosystem Function

Ellen L. Fry[1,2]*, Pete Manning[1,3], David G. P. Allen[1], Alex Hurst[1], Georg Everwand[1,4], Martin Rimmler[1,5], Sally A. Power[1,6]

1 Department of Life Sciences, Imperial College London, Ascot, Berkshire, United Kingdom, 2 Grantham Institute for Climate Change, Imperial College London, London, United Kingdom, 3 NERC (Natural Environment Research Council) Centre for Population Biology, Imperial College London, Ascot, United Kingdom, 4 Agroecology, University of Göttingen, Göttingen, Germany, 5 Department of Ecological Modelling, Bayreuth University, Bayreuth, Germany, 6 Hawkesbury Institute for the Environment, University of Western Sydney, Penrith, New South Wales, Australia

Abstract

Temperate grassland ecosystems face a future of precipitation change, which can alter community composition and ecosystem functions through reduced soil moisture and waterlogging. There is evidence that functionally diverse plant communities contain a wider range of water use and resource capture strategies, resulting in greater resistance of ecosystem function to precipitation change. To investigate this interaction between composition and precipitation change we performed a field experiment for three years in successional grassland in southern England. This consisted of two treatments. The first, precipitation change, simulated end of century predictions, and consisted of a summer drought phase alongside winter rainfall addition. The second, functional group identity, divided the plant community into three groups based on their functional traits- broadly described as perennials, caespitose grasses and annuals- and removed these groups in a factorial design. Ecosystem functions related to C, N and water cycling were measured regularly. Effects of functional group identity were apparent, with the dominant trend being that process rates were higher under control conditions where a range of perennial species were present. E.g. litter decomposition rates were significantly higher in plots containing several perennial species, the group with the highest average leaf N content. Process rates were also very strongly affected by the precipitation change treatment when perennial plant species were dominant, but not where the community contained a high abundance of annual species and caespitose grasses. This contrasting response could be attributable to differing rooting patterns (shallower structures under annual plants, and deeper roots under perennials) and faster nutrient uptake in annuals compared to perennials. Our results indicate that precipitation change will have a smaller effect on key process rates in grasslands containing a range of perennial and annual species, and that maintaining the presence of key functional groups should be a crucial consideration in future grassland management.

Editor: James F. Cahill, University of Alberta, Canada

Funding: EF was supported by a PhD studentship funded by the Grantham Institute for Climate Change at Imperial College, URL http://www3.imperial.ac.uk/climatechange. The work was further supported by UK POPnet, Centre for Population Biology and the UK Big Lottery Fund. The funders had no role in study design, data collection and analysis, decision to publish, or preparation of the manuscript.

Competing Interests: The authors have declared that no competing interests exist.

* E-mail: ellen.l.fry@gmail.com

Introduction

Grasslands provide an important range of ecosystem services, including forage production and carbon storage [1], but are often managed for food production with little consideration for biodiversity conservation, resulting in widespread declines in their species richness [2]. They are also threatened by climate change, including changes in precipitation patterns. For example, in southern England summer rainfall is projected to decrease in volume but occur in more extreme downpours, with more severe interim droughts, accompanied by increased winter rainfall [3]. Because grasslands respond directly to the volume, frequency and duration of precipitation, such changes will affect their species composition, rates of nutrient and carbon cycling and water relations, and could see them shift from carbon sinks to sources [4],[5]. Additionally, extended periods of soil drying and wetting

can be detrimental to soil microbial communities. In severe cases this may lead to cell lysis and nutrient exudation, followed by leaching and reduced soil fertility. All these changes mean that climate change may ultimately result in further diversity loss in grassland communities [6],[7].

In the last twenty years, experiments that explore the interaction between precipitation change and plant functional diversity loss have demonstrated that species richness is positively correlated with drought resistance and rates of important ecosystem processes such as respiration and soil nutrient availability [8]–[][10], but the underlying causes of this relationship are not fully understood. Meanwhile, in other areas of ecological research there have been numerous attempts to discover which functional traits drive ecosystem functioning [11]–[15]. Currently the links between these two fields of enquiry are not explicit, but making them so may yield a greater understanding of how ecosystems respond to climate change. Some climate manipulation studies demonstrate long-term effects of climate change upon function, even after the removal of stresses (e.g. drought) [6],[16],[17], whilst in others, recovery is rapid and has few long term effects [7]. This variability in ecosystem resilience may be caused by differences in the functional trait distributions of the systems measured, with certain combinations of functional effects trait values (including rooting depth, relative growth rates and nutrient turnover rates) offering greater resistance and resilience to climate stresses [18],[19]. For example, ecosystems containing high plant functional trait diversity could exhibit smaller changes to ecosystem function in response to changes in rainfall patterns than less diverse assemblages. One reason for this is that trait diversity both above- and belowground is likely to offer a greater variety of plant resource and water capture strategies, and a greater diversity of rhizosphere niches for soil microbes [20],[21]. The probability of including traits that directly provide resistance or resilience to drought (e.g. traits conferring drought tolerance or rapid regeneration) is also increased in diverse communities [19],[22]. Furthermore, under altered abiotic conditions, hitherto subdom- inant species may increase in abundance and offer higher resilience to adverse conditions (the 'insurance effect') [23].

Differences in the response of species and functional groups to climate change are likely to ultimately lead to changes in ecosystem function. Dominant species (particularly perennial- dominated communities) are possibly more vulnerable to climate changes because their resources are allocated to maintaining competitive superiority over other species rather than to resisting environmental perturbations [23]–[25]. In contrast, annual species with their short life cycles, rapid growth and prolific reproductive output are potentially more resilientand able to recover from extreme weather events [15],[26]. These functional groups are also likely to differ in the way in which they influence ecosystem function, and so ecosystems in which they are rare or absent are likely to function differently under climate change. However, such relationships remain hypothetical.Therefore, we established an experiment that combined the manipulation of a precipitation regime (as opposed to a drought event) with a diversity manipulation based on functional groups with known functional traits. Most grassland climate change studies to date have focussed on primary productivity, so we addressed a knowledge gap by placing a greater emphasis on changes in plant species composi- tion, gas fluxes and nutrient cycling [27]–[29].

Functional identity was manipulated by selectively removing functional groupings of plant species to create a gradient of functional diversity. By classifying plant species into functional types based upon effects traits, and removing the groups in factorial combination, we aimed to investigate how the presence of certain trait suites canmodify the response ofecosystem function to an altered precipitation regime.

Methods

Study site

The experiment, which is known as DIRECT -DIversity, Rainfall and Elemental Cycling in a Terrestrial ecosystem- is located in south east England, in Silwood Park, Berkshire, UK $(0°35'W, 51°25'N)$. The site containsa lowland mesotrophic *Holcus mollis-Agrostis capillaris* grassland (EUNIS code E2, (European Nature Information System, http://eunis.eea.europa.eu)) on a loamy sand soil. There are no protected or threatened species present. It is surrounded by a rabbit-proof fence, although there is some roe deer (*Capreolus capreolus* L.) browsing and mole (*Talpa europaea* L.) activity. The climate is temperate: rainfall averaged

833 mm yr^{-1} between 2000 and 2010, and temperatures averaged 4.8°C in January and 17.2°C in July over the same period [30]. The field was ploughed in October 2007, which removed most standing biomass, and was left to regenerate naturally.

Experimental design

The experiment began in June 2008, when the roofs of the precipitation change treatment were first raised. Measures were taken regularly, from October 2008 to September 2010. The experiment had a factorial, randomised block design consisting of two levels of rainfall (precipitation change rainfall and control) combined with seven combinations of three plant functional trait group (present/absent). The latter comprised every possible combination except none present (bare earth). This generated a diversity gradient of 1–3 groups. Four blocks, each containing one replicate of each treatment combination, were arranged in a row from east to west, resulting in 56 plots. This blocking accounted for a shallow incline across the site (Figure S1). Each plot was 2.4 m×2.4 m, with a 70 cm buffer zone to account for lateral drift of rain; ecosystem function measures were taken in a central 1 m×1 m central area within each plot.

Precipitation change treatment

The rainfall treatment was based upon end of the century predictions from climate models using A2 scenarios from the IPCC 4th Assessment Report [31]; these project that by 2080– 2099 south-east England will experience a reduction of ~30% rainfall volume during the summer months (June, July, August; JJA) relative to the 1961–1990 baseline.These rainfall events are also likely to become less frequent, and concentrated into more intense downpours [3]. All 56 plots were covered with a rain shelter from June 1st to August 31st each year (Figure S2). The shelters were open sided, and covered with transparent corrugated Corolux PVC, 0.8 mm thick. All the rain was removed from the precipitation change plotsand collected in individual water butts. In the control plots, roofs had approximately 100 2.5 cm diameter holes to allow rainwater to pass through. In the precipitation change treatment, if less than 20 mm fell in 24 hours, 50% of the water was reapplied manually and the rest discarded. If more than 20 mm fell, the full amount was reapplied. Based on historic rainfall data for the site, this was estimated to approximate to a net reduction of 30% volume over the growing season. Projections for the winter (Dec, Jan, Feb; DJF) consist of a 10–15% volume increase for southern England, with frequency and intensity remaining approximately the same as at present. DJF rainfall treatments were applied by collecting control rainfall in weather- resistant water trays placed adjacent to each precipitation change treatment plot, with surface area of 15% of plot size (approx. 8640 cm^2). The water collected was reapplied to all precipitation change treatment plots after every rainfall event from December 1st to February 28th each year. The PVC roofs led to an overall reduction of 34% photosynthetically active radiation (PAR) in all plots, but this was the same in both treatments as the holes had little effect upon light transmission (Analysis of variance of comparable readings under control and precipitation change roofs $F_{1,53} = 0.79$, $p = 0.377$). There was a <1% increase in temperature under the shelters compared with outside but humidity was unaffected.

Functional group identity treatment

For the functional group identity treatment three plant trait groups were derived using a divisive hierarchical cluster analysis based on functional effects trait data. These data were obtained by growing all common grassland species from the local species pool to maturity in a greenhouse and measuring above and below-ground biomass (AGB and BGB respectively), leaf nitrogen (N) content (LNC), specific plant area (plant area/AGB) (SPA), photosynthetic rate (A) and evapotranspiration rate. Additional trait data on plant lifespan and N fixing capacity were obtained from the USDA plants database [32]. Relationships between greenhouse and field traits have been the subject of some contention, but there is now compelling evidence that the two are closely related [33]. Accordingly, greenhouse-derived trait measures can be used as a measure of relative differences between species trait values in field conditions with reasonable confidence. The cluster analysis was set to divide the species into three groups (Table S1). The first of these comprised perennial grasses, forbs and legumes (hereafter FG1), whose distinguishing traits included higher SPA, LNC and a more perennial growth habit than the other groups (Table 1), characteristics which are expected to result in faster ecosystem process rates and higher and more continuous net turnover of plant material [21],[34],[35]. The second group (FG2) consisted of caespitose grasses and two forbs, with very high AGB and BGB, and low LNC. Presence of this group is expected to result in large amounts of poor quality litter inputs to the soil. The third group (FG3) consisted of annual forbs, grasses and legumes, with low SPA and biomass but high LNC. Presence of these traits, coupled with their short lifespan, could potentially result in tolerance of environmental stress as well as rapid growth and recovery from drought. Strong seasonal trends in function were expected where FG3 is present with senesced material decomposing rapidly in autumn and high nutrient and CO_2 flux rates in the spring and early summer when germination and growth occur.

The three groups were combined into every possible combination(except for the absence of all) - three individual groups, three combinations of two, one combination of three. Plant functional group identity treatments were implemented by weeding out unwanted species. All plots also contained the dominant of the site *Holcus mollis* (FG1), as its removal would have caused such significant disturbance that the functional group identity treatment would be highly confounded with this. In the absence of *H. mollis* effects of FG1 removal may have been stronger but there is no evidence that FG1 possesses any unique functional traits and therefore the removal of all except one FG1 species should be viewed as an alteration of the distribution of traits within the community. Weeding took place throughout the experiment, with

major efforts in August 2008, June 2009 and May 2010. Vegetation was surveyed before the initial weeding effort, and non-*Holcus mollis* cover was comprised of 87% FG1, 5% FG2 and 8% FG3. Biomass removal was initially large (up to 13.2 kg per plot where FG1 was removed in August 2008) but declined substantially throughout the experiment as adult plants were no longer present. Subsequent weeding efforts only required the removal of invading seedlings so did not appreciably affect total cover, which had recovered by May 2009 in all functional group treatments (Figure S3). Post-weeding total cover in September 2010 was 76% when FG1 was present and 69% when it was absent ($F_{1,42} = 4.29$, p = 0.05, see below for statistical methods, and Table S2 and Figure S3 for more complete cover data).

Field measures

Rainfall data for the duration of the experiment were obtained from an onsite Vantage Pro wireless weather station (Davis Instruments, USA) and daily measures were taken from a rain gauge to determine the amount of rainfall to be applied to treatment plots. The average soil moisture content of each plot to 10 cm depth was measured weekly using a ThetaProbe Soil Moisture Meter HH2 with ML2x probe (Delta-T, UK) at a distance of 1 m from the plot edge on all four sides.

Vegetation surveys were carried out in October 2008, May 2009, September 2009, May 2010, July 2010 and September 2010. Visual estimates of percentage cover of each species were taken from the central 1 m^2 of each plot in order to determine whether there was an effect of the treatments (both precipitation and functional group removals) upon species richness, individual species abundance and total vegetation cover (Figure S3). The cover of bare ground and dead plant material was also recorded.Total vegetation cover was derived from the sum of individual species cover estimates, and was used as a proxy, non-destructive measure of aboveground biomass.

Decomposition rate measurements began in December 2008. Two grams of dried (80°C for 24 h), cut leaf samples of *Holcus mollis* were placed in 8 cm×8 cm mesh bags (1 mm mesh size, Normesh, Oldham, UK) and secured to the soil surface in each plot. Three bags were placed in each plot and one was removed in each of March, June and September 2009. On collection, new biomass growing through the mesh was removed and the remaining material was dried at 80°C for 24 hours and weighed to determine relative mass loss.

CO_2 and water flux rates were measured using a transparent Perspex chamber (area 300 cm^2, volume 9000 cm^3) attached to a CIRAS-1 infra-red gas analyser (IRGA), (PP Systems, Hitchin, UK), which was clipped onto PVC ring collars inserted into the soil to a depth of 5 cm (20 cm diameter, 10 cm long) to create a sealed area over the plants and soil. In light conditions, the returned values were net ecosystem CO_2 exchange (NEE), (mg CO_2 m^{-2} s^{-1}). This was repeated with an opaque cover to obtain an estimate of dark ecosystem respiration (R_{eco}). To obtain ecosystem photosynthetic rate (A), NEE was subtracted from R_{eco}. These measures were taken monthly during the summerand in alternate months through the winter, between March 2009 and September 2010. Soil moisture, PAR; (Skye Instruments, Wales) and soil temperature (Hanna, Bedfordshire, UK) were measured concurrently as covariates.

Extractable soil ammonium (NH_4^+), nitrate/nitrite (NO_3^-/NO_2^-) and phosphate (PO_4^{3-}) concentrations of fresh soil were determined in December 2008 and 2009, March 2009 and 2010, and monthly from May–September in 2009 and 2010. Soil samples were taken (0–5 cm depth) from five separate areas in each plot and mixed to create a composite sample. Soils were then

Table 1. Trait means for each functional group.

Trait	FG1	FG2	FG3
Plant height (cm)	34.3±3.8	65.9±13.3	51.2±5.4
Root depth (cm)*	100+	Variable	0–10
Aboveground biomass (g)	1.9±0.2	9.0±1.6	2.8±0.3
Belowground biomass (g)	1.9±0.5	5.0±1.0	1.0±0.20
Specific plant area (mm^2 mg^{-1})	17.4±1.3	13.8±3.4	11.9±1.1
Leaf N content (mg kg^{-1})	2359±229	1357±152	2203±160
Leaf N:P ratio	8.1±3.0	4.9±0.6	7.8±0.6

Trait means ± standard error for the functional groups from plants grown in a greenhouse on mesotrophic acid soil. FG1 is dominated by perennial forbs and grasses, FG2 is dominated by caespitose grasses, while FG3 has annual grasses, forbs and legumes.
*information taken from the Ecoflora database (http://www.ecoflora.co.uk/).

extracted using 1 M potassium chloride (75 ml KCl: 20 g fresh soil) solution for NH_4^+ and NO_3^-/NO_2^- and Truog's solution for PO_4^{3-} (150 ml Truog's solution: 10 g soil), [36], and analysed colourimetrically using a Skalar SAN^{++} auto-analyser (Skalar, York, UK). Precision was verified by repeating 5% of the samples as analytical replicates and including one matrix blank per 20 samples. Soil moisture was determined for each sample in order to express values as mg kg^{-1} dry weight. NO_2^- concentrations were negligible so oxidised N will be referred to as NO_3^- hereafter.

Statistical analysis

The effect of the rainfall treatment on light interception by the shelters (see above) and soil moisture content was tested by one-way analysis of variance (ANOVA) using R2.12.0 [37] on averaged plot level data at each time point, with block as an error term [38]. Vegetation cover was analysed in order to examine the effect of the treatmentsover time. We used a linear mixed effects model (LME) across all time points. Block (four levels) and plot were included asrandom effects, and the main effects of sampling month, precipitation change treatment (two levels) and a binary presence/absence term for each functional group were calculated alongside two and three-way interaction terms between the treatments. The models excluded three-way interactions between all three functional groups, which were not possible to estimate due to the design of our experiment. This technique was also used to evaluate treatment effects upon species richness, and all individual species found in the plots.

Decomposition rates (arcsine transformed percentage mass loss) were analysed using a LME model with main effects of precipitation, presence/absence of each functional group and first order interactions between all main effects, and block as a random term at each timepoint measured.We did not analyse these data with repeated measures methods (see above) as litter bags were all installed at the same time and sub-sets harvested sequentially.

Repeated measures LMEs were carried out to test whether there was an effect of rainfall regime and functional group identity upon ecosystem functions (photosynthetic rate, ecosystem respiration, evapotranspiration, and extractable N and P) over the course of the whole study. These models included block and plot as random effects. An ANOVA was performed on each of the LME models. Following the repeated measures analysis, each time point was evaluated separately using LME so that the timing of significant effects could be evaluated.These models were identical in structure to those for vegetation cover.

Results

Soil moisture content

The winter and summer phases of the precipitation change treatment had clear measureable effects on soil moisture (Figure 1). In the winter these effects were manifested with a time delay of around 60 days from the change in rainfall pattern. In the first winter of the experiment (2008–2009) the precipitation change treatment plots received 15% more rain than the control plots (Table 2). This resulted in significantly wetter soils in February of 2009, though in general effects were small throughout the rest of the period. In the second summer (2009), precipitation change treatment plots received 38.1% less rainfall than control plots and had significantly lower soil moisture levels all the way through until November, when high natural rainfall volumes raised soil moisture contents in both treatments. The second winter (2009–2010) was exceptionally cold and wet (Figure 1, Table 2). There was a clear lag-time between the high rainfall in December and a corresponding change in soil moisture, with a peak seen in February.

After this, the soils dried rapidly, though carry-over effects of the winterprecipitation change treatment were still evident throughout the spring of 2010. The third summer (2010) was relatively dry but had three heavy rain events rather than two, so a higher proportion of total rainfall was applied to the precipitation change treatment plots (75% of 136 mm = 103 mm). As with 2009, the summer of 2010 showed highly significant treatment effects on soil moisture, with the precipitation change treatment being much drier throughout the summer. Unlike in the winter periods, a lagged effect on moisture was not apparent. No significant effects of functional group treatments were apparent for soil moisture throughout the experimental period.

Plant community composition

The precipitation change treatment had significantly lower vegetation cover than the control throughout the experiment (Table 3). Plots where FG1 species were removed had significantly lower vegetation cover, although this appears to be mainly caused by a large difference in October 2008 following the first weeding occasion (Table 3); after this the difference in cover was small compared with other FG treatments (Figure S3). Note that when FG1 is described as absent or removed, this does not include the dominant at the site H. mollis, which belonged to FG1 but was allowed to remain in all plots. Later in the experiment significant effects of FG1presence on cover were observed in the May vegetation surveys but these were likely to be due to overwintering of perennials and dieback of annuals because weeding efforts were very small (Table S2). FG2 and FG3 removal did not significantly lower total plot cover.

Species richness was not significantly affected by the treatments, nor did it change over time (Table 3), averaging seven species per m^2 throughout. Holcus mollis, while initially relatively abundant in all plots, averaging 45.5% cover, declined consistently through the experiment, and by September 2010 averaged 12% cover (Table S3, Figure S4).

When each species was tested individually for sensitivity to precipitation change over time, only one out of the 52 species recorded at the site over the duration of the experiment was significantly affected. Rumex acetosella, while always having very low cover, was almost completely lost in precipitation change treatment plots (0.12±0.04% cover compared with 0.63±0.12% in control plots, $F_{1,51} = 8.53$ p = 0.005).

Decomposition rates

Functional group identity was a significant driver of litter mass loss, although there was no significant effect of precipitation change. Decomposition rates were consistently higher when FG1 species were present throughout the nine months of measurement (Dec–Mar = 29.1%, Dec–Jun = 14.6%, Dec–Sept = 19.6% higher when present, Table 4; Figure 2a–c). This group is characterised by short, N-rich species with deep roots (Table 1). However, there was an interaction between the presence of FG1 and FG2 in the Dec–Mar period (Table 4; Figure 2a). When FG2 was absent, and FG1 was present, decomposition was close to the average in this period.When FG2 was present in the plots but FG1 was absent, decomposition was slow, possibly as a result of the production of large amounts of low quality litter; FG2's distinguishing traits include a very low LNC and high shoot and root biomass (Table 1). This effect was overwhelmed by the more abundant FG1 where it was present; when FG1 and FG2 were both present, decomposition was high.

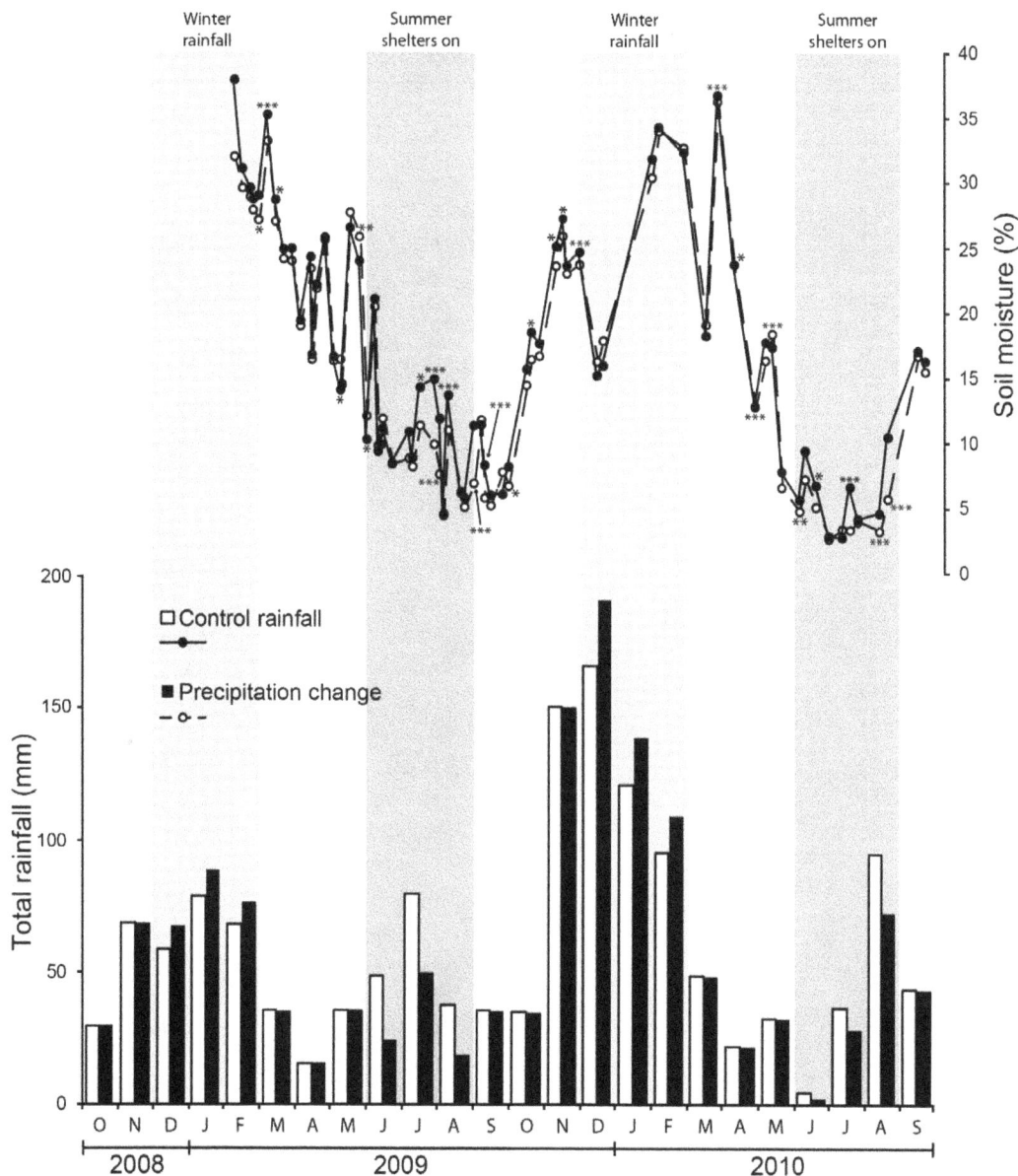

Figure 1. Rainfall applied to treatments during the measurement period of the experiment. Significant differences in soil moisture between precipitation treatments are represented by asterisks $* = p < 0.05$, $** = p < 0.01$, $*** = p < 0.001$.

Ecosystem CO_2 and water fluxes

Changes to C fluxes caused by precipitation change were strongly modified by plant functional group identity. The responses of R_{eco} and A were very similar, indicating that both fluxes were driven primarily by plant community activity. Under ambient conditions, significantly higher flux rates were associated

Table 2. Total seasonal rainfall inputs throughout the experiment.

Year	Season	Precipitation change rainfall volume (mm)	Control rainfall volume (mm)
2008	Summer	128	222
2008–9	Winter	233	206
2009	Summer	93	166
2009–10	Winter	440	382
2010	Summer	103	136

Table 3. Results of linear mixed effects models testing precipitation change (PC) and functional group (FG) treatment effects upon vegetation cover and species richness.

Treatment	d.f	Vegetation cover		Species richness	
		F	p	F	P
Intercept	1	**889.05**	**<0.001**	**258.50**	**<0.001**
PC	1	**94.88**	**<0.001**	0.12	0.728
FG1 present	1	**32.54**	**<0.001**	0.00	0.990
FG2 present	1	0.01	0.925	0.35	0.556
FG3 present	1	0.14	0.713	1.99	0.166
PC x FG1	1	0.81	0.372	1.91	0.175
PC x FG2	1	0.04	0.848	0.29	0.596
PC x FG3	1	0.85	0.362	1.54	0.222
FG1 x FG2	1	**5.22**	**0.028**	0.14	0.710
FG1 x FG3	1	0.22	0.642	1.57	0.218
FG2 x FG3	1	0.90	0.348	0.01	0.926
Residuals	42				
Month	5	**47.48**	**<0.001**	2.08	0.069
FG1 x Month	5	**16.39**	**<0.001**	1.72	0.130
FG2 x Month	5	**2.49**	**0.033**	0.62	0.683
FG3 x Month	5	0.38	0.862	1.35	0.245
PC x Month	5	30.01	**<0.001**	0.90	0.481
PC x FG1 x Month	5	1.47	0.202	1.26	0.281
PC x FG2 x Month	5	0.49	0.781	0.47	0.795
PC x FG3 x Month	5	0.77	0.574	1.62	0.157
FG1 x FG2 x Month	5	2.21	0.055	0.22	0.953
FG1 x FG3 x Month	5	1.13	0.345	**2.38**	**0.040**
FG2 x FG3 x Month	5	0.75	0.586	0.01	1.000
Residuals	225				

Table 4. Summary of treatment effects upon decomposition from mixed effects models.

Date	Treatment	d.f.	F	p
Dec–Mar 09	Intercept	1	**2948.92**	**<0.001**
	PC	1	0.92	0.342
	FG1	1	**33.93**	**<0.001**
	FG2	1	0.00	0.963
	FG3	1	0.25	0.623
	PC x FG1	1	0.21	0.647
	PC x FG2	1	0.87	0.358
	PC x FG3	1	1.59	0.214
	FG1 x FG2	1	**4.69**	**0.036**
	FG1 x FG3	1	0.23	0.638
	FG2 x FG3	1	1.17	0.287
	Residuals	42		
Dec–Jun 09	Intercept	1	**4194.75**	**<0.001**
	PC	1	1.21	0.278
	FG1	1	**40.05**	**<0.001**
	FG2	1	3.62	0.064
	FG3	1	0.61	0.440
	PC x FG1	1	1.55	0.220
	PC x FG2	1	0.00	0.960
	PC x FG3	1	0.95	0.335
	FG1 x FG2	1	1.31	0.260
	FG1 x FG3	1	0.50	0.486
	FG2 x FG3	1	0.01	0.921
	Residuals	42		
Dec–Sept 09	Intercept	1	**1641.20**	**<0.001**
	PC	1	0.13	0.721
	FG1	1	**20.94**	**<0.001**
	FG2	1	1.11	0.298
	FG3	1	1.79	0.188
	PC x FG1	1	1.74	0.194
	PC x FG2	1	2.26	0.140
	PC x FG3	1	0.00	0.989
	FG1 x FG2	1	1.95	0.170
	FG1 x FG3	1	1.59	0.215
	FG2 x FG3	1	0.05	0.820
	Residuals	42		

FGx refers to the presence of the functional group in question, PC to the precipitation change treatment.

with the presence of FG1. Across the experiment we observed average photosynthetic rates of 1.04 ± 0.08 mg CO_2 m^{-2} s^{-1} when FG1 was present compared with 0.89 ± 0.07 mg CO_2 m^{-2} s^{-1} in its absence (Table 4). The main significant effects occurred in the first year after the winter and summer treatments ceased (March and November 2009, Figure 3a–b), and during the summer rainfall treatment in the second year (June and July 2010, Figure 3c–d). For the most part, plots containing FG1 significantly differed in their photosynthesis to those without it, and there was a significant precipitation change effect in November 2009 and July 2010. In March 2009, photosynthetic rate was lower in plots where perennial species were present than in plots where there were mainly germinating annuals and caespitose grasses (Figure 3a). This was also the case in November, after the summer drought treatment had ended, and this was associated with a precipitation change effect. Plots with perennial FG1 species that had been exposed to the 2100 treatment had lower overall photosynthetic rate (Figure 3b). In June 2010, a particularly dry month at the beginning of the summer precipitation treatment (Figure 1), effects of FG1 had been superseded by FG2 (Figure 3c). The average photosynthetic rate was higher when FG2 species were present in the plots, although this appeared to be largely driven by very high photosynthetic output in plots where FG1 and FG2 were present together. However, in July 2010 the interaction between FG1 and precipitation change had returned, following a similar pattern to November 2009 (Figure 3d). The different groups had different root allocations (Table 1), which indicate that most root biomass for FG1 was distributed deeper in the soil than the other groups, and was likely to result in less optimal use of small rainfall inputs.

R_{eco} averaged 0.61 ± 0.05 mg CO_2 m^{-2} s^{-1} when FG1 was present, and 0.46 ± 0.04 mg CO_2 m^{-2} s^{-1} when absent. This pattern was reversed under the precipitation change treatment, resulting in significantly lower rates of both fluxes when FG1 was present (0.7 ± 0.09 mg CO_2 m^{-2} s^{-1} (A), 0.41 ± 0.05 mg CO_2

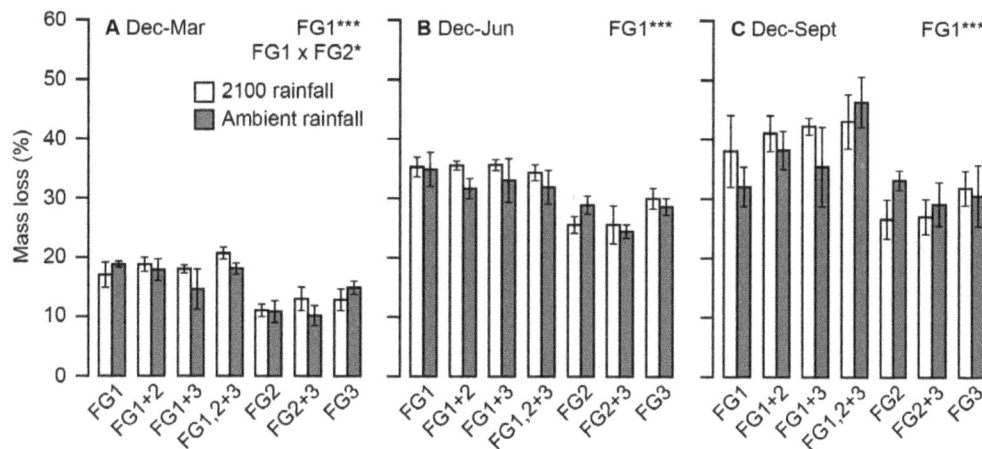

Figure 2. Effect of precipitation change and functional identity on decomposition of *Holcus mollis* at different time points. Decomposition of *H. mollis* litter in 2009 for all fourteen treatments. a) Dec–Mar $F_{1,49} = 32.87$, p<0.001, b) Dec–Jun $F_{1,49} = 40.23$, p<0.001, c) Dec–Sept $F_{1,49} = 20.31$, p<0.001. Error bars represent ± 1 SEM.

$m^{-2} s^{-1}(R_{eco})$ under the precipitation change treatment, compared to values of 1.04 ± 0.08 mg CO_2 $m^{-2} s^{-1}$ and 0.61 ± 0.05 mg CO_2 $m^{-2} s^{-1}$ for A and R_{eco}, respectively, in control plots). The dominant trends showed that FG1 was associated with higher levels of R_{eco} than the other two groups, although there was often a significant effect of the precipitation change treatment, leading to reduced R_{eco}, particularly through the summer months (Figure 4a,d–f). The precipitation change treatment was not a very strong driver on its own, and often only showed its effects in the presence or absence of certain functional groups.

Evapotranspiration rates were not strongly affected by functional group identity, but were significantly lower in the precipitation change treatment throughout the experiment (Table 5).

Extractable nutrient concentrations

Soil extractable NH_4^+ levels were not significantly altered by precipitation change or functional group identity (Table 6) throughout the experiment. Extractable NO_3^- was affected by an interaction between precipitation change and FG1; NO_3^- was slightly lower in plots where FG1 was absent. In contrast, extractable P concentrations were very strongly affected by precipitation change and presence of various functional groups across the seasons (Table 6; Figure 5a–d). Concentrations of extractable P were generally low throughout the experiment, with trace (<0.01 mg kg^{-1}) amounts in the soil in February 2009, increasing to 43.1 mg kg^{-1} in September 2010, although average P concentrations for the experiment were low, at 3.56 mg kg^{-1}. There was a highly significant interaction between FG2 presence and precipitation change. During spring, if FG2 was absent there was a higher concentration of P in the soil of precipitation change plots (i.e. those which had received higher winter rainfall), (Figure 5a). In the summer and autumn months P availability was not significantly affected by the treatments (Figure 5b,c). In the winter, soil P availability was once again affected by the nutrient-poor FG2 species and an interaction with the precipitation change treatment (Figure 5d). The wetter 2100 treatment was associated with almost total loss of P from the system when FG2 was present, although higher concentrations were found in ambient plots. Overall, FG2 presence and the precipitation change treatment appeared to affect P concentrations

during the wetter months, and have no effect during the warmer summers.

Discussion

This study showed that removal of plant functional groups based upon traits hypothesised to affect carbon and nutrient cycling altered the response of several ecosystem processes and properties to precipitation change. Our results also provide evidence that plant functional groups have complex and interactive roles in driving function in both control and altered climate conditions. Due to our sample size, caution must be used when interpreting the results, so the results we discuss here are based upon p<0.05 rather than 0.1, a potential solution for the Type II errors that may occur in small experiments [39]. As a result, we are confident that we are reporting real effects, although some may have been overlooked. Where F values in the tables exceed 2, there is a high likelihood that a significant result would have been obtained with more replicates, e.g. the effects of FG2 upon decomposition in the second period returned $F_{1,42} = 3.62$, which would have been significant with higher replication, but in this case the p value was 0.064 (Table 4). Other examples of this occurred in the gas flux measures (Table 5) and extractable nutrients (Table 6). Accordingly we have underestimated the impact of functional group presence and precipitation change on grassland ecosystem function. However, we have only discussed those with p values of lower than 0.05 in this study so as to reduce the likelihood of the more serious Type I error.

Effects of precipitation change upon ecosystem processes

The precipitation change rainfall treatment resulted in very low soil moisture levels in the summer growth period and waterlogged, possibly anoxic, conditions in the winter. This was associated with reduced rates of ecosystem A and R_{eco}, and a similar magnitude of decrease in both processes suggests that plants are the key drivers of these responses. While A and R_{eco} under the precipitation change treatment showed a variety of responses depending on the functional groups present and the season, this study supports the conclusions of a meta-analysis by Wu et al. [40] which found that the net balance of A and R_{eco} responses to rainfall manipulations (both increased and decreased) was close to neutral, thus indicating

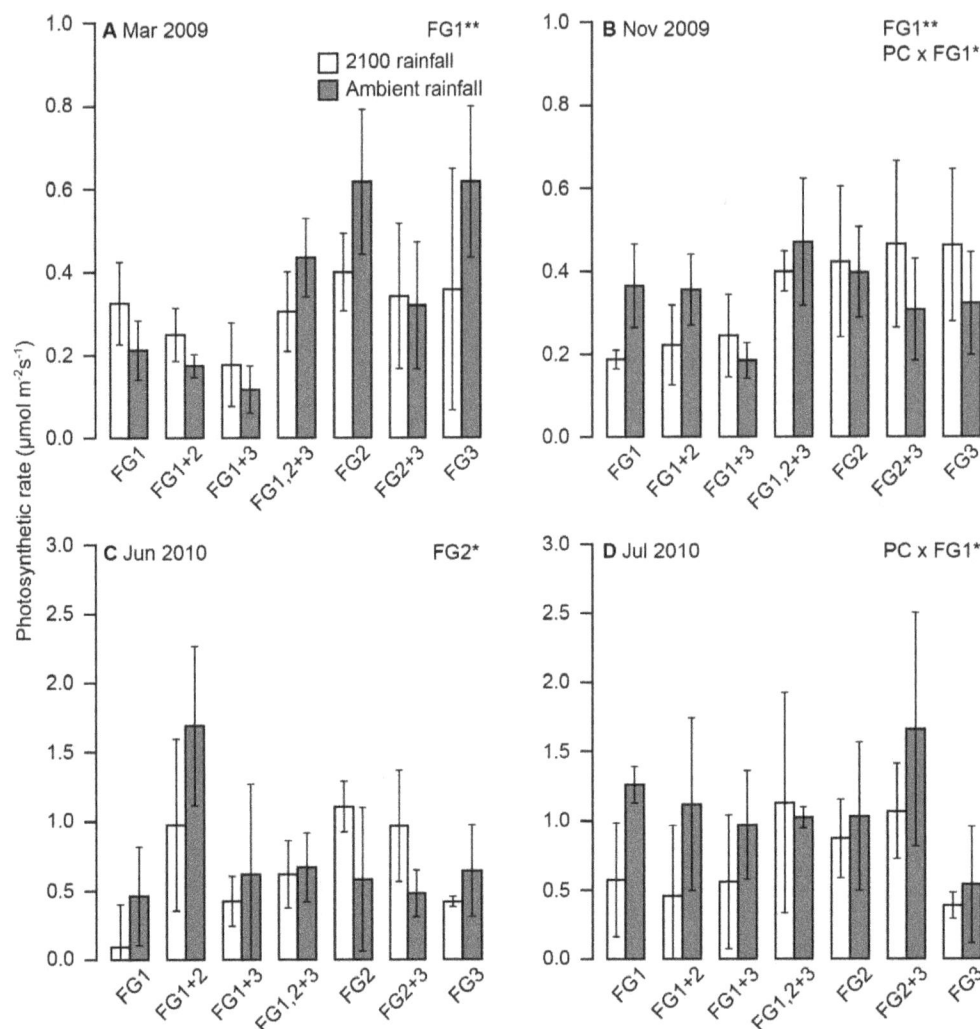

Figure 3. Effect of precipitation change and functional identity on photosynthetic rate at different time points. The response of photosynthetic rate to precipitation change (PC) and functional group (FG) identityat four time points through the experiment. a) March 2009 (presence/absence of FG1, $F_{1,42} = 9.152$, $p = 0.004$), b) November 2009 (interaction between PC and FG1, $F_{1,42} = 4.831$, $p = 0.033$), c) June 2010 (presence/absence of FG2, $F_{1,42} = 4.610$, $p = 0.037$), d) July 2010 (interaction between PC and FG1, $F_{1,42} = 5.552$, $p = 0.004$). Error bars represent ± 1 SEM.

that projected changes in rainfall patterns might not be as detrimental to soil carbon stocks as feared [31]. During the 2.5 years of this experiment, no processes were affected solely by the precipitation change treatment, thus showing that all precipitation change effects were dependent upon composition.

An unexpected result was the lagged response of soil moisture to changes in rainfall; treatment effects of the precipitation change treatment were delayed by up to six weeks, leading to stronger treatment effects on function in the spring and autumn than in the summer and winter. This highlights the importance of monitoring ecosystem function throughout the year in studies of this type. Some researchers have noticed surprisingly inconsistent relationships between ecosystem functions such as productivity [41],[42] and drought or drought alleviation, phenomena which could be explained by the lag time seen here. Studies on mixed grass prairie have demonstrated responses to a seasonal precipitation change as late as two seasons later; for example snow accumulation and melt associated with drift fences preserved ecosystem respiration levels under summer droughtby maintaining moisture in deep soil levels [27],[43].

Effects of changes in plant functional group composition on ecosystem processes

The strong observed effects of functional group composition on ecosystem properties and process rates lends support to the view that functional group identity is instrumental in driving a range ofecosystem functions in grassland systems [44],[45]–[48]. The importance of functional group identity over species richness in driving function is supported by the finding that changes in species richness throughout the main sampling seasons of the experiment were non-significant. Despite the fact that all plots contained a perennial species from FG1, *Holcus mollis*, the dominant trend that appeared over the 2.5 years of experimentation was that process rates were higher under control conditions where a range of other perennial species (FG1) were present. This indicates that several species of this group are required to maximise function, not just a single dominant. However, when several of these species were present ecosystem processes (especially carbon flux rates) were more strongly affected by precipitation change. In contrast, carbon flux rates were much less affected by the precipitation change

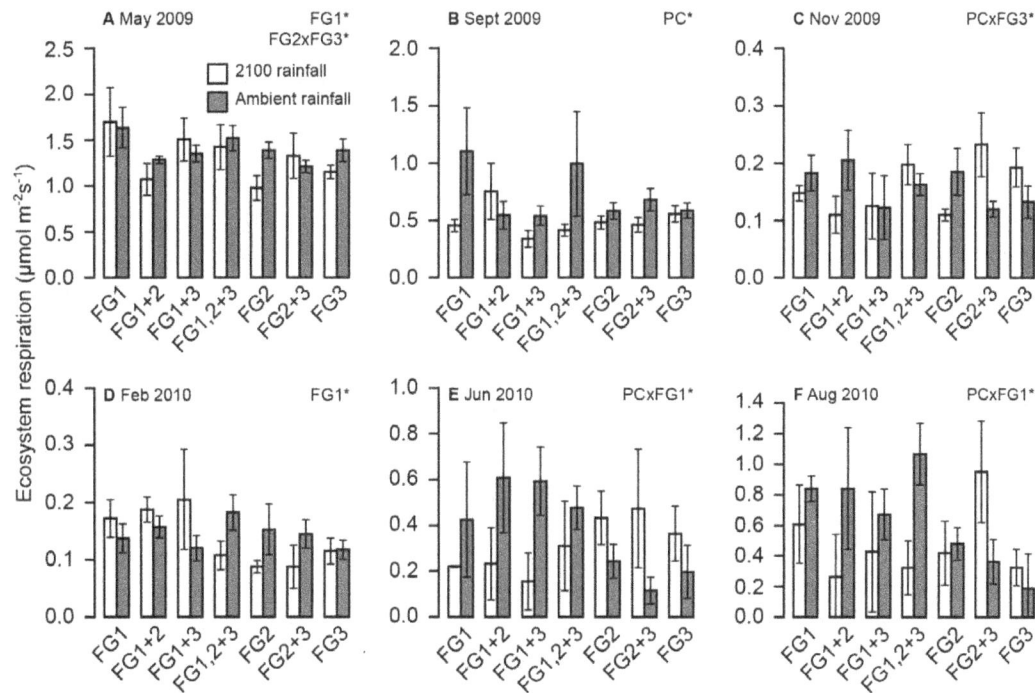

Figure 4. Effect of precipitation change and functional identity on ecosystem respiration at different time points. The response of ecosystem respiration to precipitation change (PC) and functional group (FG) identity at six time points through the experiment. a) May 2009 (presence/absence of FG1 $F_{1,42} = 4.72$, $p = 0.036$, interaction between FG2 and FG3, $F_{1,42} = 5.031$, $p = 0.03$), b) September 2009 (PC $F_{1,42} = 4.596$, $p = 0.038$), c) November 2009, (interaction between PC and FG3 $F_{1,42} = 7.165$, $p = 0.010$), d) February 2010 (FG1, $F_{1,45} = 4.521$, $p = 0.039$), e) June 2010 (interaction between PC and FG1, $F_{1,45} = 5.80$, $p = 0.020$) f) August 2010 (interaction between PC and FG1 $F_{1,45} = 6.46$, $p = 0.015$).

treatment where most of the FG1 species were absent and the other two groups were present. Morecroft and colleagues noted very similar trends in their grassland system [49]. They hypothesised that the lack of effect of summer drought on productivity in their experiment was due to gap-filling of annuals during autumn and winter, with a recovery of productivity (and consequently function) such that treatment effects were not seen in their autumn harvest. The caespitose grass and tall forb group (FG2) had consistently low abundances at our site, but the strong response of more than one ecosystem function to their removal contrasts with Grime's mass-ratio hypothesis [50], which states that species effects on ecosystem function are proportional to their biomass.

The presence of several perennial (FG1) species was a key driver of many ecosystem functions, including decomposition rates and carbon fluxes in this study. Their importance in these processes may be linked to their longevity, high LNC and deep and sparse root structures (as evidenced by their having a very similar root biomass to the much shallower rooted annuals), and the thick, dense layer of short-statured plants and associated litter they create. A more humid microclimate at the soil surface is likely to have been generated by their presence, and this combined with substantial litter inputs may have boosted microbial activity.

Decomposition responded strongly to the presence of multiple perennial species (FG1) with clear group-specific effects. Decomposition is known to be closely associated with R_{eco}, and a high abundance of FG1 perennial species increased both of these processes under control conditions in the current study [51],[52]. There is also evidence that some plant species harbour species-specific microbial communities in their rhizospheres [53]. This may in turn result in greater substrate utilisation and greater R_{eco} under certain combinations of plant functional groups [54].

Functional group identity as a regulator of ecosystem response to precipitation

Our results demonstrate that the effects of changes in rainfall on ecosystem processes can be modified by plant community composition. More specifically they show that the effects of summer drought on ecosystem processes are likely to be more substantial for communities with a high abundance of FG1 perennial species, compared to annuals, a finding which is consistent with other studies of plant community response to precipitation change [25],[55].This then has cascading effects on net photosynthetic rates and other ecosystem processes. The effect of the precipitation change treatment on soils under perennial dominated communities in the current study was smaller than that seen in some precipitation change experiments [28],[56]. This may be due to the increased winter rainfall element of our climate manipulation treatment, allowing deeper rooted species to continue to function throughout the summer drought period [14],[27].

Photosynthesis was strongly affected by a combination of the presence of multiple FG1 perennials and reducedrainfall, especially at the end of the growing season. While a reduction in process rates in response to drought is predictable, the particular response of FG1 plants is less so. When compared with the few effects of precipitation change upon the process rates of communities containing FG2 and FG3 plants, it indicates that changes in the activity, not abundance, of FG1 were responsible for the observed effects. Overall plant cover inplots containing FG1changed throughout the experiment more than in those containing the other groups, but this appeared to be a seasonal effect not a precipitation one, as shown by the lack of a three way interaction between the two treatments and time. Therefore, there

Table 5. Results of linear mixed effects models, testing precipitation change (PC) and functional group(FG) treatment effects upon carbon and water fluxes.

Treatment	d.f.	Photosynthetic rate (A)		Ecosystem respiration (R_{eco})		Evapotranspiration (ET)	
		F	p	F	p	F	P
Intercept	1	**875.59**	**<0.001**	**1016.98**	**<0.001**	**314.19**	**<0.001**
PC	1	**6.88**	**0.012**	**7.82**	**0.008**	**4.82**	**0.034**
FG1 present	1	0.56	0.458	3.83	0.057	0.19	0.669
FG2 present	1	0.65	0.425	0.15	0.700	0.02	0.881
FG3 present	1	0.16	0.688	0.73	0.398	0.83	0.367
PC x FG1	1	**7.21**	**0.010**	**12.02**	**0.001**	0.67	0.417
PC x FG2	1	0.11	0.742	0.17	0.679	0.51	0.478
PC x FG3	1	2.25	0.141	0.27	0.607	1.87	0.179
FG1 x FG2	1	2.27	0.139	0.32	0.575	1.05	0.311
FG1 x FG3	1	0.37	0.549	0.19	0.664	0.01	0.936
FG2 x FG3	1	0.04	0.843	2.87	0.098	2.10	0.155
Residuals	42						
Month	11	**42.19**	**<0.001**	**49.60**	**<0.001**	**21.85**	**<0.001**
FG1 x Month	11	1.01	0.439	0.50	0.901	0.48	0.917
FG2 x Month	11	**2.10**	**0.019**	0.77	0.675	1.59	0.097
FG3 x Month	11	0.52	0.893	0.23	0.995	0.57	0.858
PC x Month	11	0.73	0.712	0.71	0.728	0.70	0.735
PC x FG1 x Month	11	1.19	0.294	1.64	0.085	1.24	0.260
PC x FG2 x Month	11	0.28	0.990	0.66	0.780	0.82	0.624
PC x FG3 x Month	11	0.64	0.798	0.57	0.855	0.81	0.634
FG1 x FG2 x Month	11	0.91	0.534	0.60	0.830	0.70	0.742
FG1 x FG3 x Month	11	0.58	0.846	0.97	0.472	0.75	0.688
FG2 x FG3 x Month	11	0.61	0.821	0.97	0.475	0.93	0.509
Residuals	495						

is no clear link between coverand photosynthetic rate. The lack of recovery of these fluxes after drought suggests that lower soil moisture in the precipitation change treatment was associated with stomatal closure, to the detriment of photosynthetic rates in these species [57]. However, there are few studies on the effect of plant community composition on evapotranspiration in the literature, and those there are describe a positive relationship between evapotranspirationand functional group richness, possibly indicating that more diverse assemblages are less economical with water [57], though it may simply be due to higher biomass in such communities. This does not seem to be the case in our study as we found no link between evapotranspiration and the inclusion of multiple or individual functional groups. The similar magnitude of change of respiratory and photosynthetic CO_2 flux rates over the 18 month measurement period in this study indicate that neither annual- nor perennial-dominated temperate grasslands are likely to suffer a net shift in carbon sequestration as a result of the type of rainfall changes simulated here.

The differences in gas exchange noted in this experiment were not observed for soil N levels. In control plots containing multiple FG1 species, the large N inputs from decomposing litter and increases in microbial activity, as suggested by higher R_{eco} and decomposition rates, did not appear to result in significant changes to soil extractable N. This could indicate that highly competitive perennial species balance their higher N inputs with high uptake, or possibly that their deeper, sparser root structure results in a

weaker ability to prevent N leaching losses compared to annuals [58]. However, it is possible that if the experiment was better replicated or run for longer then, an effect would become apparent.

Concentrations of extractable soil P were low but significantly affected by precipitation, functional group identity and the interaction between these throughout the year. This grassland is co-limited by N and P [59] and our results indicate that it is the P cycle that is more sensitive to changes in precipitation and plant community composition. In general where several perennial (FG1) species were present P availability was higher. This may be due to the low abundance of legumes in this group, which have a high P demand and are likely to reduce soil P concentrations more than other species [60]–[][62]. Additionally, there is some evidence that deeper rooted species are able to increase the net labile P pool by taking up P from deeper soil layers, which could account for the more P-rich soils under perennial-dominated communities, and the overall increase in soil P in the soils as the experiment continued [63].

In general P was lower when caespitose grasses (FG2) were present, and the precipitation change treatment only altered P when this group was present. This suggests that the mechanisms of P uptake and availability in plots containing FG2 were strongly affected by soil moisture levels or poor ability to prevent leaching. Species in this group have a low foliar N:P ratio but a high biomass and they are notable for their high nutrient uptakes and fast

Table 6. Results of linear mixed effects models testing precipitation change (PC) and functional group (FG) treatment effects uponseasonal soil extractable nutrient concentrations.

Treatment	d.f.	Extractable NH_4^+		Extractable NO_3^-		Extractable PO_4^-	
		F	p	F	p	F	p
Intercept	1	1305.27	<0.001	876.46	<0.001	0.08	0.774
PC	1	0.02	0.901	1.56	0.218	1.18	0.284
FG1 present	1	1.65	0.206	0.89	0.351	1.89	0.177
FG2 present	1	1.28	0.265	0.00	0.952	1.17	0.286
FG3 present	1	0.00	0.959	0.61	0.440	3.60	0.065
PC x FG1	1	0.20	0.658	6.12	0.018	0.50	0.483
PC x FG2	1	1.52	0.224	1.83	0.183	6.17	0.019
PC x FG3	1	1.91	0.174	0.67	0.420	0.94	0.337
FG1 x FG2	1	2.39	0.130	0.87	0.355	2.48	0.123
FG1 x FG3	1	0.32	0.576	0.83	0.366	1.19	0.281
FG2 x FG3	1	0.00	0.965	0.08	0.780	0.08	0.776
Residuals	42						
Month	11	147.85	<0.001	154.73	<0.001	67.67	<0.001
FG1 x Month	11	1.39	0.173	0.46	0.929	0.56	0.865
FG2 x Month	11	1.09	0.371	0.57	0.858	0.75	0.692
FG3 x Month	11	1.02	0.424	0.32	0.982	3.56	<0.001
PC x Month	11	1.41	0.167	0.75	0.687	1.65	0.083
PC x FG1 x Month	11	0.90	0.544	0.83	0.608	0.34	0.977
PC x FG2 x Month	11	1.25	0.250	0.44	0.939	4.90	<0.001
PC x FG3 x Month	11	1.63	0.086	0.87	0.570	0.90	0.537
FG1 x FG2 x Month	11	1.04	0.414	0.44	0.939	2.44	0.006
FG1 x FG3 x Month	11	1.36	0.188	0.86	0.583	1.60	0.095
FG2 x FG3 x Month	11	1.13	0.337	1.02	0.429	0.15	0.999
Residuals	495						

growth rates, particularly *Dactylis glomerata* and *Lolium perenne* [64],[65]. They are also known for their resilience to drought [66]. However, our findings suggest that the presence of this group could result in very P-limited rhizospheres, particularly during the spring when the germination and establishment of annuals occurs, and this could impede colonisation by other species.

As there were relatively few significant, and/or synergistic interactions between the presence of functional groups there was little support for ageneral positive relationship between ecosystem process rates and the number of functional groups present, which is unexpected given the vast literature on the subject [11],[60],[67]. The clearest relationships have traditionally been with productivity, which we did not measure directly in this study (see [68] for a comprehensive review). The disparity of these findings may be due to a number of factors. These include the use of trait-defined functional groups as opposed to arbitrary groupings such as the commonly used grass/legume/herb classification [22],[44],[67]. Categorising species into non-arbitrary functional groups defined by traits that are likely to influence the measured functions may increase the relative importance of functional group identity effects and reduce the strength of the diversity *per se* effect. A second reason may be due to methodology; instead of constructing communities from random sets of species, as is the norm in biodiversity-ecosystem function studies [11],[15], we removed species from natural communities while leaving the dominant species present in all plots. Such an approach, which

better reflects real extinction scenarios, has been seen to result in a weaker relationship between diversity and function than is seen in artificially assembled communities [69].

Semi-natural grasslands in a changing world

From our results it can be inferred that, while the perennial plants of FG1 are important drivers of ecosystem function and theirrelationship with it is sensitive to climate change, annual plants may help maintain function during periods of water stress. This is particularly apparent for ecosystem C fluxes, which are substantially reduced when water is limiting [39]. There is some evidence, both from the current study and more generally, to suggest that traits characteristic of perennial species may make them susceptible to future drought [25]. From our results, it seems possible that while cover of these species would not change, function would be reduced. In particular, the allocation of roots to deeper soil layers could have prevented them from optimising water capture in the rainfall scenario in this experiment, thus explaining their reduced process rates under altered rainfall scenarios. An associated competitive release of annual forbs may also increase species diversity under future climate scenarios [14]. This seems to contradict the widespread support for the idea that deeper rooted species are more drought resistant, most commonly demonstrated in arid or semi- arid landscapes [70],[71]. However, we hypothesise that in our precipitation change treatment the

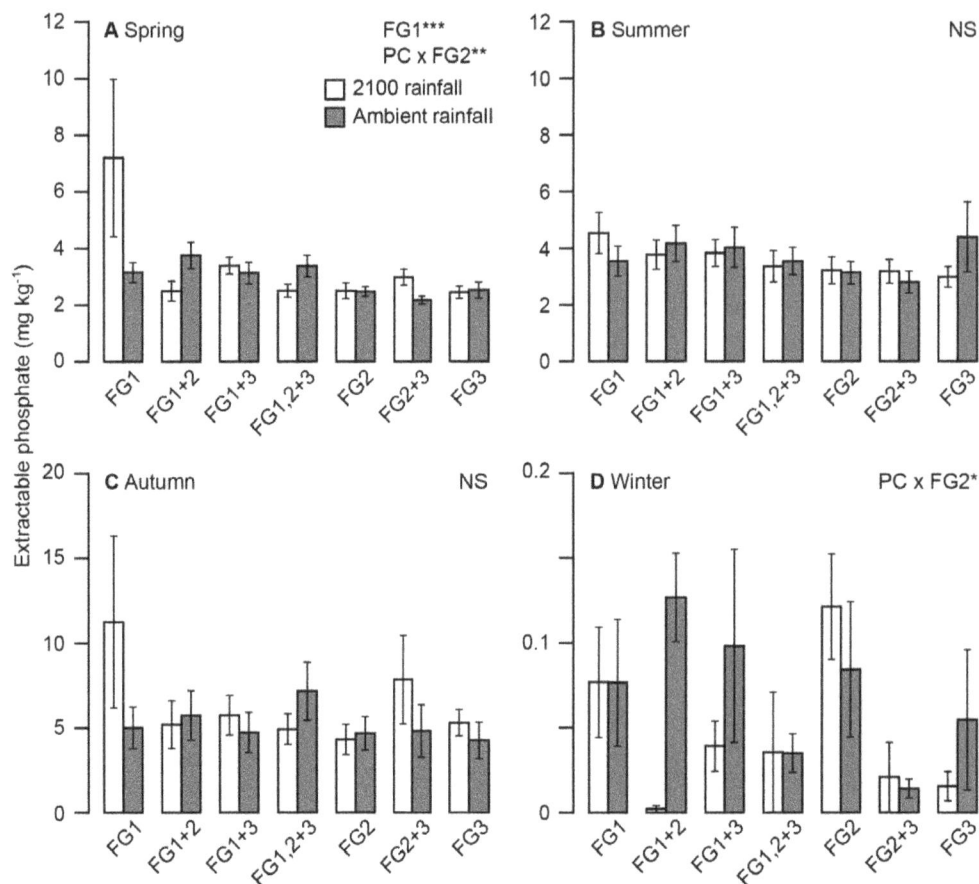

Figure 5. Effect of precipitation change and functional identity on extractable soil P in different seasons. The response of soil extractable P to precipitation change (PC) and functional group (FG) identity in each season. a) Spring (presence/absence of FG1, $F_{1,154} = 14.1$, $p = 0.0003$, interaction between PC and FG2, $F_{1,45} = 6.88$, $p = 0.0096$), b) Summer (NS), c) Autumn (NS), d) Winter (interaction between PC and FG2 $F_{1,45} = 5.005$, $p = 0.0303$).

contrast of very small rainfall pulses with sporadic high rainfall favoured the morphology of shallow rooted species which could utilise small volumes before they were lost to evaporation [6]. The lack of observed precipitation change effects on process rates in systems containing annuals could, therefore, indicate that temperate grassland systems containing this group may be more resistant to future climate change than previously thought. It should be noted, however, that the results presented here are from only 2.5 years of modest, though realistic, reductions in summer rainfall; more extreme changes in rainfall patterns, over longer timescales, or extreme weather events could have much greater ecological consequences.

Many global change drivers are known to affect grassland community composition, and large changes can be expected throughout the coming century [2],[31]. Perennial species generally dominate grasslands in temperate Europe, but these systems may be more vulnerable to changes in water inputs than systems dominated by annual species and/or caespitose grasses, which offer different life histories and strategies to cope with changing patterns of water availability. By grouping species in terms of trait complexes, differing responses to future changes in precipitation patterns can be shown in terms of gas fluxes and nutrient cycles. Our results indicate that future grassland management should aim to accommodate both perennial and annual species. The latter are often in low abundance in the improved (fertilised and sown) grasslands that are common in

Europe [72], but may help to maintain ecosystem function and the associated delivery of ecosystem services in future climates.

Supporting Information

Text S1 Supplementary methods.

Figure S1 Map, plot schematic and preliminary site characterisation of the DIRECT field site. Image of the field site of the DIRECT experiment in June 2009, accessed from Google maps 11/11/10, co-ordinates 51.4091°N, 0.6378°W. The dark blue and red rectangles delimit ongoing experiments in 2006 and 2007. The road runs parallel to the 105.8 m long periphery of the field, ~20 m away. The shelters slope into the prevailing wind. Block 1 is surrounded by light blue, through to block 4 in green. Plots without roofs belong to a related experiment.

Figure S2 Schematic of the dimensions of the rainout shelters.

Figure S3 Vegetation cover by functional diversity treatment throughout the experiment. Weeding was carried out in August 2008, May 2009 and June 2010.

Figure S4 Effect of treatments upon *H. mollis* throughout the experiment. The FG in the legend refers to presence of these functional groups in the plots.

Table S1 The list of the plant species in the field site in their allocated functional groups.

Table S2 Percentage cover estimates of plots containing each functional group in turn at the beginning and end of the experiment. Due to overlap of species the values may exceed 100%.

Table S3 Linear mixed effects model evaluating effect of treatments on *H. mollis* coverage over time. FGx refers

to the presence of the functional group in question, PC to the precipitation change treatment.

Acknowledgments

The authors are grateful for help during fieldwork fromB. Das, T. Sloan, K. Faulkner, A. Margeridas and N. Prill. We also acknowledge assistance in figure preparation by C. Tang and helpful comments from J.F. Cahill and two anonymous reviewers.

Author Contributions

Obtained permission for field site use: SAP PM. . Conceived and designed the experiments: SAP PM. Performed the experiments: ELF DGPA GE AH MR. Analyzed the data: ELF DGPA AH GE MR. Contributed reagents/materials/analysis tools: SAP PM ELF. Wrote the paper: ELF SAP PM.

References

1. Lee M, Manning P, Rist J, Power SA, Marsh C (2010) A global comparison of grassland biomass responses to CO_2 and nitrogen enrichment. Philos Trans R Soc Lond B BiolSc 365: 2047–2056.
2. Millennium Ecosystem Assessment (MEA) (2005) Current state and trends. Washington, DC. 50 p.
3. Murphy JM, Sexton DMH, Jenkins GJ, Booth B, Brown C, et al. (2009) UK Climate Projections Science Report: Climate change projections. Met Office Hadley Centre, Exeter.
4. Fay PA, Blair JM, Smith MD, Nipper JB, Carlisle JD, et al. (2011) Relative effects of precipitation variability and warming on grassland ecosystem function. Biogeosciences Discussions 8: 6859–6900.
5. Zhang L, Wylie BK, Ji L, Gilmanov TG, Tieszen LL, et al. (2011) Upscaling carbon fluxes over the Great Plains grasslands: Sinks and sources. J Geophys Res G: Biogeosci 116 G00J03.
6. Schwinning S, Sala OE (2004) Hierarchy of responses to resource pulses in arid and semi-arid ecosystems. Oecologia 141: 211–220.
7. Borken W, Matzner E (2009) Reappraisal of drying and wetting effects on C and N mineralisation and fluxes in soils. Glob Change Biol 15: 808–824.
8. Tilman D (1994) Competition and biodiversity in spatially structured habitats. Ecology 75: 2–16.
9. Knapp AK, Fay PA, Blair JM, Collins SM, Smith MD, et al. (2002) Rainfall variability, carbon cycling and plant species diversity in a mesic grassland. Science 298: 2202–2205.
10. Aanderud Z, Schoolmaster D, Lennon J(2011) Plants mediate the sensitivity of soil respiration to rainfall variability. Ecosystems 14: 156–167.
11. Hooper DU, Vitousek PM (1998) Effects of plant composition and diversity on nutrient cycling. Ecol Monogr 68: 121–149.
12. Lavorel S, Garnier E (2002) Predicting changes in community composition and ecosystem functioning from plant traits: revisiting the Holy Grail. Funct Ecol 16: 545–556.
13. Fortunel C, Garnier E, Joffre R,Kazakou E, Quested H, et al. (2009) Leaf traits capture the effects of land use changes and climate on litter decomposability of grasslands across Europe. Ecology 90: 598–611.
14. Debinski DM, Wickham H, Kindscher K, Caruthers JC, Germino M (2010) Montane meadow change during drought varies with background hydrologic regime and plant functional group. Ecology 91: 1672–1681
15. Roscher C, Weigelt A, Proulx R, Marquard E, Schumacher J, et al. (2011) Identifying population- and community-level mechanisms of diversity–stability relationships in experimental grasslands. J Ecol 99: 1460–1469.
16. Xu L, Baldocchi DD, Tang J (2004) How soil moisture, rain pulses, and growth alter the response of ecosystem respiration to temperature. Global Biogeochem Cycles 18: 4002–4012.
17. Walter J, Nagy L, Hein R, Rascher U, Beierkuhnlein C, et al. (2010) Do plants remember drought? Hints toward a drought-memory in grasses. Environ Exp Bot 71: 34–40.
18. Leps J, Osbornova-Kosinova J, Rejmanek M (1982) Community stability, complexity and species life-history strategies. Vegetatio 50: 53–63.
19. MacGillivray CW, Grime JP, The Isp Team (1995) Testing predictions of resistance and resilience of vegetation subjected to extreme events. Functional Ecology 9: 640–649.
20. Wardle DA, Bardgett RD, Klironomos JN, Setälä H, Van der Putten WH, et al.(2004)Ecological linkages between aboveground and belowground biota. Science 304: 1629–1633.
21. Orwin KH, Buckland SM, Johnson D, Turner BL, Smart S, et al. (2010) Linkages of plant traits to soil properties and the functioning of temperate grassland. J Ecol 98: 1074–1083.
22. Tilman D, Downing JA (1994) Biodiversity and stability in grasslands. Nature 367: 363–365.
23. Walker B, Kinzig A, Langridge J (1999) Plant attribute diversity, resilience, and ecosystem function: the nature and significance of dominant and minor species. Ecosystems 2: 95–113.
24. Grime JP (1985) Towards a functional description of vegetation. In: White J, editor. The Population Structure of Vegetation. Dordrecht, The Netherlands. pp. 501–514
25. Craine JM, Nippert JB, Towne EG, Tucker S, Kembel SW, et al. (2011) Functional consequences of climate change-induced plant species loss in a tall-grass prairie. Oecologia165: 1109–1117.
26. Hooper DU (2011) Biodiversity, ecosystem functioning, and global change. In: Harrison S, Rajakaruna N, editors.Serpentine: The Evolution and Ecology of a Model System.University Presses of California, Columbia, & Princeton, Ltd., Sussex, UK. pp. 329–357.
27. Chimner RA, Welker JM (2005) Ecosystem respiration responses to experimental manipulations of winter and summer precipitation in a Mixed grass Prairie, WY, USA. Biogeochemistry 73: 257–270.
28. Yahdjian L, Sala OE (2006) Vegetation structure constrains primary production response to water availability in the Patagonian Steppe. Ecology 87: 952–962.
29. Signarbieux C, Feller U (2011) Non-stomatal limitations of photosynthesis in grassland species under artificial drought in the field. Environ Exp Bot 71: 192–197.
30. Met Office (2011) Regional climate values in the UK from 1910–2010, Available: http://wwwmetofficegovuk/climate/uk/datasets/Accessed 16 July 11.
31. IPCC (2007) Climate change 2007: the physical science basis. In:Solomon S, Qin D, Manning M, et al., editors. Contribution of Working Group I to the Fourth Annual Assessment Report of the Intergovernmental Panel on Climate Change. Cambridge University Press, Cambridge, UK, pp. 996.
32. USDA, NRCS (2010) The PLANTS Database. National Plant Data Center, Baton Rouge, LA 70874-4490 USA Available: http://plants.usda.gov/java/Accessed 2nd June 2008.
33. Mokany K, Ash J (2008) Are traits measured on pot grown plants representative of those in natural communities? Journal of Vegetation Science 19: 119–126.
34. Cornelissen JHC, Pérez-Harguindeguy N, Diaz S, Grime JP, Marzano B (1999) Leaf structure and defence control litter decomposition rate across species and life forms in regional floras on two continents. New Phytol 143: 191–200.
35. Wright IJ, Reich PB, Westoby M, Ackerly DD, Barush Z, et al. (2004) The worldwide leaf economics spectrum. Nature 428: 821–827.
36. Allen SE (1989) Chemical analysis of ecological materials. Blackwell Scientific Publications, Oxford, 368 pp.
37. R Development Core Team (2009) R: A Language and Environment for Statistical Computing R Foundation for Statistical Computing, Vienna, Austria. Available: http://www.R-project.org Accessed 25 October 2009
38. Crawley MJ (2007) The R Book. Wiley Press, UK 950 pp.
39. Murphy KB, Myers B, Wolach K (2009) Statistical Power Analysis: A Simple and General Model for Traditional and modern hypothesis tests. 3rd Ed. Routledge Group, Oxon, UK. 224pp.
40. Wu Z, Dijkstra P, Koch GW, Peñuelas J, Hungate BA (2011) Responses of terrestrial ecosystems to temperature and precipitation change: a meta-analysis of experimental manipulation. Glob Change Biol 17: 927–942.
41. Haddad NM, Tilman D, Knops JMH (2002) Long-term oscillations in grassland productivity induced by drought. Ecol Lett 5: 110–120.
42. Jentsch A, Kreyling J, Elmer M, Gellesch E, Glaser B, et al. (2011) Climate extremes initiate ecosystem-regulating functions while maintaining productivity. J Ecol 99: 689–702.
43. Chimner RA, Welker JM, Morgan J, LeCain D, Reeder J (2010) Experimental manipulations of winter snow and summer rain influence ecosystem carbon cycling in a mixed-grass prairie, Wyoming, USA. Ecohydrology 3: 284–293.

44. Minns A, Finn J, Hector A, Caldeira M, Joshi J, et al. (2001) The functioning of European grassland ecosystems: potential benefits of biodiversity to agriculture. Outlook Agric 30: 179–185.

45. Tilman D, Reich PB, Knops J, Wedin D, Mielke T, et al. (2001) Diversity and productivity in a long-term grassland experiment. Science 294: 843–845.

46. Zavaleta ES, Hulvey KB (2004) Realistic species losses disproportionately reduce grassland resistance to biological invaders. Proc Natl Acad Sci USA 306: 1175–1177.

47. Diaz S, Cabido M (1997) Plant functional types and ecosystem function in relation to global change. J Veg Sci 8: 463–474.

48. Mokany K, Ash J, Roxburgh S (2008) Functional identity is more important than diversity in influencing ecosystem processes in a temperate native grassland. J Ecol 96: 884–893

49. Morecroft MD, Masters GJ, Brown VK, Clarke IP, Taylor ME, et al. (2004) Changing precipitation patterns alter plant community dynamics and succession in an ex-arable grassland. Funct Ecol 18: 648–655.

50. Grime JP (1998) Benefits of plant diversity to ecosystems: immediate, filter and founder effects. J Ecol 86: 902–910.

51. Ryan MG, Law BE (2005) Interpreting, measuring and modelling soil respiration. Biogeochemistry 73: 3–27.

52. Högberg P, Nordgren A, Buchmann N, Taylor AFS, Ekblad A, et al. (2001) Large-scale forest girdling shows that current photosynthesis drives soil respiration. Nature 411: 789–792.

53. Lamb EG, Kennedy N, Siciliano SD (2011) Effects of plant species richness and evenness on soil microbial diversity and function. Plant Soil 338: 483–495.

54. Berg G, Smalla K (2009) Plant species and soil type cooperatively shape the structure and function of microbial communities in the rhizosphere. FEMS Microbiol Ecol 68: 1–13.

55. Laporte M, Duchesne LC, Wetzel S (2002) Effect of rainfall patterns on soil surface CO_2 efflux, soil moisture, soil temperature and plant growth in a grassland ecosystem of northern Ontario, Canada: implications for climate change. BMC Ecol 2: 1–10.

56. Sternberg M, Brown VK, Masters GJ, Clarke IP (1999) Plant community dynamics in a calcareous grassland under climate change manipulations. Plant Ecol 143: 29–37.

57. Van Peer L, Nijs I, Reheul D, De Cauwer B (2004) Species richness and susceptibility to heat and drought extremes in synthesized grassland ecosystems: compositional vs physiological effects. Funct Ecol 18: 769–778.

58. Tilman D, Wedin D (1991) Plant traits and resource reduction for five grasses growing on a nitrogen gradient. Ecology 72: 685–700.

59. Edwards GR, Bourdôt GW, Crawley MJ (2000) Influence of herbivory, competition and soil fertility on the abundance of Cirsium arvense in acid grassland. J Appl Ecol 37: 321–334.

60. Janssens F, Peeters A, Tallowin JRB, Bakker JP, Bekker RM, et al. (1998) Relationship between soil chemical factors and grassland diversity. Plant Soil 202: 69–78.

61. Spehn EM, Scherer-Lorenzen M, Schmid B, Hector A, Caldeira MC, et al. (2002) The role of legumes as a component of biodiversity in a cross-European study of grassland biomass nitrogen. Oikos 98: 205–218.

62. Pang J, Tibbett M, Denton MD, Lambers H, Siddique KHM, et al. (2011) Soil phosphorus supply affects nodulation and N:P ratio in 11 perennial legume seedlings. Crop Pasture Sci 62: 992–1001.

63. McCulley RL, Jobbagy EG, Pockman WT, Jackson RB (2004) Nutrient uptake as a contributing explanation for deep rooting in arid and semi-arid ecosystems. Oecologia 141: 620–628.

64. Ryser P, Lambers H (1995) Root and leaf attributes accounting for the performance of fast- and slow-growing grasses at different nutrient supply. Plant Soil 170: 251–265.

65. Markham JH, Grime JP, Buckland S (2009) Reciprocal interactions between plants and soil in an upland grassland. Ecological Research 24: 93–98.

66. Turner LR, Holloway-Phillips MM, Rawnsley RP, Donaghy DJ, Pembleton KG (2012) The morphological and physiological responses of perennial ryegrass (Lolium perenne L), cocksfoot (Dactylis glomerata L) and tall fescue (Festuca arundinacea Schreb; syn Schedonorus phoenix Scop) to variable water availability. Grass and Forage Science (In Press).

67. Hector A, Beale AJ, Minns A, Otway SJ, Lawton JH (2003) Consequences of the reduction of plant diversity for litter decomposition: effects through litter quality and microenvironment. Oikos 90: 357–371.

68. Hooper DU, Chapin FS, Ewel JJ, Hector A, Inchausti P, et al. (2005) Effects of biodiversity on ecosystem functioning: A consensus of current knowledge. Ecol Monogr 75: 3–35.

69. Smith MD, Knapp AK (2003) Dominant species maintain ecosystem function with non-random species loss. Ecol Lett 6: 509–517.

70. Volaire F, Thomas H, Lelièvre F (1998) Survival and recovery of perennial forage grasses under prolonged Mediterranean drought: I. Growth, death, water relations and solute content in herbage and stubble. New Phytol 140: 439–449.

71. Padilla FM, Pugnaire FI (2007) Rooting depth and soil moisture control Mediterranean woody seedling survival during drought. Funct Ecol 21: 489–495.

72. Carey PD, Wallis S, Chamberlain PM, Cooper A, Emmett BA, et al. 2008 Countryside Survey: UK Results from 2007. NERC/Centre for Ecology & Hydrology, 105pp. (CEH Project Number: C03259).

20

Weather Indices for Designing Micro-Insurance Products for Small-Holder Farmers in the Tropics

Jacqueline Díaz Nieto[1], Myles Fisher[2], Simon Cook[2], Peter Läderach[2]*, Mark Lundy[2]

1 Catchment Science Centre, Kroto Research Institute, University of Sheffield, North Campus, Sheffield, United Kingdom, **2** Decision and Policy Analysis (DAPA), Centro Internacional de Agricultura Tropical (CIAT), Cali, Colombia

Abstract

Agriculture is inherently risky. Drought is a particularly troublesome hazard that has a documented adverse impact on agricultural development. A long history of decision-support tools have been developed to try and help farmers or policy makers manage risk. We offer site-specific drought insurance methodology as a significant addition to this process. Drought insurance works by encapsulating the best available scientific estimate of drought probability and severity at a site within a single number- the insurance premium, which is offered by insurers to insurable parties in a transparent risk-sharing agreement. The proposed method is demonstrated in a case study for dry beans in Nicaragua.

Editor: Alex J. Cannon, Pacific Climate Impacts Consortium, Canada

Funding: The authors acknowledge partial funding from the German Federal Ministry for Economic Cooperation and Development (BMZ) and support from the Climate Change, Agriculture and Food Security Program (CCAFS/http://ccafs.cgiar.org). The funders had no role in study design, data collection and analysis, decision to publish, or preparation of the manuscript.

* E-mail: p.laderach@cgiar.org

Introduction

Agriculture is inherently risky, a review of rural poverty identified exposure to risk as a major modifiable reason for chronic poverty, noting the widespread evidence that correlates risk with poverty [1]. Production risks include, but are not limited to climatic hazard, which of all the hazards agriculture faces is perhaps the most difficult one for agriculturalists to manage. Drought is the most serious of the natural hazards globally in terms of loss of life, accounting for 44% of reported deaths in the period 1974–2003 [2].

The mere expectation of drought is sufficient in some cases to reduce agricultural production. Nearly 80% of farmers interviewed in Ethiopia cited harvest failure caused by drought and other natural hazards as the event that caused them most concern [3]. Pandey *et al.* [4] revealed a huge drop in income for rice farmers in Orissa state in India as a result of drought. The impacts of drought extend beyond the loss of production. Sakurai and Reardon [5] include increases in local interest rates due to a rise in households seeking credit, a decline in farm labor demand, a reduction in local wages due to greater numbers seeking off-farm employment, drops in livestock prices due to distress sales of livestock and increases in food prices coinciding with low financial resources.

Additionally to the risk drought prone farmers face there is growing interest for weather insurance schemes for poor farmers to balance their risk as shown in a recent studies in Africa [6] and China [7]. Insurance schemes have been developed, for example, for east Africa [8] and Central America [9]. The reasons for the low uptake of index-based insurance schemes that is cited in the literature is the lack of understanding of the core concepts [6], and the lack of trust in the schemes and in insurance companies

[10,11]. An additional hindrance for wider uptake may be the low spatial resolution of climate data and lack of suitable crop yield data, which leads to high basis risk, which the farmers assume, and which makes the insurance unattractive for farmers. It is this aspect that we address in this study.

In this paper we introduce index rainfall insurance methodology as a tool that can help smallholder farmers manage the risk of drought. We then briefly recapitulate on a previous paper [12] where we looked at the possibility of using a weather generator to provide data to simulate crop yields used to design an indexed rainfall insurance instrument for smallholder drybean growers in Honduras. We then extend the method to determine the probabilities of damaging drought over the area in the north-central mountains of Nicaragua where drybeans are the main food crop. We finally discuss the need for insurance instruments that reduce basis risk by taking account of site specificity, crop variety and soil to design crop insurance instruments with emphasis on smallholder farmers.

1. Drought, Risk and Smallholder Farmers

Aside from drought, farmers face other environmental hazards such as hail, floods and frosts. In the north-central mountains of Nicaragua, where most of the drybeans are produced as a food staple by smallholder farmers, drought is by far the most common hazard and according to climate predictions it is going to get even dryer in the future [13]. Flood rains from tropical hurricanes do occur of course, but much less frequently than drought. Hail is rare and frost at altitudes lower than 1200 m does not occur at this latitude.

Drought is an especially serious problem for small-scale producers, most of whom do not have access to irrigation. For example, in Nicaragua only 8% of the land is irrigated [14], and almost none of this is in the central-north region where bean growers are located.

Droughts cause food and income insecurity through both acute effects and chronic secondary effects. Acute effects are immediate crop failure, which in extreme cases leads to hunger and even starvation. Secondary consequences of drought include increases in local rates of interest due to an increase in the number of households seeking credit and a decline in the demand for farm labor leading to a reduction in local wages due to greater numbers seeking off-farm employment. Livestock also suffer hunger and starvation leading to falling prices due to distress sales. Food prices increase coincidental with falling financial resources available to rural households as sources of income dry up [5].

The rural poor are often, indeed usually, found on lands that are marginal for one reason or another, such as low fertility soils, steep slopes and remoteness. They are especially vulnerable to drought. Large numbers of people are affected. Numerous studies have shown a strong link between risk, vulnerability and poverty [3,15,16,17]. Poor households lack resources with which to absorb the shocks of natural hazards.

Even small disruptions in the flow of income can have serious implications for them, so poor farmers commonly use informal and self-insurance measures to avoid risk. As discussed in more detail below, while these measures can help survival (e.g. [18]), most studies conclude that they are not the most effective tools for risk management, since they reduce the impact of a hazard at the expense of more profitable activities [19,20,21]. Although any risk-management strategy has a cost, the poor often have no other options besides informal methods because insurance is rarely available to them. If the insurance is more attractive than the informal methods, our consultations with smallholders in Nicaragua suggests that they would welcome the opportunity to participate.

2. Risk and Insurance

2.1 Strategies for coping with risk and their effects on livelihoods.
Most of the modern measures to mitigate risk are not readily available in developing countries, hence farmers in these regions are obliged to adopt traditional informal risk coping mechanisms [22] (Table 1).

The implicit costs associated with informal strategies can be quite high [15], which many argue are a barrier to poverty alleviation and indeed reinforce poverty [21,23]. If it were possible to accomplish the same risk reduction or risk transfer at lower cost

using formal insurance, then this could increase household profits and reduce poverty. Traditional risk-coping mechanisms are also risk-averse strategies that use resources inefficiently and fail to exploit more productive investments and technologies that in the long term would result in more productive systems [14,24]. For example, when faced with the possibility of losing an entire crop due to drought, farmers may lessen risk by minimizing investment in the crop by not applying fertilizer. They do this because making the additional investment increases their loss should the crop fail.

2.2 Risk sharing through insurance is an option but has traditionally not been available to the poor.
Formal insurance has provided benefits to individual consumers for centuries and in the last few years has also been suggested as a pro-poor tool for managing risk [25]. A growing number of micro-insurance products (products offered to insure items in the range of a few hundreds of dollars) are now being offered in poor countries in the areas of life, health and property insurance and in some cases, schemes for crop insurance. This growing interest in micro-insurance products as development tools is associated with the expansion of micro-credit schemes [20]. There is also a growing recognition of the mutual benefits of risk management as a tool for poverty alleviation. Micro-insurance is not only justified on the basis of humanitarian need.

Insurance can be thought of as exchanging the irregular uncertainty of large losses for regular small premium payments. A general rule of thumb seems to be that the larger the proportional loss in assets and income to the household, the fewer alternatives there are to recover from the loss [23]. Insurance is one of the few viable options for poor people to manage uncertain events that can cause large losses.

2.3 Previous experience with insurance has not been good.
Although we have made the case for crop insurance above, crop-insurance schemes in general in the tropics have a sorry record [26]. Several governments have developed crop insurance schemes. To date, most agricultural insurance has been either fully publicly owned or has involved large government subsidies to schemes operated by private companies. Unfortunately most of them have failed.

The main reason for failure of publicly-owned insurance schemes is because they were either multiple-peril or all-risk programs [26]. This means that virtually any cause of crop failure has been insured, which results in moral hazard where there is no incentive for the insured to use the best possible practices to avoid yield loss. Moreover, risks are widely correlated or systemic, that is a weather risk event affects many crops at the same time over an extensive geographic area [27]. Further problems are adverse

Table 1. Risk management tools.

Self insurance measures	Modern risk avoidance measures
Crop diversification	Production contracting
Maintaining financial reserves	Marketing contracting
Reliance on off-farm employment	Forward pricing
Other off-farm income generation	Futures options contracts
Selling family assets (e.g. cattle)	Leasing inputs
Avoidance of investments in expensive processes such as fertilizing (especially in high-risk years)	Invest in fertilizer, use long-term forecasts
Accumulation of stocks in good years	Acquiring crop and revenue insurance
Removal of children from education to work on farm	Custom hiring

(Source: Wenner and Arias, 2003; Skees et al., 2001; Hess, 2003).

selection (the insured knows more about the risk than the insurer) and the high transaction costs associated with sales, underwriting (to control adverse selection), and monitoring (to control moral hazard). A benefit of index insurance is that there is no need to underwrite each policy individually or to monitor for moral hazard, which greatly reduce transaction costs. Nevertheless, there are still sales costs, which are much proportionally much higher for the small policies sold to smallholders.

The Nicaraguan Institute for Insurance and Reinsurance (INISER) has developed a index insurance for groundnut (http://sagropecuarios.org/guide.php?p = 9). A regional risk and vulnerability assessment was conducted to determine a unique insured sum, premium and indemnization per region. The INISER insurance scheme uses meteorological stations to calculate a unique basis risk per region and a non-crop specific algorithm to calculate the loss in yield due to a shortfall of precipitation. This approach may have the limitation that the basis risk is inadequately calculated in remote areas where there are few data and details of crop management available.

2.4 Principles of weather insurance. Weather micro-insurance has been proposed as a viable tool to help farmers manage weather risk, which translates into crop production risk. The principles behind weather insurance have been widely discussed [26,28,29,30,31]. A review of the principles and experience of the insurance processes follows.

A number of factors govern the viability of insurance. Risk-sharing can only occur when both parties (the insurer and insured) have accurate information about a hazard and its likelihood. This has been the basis of insurance for over three centuries and Skees [32] maintains that a sound weather insurance product is transparent thus eliminating both moral hazard and adverse selection. Risk sharing must be broad enough to overcome co-variate risk (the risk that all crops insured in a scheme are affected), given that major weather events typically have broad geographic coverage. Many other factors are also important such as consumer demand, data availability, acceptably low delivery costs, capacity of local insurers, and an enabling legal and regulatory environment.

The probabilities of occurrence of adverse weather events that reduce crop yield can usually be estimated from historical weather data, provided that the available data captures the innate variability of the weather. In developing countries, this is rarely the case. Moreover, some areas are riskier than others. In an insurance scheme the probability of occurrence must be identified for specific areas and be agreed by both parties (symmetry of information).

Insurance based on weather indices is a relatively recent development, in which weather events, not yield, are the basis for determining indemnity payment. Compared to area-average indices, weather-based indices have the advantage that weather data are generally more accessible and reliable than yield data. This is especially the case in developing countries [32]. Weather-related crop insurance products succeed or fail on their ability to present accurate information about weather-related risks that are specifically associated with yield loss. The critical step is to identify the relationship between an insured weather event and consequent crop loss. There are those who argue that more generalized weather indices should be developed that can be used to protect households from the variety of losses that occur due to extreme weather events [33,34,35,36], but we are not in a position to consider all these many components. We restrict ourselves to formulating approaches that might be useful to smallholders to confront the loss of one or more key staple crops

due to unfavorable weather, rather than extreme weather events, which are very different.

A key attribute of weather-based index insurance is its simplicity and transparency, which makes them more attractive to global insurance markets [37]. Weather-index insurance also provides a hedge against the cause of the yield loss, rather than its cost, which is the underlying concept of insurance against yield reduction. This removes the need to estimate prices [26,38], a critical component of many of the traditional yield-triggered insurance schemes.

Results

1. Nicaragua Study Site

The main drybean-producing departments in Nicaragua are Matagalpa, Jinotega, Estelí and Nueva Segovia [41] in the north-central mountains. Most drybeans are produced on hilly to steep slopes [42].

For the baseline study, we chose San Dionisio, in Matagalpa Department, which is one of the major drybean producing areas of Nicaragua [43]. At San Dionisio drybeans and maize are generally grown at altitudes 500–800 meters on steep slopes; 67% of the area has slopes greater than 30%.

Nicaragua has a well-defined dry season from December to May and a rainy season from June to November. The rainy season is long enough to allow two successive crops to be grown known as the *primera* and *postrera*, separated by a short drought that usually occurs in July or August [44] called the *canicula*. Although the *primera* and *postrera* cropping periods are well defined, the onset of the rains is highly variable so that sowing date is of great importance to make the best use of both the *primera* and the *postrera* cropping periods.

The drybean varieties grown in Nicaragua are adapted to temperatures of between 17 and 24°C [42] and have a life cycle of 60–75 days. Farmers generally prefer small- and medium-seeded black and red types [45].

Temperature and solar radiation vary little during the growing season for any particular site in Nicaragua; it is rainfall that has the greatest climatic influence on drybean production. The optimum rainfall is between 300 and 400 mm while Jaramillo [46] quoted by Rios and Quiros [47] found that the maximum yields were obtained with 400 mm precipitation distributed according to the water requirements of the crop.

2. Methodology

We selected the 151 10-arc minute pixels that covered the departments of Matagalpa, Jinotega, Estelí and Nueva Segorvia where drybeans are grown (Figure 1). We generated 99 years of weather data in MarkSim using the coordinates of the geographical center of each pixel. For each pixel, we input these data into the Decision Support System for Agrotechnology Transfer [40] drybean model to simulate yields for the 99 years for eight generic soils with textures ranging from sand to silty clay and either deep or shallow profile from the DSSAT soil database. We used the genetic coefficients for the variety Rabia de Gato, whose physiological characteristics are similar to the traditional varieties grown in the region. In total we simulated almost 120,000 separate crops of drybeans.

For each soil within each pixel (called a "run"), we established the minimum water requirement (MWR, as rainfall) for each dekad below which there was a yield reduction, We tabulated the rainfall data for each dekad with the simulated yield and for each dekad we estimated plausible values for the minimum MWR. We subtracted these MWRs from the observed rainfall for each dekad to calculate deficits, that is, we ignored positive values. The total

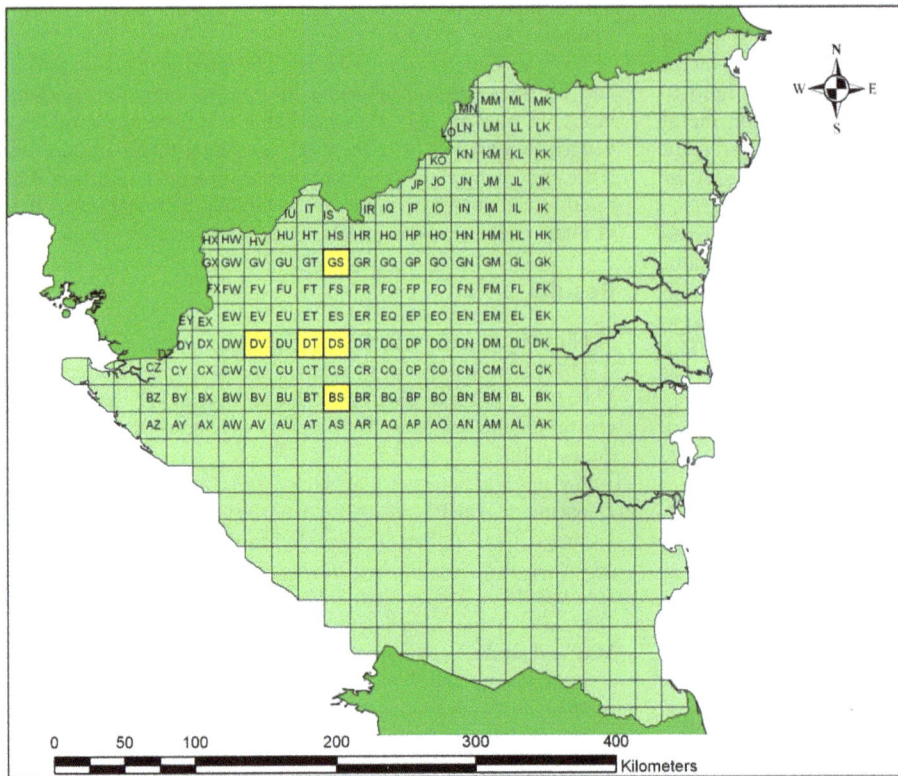

Figure 1. Two letter codes of each pixel used to identify the generated weather data.

rainfall deficit for the growing period was therefore the sum of all the deficits. Note that the MWR is a simply a plausible starting value, which is subsequently adjusted in the optimization procedure in the next step.

We selected the lowest quartile of each run and calculated total rainfall deficits from day −10 to day +70 for each simulation within this subset. We then calculated the regression coefficient of total deficit on crop yield. We optimized the estimates of MWR for each dekad to maximize the correlation coefficient using the Solver procedure of Excel with the constraint that MWR for each dekad ≥0. The upper and middle quartiles of yield have rainfall deficits of zero, and therefore were not relevant to establish MWRs. We then calculated the rainfall index for each run as the sum of the MWRs.

The procedure for the deep loam for the pixel BS (Figure 1), which contains the locality of San Dionisio (12° 45′, 85° 51′W), is summarized in Table 3.

3. Results

We applied the method to each soil-profile depth combination of the 151 pixels, but the rainfall indices for soils differed little so we present means.

The correlation of the rainfall index with crop yield was in general satisfactory with R^2 0.7–0.9 and higher (Figure 2). Soil texture and slope affected the R^2 values because there is more runoff of rainfall on the heavier soils and particularly on sloping land.

We used a range of generic soils with both deep and shallow profiles. As expected, sandy soils were much droughtier than heavier-textured soils and especially if they were shallow.

Using the relation of total rainfall deficit against yield we set levels of deficit that would trigger an indemnity payout in a hypothetical insurance instrument. The probabilities of reaching a

given level of deficit were then calculated for each of the eight soils for each pixel. The probabilities of reaching deficits of 50 and 70 mm, averaged over all eight soils for simplicity, are presented in Figure 3.

Based on these data, it was then straightforward to design an insurance instrument for each soil within each pixel. The details of a hypothetical contract are shown in Table 3. Tables 4 and 5 show hypothetical growing seasons that do not reach, and do reach, respectively, the trigger level. In designing an insurance instrument based on modeling as described above, it is relatively simple matter to obtain the information necessary for the actual soils in question and adjust the index criteria accordingly.

This exercise shows that it is feasible for any given location to simulate the yield of any particular crop for which there is a simulation model in the DSSAT series.

Discussion

Sound insurance requires best estimates of hazard probability. It also requires agreement about the likelihood of the hazard occurring. Errors in estimation of the hazard can be due to three sources:

- An incomplete model in which the weather event cannot be related to the loss,
- (a) Spatial and (b) temporal variation in which the model is complete, but data are incomplete, and
- Basis risk.

We discuss each of these as they apply to the Nicaragua case study.

Table 2. Summary of main challenges that need to be addressed and possible areas of action.

Basis risk	Details	Solutions
Temporal risk	The level of impact of a weather phenomenon will vary according to the time at which it occurs during the crop cycle. E.g. a shortage of rainfall at just before maturity may kill a crop, whereas just after seeding may have little effect.	Indices that represent the temporal variability in sensitivity to rainfall deficit.
Spatial risk	A rainfall deficiency may occur at one location causing crop losses, but this rainfall deficiency did not occur at the recording location and so no payment is triggered.	Offset the risk by offering site-specific contracts that account for spatial variability.
Crop specific risk	A rainfall deficiency may kill a drought sensitive crop, whereas a drought resistant crop will survive through longer periods of drought.	Offset the risk by tailoring the insurance to specific crops.

(Source: World Bank, 2001).

1. Incomplete Model. Exclusion of Major Factor Such as Soil and Crop Cultivar

1.1 Soil specificity. The effectiveness of rainfall is strongly influenced by soil characteristics. In soils that have low water-storage capacity, the impact of rainfall shortages will be felt much sooner than in the case of soils with high water-storage capacity. Conversely, when soils are dry, small falls of rain can be more effective on sandy soils compared with clay soils, which require more water to "wet up". Soil texture, soil depth and water-holding capacity are key factors to take into account in designing an effective insurance scheme. Farmers growing crops on very risky soils will need indemnity payments more often than farmers on less risky soils, which must be reflected in both a soil-specific payout structure and in the cost of the insurance coverage.

1.2 Cultivar specificity. Rainfall requirements will also vary greatly from crop to crop and within the same crop depending on the cultivar. Drought-tolerant varieties will naturally withstand rainfall deficits more successfully than drought-sensitive varieties. Therefore in order to improve the relationship between the rainfall weather index and crop losses, the rainfall indices need to be tailored specifically to the crop variety.

Table 3. Sample insurance contract.

RAINFALL INSURANCE CONTRACT

REFERENCE WEATHER STATION	*(e.g.)* San Dionisio INETER weather station
Crop	*(e.g.)* Dry beans – drought tolerant type
Reference soil type	*(e.g.)* Deep sand
Sowing window	*(e.g.)* 15 May to 15 June
Sowing date rule	*(e.g.)* First day after 5 consecutive rainy days over 5 mm each
Trigger value	*(e.g.)* −70 mm
Premium price	*(e.g.)* US$3
Indemnity	*(e.g.)* US$5 for every mm of rainfall deficit after the trigger value

Minimum rainfall requirements (given crop and soil stated above)

	Day 1 to 10	Day 11 to 20	Day 21 to 30	Day 31 to 40	Day 41 to 50	Day 51 to 60	Day 61 to 70	Day 71 to 80	Day 80 to 90
MIN	0	10	10	25	40	40	40	30	0
RAIN									
DEF									

a. TOTAL Rainfall deficit

Calculation of indemnity payments:

1. MIN is the minimum rainfall that is required for your crop in each of the 10 day windows.

2. RAIN is the rainfall observed at the reference weather stations (you may enter this into the RAIN box, however it is the official rainfall recorded at the weather station that determines whether you are entitled to an indemnity payment).

3. DEF is the rainfall deficit. This is calculated by subtracting MIN from RAIN (only negative values are taken into account).

4. Indemnity payments occur when the TOTAL rainfall deficit is equal to or less than the trigger value.

5. The rainfall deficit is the sum of the 10 day rainfall deficits.

■ Rainfall from planting to harvest

□ Summed rainfall deficit

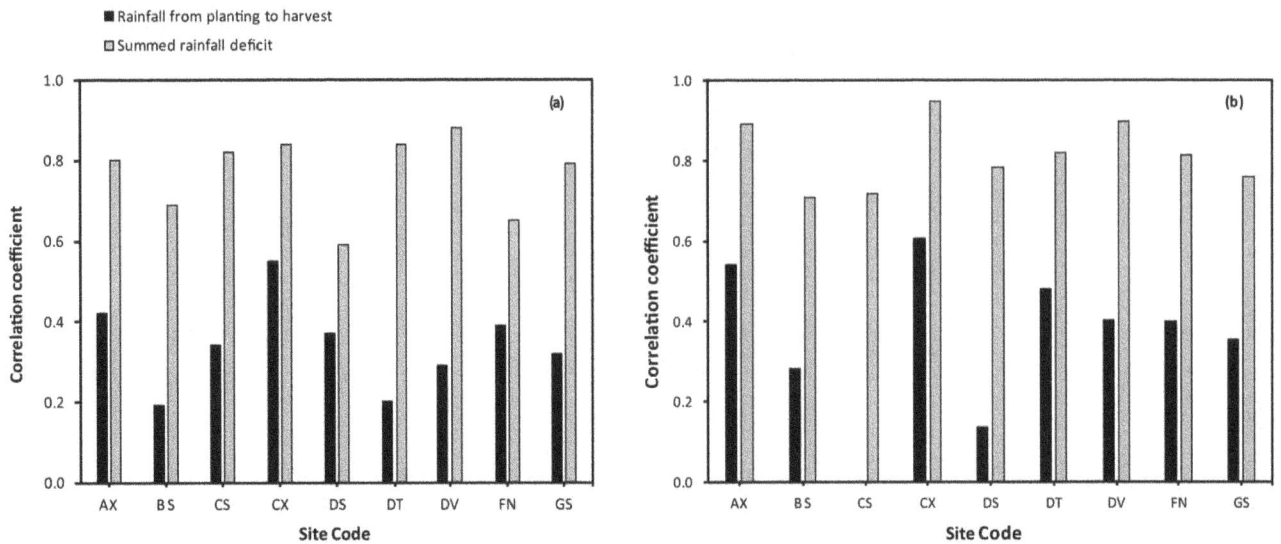

Figure 2. Correlation coefficients of total rainfall deficit and rainfall on yield of drybeans simulated by the DSSAT drybean model on contrasting soils for a selection of sites in north-central Nicaragua. Soil textures are (a) sand, and (b) silty clay. The rainfall for each cell was generated using the MarkSim procedure.

The implications of this for modeling are that the genetic coefficients must be known for the cultivar or cultivars in question. Ideally these should be the outcome of carefully-designed experiments. Nevertheless, it is possible to make some informed guesses as to what the coefficients should be, based on phenological data from different latitudes for the cultivar in question. But the guessing should only be undertaken by experts with a clear understanding of how the particular model represents physiological factors such as photoperiod response and the thermoregulation of plant development.

1. 3 Planting date. In rain-fed agriculture, which is implicit in designing a drought index, sowing date varies from season to season depending on the onset of rain at the start of the growing season. Since weather insurance schemes will be sold in advance when there is no information about what the weather will be, a transparent system is needed that incorporates variable planting dates into the insurance products. Both insurer and insured will need to know the exact start and end dates within which the observed rainfall will be taken into account for determining indemnity payments. To maximize the effectiveness of the insurance product, the method used to establish the sowing date used in the product must reflect the actual planting date as closely as possible.

2. Spatial Error

Crop yields from research stations are typically 30%, or more, higher than those of farmers' fields [48], so that using them as the basis for estimating the effect of a given weather event on farmers' yields is dangerous. Moreover, weather risk varies spatially. To reflect this spatial variation of risk in the premium, methods to estimate it in risk evaluation are needed so that the insured pays the price of the risk they actually confront.

The spatial limitations of MarkSim's weather surface, 2.5 arc minutes for Asia (4 km near the Equator), 10 arc minutes elsewhere (18 km), are now irrelevant with the availability of the WorldClim surface [49]. WorldClim's surface has a resolution of 30 arc seconds, or about 1 km at the Equator and it is a simple procedure to extract data from it and use these as external input to MarkSim. This permits further lessening of basis risk, in all but

extreme terrain, where it is unlikely that insurance would be considered.

3. Temporal Error, Estimating Extreme Events from Short-run Data

It is common to think that 50 years' (or so) weather data is sufficient to estimate yield variation in crops. We caution that this is a dangerous assumption. Engineers design structures and other works to withstand a given frequency of extreme weather, for example, a river levy to withstand a one in 100 year flood, termed more simply a 100-year flood. Clearly, a short run of historical data (50 years or even less) is only a limited sample of a very large population. Using such limited data alone to generate probabilities of climate risk will lead to seriously under- or over-estimated risk since by definition, only the extremes encompassed by the actual data are represented.

A different component of temporal factors is some method of incorporating the El Niño-Southern Oscillation (ENSO) phenomenon. Recent studies have shown that the ENSO has a profound effect on weather, not only in the eastern Pacific but more generally globally. Although this may make long-term forecasts more reliable, it is not yet clear how this can be applied in practical terms. MarkSim does not attempt to identify the ENSO phenomenon, although it does include its effect in the temporal variation it represents.

No weather simulator will forecast extreme events, so the method presented here will need to be modified to take account of their historical frequency if that is deemed necessary [35]. As it stands, the method does not address this issue. Typically, engineers use a Pearson function (logarithmic extrapolation) based on historical data, but consideration of this approach is outside the scope of this paper.

4. Consequences of Basis Risk

As the checkered history of insurance shows, commercial viability is essential to ensure a self-sustaining insurance process. Viability of insurance is determined by the design of the insurance process, which encourages risk-sharing on the basis of transparent

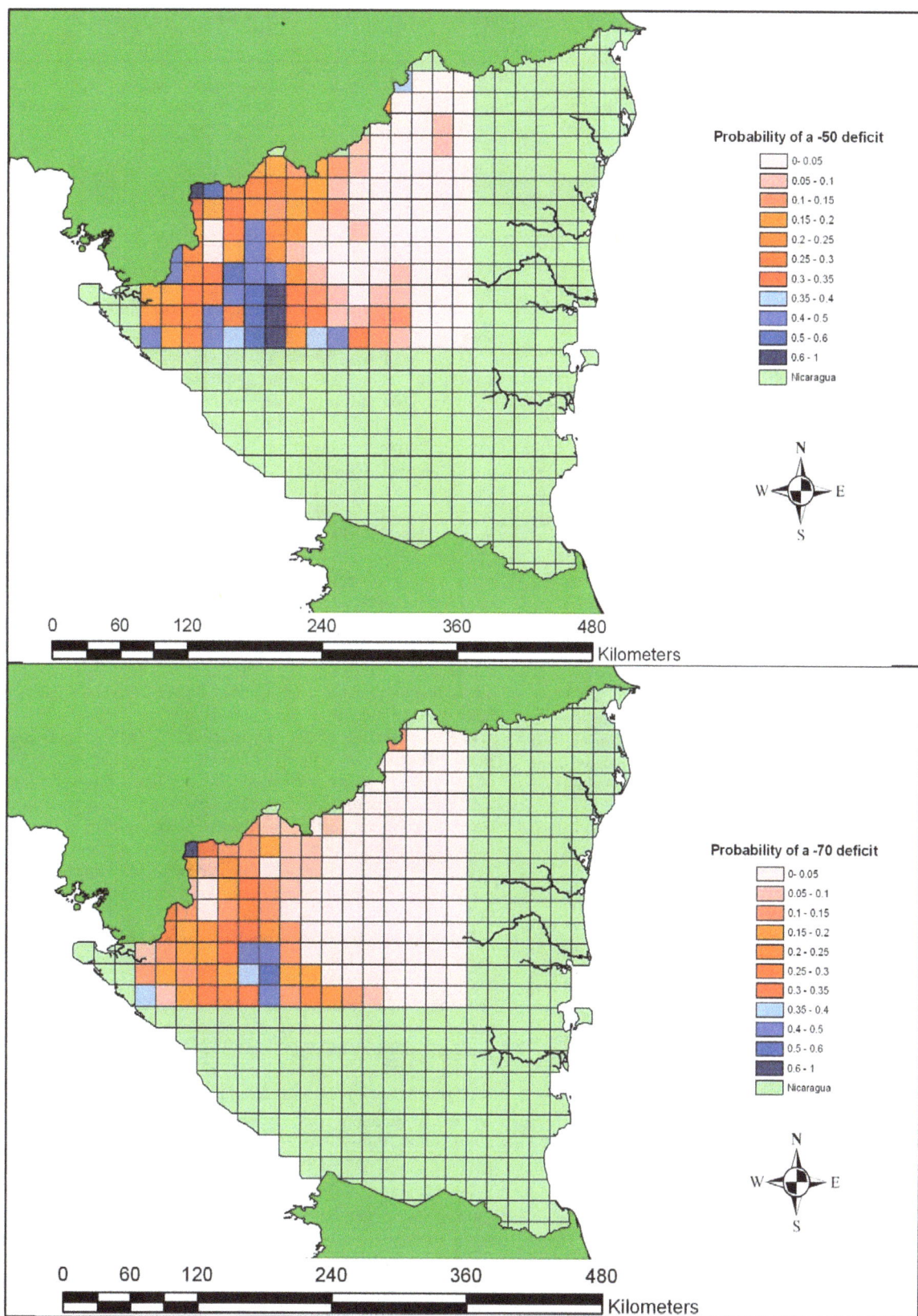

Figure 3. Probability of accumulated rainfall deficits of 50 and 70 mm during the growth of dry beans during the first growing season in north central Nicaragua.

Table 4. Example of a season not entitled to an indemnity payment (total rainfall deficit does not reach the trigger value of −70 mm).

	Day 1 to 10	Day 11 to 20	Day 21 to 30	Day 31 to 40	Day 41 to 50	Day 51 to 60	Day 61 to 70	Day 71 to 80	Day 80 to 90
MIN	0	10	10	25	40	40	40	30	0
RAIN	34.9	22.4	0.6	33.8	0	57.6	73.4	161.8	112.9
DEF			−9.4		−40				
a. TOTAL Rainfall deficit									−49.4

agreements between the insurer and the insured about drought probabilities. A key part of this agreement is the provision of accurate estimates, and in this respect we have concerns about potentially imprudent application of insurance. Insurance with excessive basis risk will be expensive or, worse, may invoke moral hazard since farmers will believe themselves to be protected whereas in fact they are not.

Index-based schemes seem particularly vulnerable to basis risk, since their prime attraction is cost reduction through insuring weather events rather than actual inspectable loss. We discuss this point in more detail in Diaz Nieto [12], but briefly the actuarial component of the index instrument is calculated on the basis of the data recorded at a particular meteorological station, which is also where the current rainfall is measured on which a payout will be assessed. Any gradient in the actual climate surface from the station to a farmer's fields constitutes basis risk, *which is borne by the farmer*. Proponents of indexed insurance and assessments of unsuccessful pilot schemes consistently ignore this, in the latter case often expressing bewilderment that farmers are unwilling to avail themselves of the offered instrument. We have seen no case where the reasons for the farmers' unwillingness to buy have been disaggregated to include farmers' perception of the suitability of the instrument for *their own farms*.

We believe that basis risk is a key issue and minimizing it is a major advantage of the scheme we propose here. There can be as many insurance instruments as are necessary to provide coverage that individual groups of farmers perceive to be relevant to them as the procedures we describe here can generate pseudo-historical data of both weather and crop yield for any point for which they are needed. The only requirement is that each will be required to have its own rain gauge on which to determine any payout. In a successful scheme for maize farmers in western Kenya in which we had some involvement, cell-phone masts were the sites of choice for the rain gauges, with the data being recorded in near-real time, astonishingly, in Austria, from where it was readily available to both the insurer and the insured.

5. Practical Implications: Technical Considerations in the Design of an Effective Weather Insurance Scheme

A weather-index insurance scheme should ideally take into account the following scientific and technical details:

5.1 Payable index. Several models, typified by the DSSAT series, are available to simulate crop yield. The minimum climatic variables required as key drivers are daily maximum and minimum temperatures, solar radiation and rainfall. In principle, such models could be used to determine whether farmers receive an indemnity or not, by inputting the current weather data into the model as they become available. Although this approach is scientifically sound, it is unlikely to be thought transparent by either the insured or the insurer. The requirement of a weather index simply means that a complex relationship between one climatic variable, such as rainfall in the case of drought, and crop yield must be converted into a simple index. Moreover, the index must be easily understood by all parties so that the trigger event for an indemnity payment and its magnitude is clearly defined.

5.2 Accurate estimation of payment probabilities. Insurance companies will need to know how often they will be paying out indemnities based on each of the weather stations they are using as a reference for payments. In some cases these weather stations will not have the necessary historical data to determine this probability. A method therefore needs to be established that will enable accurate estimation of the probability at points where the historical data are inadequate or lacking.

5.3 Weather insurance package or stand alone solution crop solution. The kind of weather index method presented in this paper is applicable to any crop included in DSSAT, which are the main staple crops grown. Furthermore, the index developed for any one component may be part of a broader insurance package, which includes, for example, excessive precipitation and other risks, or as a stand-alone solution. The type of instrument offered will depend on the geographical location and the risks that farmers there face.

Of course, the approach that we propose must be validated in the real world, and we present it as a paradigm that can address

Table 5. Example of season resulting in an indemnity payment (total rainfall deficit exceeds the trigger value of –70mm).

	Day 1 to 10	Day 11 to 20	Day 21 to 30	Day 31 to 40	Day 41 to 50	Day 51 to 60	Day 61 to 70	Day 71 to 80	Day 80 to 90
MIN	0	10	10	25	40	40	40	30	0
RAIN	5.8	3.6	0	9.5	4.1	23.5	12.6	2	96.1
DEF		−6.4	−10	−15.5	−35.9	−16.5	−27.4	−28	
a. TOTAL Rainfall deficit									−139.7

the needs of smallholders who lie outside the ambit of current schemes. These validations can only be done in the future (unless there is some trove of data of which we are not aware), with a well-distributed set of weather stations and reliable data of farmers' yields in the vicinity of each. A scheme in process by the Syngenta Foundation in Kenya collects weather data in real time from sensors on cellphone masts for farmers located in the coverage area of the tower. Yield data from nearby farmers' fields could provide the validation that we seek over several years.

5.4 Methodological issues. We discussed the suitability of the DSSAT models to formulate indices for crop weather insurance in Diaz Nieto et al. [12], especially the criticism that 'DSSAT results are calibrated to a very specific and idiosyncratic situation' [8]. We disagreed with this assessment, pointing out that, 'far from a weakness, this is [DSSAT's] great strength'. We argued that DSSAT allowed us to reduce spatial risk in a 'transparent and logically consistent manner ..., which is impossible in the statistical approaches advocated by others.' We went on to cast doubt on the DSSAT modeling that Osgood et al. [8] did, concluding that, 'In our experience, the results that Osgood et al. [8] report are so bad that we wonder whether the simulations were set up correctly. Certainly DSSAT can give bad results if the models are not set up with some basic understanding of crop agronomy.'

We note that Gianini et al. [9], in examining artificially-generated weather for sites in Central America, used the WGEN routine, which uses a first-order Markov model. As Jones and Thornton [50] point out, first-order Markov simulates temperate weather, which is controlled by a more-or-less orderly procession of weather systems from west to east, relatively well, but it fails to capture the very different synoptic situations of the tropics. Jones and Thornton [50] showed that a third-order Markov model was required to simulate the different patterns of rainfall in the tropics, producing MarkSim [39], which we have used here. We urge others to make use of this tool that more closely reflects the behaviour of tropical systems.

No weather simulator will forecast extreme events, so the method presented here will need to be modified to take account of their historical frequency if that is deemed necessary [35]. As it stands, the method does not address this issue. Typically, engineers use a Pearson function (logarithmic extrapolation) based on historical data, but consideration of this approach is outside the scope of this paper.

The task of producing index insurance instruments for smallholders anywhere in the tropics is frustrated by two realities: there are few long-term sets of meteorological data (and let us not even think about how reliable they might be), and data of farmers' yields are similarly sparse and unreliable. Rather than treat the problem as intractable and ignore the needs of smallholders that conventional approaches regard as uninsurable, which most insurers and especially reinsurers do, we propose an alternative. MarkSim does reliably represent the climate variability in the tropics, especially when combined with the WorldClim database at 1-km resolution. The DSSAT suite of crop models incorporates understanding of crop physiology, biochemistry, and agronomy, developed over more than 30 years and have been widely documented in the literature over the last 30 years. Thus we feel that it is realistic to combine modeled weather data extrapolated from the best available meteorological data in combination with a tested crop model to generate reasonable predictions of risk in areas that are currently unserved by commercial approaches.

Conclusion

We present methods of providing low-cost, site-specific drought insurance products for most crops in any location in the tropics. We explain the benefit of insurance to risk takers, and especially those with minimal resources, from which it should become apparent that the major contribution this innovation offers is that it streams best available science about natural hazards directly to decision makers, through the medium of commercially-viable insurance products. Insurance provides decision-support to manage drought risk. The basis of the method, the insurance premium, transmits the best-available estimate of drought probabilities.

Estimates are only as accurate as the predictive model that produces them and we reflect here on three sources of basis risk that are likely to occur when modeling crop drought risk: structural uncertainty of the model; spatial error and temporal error. Structural uncertainty increases when the model fails to represent processes that significantly influence drought risk. In this respect, a model that depends solely on correlation between rainfall and yield will not represent systematic and significant yield variations that are caused by temperature, soil, crop variety or a number of other factors.

Spatial error introduces a second major source of basis risk, since it is rare that weather data, and even more so, yield data, are available with sufficient density to enable simple interpolation over large areas. Even where dense networks of weather stations exist the degree of bias towards non-marginal sites is unknown, hence its ability to represent higher risk in marginal areas. Thirdly, error can occur due to unexplained temporal error caused by inadequate data runs. A purely empirical estimation of low probability events requires long runs of data.

We do caution that the method should not be applied uncritically as illustrated by the effect of soil texture and slope on soil water recharge, and the influence of temperature on growth and hence yield at higher altitudes.

Methods

1. The Main Challenge in Developing Weather Insurance: Basis Risk

The greatest challenge facing weather-based insurance products is basis risk [14,26,32,37,38]. Basis risk occurs when the insurance index does not accurately represent loss: a weather index may not trigger a payment when there has indeed been a loss; or payment may occur without serious loss. The insurance product will not be attractive to potential customers if they think that the basis risk is too high [26].

A feasibility study of rainfall indices for Nicaragua concluded that even within departments a single index did not adequately represent the spatial variability of risk [14]. In each department there was at least one weather station where the data were markedly different from the others. A study by Diaz-Nieto et al. [12] using simulated data for Honduras also revealed that a single weather index was not appropriate for a country the size of Honduras.

Basis risk is caused by the need to model complex heterogeneous systems within a single index. There are three sources of basis risk (Table 2).

Specialized contracts can be designed to offset much of temporal, spatial and crop-specific basis risk [37]. However, doing so may increase administrative costs and, more importantly, increase the complexity involved in marketing and distribution. An alternative to overcome basis risk is a larger number of standard contracts that cover all possibilities and priced accordingly, and allow the insured to select the contract they consider most appropriate [38].

2. Establishing the Correlation between Crop Yield and the Rainfall Index

The fundamental requirement of a rainfall index is that rainfall must explain a large proportion of the variability in yield [26,30,32,38]. As a first step, it is essential to establish the cause and effect relationship [38], so that the index represents critical rainfall deficits that account for crop yield losses. It is not sufficient, for example, to posit that a rainfall deficit of 30% of the long-term average will trigger payment because this provides no information about the timing of rainfall in relation to crop demands at different growth stages.

Defining the weather events that cause the most serious yield losses and that cover as many of the loss-causing events as possible requires a considerable investment in research [26]. Furthermore it is critically important that both parties agree that the weather index adequately explains the variability in crop yields [30]. Few customers would be inclined to purchase insurance that they did believe protected them against risk.

3. Limited Availability of Yield and Climate Data on which to Base Indices

Stoppa and Hess [30] suggested that to develop effective weather-index insurance the weather variable must not only be measurable but adequate historical weather records must be available from which to estimate probabilities of a risk event occurring and its magnitude. In spite of this, many of the feasibility studies into the use of weather-based indices in developing countries provide indices based on relatively few data. Reliable long-term datasets of weather in developing countries are very limited and this presents a major potential challenge. It is noteworthy that countries with poor infrastructure are amongst those places where an effective insurance product could have most impact. The danger is that poor regions, which have greatest need for insurance, are those which are excluded, precisely for reasons of poor infrastructure associated with poverty.

An alternative approach, which we describe below, is to use statistical models and process-based simulation models, based on decades of scientific analysis, to generate 'pseudo-historical' data of climate and yield. Where possible these pseudo-historical data can be complemented with such weather data as are available.

4. Payout Index Highly Correlated with Yield Loss

In a weather insurance scheme it is not the actual crop loss that is insured but the loss-causing event, which in this case is a specified adverse weather event. Therefore the way in which the relationship between weather and crop losses is expressed in an insurance index needs to be carefully thought out and appropriately designed. A producer will be interested in a weather-insurance scheme that is highly likely to pay out when (s)he does indeed suffer a crop loss. Ideally the relationship between weather and crop yield can be extracted from long historical records of both. In practice, as in the case of drybean yields in Nicaragua, data are typically very scarce. It was therefore necessary to design a methodology that allowed weather insurance to be developed in these circumstances.

5. Summary of the Honduras Study

Díaz Nieto et al. [12] proposed a method for an indexed insurance instrument for tropical sites for which historic data were not available for either rainfall or crop yield. Briefly, they combined the MarkSim weather generator [39] with the drybean simulation of the DSSAT series of crop models [40] for six sites in central Honduras. Because there was a low frequency of drought for some of the sites, they randomly imposed droughts for ten-day periods (dekads) during crop growth. By comparing the simulated yields of the droughted crops with those with no drought, they determined sensitivity coefficients of the crop to drought at different stages of growth for each site. Rainfall for each dekad of crop growth was weighted by the crop sensitivity coefficient and the total weighted rainfall was expressed as a percentage of the long term mean. They selected an arbitrary "strike" value for each site for years with rainfall deficit greater than 65%. They based payout on the percentage deficit and calculated notional premiums based on the frequency and amount of payout.

Because the north-central mountains of Nicaragua are drier than the six sites in Honduras chosen by Díaz Nieto et al. [12], we used a different method to determine the crop sensitivity coefficients, described below.

Author Contributions

Conceived and designed the experiments: JD MF SC. Performed the experiments: JD. Analyzed the data: JD MF SC PL ML. Contributed reagents/materials/analysis tools: JD MF SC PL ML. Wrote the paper: JD MF SC PL ML.

References

1. Bird K, Hulme D, Moore K, Shepherd A (2002) Chronic poverty and remote rural areas, CPD Working Paper No.13. Birmingham: University of Birmingham.
2. EM-DAT (2004) The OFDA/CRED (Office of U.S. Foreign Disaster Assistance/WHO Collaborating Centre for Research on the Epidemiology of Disasters). International Disaster Database. Available: www.em-dat.net. Accessed April 2012.
3. Dercon S (2001) Income risk, coping strategies and safety nets. Center for the Study of African Economics, Oxford University.
4. Pandey S, Behura D, Villano R, Naik D (2001) Drought risk, farmers' coping mechanisms, and poverty: A study of the rainfed rice system in eastern India. Rice Research for Food Security and Poverty Alleviation 67–274.
5. Sakurai T, Reardon T (1997) Potential demand for drought insurance in Burkina Faso and its determinants. American Journal of Agricultural Economics 79: 1193–1207.
6. Patt A, Suarez P, Hess U (2010) How much do smallholder farmers understand and want insurance? Evidence from Africa. Global Environmental Change 20: 153–161.
7. Liu B, Li M, Guo Y, Shan K (2010) Analysis of the Demand for Wather Index Agricultural Insurance on Household level in Anhui. International Conference on Agricultre Risk and Food Security 1: 179–186.
8. Osgood D, Mclaurin M, Carriquiry M, Mishra A, Fiondella F, et al (2007) Designing Weather Insurance Contracts for Farmers in Malawi, Tanzania and Kenya. Final report to the World Bank Commodity Management Group, ARD, Word Bank, IRI Technical report 07-02.
9. Giannini A, Hansen J, Holthaus E, Ines A, Kaheil Y, Karnauskas K, Mclaurin M, Osgood D, Robertson A, Shirley K, Vicarelli M (2009) Designing index-based weather insurance for farmers in Central America. International Research Institute for Climate and Society, Palisades: Project and report to the World Bank Commodity Management Group, ARD, Word Bank. IRI Technical report 09-01. 19p.
10. Cohen M, Sebstad J (2006) The demand for microinsurance. Churchill, C. (Ed.), Protecting the Poor: A Microinsurance Compendium. Geneva: International Labour Organization. pp 25–44.
11. Dercon S, Kirchberger M, Gunning J, Platteau J (2008) Literature Review on Microinsurance. Geneva: International Labour Organization.
12. Díaz Nieto J, Cook SE, Läderach P, Fisher MJ, Jones PJ (2010) Rainfall index insurance to help small-holder farmers manage drought risk. Climate and Development 2: 233–247.
13. Läderach P, Lundy M, Jarvis A, Ramires J, Emiliano Pérez PE, et al (2011) Predicted impact of climate change on coffee-supply chains. The Economic, Social and Political Elements of Climate Change Chapter 43. Berlin: Springer Verlag. 19 p.
14. World Bank (2001) Innovations in managing catastrophic risks: How can they help the poor? Washington, DC: Workshop.

15. Rosenzweig MR, Binswanger HP (1993) Wealth, weather risk and the composition and profitability of agricultural investments. The Economic Journal 103: 56–78.

16. Mosley P, Krishnamurthy R (1995) Can crop insurance work? The case of India. Journal of Development Studies 31: 428–450.

17. World Bank (2000) Managing catastrophic risks using alternative risk financing and insurance pooling mechanisms. Washington: Report No.20448-LAC.

18. Webb P, Reardon R (1992) Drought impact and household response in East and West Africa. Quarterly Journal of International Agriculture 31: 230–246.

19. Morduch J (1995) Income smoothing and consumption smoothing. Journal of Economic Perspectives 9: 103–114.

20. Morduch J (1999) Between the state and the market: Can informal insurance patch the safety net? The World Bank Research Observer 14: 187–207.

21. Barrett CB, Reardon T, Webb P (2001) Non farm income diversification and household livelihood strategies in rural Africa: Concepts, dynamics and policy implications. Working paper. Available: http://www.inequality.com/publications/working_papers/Barrett-Reardon-Webb_IntroFinal.pdf.

22. Wenner M, Arias D (2003) Agricultural insurance in Latin America: Where are we? Paving the Way Forward for Rural Finance: An International Conference on Best Practices, Washington DC. Case Study. Available: http://www.basis.wisc.edu/live/rfc/cs_03b.pdf.

23. Brown W, Churchill C (1999) Providing insurance to low-income households Part I: A primer on insurance principles and products. USAID - Microenterprise Best Practices (MBP) Project. 91p.

24. Hazell P, Larson D, Mac-Isaac D, Zupi M (2000) Weather Insurance Project Study: Ethiopia. The World Bank/Fondazione Salernitana Sichelgaita/IFPRI.

25. van Oppen C (2001) Insurance: A tool for sustainable development. J. Insur. Res. Prac.19 p.

26. Skees JR, Gober S, Varangis P, Lester R, Kalavakonda V (2001) Developing rainfall based index insurance in Morocco. Policy, Research working paper. WPS 2577. Washington: The World Bank. Available: http://wdsbeta.worldbank.org/external/default/WDSContentServer/IW3P/IB/2001/04/27/0000949.

27. Miranda MJ, Glauber JW (1997) Systematic risk, reinsurance, and the failure of crop insurance markets. American Journal of Agricultural Economics 79: 206–215.

28. Bryla E, Dana J, Hess U, Varangis P (2003) The use of price and weather risk management instruments. Risk Management: Pricing, Insurance, Guarantees. Washington: The World Bank. Available: http://www.basis.wisc.edu/live/rfc/cs_03a.pdf.

29. Hess U (2003) Innovative financial services for rural India: Monsoon-indexed lending and insurance for small holders. Agriculture and Rural Development Department Working Paper 9, Washington: World Bank.

30. Stoppa A, Hess U (2003) Design and use of weather derivatives in agricultural policies: The case of rainfall index insurance in Morocco. International Conference "Agricultural policy reform and the WTO: where are we heading?" Capri, Italy, June 23–26.

31. Varangis P, Skees J, Barnett B (2003) Weather indexes for developing countries (CHP).

32. Skees JR (2003) Risk management challenges in rural financial markets: Blending risk management innovations with rural finance.

33. Skees JR, Collier B (2008) The potential of weather index insurance for spurring a green revolution in Africa.

34. Leftley R (2009) Micro-insurance for health and agricultural risks. In Innovations in Insuring the poor. Focus 17, Brief 4.

35. Collier B, Barnett B, Skees J (2010) State of knowledge report – data requirements for the design of weather index insurance.

36. Binswanger-Mkhize HP (2011) Is there too much hype about index-based agricultural insurance? Journal of Development Studies.

37. Miranda MJ, Vedenov DM (2001) Innovations in agricultural and natural disaster insurance. American Journal of Agricultural Economics 83: 650–655.

38. Turvey CG (2001) Weather derivatives for specific event risks in agriculture, Review of Agricultural Economics. Agricultural Economics 23: 333–351.

39. Jones PG, Thornton PK, Díaz W, Wilkens PW (2002) MarkSim: A computer based tool that generates simulated weather data for crop modelling and risk assessment. CD-ROM Series, Cali: CIAT. 88 p.

40. Jones JW, Hoogenboom G, Porter CH, Boote KJ, Batchelor WD, et al (2003) The DSSAT cropping system model. European Journal of Agronomy 18: 235–265.

41. Ministerio Agropecuario y Forestal (2001) Atlas Rural de Nicaragua. Managua, Nicaragua.

42. Quintana JO, Caceres VH (1983) Suelos y fertilización del frijol Phaseolus vulgaris L. en las zonas de producción de Nicaragua.

43. Baltodano ME (2001) Línea Base del Sitio de Referencia, San Dionisio, Nicaragua.

44. Magaña V, Amador JA, Medina S (1999) The midsummer drought over Mexico and Central America. Journal of Climate 12: 1577–1588.

45. Voysest O, Dessert M (1991) Bean cultivars: Classes and commercial seed types. pp 119–162. Common Beans: research for crops improvement.C.A.B. Int Wallingford, reuno unido & CIAT, Cali, Colombia.

46. Jaramillo PM (1989) El cultivo del fríol. Federación Nacional de Cafeteros.

47. Ríos MJ, Quirós JE (2002) El Frijol (Phaseolus vulgaris L.): Cultivo, Beneficio y Variedades.

48. Davidson BR (1965) The Northern Myth: A Study of the Physical and Economic Limits to Agricultural and Pastoral Development in Tropical Australia. Carlton: Melbourne University Press. 283p.

49. Hijmans RJ, Cameron SE, Parra JL, Jones PG, Jarvis A (2005) Very high resolution interpolated climate surfaces for global land areas. International Journal of Climatology 25: 1965–1978.

50. Jones PG, Thornton PK (2000) MarkSim: software to generate daily weather data for Latin America and Africa. Agronomy Journal 93: 445–453.

21

An Ensemble Weighting Approach for Dendroclimatology: Drought Reconstructions for the Northeastern Tibetan Plateau

Keyan Fang[1,2,3]*, Martin Wilmking[4], Nicole Davi[5,6], Feifei Zhou[1], Changzhi Liu[2]

1 Key Laboratory of Humid Subtropical Eco-Geographical Process (Ministry of Education), College of Geographical Sciences, Fujian Normal University, Fuzhou, Fujian Province, China, **2** Key Laboratory of Western China's Environmental Systems (Ministry of Education), Research School of Arid Environment and Climate Change, Lanzhou University, Lanzhou, Gansu Province, China, **3** Department of Geosciences and Geography, University of Helsinki, Helsinki City, Helsinki, Finland, **4** Institute of Botany and Landscape Ecology, Greifswald University, Greifswald, Mecklenburg-Vorpommern, Germany, **5** Tree-Ring Lab, Lamont-Doherty Earth Observatory, Columbia University, Palisades, New York, United States of America, **6** Department of Environmental Science, William Paterson University, Wayne, New Jersey, United States of America

Abstract

Traditional detrending methods assign equal mean value to all tree-ring series for chronology developments, despite that the mean annual growth changes in different time periods. We find that the strength of a tree-ring model can be improved by giving more weights to tree-ring series that have a stronger climate signal and less weight to series that have a weaker signal. We thus present an ensemble weighting method to mitigate these potential biases and to more accurately extract the climate signals in dendroclimatology studies. This new method has been used to develop the first annual precipitation reconstruction (previous August to current July) at the Songmingyan Mountain and to recalculate the tree-ring chronology from Shenge site in Dulan area in northeastern Tibetan Plateau (TP), a marginal area of Asian summer monsoon. The ensemble weighting method explains 31.7% of instrumental variance for the reconstructions at Songmingyan Mountain and 57.3% of the instrumental variance in the Dulan area, which are higher than those developed using traditional methods. We focus on the newly introduced reconstruction at Songmingyan Mountain, which showsextremely dry (wet) epochs from 1862–1874, 1914–1933 and 1991–1999 (1882–1905). These dry/wet epochs were also found in the marginal areas of summer monsoon and the Indian subcontinent, indicating the linkages between regional hydroclimate changes and the Indian summer monsoon.

Editor: Eryuan Liang, Chinese Academy of Sciences, China

Funding: This research is financed by National Basic Research Program of China (2012CB955301), the National Science Foundation of China (41001115 and 41171039), and the Minjiang Special-term Professor fellowship. Support for Davi comes from NSF AGS # 1137729. The funders had no role in study design, data collection and analysis, decision to publish, or preparation of the manuscript.

Competing Interests: The authors have declared that no competing interests exist.

* E-mail: kujanfang@gmail.com

Introduction

Global warming has brought long-term climate data inferred from proxies into focus for both the scientific and public communities. These proxies enable us to assess recent climate change in the context of hundreds to thousands of years and to evaluate changes prior to any anthropogenic influences [1,2]. In addition, the availability of large-scale paleoclimate data increases the robustness of the analyses of regional climate regimes in relation to external forcings and internal feedback loops [3]. Because tree-ring records are climate sensitive and can be exactly dated, they have been widely used to extend short-term instrumental climate by centuries to millennia from regional to global scales [4]. A variety of detrending methods have been developed to isolate and extract the climate signal from tree-ring series, such as negative exponential or straight line splines [5]. However, the too flexible forms result in the inevitable loss of longer-timescale climate signal. In addition, the medium-frequency (e.g. decadal/multi-decadal scales) variations can bias the final chronology [6]. This is referred to as the "trend distortion" problem, which can be mitigated by the

"signal-free" method [6]. In addition, potential bias in the traditional methods can arise when setting the mean value of the tree-ring indices to 1 for different tree-ring indices covering different time intervals (Figure 1). The mean values of individual tree-ring series can be different in different temporal intervals, therefore low (high) tree-ring indices can be increased (decreased) when assigning the same value to these indices (Figure 1). Third, the tree-ring chronology can be biased when including some less climate-sensitive tree-ring series or those showing a different climate-growth relationship [7,8]. In order to mitigate the three potential biases, we propose a method, termed the "ensemble weighting method", to iteratively weight individual tree-ring series according to their mean climate values and by the sensitivity of each series to climate.

We use this new ensemble weighting method to develop the first tree-ring chronology from the Songmingyan Mountains in the Linxia district of the northeastern Tibetan Plateau (TP) and to ̄ a shoulder region of Asian summer monsoon [9]. Understanding hydroclimate dynamics in regions that are only marginally affected by Asian summer monsoon is highly needed due to the sensitivity of these regions to large-scale atmospheric

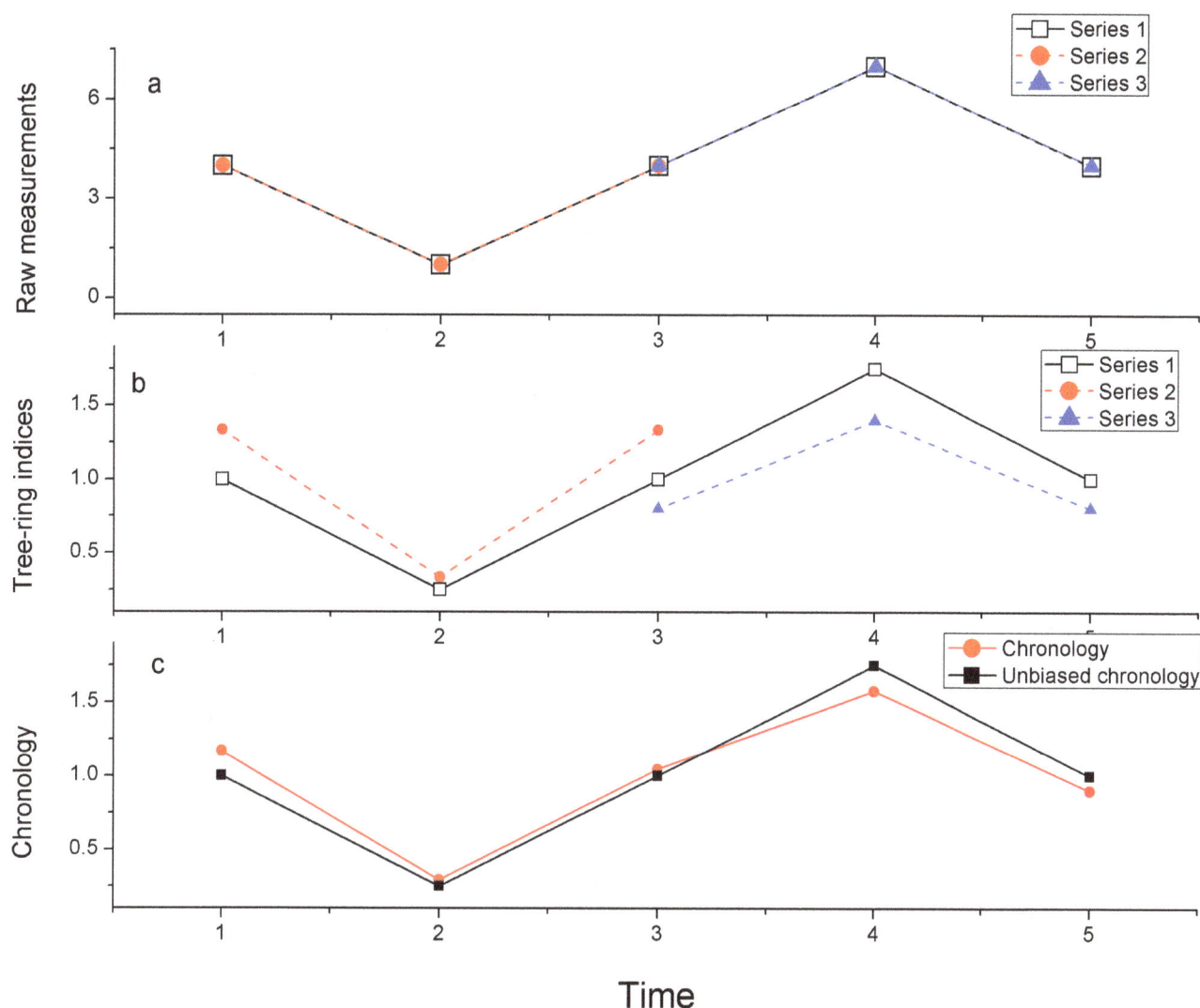

Figure 1. Schematic illustration of the distortion in chronology developments. (a) raw tree-ring measurements, (b) the standardized tree-ring indices calculated between the raw measurements and their mean values, and (c) the comparisons between chronology calculated as the mean among the tree-ring indices and the unbiased chronology.

changes and the potential impacts of water variability. Most instrumental records from these regions begin after the 1950s and tree-ring based reconstructions are sparse, limiting our ability to understand large-scale monsoon dynamics. Although there are many climate-sensitive tree-ring chronologies available for neighboring regions in northeastern TP [10,11,12,13,14,15], there are still no chronologies published from the Linxia district largely due to the limited availability of old-growth forests. In addition, we also test the use of this method for the development of a 1835-year tree-ring chronology in a nearby region in Dulan, which has been used for a hydroclimate reconstruction in the previous study [12,13,14,15]. Therefore, the goals of this study are to (1) introduce the "ensemble weighting method", (2) test its applicability by developing the first tree-ring chronology in the Linxia area and the tree-ring chronology in Dulan area, and (3) provide climate reconstructions for the northeastern TP and to detect its linkages with monsoon dynamics.

Data and Methods

The Instrumental Data

Monthly mean temperature peaks in July and monthly total precipitation peaks in August when the monsoon front reaches its northern shoulder region in the northeastern TP [16]. Old-growth forests in this region can only be found on the TP and on some of the mountain ranges on the arid or semi-arid Chinese Loess Plateau [17] (Figure 2). The newly introduced sampling site is located in the Songmingyan Mountain area (35.23°N, 103.39°E, 2589 m a.s.l.) in Linxia (Figure 2), in the northeastern TP. Total annual precipitation is 531 mm and annual mean temperature is 7.2°C according to the nearest meteorological station at Lintao (35.33°N, 103.82°E, 1891 m a.s.l., WMO NO. 52986). Pines and/or spruces growing near these temples are often well protected and can grow very old and can be used to develop long-term tree-ring chronologies [10,18,19]. The tree-ring samples used here are from Wilson's spruce (*Picea wilsonii*) growing in the sacred green island region on Songmingyan Mountain, the biggest national forest park in the Linxia district (Figure 2). The second tree-ring site at Dulan area is located near the Shenge town, northern to the Songmingyan Mountain area with a drier and colder climate than Songmingyan Mountain [15]. Total annual precipitation is 188 mm and the mean annual temperature is 3°C in Dulan meteorological station (36.00°N, 98.00°E, 3800 m a.s.l., WMO NO. 52986).

Tree-ring Data and Traditional Methods

We collected 30 cores from 20 *Picea wilsonii* trees at the Songmingyan Mountain site with permission from the Songmingyan Forest Park. 132 cores of 60 *Sabina przewalskii* trees were from Shenge site were downloaded from the International Tree-ring Data Bank (http://www.ncdc.noaa.gov/paleo/treering.html) [15]. These samples were mounted, air dried, polished and crossdated according to a skeleton plotting scheme by visually comparing the extremely narrow and wide rings [20]. The crossdated tree rings were measured to 0.001 mm accuracy and quality checked with moving correlations using the program COFECHA [21]. The growth trend in the raw measurements is removed by fitting an age-related growth curve (herein a cubic smoothing spline with a 50% cutoff at around 67% of the mean segment length) [22]. The dimensionless tree-ring indices are calculated as ratios between raw measurements and the fitted growth values, which are then averaged to produce a chronology based on a robust mean methodology [23]. In the signal-free method, the signal-free measurements are indexed as ratios between raw measurements and the initial robust mean chronology indices, which are again fitted with a growth curve (herein a spline) to create the "signal-free curve" representing the age-related growth trend. Then the tree-ring indices are calculated as ratios between raw measurements and the signal-free curve, which were used to create a new chronology. The final chronology is produced by iterating the aforementioned steps until the two latest versions of the chronologies showed only limited differences [6].

Ensemble Weighting Method

As shown in Figure 3, the ensemble weighting method for chronology development contains 3 stages and 6 steps, including:

Stage 1. Developing an initial chronology using traditional methods.

Step 1–2. Fitting growth curves and developing the chronologies using traditional methods detailed in the section above.

Stage 2. Developing the first version of the ensemble weighting chronology.

Step 3. The mean values of the initial chronology for the time period (segment length) same as the duration of a given tree-ring index are calculated. These mean values are then assigned to the associated tree-ring index by multiplying the ratio between the mean chronology value corresponding to the length of the tree-ring index and the mean chronology value during the entire period.

Step 4. The target climate variable used for reconstruction is selected based on climate-growth relationships of the initial chronology. The Pearson correlation between resulting tree-ring indices and the target climate variable is calculated for individual tree-ring series. Each tree-ring index is weighted by its correlation

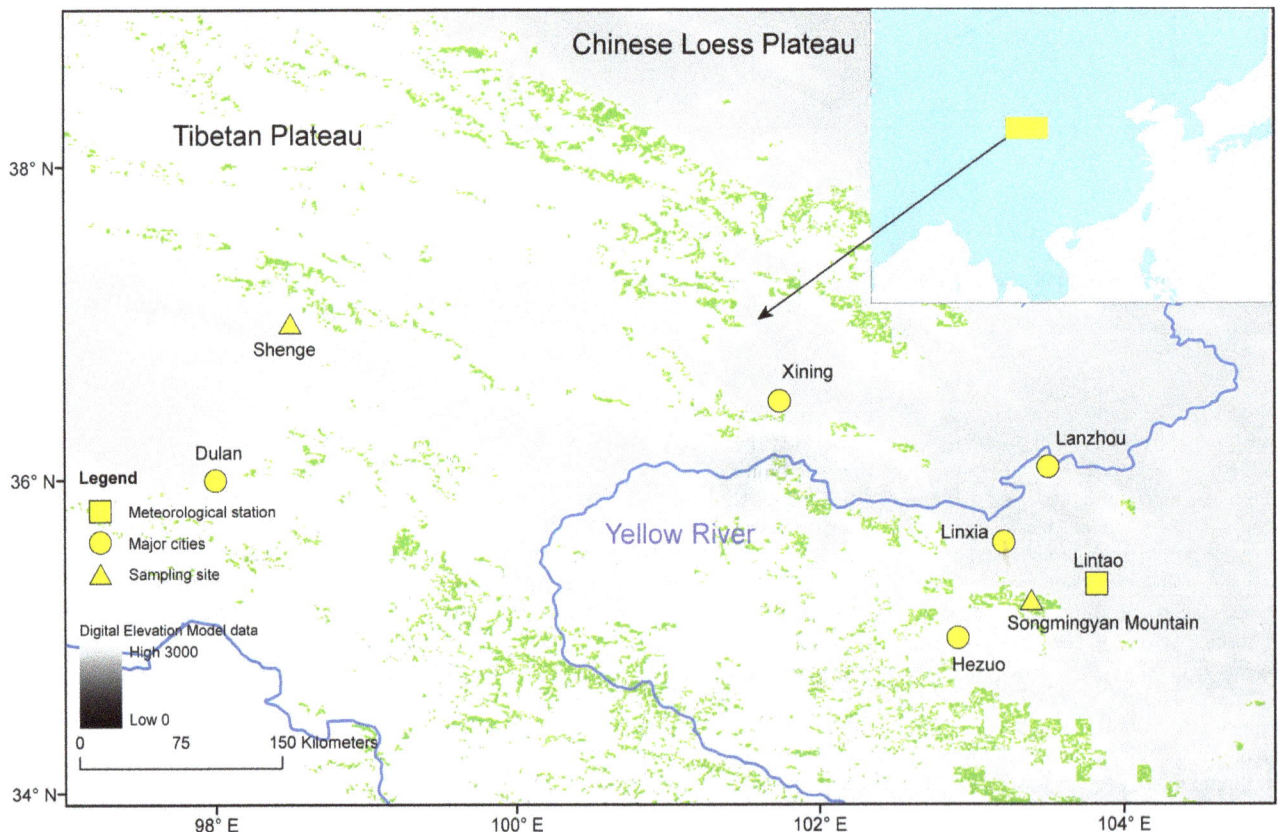

Figure 2. Map of the sampling site at the Songmingyan Mountain at Linxia area, the Shenge site at Dulan area, the meteorological station and major cities in the study region. The topographic features are indicated by digital elevation model data in grey colors and the boreal forests are shown in green color as well as the position of the study region in East Asia.

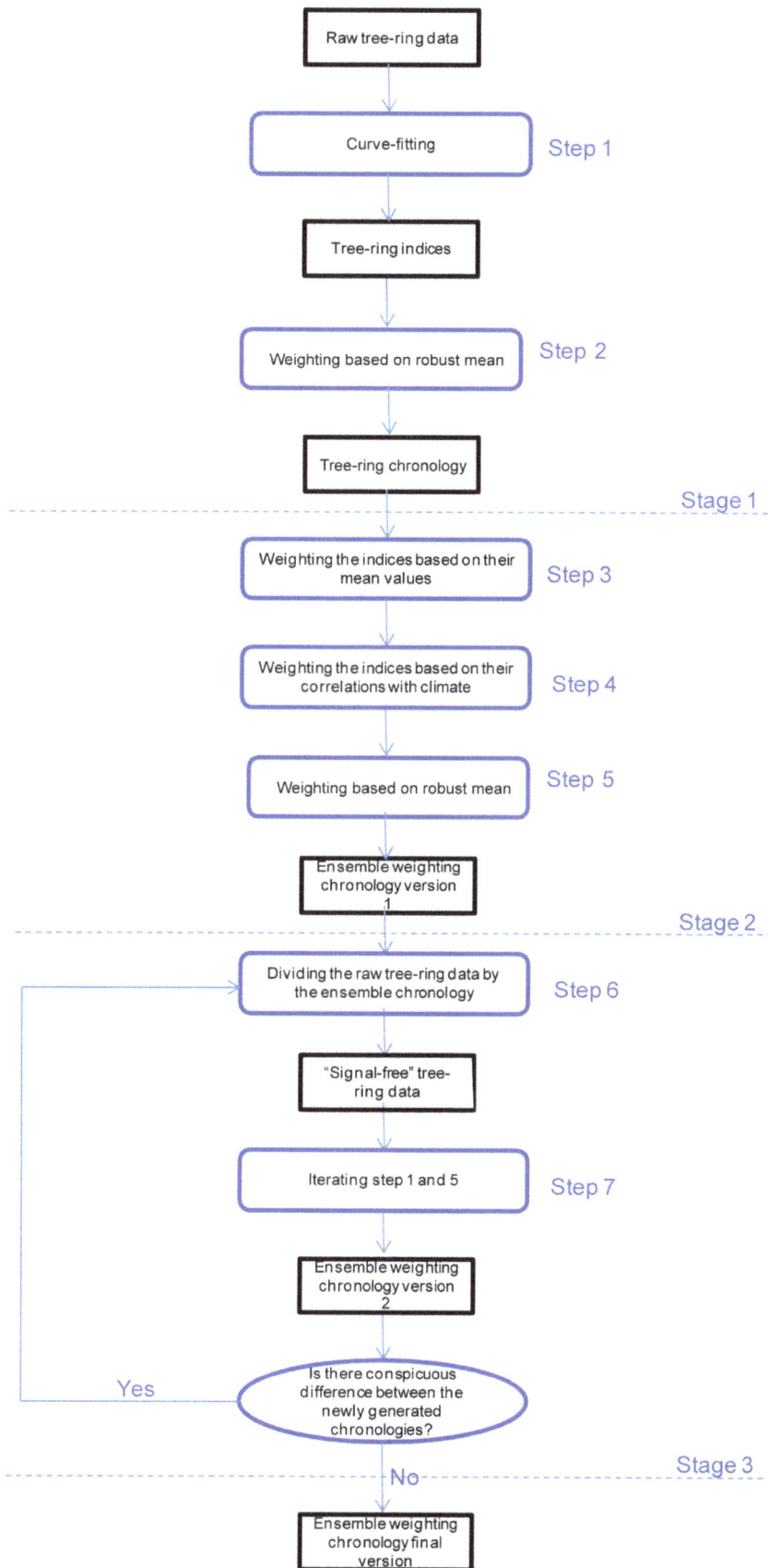

Figure 3. Flow diagram illustrating the development of an ensemble weighting chronology.

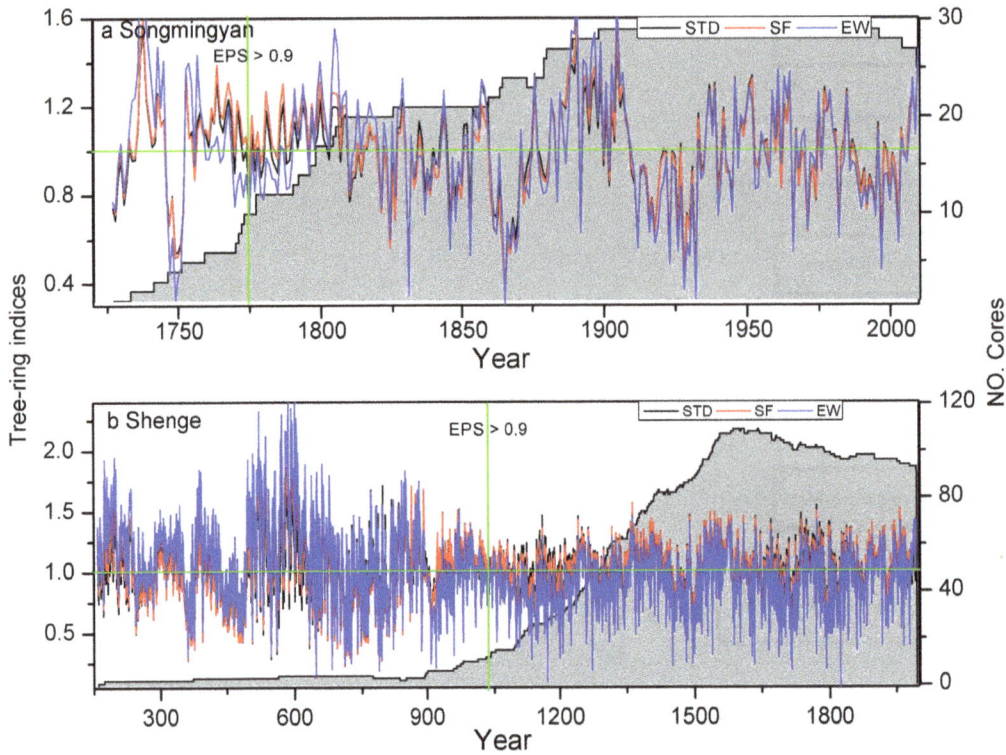

Figure 4. Comparisons of tree-ring chronologies based on the traditional method (STD), the signal-free (SF) method and the ensemble weighting (EW) method for (a) the Songmingyan Mountain site and (b) the Shenge site.

with climate at different powers (herein 0, 0.5, 1, 1.5 and 2). A power of 0 equals to an unweighted series. That is,

$$wTR = m \times r^p \times TR$$

where wTR is the weighted tree-ring index, TR is the unweighted tree-ring index, m is the mean chronology index in the sub-period of a given tree-ring series, r^p is the climate-growth correlations in different powers of p.

Step 5. The robust mean method is applied to the weighted tree-ring index to produce the chronologies (herein 5 chronologies) weighted by climate-growth correlations [8] at given powers (herein power of 5). The ensemble weighting chronology is the arithmetic mean of these chronologies, because an ensemble approach can dampen the influences of spurious climate-growth correlations due to the noise in both climate and tree ring data.

Stage 3. Developing the final chronology using signal-free iterations.

Step 6. The signal-free tree-ring measurements are produced as ratios between raw measurements and the original ensemble weighting chronology.

Step 7. The iteration procedures are the same as in the traditional signal-free method developed by Melvin and Briffa (2008) introduced above. The major difference from traditional applications of signal-free method is that the tree-ring indices during the iterations process are weighted. The iterations stop when only a minor difference between chronologies is found. The first and third stages, discussed above are similar to traditional methods. Major improvements occur in the second stage when the two weighting procedures are used to produce an ensemble of chronologies.

One limitation to this method, however, is the difficulty in determining the climate-growth correlations for sub-fossil samples that do not have any overlap with instrumental data. For such sub-fossil samples, we herein weight them based on their correlations with the master chronology and the correlations between the master chronology and climate, i.e. using a weighted multiplier between the series-chronology correlations and the chronology-climate correlations. The strength of the reconstructions was tested by linearly regressing the chronologies with the instrumental climate variable and evaluating the variance explained by each. The robustness of the reconstruction was further examined by split calibration-verification procedure [24], which calibrates the instrumental data from one sub-period and verifies the reconstruction using the remaining instrumental data. The verification sub-periods are the 1980–2008 and 1952–1979 for the Songmingyan Mountain reconstruction. Keeping in mind with the relatively short common period (1954–1993) between instrumental and tree-ring data at the Shenge site, we used a slightly longer sub-period for calibration (1954–1974 and 1973–1993) to maintain the robustness the split calibration-verification. Attention is also required for tree-ring series showing unstable climate-growth associations through the instrumental period [25], which could either be excluded or receive less weights. The samples used here generally show stable responses to climate through time as indicated by acceptable split calibration-verification statistics (detailed below).

Results

In this study, we applied 3 signal-free iterations for the Songmingyan Mountain site and 5 iterations for the Shenge site as suggested in previous studies [6]. We truncated the chronology

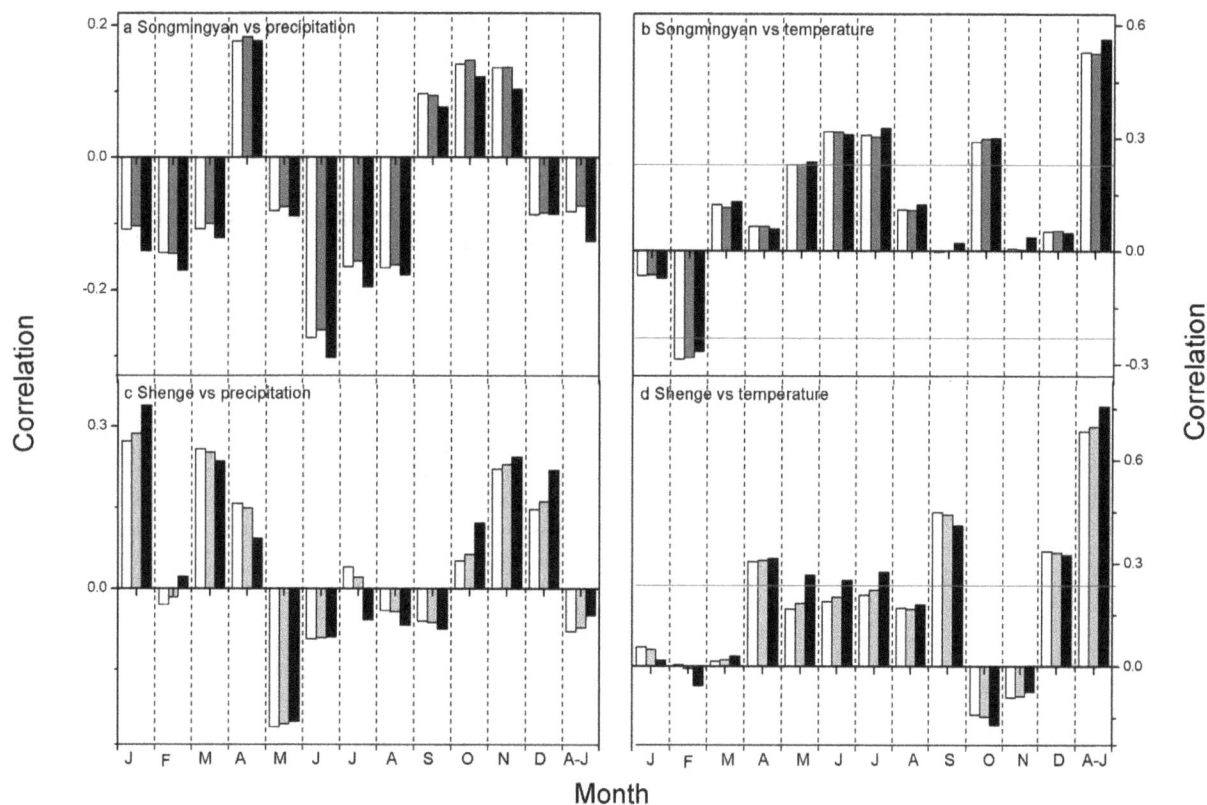

Figure 5. Correlations between tree-ring chronologies derived from traditional method (white bar), signal-free method (grey bar) and the ensemble weighting method (black bar) and the (a) monthly precipitation and annual precipitation from previous August to current July (A–J) for the Songmingyan Mountain, the (b) monthly temperature and annual temperature for the Songmingyan Mountain, the (c) monthly precipitation and annual precipitation from previous August to current July (A–J) for the Shenge site and the (d) monthly temperature and annual temperature for the Shenge site. The significance level of 0.1 is indicated by horizontal line.

at Songmingyan Mountain from 1773–2010 and Shenge chronology from 1041 to 1993 as there are sufficient replications indicated by an expressed population signal (EPS) over 0.90 [26]. Only the robust portions of the chronologies were employed in the following reconstructions. At the Songmingyan site, the signal-free method successfully mitigates the trend distortion problem by increasing (decreasing) the high (low) chronology indices in the latter half of the 18th century (during the ~1830s–1870s) (Figure 4a). At the Shenge site, the signal-free method adds back the climate signals removed in the traditional method and thus generally increase (decrease) the chronology indices when they are high (low) (Figure 4b). We additionally plotted the chronology indices derived from the three methods in the appendix (Figure S1) in a roughly 300-year interval to better illustrate the changes between them. Based on the ensemble weighting method, the chronology indices at Songmingyan Mountain increases (decreases) in the ~1790s–1820s and near 1900s (1870s–1880s and 1900s–1950s) (Figure 4a). At the Shenge site, 4 tree-ring samples (DU07A, DU12A1, DU70A and DU70B) spanning before ~800 show above average chronologies indices, which can be artificially lowered down in the traditional chronology. This problem has been mitigated in the ensemble weighting method that generally increases the chronology indices before ~800 and decreases the indices afterwards (Figure 4b). Both the signal-free and ensemble weight methods are more efficient and result in larger differences during the early periods with larger variance due to the availability of only a few tree-ring samples.

All the chronologies show positive, significant correlations with precipitation and negative correlations with temperature in the growing season (Figure 5). The highest climate-growth correlations are found with annual precipitation from the previous August to the current July for the Songmingyan Site and from previous July to currently June for the Shenge site (Figure 5). The ensemble weighting chronology shows higher correlations with precipitation than the chronologies developed using traditional and the signal-free methods (Figure 5). Therefore our later analyses are based on the reconstruction using ensemble weighting chronology since it contains a more "pure" precipitation signal and explains a higher percent of the variance. The precipitation reconstructions derived from the ensemble weighting chronologies explain 31.7% and 57.3% of the instrumental variance for Songmingyan Mountain and the Shenge site, respectively (Figure 6). The explained variance of the Shenge reconstruction is higher than that in previous reconstruction (47.8%) using traditional methods [15].

In the split calibration-verification method, the above-zero values of reduction of error (RE) and coefficient of efficiency (CE) statistics indicate that the reconstruction model has acceptable reliability in reproducing the climate signal in both sub-periods [27]. The RE and CE are acceptable for the verification sub-periods 1980–2008 (RE = 0.38, CE = 0.19) and 1952–1979 (RE = 0.31, CE = 0.06) for the Songmingyan Mountain reconstruction. For the Shenge site, RE and CE for the verifications are acceptable for verification sub-periods of 1975–1993 (RE = 0.45, CE = 0.25) and 1954–1972 (RE = 0.73, CE = 0.68). We additionally tested the robustness of the newly introduced reconstruction at

Songmingyan Mountain by comparing its extreme values with nearby reconstructions. Extremely dry years (< mean −2SD) in the reconstruction at Songmingyan Mountain are 1831, 1865, 1923, 1928, 1932 and 1997, and extremely wet (> mean +2SD) years are 1891 and 1905 (Figure 6). Some of the extreme dry years (1865, 1928 and 1997) reconstructed were also found in yearly dryness/wetness chats derived from historical documents from neighboring Lanzhou [28]. Although the extreme droughts in 1923 and 1932 were not found in the documentary records, these two extremely dry years were found in hydroclimate reconstructions in other nearby regions [10,19,29]. The extreme drought of 1997 was the most extreme in the Guiqing Mountain area [18], and is east of our study region.

Discussion

Performances of the Ensemble Weighting Method

The ensemble weighting chronologies contains two weighting procedures apart from the robust mean weighting in traditional methods [23]. This weighting procedure may have larger efficiency in modulating the chronologies for samples with both living and sub-fossil cores. This is because the living and fossil cores generally have less overlapping periods and can thus show larger differences in mean chronology values. Weighting procedure can better retain the low-frequency variations and tends to increase (decrease) the chronology indices in periods with high (low) climate signals.

Our weighting procedure is designed to put more emphasis on the most climate-sensitive samples, similar to fieldwork where scientists select sites and collect samples at sites with a high degree of climatic signal (determined by ecological conditions), at, for example, the treeline locations (a *priori* knowledge) [20]. The weighting of individual tree-ring series based on their correlations with climate provides a "quality control" to test whether individual tree-ring series contains a pure climate signal (a *posteriori* knowledge) [8]. This weighting procedure may be necessary in regions that have both arid and cold climate, such as northeastern TP. In these regions, tree growth may be sensitive to both precipitation and temperature [20] and thus it is possible to include tree-ring series with a different climate-growth relationships. In addition, this method is a more justified approach than completely excluding tree-ring samples from a chronology that have a limited climate signal. For example, the weighting of a few temperature-sensitive tree-ring series in the ensemble weighting chronology has increased (decreased) the chronology values in the warm (cold) period of the latter half of the 18th century and 19th century (early 19th century).

This ensemble weight procedure incorporates the principles of the signal-free method that has a number of iterations to mitigate the trend distortion problem. The key difference the original signal-free method gives equal weights to the tree-ring indices. From this point of view, we can consider this ensemble weighting method as a updated version of the signal-free method. Similar to the signal-free method, our method can be improve the regional curve standardization (RCS) method, a specific technique to overcome the segment length curse problem [30]. Apart from the trend distortion problem, our method can aid in producing a climate sensitive RCS chronology. However, there is no need to adjust the mean values of individual tree-ring indices in RCS, because their mean values are the ratio between mean growth measurements and the regional growth curve and thus can be different [31,32]. Therefore in the application of the ensemble weighting method in RCS, we only need to weight the tree-ring indices according to their associations with climate.

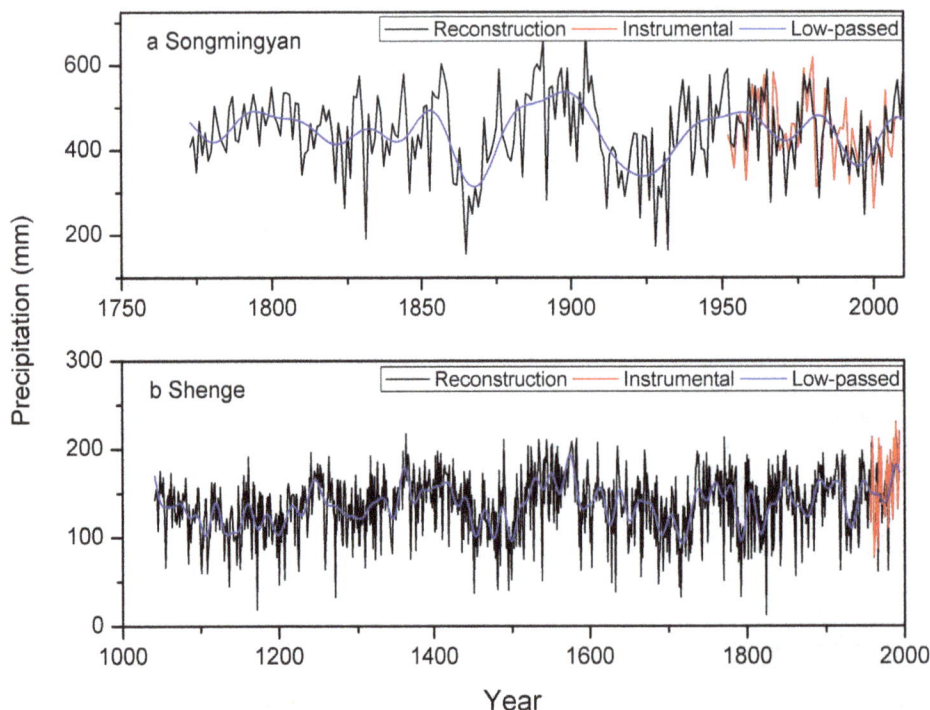

Figure 6. The reconstructed precipitations using the ensemble weighting chronology for the (a) Songmingyan and the (b) Shenge sites, and their low-passed values and the instrumental data.

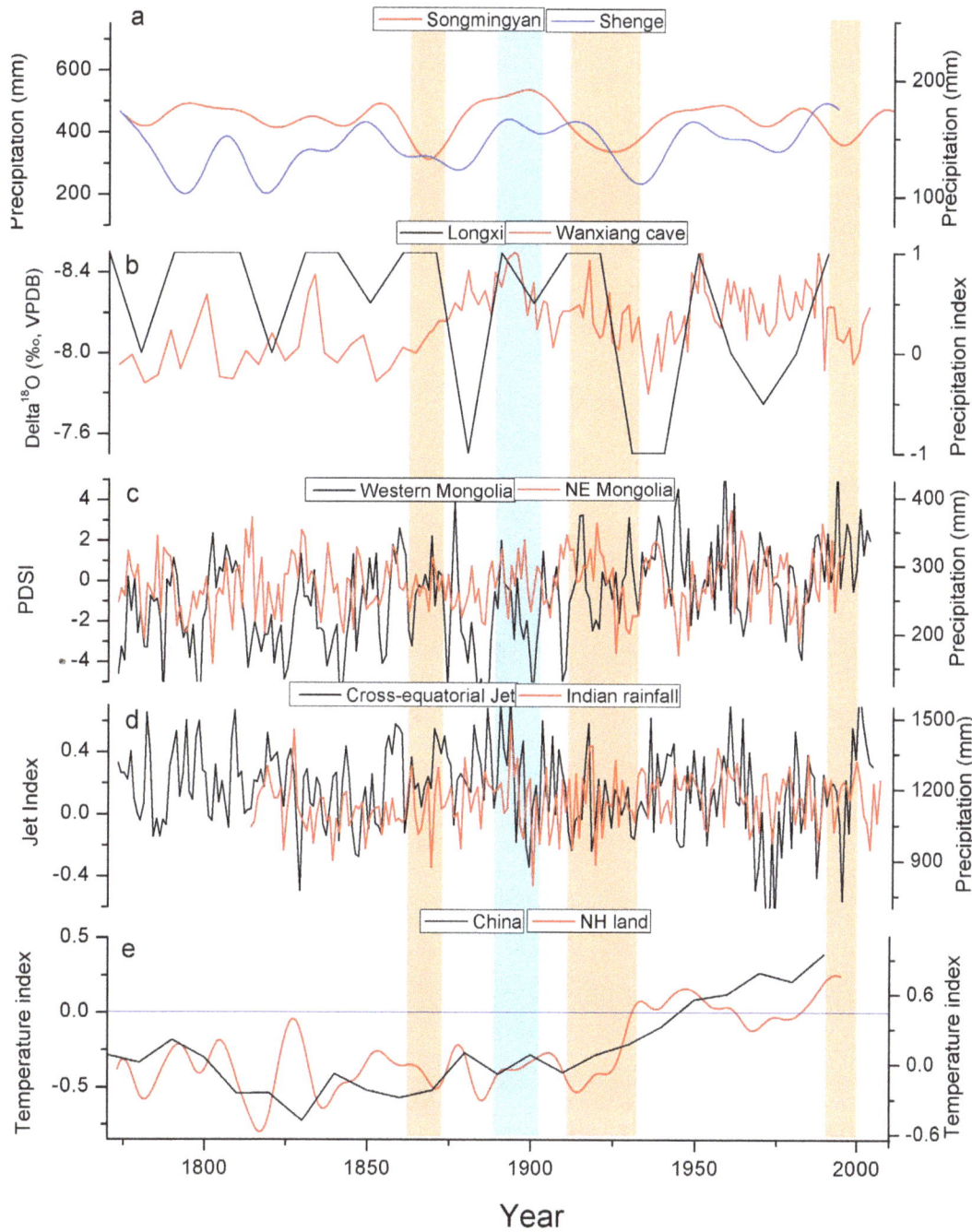

Figure 7. Comparisons among (a) precipitation reconstructions in the Songmingyan Mountain and the Shenge site, and (b) a speleothem-based monsoon record from the Wanxiang cave [38] southeastern to our study region and a precipitation reconstruction using historical documents from Longxi [39] eastern to our study region, (c) the hydroclimate reconstruction in far-western Mongolia [41] and the northeastern Mongolia [37], (d) coral-inferred variations of the low-level cross-equatorial jet in the western Indian Ocean [43] and the longest instrumental precipitation for all the India [42], and (e) the reconstructed temperature index in China derived from proxies of ice cores, tree rings, peat and historical documents [45] and the land temperature of northern hemisphere reconstructed from multiple proxies using composite-plus-scale method [1].

Precipitation Reconstruction and Monsoon Dynamics

A drought-sensitive growth pattern, negative correlations with temperature and positive correlations with precipitation, has been widely seen in arid regions [10,11,12,13,15,19], and is also found in our study region. Drought stress and less annual growth can occur when there are increases in evapotranspiration in warmer temperatures. We have found that tree growth at our study site correlates

highest with annual averaged hydroclimate conditions rather than monthly or seasonal climate data like in our study region and some sites in southeastern TP [33,34] and in the northeastern TP [12,15,35], north central China [36] and eastern Mongolia [37]. This is because tree rings can integrate monthly precipitation of consecutive months in the growing season and the precipitation from the dormant season can "compensate" for monthly water shortages

[10]. This phenomenon is more conspicuous in regions with deep soil that can retain the water in winter (non-growing season) and facilitate tree growth in the following year. Tree growth at sites with very shallow soil tends to be more sensitive only to growing season hydroclimate conditions, such as the Xiaolong Mountain [10] and the Guiqing Mountain areas [18].

Dry epochs with more than 5 continuously dry years ($<$ mean $-SD$) in the 20-year low-pass Songmingyan reconstruction are from 1862–1874, 1914–1933 and 1991–1999, and the wet epoch with over 5 continuously wet years ($>$ mean $-SD$) is from 1882–1905 (Figure 7a). For the reconstruction at Shenge site, the dry epochs are found in periods of 1100–1111, 1127–1152, 1170–1178, 1192–1207, 1449–1459, 1470–1481, 1493–1506, 1686–1698, 1709–1725, 1788–1797, 1813–1822 and 1929–1935, and the wet epochs are found in the 1241–1251, 1355–1367, 1400–1421, 1525–1538, 1543–1557, 1566–1582, 1844–1851, 1888–1899, 1906–1918, 1945–1953 and 1982–1993. Our discussions only focus on the dry and wet epochs at Songmingyan Mountain, since most of these epochs have been mentioned in previous tree-ring reconstructions at Shenge site in the Dulan area [12,13,14,15]. For the reconstruction at Shenge site, we herein paid special attention to the most extreme drought over the entire reconstruction period in 1824 (Figure 6b). This extreme drought was also recorded by locally historical documents the major city of Xining near the study site [28]. This drought was centered in the Dulan area in northeastern TP, which was the driest region in 1824 in the Monsoonal Asia Drought Atlas (Figure S2). The drought reconstruction for Songmingyan Mountain does not record this drought, because it is located outside of this drought center. The reconstructed precipitation dropped sharply from 100.5 mm in 1823 to 12.8 mm in 1824, and increased to 144.7 mm in 1825. Similarly, this extreme drought was not observed in 1823 and 1825 (Figure S2). This drought might indicate that an abnormally high pressure controlled the Dulan area in 1824, which requires future modeling studies to examine the occurrence of this anomalous high and its associations with large-scale circulation anomalies. Along the waveguide of westerlies, this anomalous high might be related to the abrupt shift from positive phase of Pacific/North America teleconnection to its negative phase in 1824 ().

The dry and wet epochs in the Songmingyan reconstruction are seen in hydroclimate reconstructions from neighboring regions (Figure 7a and 7b), e.g. the reconstruction from Shenge site in Dulan region and the speleothem records from Wanxiang cave [38]. However, some dry epochs in Songmingyan reconstruction, for example, in 1862–1874 and 1914–1933 are not found in a document-based precipitation reconstruction from Longxi [39], east of our study region. The Longxi reconstruction shows similar variations with the Dulan reconstruction. These differences may be related to varying hydroclimate regimes in different regions and the different seasons for reconstruction (e.g. summer or annual precipitation).

These dry/wet epochs were not only found in marginal areas of the Asian summer monsoon in the northeastern TP region, but also in northeastern China and northeastern Mongolia, such as the dry epoch from 1914–1933 [10,18,29,37,38,40] (Figure 7c). Our study adds additional proxy evidence that this dry epoch reached the southwestern boundary of the Songmingyan Mountain. In the northeastern TP region this drought began in the 1910s, nearly a decade before other regions, including north central China and eastern Mongolia, suggesting that this persistent drought initiated from the west. However, the hydroclimate variations have limited resemblance with the hydroclimate changes from the westerlies-dominated regions, such as in western Mongolia [41] (Figure 7c).

This indicates that hydroclimate changes in the past two centuries are more likely dominated by the monsoon and are different from the hydroclimate changes in westerlies-dominated regions, which is in agreement with previous studies [9]. Additional evidence of the influences of the monsoon in this region comes from the similar variations between hydroclimate changes and a long instrumental precipitation record from India [42] and the coral-inferred low-level cross-equatorial jet in the western Indian Ocean [43] (Figure 7d). Previous studies documented that the Asian summer monsoon, particularly the Indian summer monsoon, can reach northeastern TP along the eastern boundary of TP [44]. The dryness (wetness) in the northeastern TP and associated weakened (strengthened) Indian summer monsoon often corresponds to cold (warm) periods of the reconstructed land temperature of the northern hemisphere [10] and China [45], except for the recent monsoon failure since around the 1980s (Figure 7e). The positive relationship between temperature and the monsoon is likely a result of increased land-ocean temperature gradients [9,38] and the northward shifts of the intertropical convergence zone (ITCZ) [46]. The monsoon failure in recent decades may be caused by intensified human activities such as increases in aerosols emissions, which can weaken the land-ocean temperature gradients and the gradients between the northern and southern hemispheres [47] and the cooling of the upper troposphere [48].

Conclusions

We introduced an ensemble weighting method to alleviate two potential biases in traditional methods of chronology development. This method allows the mean value of tree-ring series to vary at different time intervals, instead of assigning a value of 1. In addition, this ensemble weighting method assigns weights to individual series depending on the strength of the climate-growth relationship. The resulting chronology is then averaged from an ensemble of chronologies with weighted individual tree-ring indices. The chronology development is iterated to adjust the mean values of individual tree-ring indices and to alleviate the trend distortion problem, similar to signal-free methods. We tested the efficiency of this method by developing a new tree-ring chronology at the Songmingyan Mountain and by recalculating a tree-ring chronology at Shenge site from a marginal area of the Asian summer monsoon in the northeastern TP. These reconstructions explain higher instrumental variance, 31.7% for the Songmingyan reconstruction and 57.3% for the Shenge reconstruction, than the reconstructions based on traditional methods. The reconstructed dry epochs range from the marginal area of the Asian summer monsoon from the northeastern TP to eastern Mongolia, as well as the monsoon dominated Indian subcontinent, indicating the linkages between regional hydroclimate changes and the Asian summer monsoon.

Supporting Information

Figure S1 Indices of the tree-ring chronologies developed from traditional method (black), the signal-free method (red) and the ensemble weighting method (blue) for the Shenge site at a roughly 300-year interval.

Figure S2 The reconstructed summer (June-August) Palmer Drought Severity Indices in years of 1823, 1824 and 1825 from the Monsoon Asia Drought Atlas (Cook et al. 2010).

Acknowledgments

We kindly thank Mingshi Zhao for his assistance in the field and laboratory. We acknowledge the Environmental and Ecological Science Data Center, West China for sharing the plant function data from boreal forests.

Author Contributions

Conceived and designed the experiments: KF. Performed the experiments: KF CL FZ. Analyzed the data: KF MW ND. Contributed reagents/materials/analysis tools: KF MW ND. Wrote the paper: KF.

References

1. Mann ME, Zhang Z, Hughes MK, Bradley RS, Miller SK, et al. (2008) Proxy-based reconstructions of hemispheric and global surface temperature variations over the past two millennia. Proceedings of the National Academy of Sciences 105: 13252–13257.

2. Cook E, Anchukaitis KJ, Buckley BM, D'Arrigo RD, Jacoby GC, et al. (2010) Asian Monsoon Failure and Megadrought During the Last Millennium. Science 328: 486–489.

3. Jones PD, Briffa KR, Osborn TJ, Lough JM, Van Ommen TD, et al. (2009) High-resolution palaeoclimatology of the last millennium: a review of current status and future prospects. Holocene 19: 3–49.

4. IPCC (2007) Climate Change 2007: The Physical Science Basis: IPCC.

5. Cook E, Briffa K (1990) Data analysis. In: Cook E, Kairiukstis L, editors. Methods of Dendrochology: Application in the Environmental Sciences. Dordrecht: Kluwer Academic Publisher.

6. Melvin TM, Briffa KR (2008) A signal-free approach to dendroclimatic standardisation. Dendrochronologia 26: 71–86.

7. Wilmking M, Juday GP, Barber VA, Zald HSJ (2004) Recent climate warming forces contrasting growth responses of white spruce at treeline in Alaska through temperature thresholds. Global Change Biology 10: 1724–1736.

8. Bunn AG, Hughes MK, Salzer MW (2011) Topographically modified tree-ring chronologies as a potential means to improve paleoclimate inference. Climatic Change 105: 627–634.

9. Chen FH, Chen JH, Holmes J, Boomer I, Austin P, et al. (2010) Moisture changes over the last millennium in arid central Asia: a review, synthesis and comparison with monsoon region. Quat Sci Rev 29: 1055–1068.

10. Fang K, Gou X, Chen F, Frank D, Liu C, et al. (2012) Precipitation variability of the past 400 years in the Xiaolong Mountain (central China) inferred from tree rings. Clim Dyn 39: 1697–1707.

11. Gou XH, Deng Y, Chen FH, Yang MX, Fang KY, et al. (2010) Tree ring based streamflow reconstruction for the Upper Yellow River over the past 1234 years. Chinese Science Bulletin 55: 4179–4186.

12. Liu Y, An Z, Ma H, Cai Q, Liu Z, et al. (2006) Precipitation variation in the northeastern Tibetan Plateau recorded by the tree rings since 850 AD and its relevance to the Northern Hemisphere temperature. Science in China Series D: Earth Sciences 49: 408–420.

13. Shao X, Huang L, Liu H, Liang E, Fang X, et al. (2005) Reconstruction of precipitation variation from tree rings in recent 1000 years in Delingha, Qinghai. Sci China 48: 939–949.

14. Zhang QB, Cheng GD, Yao TD, Kang XC, Huang JG (2003) A 2,326-year tree-ring record of climate variability on the northeastern Qinghai-Tibetan Plateau. Geophys Res Lett 30: 1739–1742.

15. Sheppard PR, Tarasov PE, Graumlich LJ, Heussner KU, Wagner M, et al. (2004) Annual precipitation since 515 BC reconstructed from living and fossil juniper growth of northeastern Qinghai Province, China. Clim Dyn 23: 869–881.

16. Ramage CS (1971) Monsoon Meteorology. New York: Academic Press.

17. Ran YH, Li X, Lu L, Li ZY (2012) Large-scale land cover mapping with the integration of multi-source information based on the Dempster–Shafer theory. International Journal of Geographical Information Science 26: 169–191.

18. Fang K, Gou X, Chen F, D'Arrigo R, Li J (2010) Tree-ring based drought reconstruction for the Guiqing Mountain (China): linkages to the Indian and Pacific Oceans. Int J Climatol 30: 1137–1145.

19. Li J, Chen F, Cook ER, Gou X, Zhang Y (2007) Drought reconstruction for north central China from tree rings: the value of the Palmer drought severity index. Int J Climatol 27: 903–909.

20. Fritts HC (1976) Tree rings and climate. New York: Academic Press.

21. Holmes RL (1983) Computer-assisted quality control in tree-ring dating and measurement. Tree-ring bull 43: 69–78.

22. Cook E, Kairiukstis L (1990) Methods of Dendrochronology Netherlands: Kluwer Academic Press.

23. Cook ER (1985) A time series analysis approach to tree ring standardization. Tucson: The University of Arizona.

24. Meko D, Graybill DA (1995) Tree-ring reconstruction of upper Gila river discharge. J Am Water Resour As 31: 605–616.

25. D'Arrigo RD, Kaufmann RK, Davi N, Jacoby GC, Laskowski C, et al. (2004) Thresholds for warming-induced growth decline at elevational tree line in the Yukon Territory, Canada. Global Biogeochem Cy 18: GB3021 doi: 3010.1029/2004GB002249.

26. Wigley TML, Briffa KR, Jones PD (1984) Average value of correlated time series, with applications in dendroclimatology and hydrometeorology. J Appl Meteorol Clim 23: 201–234.

27. Cook E, Meko DM, Stahle DW, Cleaveland MK (1999) Drought reconstructions for the continental United States. Journal of Climate 12: 1145–1162.

28. Zhang D, Li X, Liang Y (2003) Complementary Data of the Yearly Charts of Dryness/Wetness in China for the Last 500 years Period. Journal of Applied Meteorological Science 14: 379–384.

29. Liang E, Liu X, Yuan Y, Qin N, Fang X, et al. (2006) The 1920s drought recorded by tree rings and historical documents in the semi-arid and arid areas of northern China. Climatic Change 79: 403–432.

30. Cook E, Briffa KR, Meko DM, Graybill DA, Funkhouser G (1995) The'segment length curse'in long tree-ring chronology development for palaeoclimatic studies. Holocene 5: 229–237.

31. Esper J, Cook ER, Schweingruber FH (2002) Low-frequency signals in long tree-ring chronologies for reconstructing past temperature variability. Science 295: 2250–2253.

32. Briffa KR, Melvin TM (2009) A Closer Look at Regional Curve Standardization of Tree-Ring Records: Justification of the Need, a Warning of Some Pitfalls, and Suggested Improvements in Its Application. In: Hughes IMK, Diaz HF, Swetnam TW, Progress D, editors. Dendroclimatology: progress and prospects. New York City: Springer Verlag.

33. Fang K, Gou X, Chen F, Li J, D'Arrigo R, et al. (2010) Reconstructed droughts for the southeastern Tibetan Plateau over the past 568 years and its linkages to the Pacific and Atlantic Ocean climate variability. Clim Dyn 35: 577–585.

34. Fan ZX, Bräuning A, Cao KF (2008) Tree-ring based drought reconstruction in the central Hengduan Mountains region (China) since AD 1655. International Journal of Climatology 28: 1879–1887.

35. Gou X, Chen F, Cook E, Jacoby G, Yang M, et al. (2007) Streamflow variations of the Yellow River over the past 593 years in western China reconstructed from tree rings. Water Resour Res 43: W06434.

36. Fang K, Gou X, Chen F, Yang M, Li J, et al. (2009) Drought variations in the eastern part of Northwest China over the past two centuries: evidence from tree rings. Clim Res 38: 129–135.

37. Pederson N, Jacoby GC, D'Arrigo RD, Cook ER, Buckley BM, et al. (2001) Hydrometeorological Reconstructions for Northeastern Mongolia Derived from Tree Rings: 1651–1995. J Climate 14: 872–881.

38. Zhang P, Cheng H, Edwards RL, Chen F, Wang Y, et al. (2008) A test of climate, sun, and culture relationships from an 1810-year Chinese cave record. Science 322: 940–942.

39. Tan L, Cai Y, Yi L, An Z, Ai L (2008) Precipitation variations of Longxi, northeast margin of Tibetan Plateau since AD 960 and their relationship with solar activity. Climate of the Past 4: 19–28.

40. Davi N, Pederson N, Leland C, Nachin B, Suran B, et al. (2013) Is eastern Mongolia drying? A long-term perspective of a multi-decadal trend. Water Resources Research doi:10.1029/2012WR011834.

41. Davi N, Jacoby G, D'Arrigo R, Baatarbileg N, Li J, et al. (2009) A tree-ring-based drought index reconstruction for far-western Mongolia: 1565–2004. Int J Climatol 29 1508–1514.

42. Sontakke NA, Singh N, Singh HN (2008) Instrumental period rainfall series of the Indian region (AD 1813–2005): revised reconstruction, update and analysis. Holocene 18: 1055–1066.

43. Gong DY, Luterbacher J (2008) Variability of the low-level cross-equatorial jet of the western Indian Ocean since 1660 as derived from coral proxies. geophysical research letters 35: L01705, doi:01710.01029/02007GL032409.

44. An Z, Colman SM, Zhou W, Li X, Brown ET, et al. (2012) Interplay between the Westerlies and Asian monsoon recorded in Lake Qinghai sediments since 32 ka. Scientific reports 2: doi:10.1038/srep00619.

45. Yang B, Bräuning A, Johnson KR, Yafeng S (2002) General characteristics of temperature variation in China during the last two millennia. Geophys Res Lett 29: 1324. 1310.1029/2001GL014485.

46. Sachs J, Sachse D, Smittenberg R, Zhang Z, Battisti D, et al. (2009) Southward movement of the Pacific intertropical convergence zone AD 1400–1850. Nature-Geoscience: doi: 10.1038/NGEO1554.

47. Bollasina MA, Ming Y, Ramaswamy V (2011) Anthropogenic aerosols and the weakening of the South Asian summer monsoon. Science 334: 502–505.

48. Ding Y, Wang Z, Sun Y (2008) Inter-decadal variation of the summer precipitation in East China and its association with decreasing Asian summer monsoon. Part I: Observed evidences. International Journal of Climatology 28: 1139–1161.

22

Variability in Seroprevalence of Rabies Virus Neutralizing Antibodies and Associated Factors in a Colorado Population of Big Brown Bats (*Eptesicus fuscus*)

Thomas J. O'Shea[1]*, Richard A. Bowen[2], Thomas R. Stanley[1], Vidya Shankar[2,3¤], Charles E. Rupprecht[4,5]

1 United States Geological Survey, Fort Collins Science Center, Fort Collins, Colorado, United States of America, 2 Department of Biomedical Sciences, Colorado State University, Fort Collins, Colorado, United States of America, 3 Centers for Disease Control and Prevention, Atlanta, Georgia, United States of America, 4 Ross University School of Veterinary Medicine, Basseterre, Saint Kitts, West Indies, 5 The Global Alliance for Rabies Control, Manhattan, Kansas, United States of America

Abstract

In 2001–2005 we sampled permanently marked big brown bats (*Eptesicus fuscus*) at summer roosts in buildings at Fort Collins, Colorado, for rabies virus neutralizing antibodies (RVNA). Seroprevalence was higher in adult females (17.9%, n = 2,332) than males (9.4%, n = 128; $P = 0.007$) or volant juveniles (10.2%, n = 738; $P < 0.0001$). Seroprevalence was lowest in a drought year with local insecticide use and highest in the year with normal conditions, suggesting that environmental stress may suppress RVNA production in big brown bats. Seroprevalence also increased with age of bat, and varied from 6.2 to 26.7% among adult females at five roosts sampled each year for five years. Seroprevalence of adult females at 17 other roosts sampled for 1 to 4 years ranged from 0.0 to 47.1%. Using logistic regression, the only ranking model in our candidate set of explanatory variables for serological status at first sampling included year, day of season, and a year by day of season interaction that varied with relative drought conditions. The presence or absence of antibodies in individual bats showed temporal variability. Year alone provided the best model to explain the likelihood of adult female bats showing a transition to seronegative from a previously seropositive state. Day of the season was the only competitive model to explain the likelihood of a transition from seronegative to seropositive, which increased as the season progressed. We found no rabies viral RNA in oropharyngeal secretions of 261 seropositive bats or in organs of 13 euthanized seropositive bats. Survival of seropositive and seronegative bats did not differ. The presence of RVNA in serum of bats should not be interpreted as evidence for ongoing rabies infection.

Editor: Michelle L. Baker, CSIRO, Australia

Funding: Financial support was provided via a National Science Foundation Ecology of Infectious Diseases (EID) grant 0094959 (http://www.nsf.gov/funding/pgm_summ.jsp?pims_id = 5269&org = EF) to Colorado State University. RAB and TJO were supported by the Research and Policy for Infectious Disease Dynamics (RAPIDD) program of the Science and Technology Directorate (US Department of Homeland Security) and the Fogarty International Center (National Institutes of Health; http://www.fic.nih.gov/about/staff/pages/epidemiology-population.aspx). The funders had no role in study design, data collection and analysis, decision to publish, or preparation of the manuscript.

Competing Interests: The authors have declared that no competing interests exist.

* E-mail: osheat@usgs.gov

¤ Current address: U.S. Army Public Health Command, Joint Base Lewis-McChord, Fort Lewis, Washington, United States of America

Introduction

The presence of rabies virus neutralizing antibodies (RVNA) in serum of insectivorous bats of North America has been documented for over 50 years (*e.g.* [1–4]). However, initial investigations could not determine to what degree the presence of serum RVNA signaled past exposure, immunity, abortive infection, subclinical, or incubation phases of rabies [1,3,5]. Additionally, past serological surveys for RVNA in North American insectivorous bats were cross-sectional, in that wild bat populations were sampled once (sometimes terminally) and not marked for subsequent sampling. Historically, such serological surveys also concentrated on samples from small numbers of bat colonies for which there was limited ecological background information. More recent serological studies in Europe have indicated the presence of serum antibodies to other bat lyssaviruses in several species of insectivorous bats, usually at low prevalence (reviewed by Schatz et al. [6]). These latter studies included cross-sectional sampling at multiple locations and colonies [7–11], limited longitudinal sampling of marked individual bats [9,12,13] and analysis of ecological factors associated with seroprevalence [10,13].

Herein we report on both cross-sectional and longitudinal prevalence of RVNA in serum samples of big brown bats (*Eptesicus fuscus*) roosting commensally with humans in the urbanizing setting of Fort Collins, Colorado, U.S.A. Big brown bats are the most common species of bat submitted for rabies diagnostic testing in passive public health surveillance programs in the U.S. and in Colorado [14–16]. Recent complementary studies of rabies pathogenesis in captive big brown bats have included measurements and interpretations of the presence of RVNA based on laboratory experiments [17–22]. The big brown bat population we sampled roosts commensally with people in buildings and has been characterized by a number of concurrent ecological [23–25], demographic [26,27], and genetic studies [28]. This surveillance, laboratory, and field background provides additional information of potential importance for understanding the significance of serum RVNA in big brown bats (see *Methods* for more detail).

Our study focuses on testing the hypothesis that the presence of RVNA in serum of bats is indicative of past exposure of bats to rabies and is not evidence for an ongoing rabies infection per se. Our first objective in the present paper is to provide an in-depth cross-sectional profile of RVNA seroprevalence in a big brown bat population, and to test variation in RVNA seroprevalence based on sex, age, year of study, and roosting colony. Secondly, we describe longitudinal variability in the presence of RVNA in individually marked bats, and test multiple competing hypotheses about the relative importance of a number of biological and environmental factors that have potential influence on the serological status of individual bats over time. A cross-sectional study of Mexican free-tailed bats (*Tadarida brasiliensis mexicana*) roosting in large colonies in caves and bridges in Texas has reported interactions among ecological factors and variance in seroprevalence during a single year, with important associated factors including roost and season [29]. Recent studies of other diseases in unrelated species of wildlife have suggested that temporal changes in the immune status of individuals can occur in relation to a variety of environmental stress-related factors (*e.g.* [30,31]). Our longitudinal analyses attempt to explore the importance of such influences on RVNA seroprevalence in big brown bats at the Colorado study area. Our final objective was to shed additional light on the interpretation of serum RVNA in bats and on our initial hypothesis by sampling seropositive bats for evidence of rabies virus (RV) RNA in oral secretions and tissues, and to describe the survival of seropositive bats over time.

Methods

Study Area and Supporting Background Research on the Bat Population

We studied big brown bats at Fort Collins, Colorado, during summers 2001–2005. Many aspects of the big brown bat population at Fort Collins were studied intensively simultaneous to our serological sampling, and a more detailed description of the study area appears in [25]. Big brown bats are the most common bat in Fort Collins, as in many urban areas in North America (reviewed in [25]). The population of big brown bats at Fort Collins roosts only in buildings, and the bats migrate to higher elevations in the adjacent Rocky Mountains for winter hibernation in rock crevices [23,25]. Summer roost locations occurred throughout the city, and were previously mapped and character-ized [24,25]. Big brown bats generally show high fidelity to roosts, including high natal philopatry [26], but will move to neighboring roosts during periods of hot weather [32]. Numbers of adults counted emerging from roosts ranged from 10 to 219, with a geometric mean of 47 bats (95% Confidence Interval [*CI*] 39–56; [25]). Bats from multiple colonies concentrated foraging at overlapping riparian areas, and extensive dietary heterogeneity was inferred based on isotopic analysis of hair samples of many individuals from several roosts [25] [33]. Two mitochondrial DNA haplotypes of big brown bats are found at the study area but interbreed and mingle within colonies [28]. Demographic characteristics of the population include: a litter size of 1 with occasional twinning, a survival rate of 0.67 from volancy to the second summer, and breeding probabilities of 0.64 for one-year-old females and 0.95 for older females [26]. Annual adult survival averaged 0.79 with the best candidate models for explaining survival including roost and year; winter survival was higher than survival in summer, and population growth rates were positive [27].

The RVs in Colorado big brown bats were genotyped and include primarily two clades based on nucleoprotein (N) gene sequences [34]. Public health surveillance records for rabies in big brown bats in the region were analyzed and include the regular occurrence of rabid bats and exposure of humans and domestic animals to rabid bats [14,25]. About 2.5% of 199 big brown bats taken from the population eventually developed rabies when held captive for six months [21]. Big brown bats captured at the study area were experimentally subjected to aerosol exposures to one of the RV variants circulating in their wild population and showed serum antibody responses, but succumbed to subsequent intra-muscular inoculation with a different variant [20]. In a separate experiment with captive bats from the study population, most (78%) bats inoculated with varying dosages of the same variant responded by producing serum RVNA but none developed rabies [22]. Eight of 11 bats that were seronegative at capture developed rabies when inoculated with a second variant found in the population, but none of the bats that were seropositive at the time of capture developed rabies when inoculated with this second variant [22]. The survival of experimentally inoculated seropos-itive bats in these studies suggests some degree of protection is afforded by RVNAs. The dynamics of RV infection in this population were modeled mathematically by George et al. [35], who found that seasonally high adult survival in winter favors maintenance of the host population, while the virus is maintained by long incubation periods and overwinter dormancy followed by a seasonal birth pulse of immunologically susceptible young.

Our study was conducted during a long drought that varied in intensity by year (Figure 1), as measured by the Palmer Drought Severity Index (PDSI) for the warm-season period of bat activity (April-September) for Colorado Zone 11, north Front Range, and adjacent plains [36]. The PDSI is a standardized method for measuring intensity, duration, and spatial extent of drought based on precipitation, air temperature, and local soil moisture, with values ranging from −6.0 (extreme drought) to +6.0 [37]. Insect prey and demography of bats can be depressed during droughts [38–40], making year of sampling an important environmental variable. In addition to the PDSI, we also compiled and summarized daily temperature data for the summer sampling periods in each year of study using data available for Fort Collins [41].

Capture, Marking, and Sampling of Bats

We captured bats in mist nets over water at night, applied radio transmitters before releasing, and located roosting colonies by searching for radio signals while driving city streets during the day [25,42]. We captured bats haphazardly at roost entrances as they emerged after dusk using mist nets, harp traps, funnel traps, and hand-held nets (we followed guidelines of Sheffield et al. [43] and usually avoided entering maternity roosts to capture bats). We sampled bats at five roosts for each of the five summers, and sampled additional bats at 17 other roosts for 1–4 summers (bats were often excluded from roosts by building owners, preventing re-sampling every summer; see [25] for details). We repeated captures of bats at the same roosts 1–8 times each summer with a mean interval of 20.5 days (95% confidence interval [*CI*] 20.0–23.0 days) [25] but could not always recapture the same individuals. Bats were transported to the laboratory in individual cloth bags nested within disposable cups with lids. At the laboratory each bat was examined, measured, and permanently marked with passive integrated transponders (PIT tags) [44]. Volant juveniles were distinguished from adults based on lack of ossification of phalangeal epiphyses [45]. This allowed assigning of accurate ages to bats first marked as volant juveniles when they

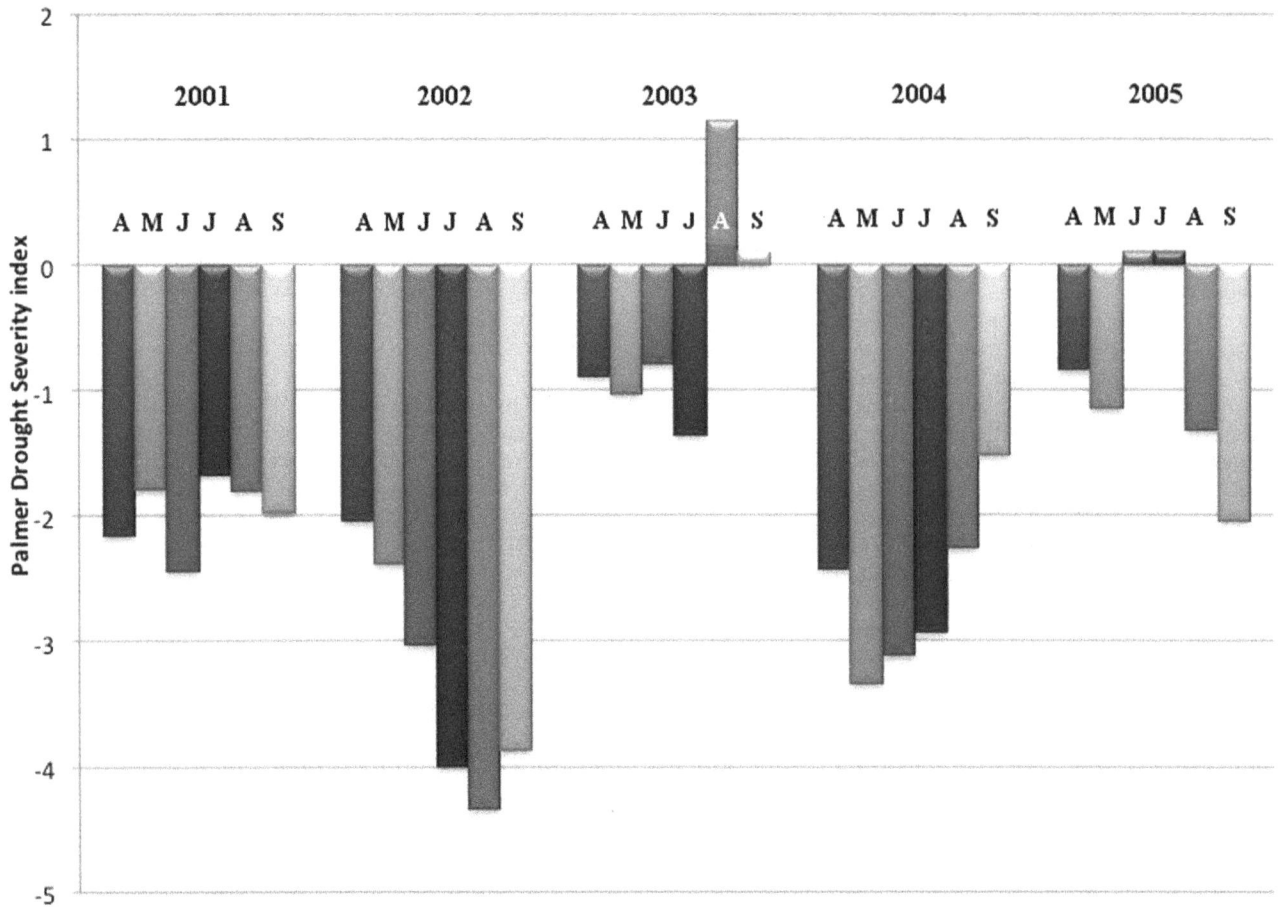

Figure 1. Palmer Drought Severity Index for the spring and summer months in the region including Fort Collins, Colorado, 2001–2005. Relative monthly drought is indicated by the depth of bars below zero.

were captured in subsequent years. Entry points of selected roosts were continuously monitored with PIT tag readers to record presence of individuals for survival analyses [42,44,46].

Colony sizes were estimated through emergence counts [47] at dusk. We provide these data for each roost to indicate relative size of colonies and base them on the highest counts of any year during the month of June, when colonies had formed but most juveniles were not volant (based on captures at roosts). The counts thus are assumed to represent adults only. Evening emergence counts can show daily variability because big brown bat colonies may occupy a number of alternate neighboring roosts (e.g. [32,48]. We did not make numerous replicate counts at roosts to quantify this variability. Instead we used evening emergence counts only to document the approximate magnitude of colony sizes.

Captured bats were selected for serological sampling randomly in 2001–2003; beginning in 2003 we gave priority to sampling bats that had been sampled previously, with additional bats sampled at random as time allowed. Blood was obtained through interfemoral vessels as described by Wimsatt et al. [44], except that anesthesia was discontinued because there was no difference in survival between anesthetized, unanaesthetized, and unbled bats [49]. Oropharyngeal secretions of bats were sampled at the laboratory using cotton-tipped swabs inserted into 0.5 ml of BA-1 medium (Minimal Essential Medium with Hanks' balanced salt solution containing 0.05 M Tris buffer at pH 7.6, 1% Bovine Serum Albumin, 0.35 g/L of Sodium bicarbonate, 50 mg/L Gentamicin,

and 2.5 mg/L Amphotericin B). Serum and oropharyngeal swabs were frozen at −80 C until shipment to the Centers for Disease Control and Prevention (CDC), Atlanta, Georgia for testing [17,50]. Capture, marking, sampling, and euthanasia of bats were approved by the Institutional Animal Care and Use Committees of Colorado State University and the United States Geological Survey. Bats were captured under authority of a scientific collecting license issued by the Colorado Division of Wildlife.

Laboratory Analyses

Serum samples were analyzed at the Rabies Laboratory, CDC, Atlanta, Georgia for determination of RVNA using the Rapid Fluorescent Focus Inhibition Test (RFFIT), a standard RV neutralization assay used in diagnostic laboratories [17,51]. RV neutralization activity in big brown bat serum is associated with the immunoglobulin G antibody fraction rather than with non-specific serum components [17]. Rabies Immune Globulin International Standard from the World Health Organization was used as positive control and a pool of unvaccinated mouse sera served as negative control in all assays. The RFFIT results were expressed as endpoint titers. Titers <5 were considered negative, and samples with RFFIT titers >5 were considered positive. Negative samples were excluded prior to statistical summarization and analysis if serum volumes were less than 12 µl due to concerns about assay sensitivity.

An aliquot (100 µL) of each oropharyngeal swab sample was tested for the presence of RV RNA using a nested RT-PCR assay targeting the N gene [17]. Samples from brain and salivary glands of seropositive bats were also assayed from euthanized bats with this technique. Brain tissue of suspected rabid bats was assayed for RV antigen using the Direct Fluorescent-Antibody (DFA) test. Bat brain impressions from frozen tissue were processed and examined for RV antigen by using a fluorescein isothiocyanate-labeled anti-rabies antibody [52,53].

Data Analysis

We provide simple summary statistics to describe seroprevalence patterns in various groupings of bats: proportion seropositive, number of bats sampled, and 95% *CI* for proportions with correction for continuity [54]. Seroprevalence at other roosts that were sampled only in 2003 are reported elsewhere [50]. We use study numbers for roosts listed in summary statistics, but to protect confidentiality of building owners and occupants we do not provide details of roost locations or names. For cross-sectional summaries by year, we included only one serological value in our computations for an individual sampled more than once in that summer. Because we were interested in numbers of individual bats that showed RVNA (were seropositive), and if serial assays of an individual included both positive and negative values within a summer we included only the positive value in the cross-sectional summaries. Similarly, because we were interested in past evidence of exposure to RV, in calculating seroprevalence in known-age adult bats we present data as cumulative over time (proportion of individuals of known age with a history that included a seropositive sample). In some cases we were interested in whether the proportion of seropositive individuals differed across groupings (*e.g.*, age classes, years). To evaluate this we used the Marascuilo procedure [55] to compute χ^2 statistics for pair-wise comparisons among proportions.

Longitudinal statistical samples involved many individuals. For illustrative purposes, we provide results of individual longitudinal serological sampling for 20 adult female bats sampled 5 or more times to show examples of temporal variability in expression of RVNA. We chose all individuals with 6 or more sampling dates (n = 7), and randomly selected 13 additional bats from among 37 individuals bled 5 times each over the course of the study. We also analyzed serological data from adult female bats using logistic regression in a generalized linear models context [56] under three distinct frameworks: 1) we modeled the serological status of the bat only on the date it was first sampled ($y = 1$ if seropositive, $y = 0$ otherwise); 2) we conditioned on bats that were seronegative on a capture date, then modeled the status of the bat on the next occasion it was sampled ($y = 1$ if it transitioned to seropositive, $y = 0$ otherwise); and 3) we conditioned on bats that were seropositive on a capture date, then modeled the status of the bat on the next occasion it was sampled ($y = 1$ if it transitioned to seronegative, $y = 0$ otherwise). Under frameworks 2 and 3 we only used data from bats that had a prior history of exhibiting RVNA as adults to ensure that we had evidence they could be capable of mounting a RVNA response. For all three sets of analyses we used a logit link to incorporate covariates for use as predictor variables, and evaluated an *a priori* set of candidate models that we believed could potentially explain variation in serological status. The variables in our candidate set of models under all three frameworks were: year of sampling; number of fresh open puncture holes in the wing that were the approximate size of big brown bat canine tooth tips (assumed to be a measure of recent exposure to bites from another bat, but not definitively known); and mean minimum and maximum ambient temperatures for the 5 days preceding

sampling. None of the explanatory variables appearing in a particular candidate model had a correlation coefficient with an absolute value greater than 0.29. Big brown bats are heterothermic (body temperature of inactive bats can vary with environmental temperature); body temperature and metabolism of heterothermic mammals can affect immune function and expression of antibodies [57–60]. We chose thermal regimes for the 5 days prior to sampling considering that longer intervals might mask the importance of recent mean maxima or minima, and because the half-lives of immunoglobulin antibodies in experimental mice are short, ranging from 0.4 days to 11.7 days (*e.g.* [61,62]). Two additional variables used in the candidate set of models were day of season (beginning 6 April each year, approximating when bat activity begins at the study area [25,35]); and reproductive status (pregnant, lactating, post-lactating or non-reproductive; [26]). Under frameworks 2 and 3, we added a variable representing the number of days elapsed between the two sampling occasions. Our candidate sets of models included those representing potential additive and interactive effects of pairs of variables. We did not include a roost variable and its interactions in our final set of candidate models, based on the low ranking of roost in preliminary exploratory analyses. We limited these analyses to the years 2002–2005 because sampling in the first year of study (2001) was biased towards dates later in the season.

We compared alternative models within the three candidate sets of models using Akaike's Information Criterion corrected for small sample size (AIC_c), and computed model weights (w) [63]. The ΔAIC_c, computed as the difference between the AIC_c value of the model of interest and the AIC_c of the best model (*i.e.*, $AIC_{c\ min}$), along with model weights (w), provided a measure of the relative strength of evidence for each model given the data and our candidate set of models. Models with $\Delta AIC_c \leq 2$ were considered to have substantial support, with the exception of those that differed from the best model by one additional parameter and which had essentially the same values of the maximized log-likelihood as the best model [63,64]. Because these latter models provide no net reduction in AIC_c [64], we excluded them from consideration. Herein we report only results for models with $\Delta AIC_c \leq 2$.

We estimated apparent annual survival (ϕ) over 2001–2005 for female bats that were seropositive as adults at any time in 2001–2003 in comparison with adult females that were only seronegative when sampled in 2001–2003. We based the survival estimates on encounters of tagged bats with PIT tag readers in fixed positions at entrances of five roosts monitored from 2001–2005 [46]. Survival was calculated using Program Mark with differences between the seropositive and negative groups determined by χ^2 goodness-of-fit tests in Program Release Test 1 [49,65]. A more detailed multi-model analysis of factors influencing adult survival that does not include serological status was previously published [27]. We also summarize numbers of seropositive bats known alive from encounters by tag readers or captures after lengthy intervals during this study, and in nets during an unrelated subsequent study [33] of the same population.

Results

Seroprevalence in Volant Juveniles and Known-Age Adults

We sampled 738 marked volant juveniles (including 74 sampled twice as juveniles at variable intervals): 75 were seropositive (10.2%; Table 1). The proportion seropositive was similar between volant juvenile females and volant juvenile males (Table 1). We sampled 388 volant juvenile females 434 times: 39 were

Table 1. Seroprevalence of rabies virus neutralizing antibodies in known-age, marked individual big brown bats (*Eptesicus fuscus*), Fort Collins, Colorado, 2001–2005.

VJF	VJM	VJ	1 yr-old F	2 yr-old F	3 yr-old F	4 yr-old F	≥5 yr-old F
10.1%	10.3%	10.2%	14.3%[a]	27.1%[ab]	20.8%[ab]	47.8[b]	24.5[ab]
7.3–13.6	7.4–14.1	8.1–12.6	9.2–21.5	18.8–37.3	11.3–34.5	27.4–68.9	16.6–34.4
388	350	738	140	96	53	23	98

Proportions with superscripts in common for adult (≥1 year old) bats are not significantly different (*P*>0.05). Values are seroprevalence (% seropositive), 95% confidence interval for seroprevalence, and sample size (number of unique individuals sampled). Abbreviations: F = female, M = male, VJ = volant juvenile.

seropositive (10.1%; Table 1). Serial sampling involved two samples each from 46 tagged volant juvenile females: 38 juvenile females were negative on both sample dates (25.8±9.5 days apart), two were positive on both dates (11 and 27 days apart), five were first positive, then negative on second sampling dates (34.4±8.4 days apart), and one was first negative, then positive on a second sampling date (32 days apart). We sampled 350 marked volant juvenile males 378 times: 36 were seropositive (10.3%; Table 1). Serial sampling involved two samples each from 28 volant juvenile males during a single year. Twenty-four were negative on both samples (17.3±7.2 days apart), 3 were first positive, then negative on second sampling dates (17, 22, and 34 days elapsed), and one was first negative, then positive (10 days elapsed).

We sampled 330 known-age (first tagged as juveniles) adult bats of both sexes 503 times. Only 19 were males, age 1 (n = 14) or 2 (n = 5) years old, none of which was seropositive. Adult females (n = 410) showed differences in cumulative seroprevalence with age (Table 1, $\chi^2 = 15.4$, 4 d.f., *P*=0.004) due primarily to the contrast between 1-year-old and 4-year-old bats (Table 1; $\chi^2 = 9.6$, 4 d.f., *P*=0.048).

Seroprevalence Patterns by Age Class, Sex, and Year of Sampling

In city-wide sampling 2001–2005, marked big brown bats ranked from highest to lowest seroprevalence in the order adult females>volant juveniles>adult males (Table 2; overall $\chi^2 = 29.3$, 2 d.f., *P*<0.0001). This pattern generally held within years. Females had higher seroprevalence than adult males ($\chi^2 = 10.0$, 2

d.f., *P*= 0.007) or volant juveniles ($\chi^2 = 31.9$, 2 d.f., *P*<0.0001). Seroprevalence did not differ between juveniles and adult males ($\chi^2 = 0.08$, 2 d.f., *P*=0.961). Seroprevalence in adult females varied among years (Table 2; $\chi^2 = 94.6$, 4 d.f., *P*<0.0001). Adult females sampled in year 2003 had higher seroprevalence than in any other year except 2002; adult females sampled in 2004 had lower seroprevalence than in 2002 ($\chi^2 = 53.1$, 4 d.f., *P*<0.0001) and 2003 ($\chi^2 = 100.1$, 4 d.f., *P*<0.0001; Table 2). Year-to-year differences in overall seroprevalence of adult males (with smaller sample sizes) also were evident ($\chi^2 = 15.3$, 4 d.f., *P*=0.004; Table 2), largely due to contrasts of seroprevalence of adult males in 2003 with seroprevalence in 2002 ($\chi^2 = 10.8$, 4 d.f., *P*=0.029) and 2004 ($\chi^2 = 10.8$, 4 d.f., *P*=0.029). Volant juveniles had higher seroprevalence during 2003 than during all years except 2005, with juveniles in no other years differing significantly among each other (Table 2; overall $\chi^2 = 53.0$, 4 d.f., *P*<0.001).

Seroprevalence in Adult Females by Roost

At the five roosts sampled in all five years total seroprevalence in adult females across years varied from 6.2 to 26.7% (Table 3; overall $\chi^2 = 31.5$, 4 d.f., *P*<0.0001). Seroprevalence at one roost (#29) differed significantly from the other four roosts (*P*≤0.017), and seroprevalence values in bats from two other roosts (#58 and #60) were significantly different from each other (P = 0.03). No other pair-wise comparisons of seroprevalence at roosts showed significant differences. This pattern appeared to be held qualitatively for seroprevalence at roosts within single years, but with widely overlapping *CIs* for seroprevalence. The pattern among

Table 2. Cross-sectional summaries of seroprevalence of RVNA by age class and sex in individual big brown bats (*Eptesicus fuscus*) by year, Fort Collins, Colorado, 2001–2005, all roosts combined.

Sex/Age	Statistic	2001	2002	2003	2004	2005	All Years
AF	%	12.3[a]	22.0[b]	26.4[b]	6.0[a]	11.9[a]	17.9[x]
	CI	8.9–16.6	18.6–25.9	23.3–29.8	3.9–9.0	9.0–15.6	16.4–19.5
	N	310	518	734	368	402	2,332
AM	%	3.6[a,b]	0.0[a]	25.8[b]	0.0[a]	10.7[ab]	9.4[y]
	CI	0.1–20.2	0–18.5	12.5–44.9	0–20.9	2.8–29.4	5.2–16.1
	N	28	22	31	19	28	128
VJ	%	3.2[a]	6.3[a]	21.1[b]	2.6[a]	10.0[ab]	10.2[y]
	CI	1.4–6.7	3.4–11.3	16.4–26.7	0.4–9.8	0.5–45.9	8.1–12.6
	N	220	174	256	78	10	738
Roosts	N	11	19	23	19	15	34

Proportions with superscripts in common for years across rows within age and sex categories, and among sex and age categories all years combined (right hand column) are not significantly different (*P*>0.05). Abbreviations: AF = adult females, AM = adult males, VJ = volant juveniles (sexes combined), *CI* = 95% confidence interval for proportion, *N* = sample size, % = per cent seropositive.

Table 3. Seroprevalence of rabies virus neutralizing antibodies in individual adult female big brown bats (*Eptesicus fuscus*) within roosting colonies by year at five roosts, Fort Collins, Colorado, 2001–2005.

Roost ID #	Statistic	2001	2002	2003	2004	2005	All Years
#58 (219)	%	10.4	19.8	31.2	7.2	10.9	16.9 [x]
	CI	4.9–20.0	12.0–30.4	23.0–40.8	3.0–15.7	5.6–19.5	13.6–20.7
	N	77	81	112	83	92	445
#60 (203)	%	28.3	32.3	34.9	7.0	18.9	26.7 [z]
	CI	17.2–42.6	21.6–45.2	25.0–46.3	1.8–20.1	8.6–35.7	21.7–32.3
	N	53	65	83	43	37	281
#29 (23)	%	6.4	7.7	10.4	0.0	0.0	6.2 [y]
	CI	1.1–22.8	1.3–26.6	3.9–23.4	0.0–17.2	0.0–22.9	3.0–11.7
	N	31	26	48	24	17	146
#44 (77)	%	16.7	20.5	34.3	0.0	22.2	20.8 [x z]
	CI	8.0–30.8	10.3–35.7	19.7–52.3	0.0–26.8	7.4–48.1	14.9–28.0
	N	48	44	35	14	18	159
#51 (126)	%	13.3	23.1	32.7	16.0	14.3	22.4 [xz]
	CI	2.3–41.6	9.8–44.1	20.4–47.6	5.2–36.9	4.7–33.6	16.0–30.3
	N	15	26	49	25	28	143
All Roosts	%	15.6 [ac]	22.3 [ab]	29.7 [b]	6.9 [c]	13.0 [ac]	19.1
	CI	11.3–21.2	17.3–28.2	24.8–35.0	3.9–11.7	8.8–18.8	16.9–21.5
	N	224	242	327	189	192	1,174

Proportions with superscripts in common for roosts within the right-hand column or for years across the bottom row are not significantly different (P>0.05). CI = 95% confidence interval for proportion, % = per cent seropositive, N = sample size. Maximum count of adults as a measure of relative colony size (see Methods) is given in parentheses following roost identifiers. Seroprevalence by roost across all years is given in the right hand column, and by year across all roosts in the bottom row, with seroprevalence across all roosts and years provided in the lower right corner.

years seen in the city-wide sampling (Table 2) also appeared to hold for this subset of five roosts (Table 3; overall $\chi^2 = 49.9$, 4 d.f., $P<0.0001$): seroprevalence was significantly higher in 2003 than during all other years ($P<0.003$) except 2002 ($P = 0.408$); adult females sampled in 2004 had lower seroprevalence than in 2002 ($\chi^2 = 22.6$, 4 d.f., $P = 0.0002$) and 2003 ($\chi^2 = 53.1$, 4 d.f., $P<0.0001$; Table 3). No other pair-wise comparisons of years showed significant differences.

Variability in seroprevalence at 17 other roosts sampled for fewer than five summers ranged from 0.0 to 41.7% during individual years (Table S1). Efforts at roosts varied among years, making comparisons across years and roosts less straightforward to interpret. However, the presence of RVNA was widespread. RVNA were detectable in each year of study, and in each roost across the years sampled (Table S1). Within individual years and roosts, seropositive bats were detected in 36 of 49 summer samplings (Table S1). Within years, CIs for seroprevalence at individual roosts all were widely overlapping. Across years sampled in common, pooled seroprevalence at individual roosts also showed widely overlapping CIs (e.g., cf. roosts in 2003–2005; Table S1).

Factors Influencing Serological Status of Individuals

The presence of RVNA in individual bats showed temporal variability, sometimes after short intervals between samples (Figure 2). In our logistic regression analysis of the factors influencing the serological status of the bat when it was first sampled, the highest ranking model in our candidate set had an intercept (β_0), main effects for year (β_{yr}) and day of season (β_{dos}), and a year x day of season interaction term (β_{yr*dos}) ($w = 0.670$, $k = 8$, $n = 1,343$ observations). No other models were competitive

(i.e., $\Delta AIC_c >8$ for all other models). Diagnostic plots suggested that the interaction was complex and varied with year (Figure 3). In comparison with 2003, which was the reference year in this analysis (i.e., β [95% CI]; $\beta_0 = -1.097$ [-1.796 to -0.391], $\beta_{yr} = 0$, $\beta_{dos} = -0.002$ [-0.011 to 0.006], and $\beta_{yr*dos} = 0$), the likelihood of a bat being seropositive in 2005 began lower and increased as the season progressed (β_0 and β_{dos} are unchanged, $\beta_{yr} = -6.741$ [-12.296 to -2.984], $\beta_{yr*dos} = 0.056$ [0.019 to 0.105]). In 2002 the likelihood of a bat being seropositive began higher than in 2003 and continued to decrease as the season progressed (β_0 and β_{dos} are unchanged, $\beta_{yr} = 1.510$ [0.106 to 2.936], $\beta_{yr*dos} = -0.019$ [-0.036 to -0.002]), whereas in 2004 the likelihood of a bat being seropositive began lower than in 2003 but also decreased as the season progressed (Figure 3; β_0 and β_{dos} are unchanged, $\beta_{yr} = -0.893$ [-3.931 to 1.866], $\beta_{yr*dos} = -0.011$ [-0.045 to 0.020]). Although overall the PDSI for the April-September period in 2005 documented drought conditions, the PDSI in months of June and July 2005 was normal, whereas in both 2002 and 2004 drought was persistent throughout the summer months (Figure 1).

In our logistic regression analysis of the likelihood of an adult female transitioning to seronegative from a previously seropositive state, the highest ranking model in our candidate set depended only on year ($w = 0.906$, $k = 4$, $n = 169$ observations). Parameter estimates [95% CI] indicated that the likelihood of this transition was higher in 2004 ($\beta = 1.280$ [0.464 to 2.143]) and 2005 ($\beta = 1.727$ [0.899 to 2.624]) relative to 2003, but not in 2002 ($\beta = -0.713$ [-2.303 to 0.633]). In our logistic regression analysis of the likelihood of an adult female transitioning from seronegative to seropositive, the highest ranking model in our candidate set depended only on day of the season ($w = 0.410$, $k = 2$, $n = 52$

Bat

170
245
331
359
388
499
531
618
682
733
748
871
909
917
1218
1378
1386
1509
2870
2873

M J J A M J J A M J J A M J J A M J J A
2001 2002 2003 2004 2005

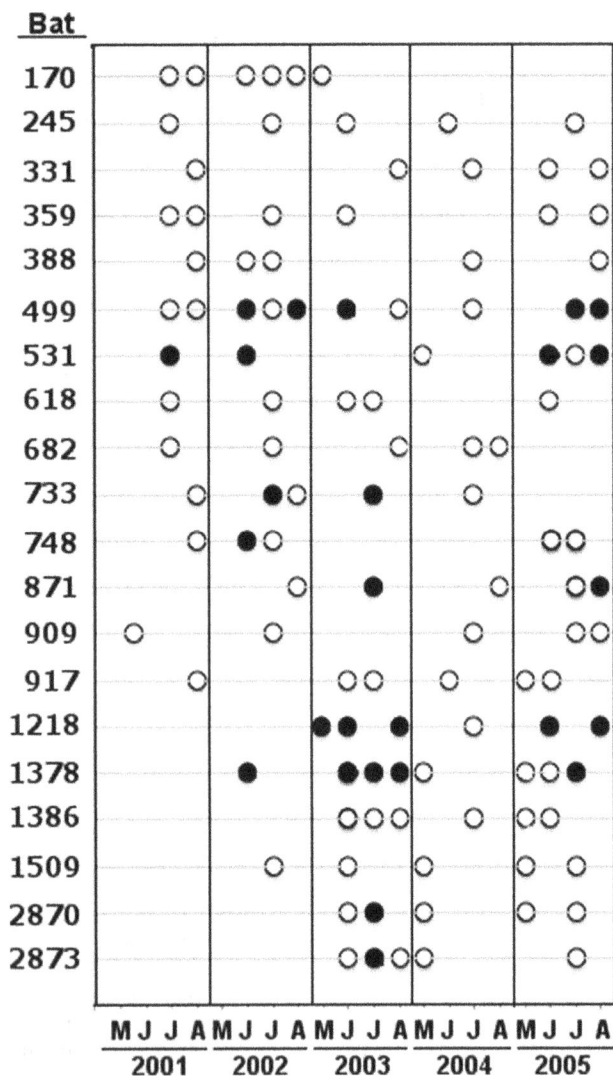

Figure 2. Examples of temporal variability in RVNA serological status of 20 wild big brown bats (*Eptesicus fuscus*) sampled 5 or more times, Fort Collins, Colorado 2001–2005. Open circles denote seronegative, closed circles denote seropositive at date shown at the bottom of the figure (vertical lines separate years). M-A are months of May through August each year. Individual bat identification number is given at the left-hand column.

observations). This likelihood increased as the season progressed (Figure 4; $\beta = 0.036$ [0.011 to 0.065]).

Lack of Rabies Viral RNA in Bats

We tested 467 samples of oropharyngeal secretions from 363 individual bats for the presence of RV RNA. Only one sample was positive, from a bat that was seronegative on the day of sampling. This bat was never captured again, and was never detected by an automated tag reader in place at its roost each summer for the next four years. RV genome-specific N gene RNA was not detected in any of 293 oropharyngeal secretions from a subsample of 261 individuals that were all seropositive on the dates of testing (32 individuals were sampled on >1 date).

The remaining negative samples were from bats that were seronegative at the time of sampling (106 oropharyngeal secretion samples from 98 individual bats) or did not have serological status

determined on the date of sampling (67 secretion samples from 66 individuals). Among the bats in these latter two groups were individuals that had negative oropharyngeal samples but were seropositive for RVNA earlier during the same summer (34.1 ± 19, range 13–73 days prior to oropharyngeal sampling, $n = 21$ intervals from 19 individuals) or during the summer of the previous year (8 intervals from 6 individuals).

Samples of total RNA from salivary glands and brains of 13 adult female bats that were seropositive during summer 2001 were assayed after recapture during 2002. None of the brain or salivary gland preparations from these bats was positive for RV genome-specific N gene RNA. Eleven of the 13 were also sampled for serology when recaptured in 2002, with nine seropositive on recapture (1 seronegative bat recaptured in 2002 was a volant juvenile when sampled in 2001). Similarly, an additional bat was seropositive in June 2002 and again when recaptured 34 days later, but had no detectable RV genome-specific N gene RNA in brains or salivary glands at the date of final recapture. In addition to these samples, two sick bats found near roosts were diagnosed as rabid based on DFA analysis of brain tissue: one was seronegative and one was seropositive on the day they were euthanized.

Survival of Seropositive Bats

The annual survival estimate of adult female bats during 2001–2005 that were seropositive during 2001–2003 ($\phi = 0.85$ [95% CI 0.80 to 0.89], $n = 360$ bats) showed no significant difference ($P = 0.783$, $\chi^2 = 4.0$, 7 d.f.) with that of bats only known to be seronegative during this period ($\phi = 0.83$ [95% CI 0.81 to 0.86], $n = 106$ bats).

Ten of 15 bats at regularly monitored roosts that were seropositive in 2001 were known to be alive in 2005, and 39 of 55 bats seropositive in 2002 were known to be alive in 2005. Bats captured haphazardly in nets during limited efforts for the unrelated study in 2007 and 2008 included 10 individuals that were known to be seropositive as adults 5 years earlier, 4 individuals seropositive as adults 6 years earlier, and one bat seropositive as an adult 7 years earlier.

Discussion

Seroprevalence in Volant Juveniles and Known-Age Adult Females

It is likely that many of the seropositive volant juveniles exhibited passive immunity from maternally derived antibodies. This also was reported for juvenile Mexican free-tailed bats [2]. In big brown bats at Fort Collins, one-year-old females had 14% seroprevalence, whereas RVNA were not detected in one-year-old males. Unlike breeding adult females, adult male big brown bats typically do not roost in clustering groups [66] that might foster greater exposure to RV due to higher densities of bats. Cumulative evidence for RVNA in serum ranged from 21 to 48 percent in adult females that were 2 years old or older, indicating that exposure of big brown bats to RV is common. The proportion seropositive was significantly lower in one-year olds compared to four-year olds; a tendency for increased seroprevalence with age is consistent with the finding that in this population one-third of one-year-old females do not breed, whereas nearly all older females breed [26]. Although occupying the same maternity roosts as breeders, non-breeding one-year olds may not exhibit a strong tendency to engage in clustering with its presumed increased likelihood of exposure to RV.

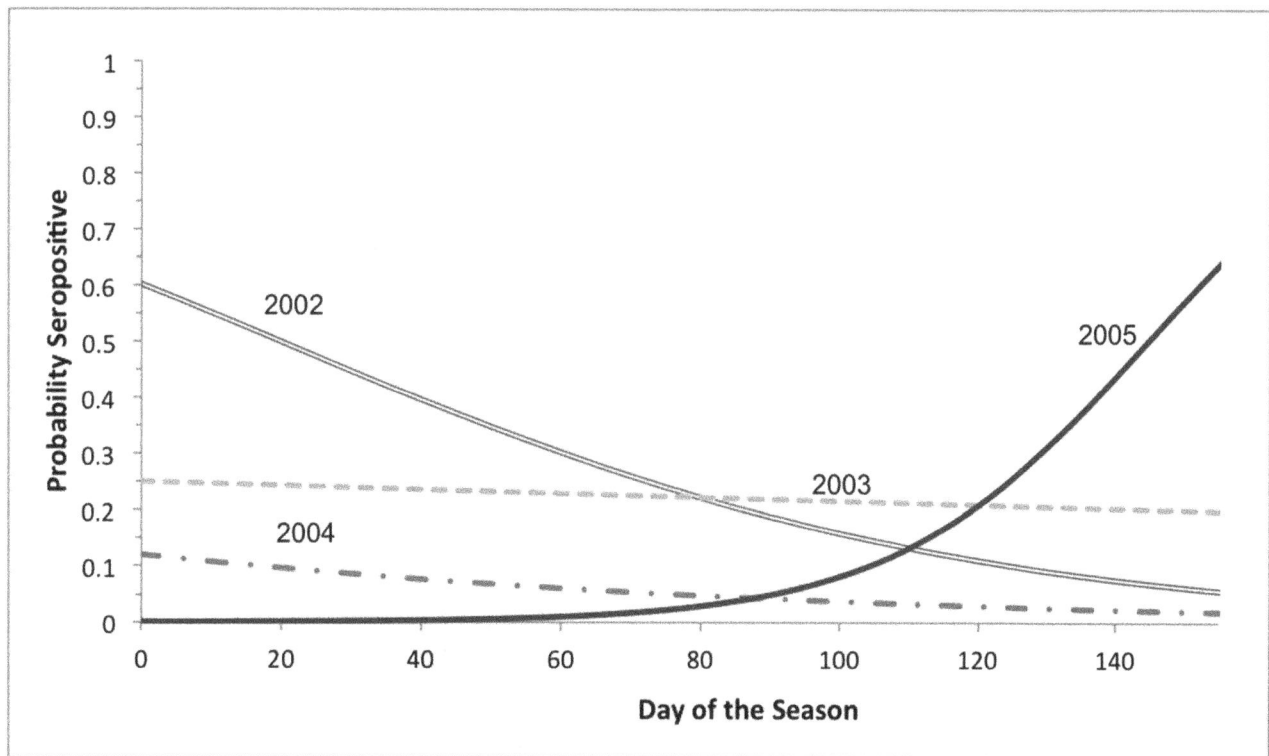

Figure 3. Relationship between day of the season (horizontal axis) and probability of an adult female big brown bat (*Eptesicus fuscus*) being seropositive for RVNA on first sampling in three different years of study. The reference year is 2003.

Seroprevalence Patterns by Sex, Age Class, and Year of Sampling

Adult females consistently showed higher seroprevalence than adult males. This pattern generally held within years, and is concordant with the greater degree of coloniality and likely exposure to RV in females. However, differences in seroprevalence between males and females have not been observed consistently in the few other species of North American insectivorous bats that have been sampled for RVNA. Differences in seroprevalence between sexes were not apparent in adult Mexican free-tailed bats at colonies in Texas or at Carlsbad Caverns, New Mexico [2,29]. Adult females may tend to have higher seroprevalence than adult males in hoary bats (*Lasiurus cinereus*), but not in silver-haired bats (*Lasionycteris noctivagans*), little brown bats (*Myotis lucifugus*), or long-legged myotis (*M. volans*) [50].

Seroprevalence in adult female big brown bats varied among years in city-wide sampling. Most notably, adult females sampled in 2003 had highest seroprevalence (but not significantly different from 2002), and those sampled in 2004 had the lowest seroprevalence (but not significantly different from 2001 and 2005). This pattern also held at the five roosts sampled each year. Volant juveniles also had higher seroprevalence during 2003 (but similar to 2005) in city-wide sampling, presumably because a greater proportion expressed maternally derived antibodies, paralleling higher adult female seroprevalence. Year-to-year differences in overall seroprevalence within adult males (with much smaller sample sizes) also followed this pattern in part, with higher seroprevalence in 2003 (but not significantly different than 2005) and lowest in 2004 (but not significantly lower than during years other than 2003).

The year-to-year pattern in seroprevalence roughly followed the annual variation in drought severity affecting the Fort Collins region, suggesting that lower production of RVNA in times of environmental stress may contribute to variability in RVNA seroprevalence. The year 2003 had reduced drought conditions during the April-September period of bat activity compared to other years, although in 2005 June and July were more favorable (Figure 1). Drought occurred to a greater degree in all other years, with 2002 the greatest and 2004 the next greatest in drought severity. Although seroprevalence was higher in 2002 than 2004, we speculate that the lower seroprevalence in 2004 compared to 2002 may have been exacerbated by other factors with potential stress on the big brown bat population. In 2004, areas used for foraging by big brown bats [25] were subject to much more intensive mosquito control targeting vectors in a novel West Nile virus epizootic that had recently reached Colorado [67]. This included use of pyrethroid and organophosphate insecticides as well as mosquito larvicides [68,69]. Effects of modern insecticides on bats or their specific food supplies are poorly known [70], but insect control may have been an additional environmental stress in 2004 that was not present in 2002. Dietary analysis shows that big brown bats consumed prey from different orders of insects in 2004 than in other years, including a large proportion of small midges instead of more typical dietary items such as large beetles (E. Valdez, U.S. Geological Survey, unpublished observations). Also in 2004, the month of July was notably cooler than in other years (Table 4), with nursing bats and suckling young found in torpor during sampling at roosts (unpublished observations). We are uncertain why seroprevalence in the drought year of 2002 was not significantly lower than seroprevalence during 2003, although temperatures were considerably warmer in June 2002 than in June

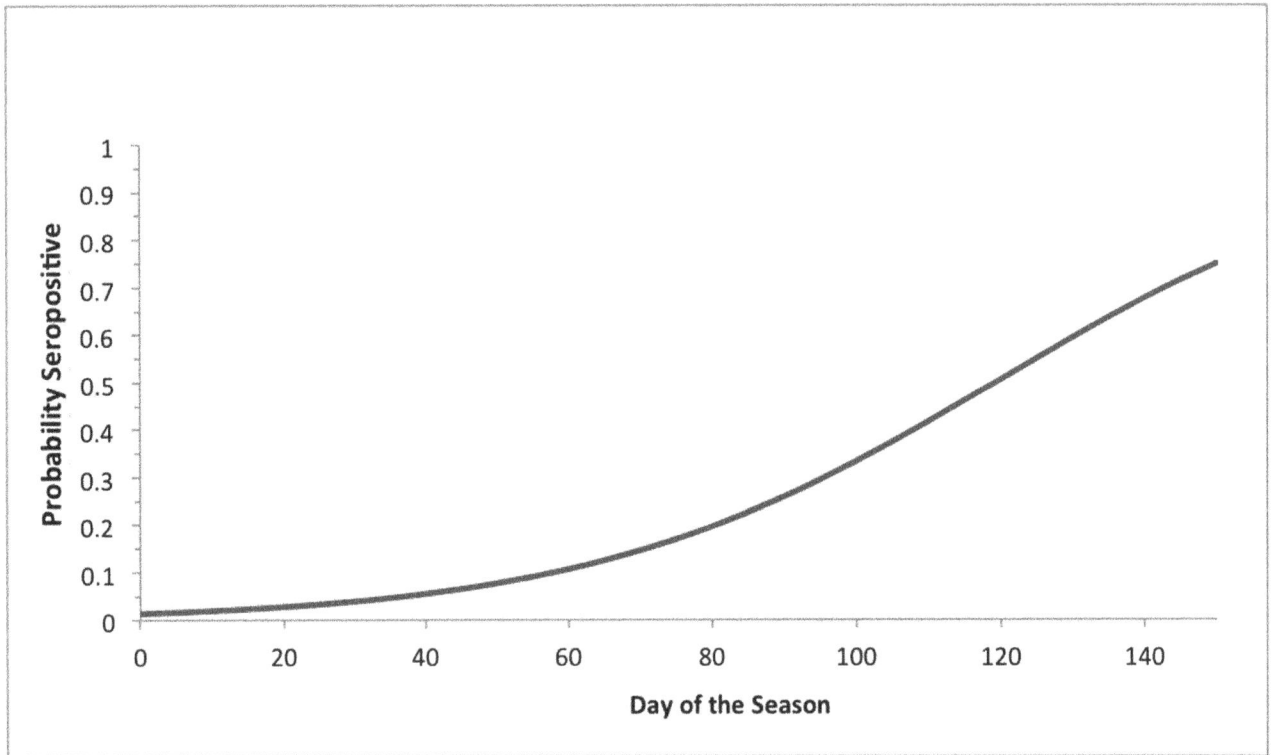

Figure 4. Probability of an adult female big brown bat (*Eptesicus fuscus*) transitioning from seronegative to seropositive for RVNA with the advancing season of bat activity (early April to September), conditional that the bat was seronegative on the previous sampling (see Methods).

of other years (Table 4) and warmer conditions can positively affect expression of antibodies in bats [57].

Seroprevalence in Adult Females by Roost

Results of sampling at 17 roosts for variable numbers of years support previous conclusions [14,50] that the presence of antibodies and exposure to RV are widespread among big brown

Table 4. Summary statistics for ambient temperatures (°C) during June and July 2001–2005, Fort Collins, Colorado: means of daily maximum temperatures ($\bar{X}_{max}$) and their lower and upper 95% confidence limits (CL_{max}), the range of daily maxima (range$_{max}$, and the number of days each month with temperatures exceeding 32°C (N days >32°C).

	2001	2002	2003	2004	2005
JUNE $\bar{X}_{max}$	27.6	29.5	23.1	23.3	25.8
JUNE CL_{max}	25.6, 29.6	27.9, 31.2	21.7, 24.6	21.3, 25.4	23.9, 27.8
JUNE range$_{max}$	16.4–35.4	17.2–34.9	13.4–29.7	10.2–35.9	14.3–32.4
JUNE N days >32°C	10	15	0	3	5
JULY $\bar{X}_{max}$	30.7	32.2	32.7	27.9	32.0
JULY CL_{max}	29.6, 31.7	31.2, 33.2	31.6, 33.8	26.0, 29.7	30.6, 33.4
JULY range$_{max}$	26.0–35.6	25.8–37.4	26.7–37.2	14.6–35.5	19.1–38.5
JULY N days >32°C	15	21	22	11	22

bat colonies in the region. Differences in seroprevalence among most roosts were not readily apparent. However, at the five roosts sampled in each of the five years of study, seroprevalence was consistently and significantly lower at one roost (#29, Table 3). We are uncertain why this was so. However, bats at this roost also had lower adult and first-year survival [26,27]. They also occurred in a smaller colony than at the other four roosts, perhaps reducing chances of exposure to RV. The annual variation in seroprevalence at the five roosting colonies also followed the year-to-year patterns seen in the city-wide sampling: highest seroprevalence in 2003 and lowest in 2004.

Ranges of seroprevalence at roosts of big brown bats in Fort Collins were similar to the reported ranges of seroprevalence for this species from colonies in New York in 1973–1976, but with higher overall seroprevalence (0.0 to 40%, overall 9.6% in New York; [3]), and similar to seroprevalence at other roosts in Colorado during 2003 (5.6 to 54.6%; [50]). Seroprevalence in big brown bats was higher than that reported for little brown bats (*Myotis lucifugus*) at colonies in New York (0.0 to 6.9%, overall 2.4%; [3]) and in little brown bats captured at foraging areas in Colorado (0.0 to 8.3%, overall 2.9%; [50]). Seroprevalence can be as high as 80% in adult female Mexican free-tailed bats [5,29,71]. Seroprevalence of RVNA in adult females of the largely solitary hoary bat was 31.8%, and seroprevalence in silver-haired bats was 6.8% in Colorado and New Mexico [50].

Factors Influencing Serological Status of Individuals

Relatively high fidelity of big brown bats to roosts within core areas [25,48] facilitated repeated serological sampling of individual bats. The degree of waxing and waning of detectable RVNA observed in repeatedly recaptured individuals was not anticipated.

It suggests that many adult big brown bats may be immunologically primed against RV due to past exposures, are not actively producing antibodies at the time of sampling, but may exhibit anamnestic responses on subsequent exposures. If this is true, then cross-sectional surveys underestimate the amount of exposure to RV that is encountered by bats. An absence of RVNA production also may be due to other demands on the immune system, or environmental and physiological factors as suggested for other wildlife [30,31]. Although re-exposure may be a trigger to renewed RVNA production in big brown bats, we were not able to document or measure such events in the field. However, the likely capability of many seemingly seronegative bats to exhibit anamnestic responses [72] on subsequent exposures to RV would provide some immunity and result in a lack of observations of epizootic increases in rabies deaths among big brown bats during this and other studies [14,25,73] (see also "Comparisons with Captive Exposure Experiments" below).

The environmental and physiological factors in our models for serological status on first sampling did not implicate female reproductive condition, recent biting of wings, short term ambient temperatures prior to sampling, and days elapsed since prior sampling as factors associated with serological status. However, an interaction of year and day of the season at the time of sampling influenced the likelihood of an adult female being seropositive on first capture. When compared to 2003 (near normal temperature and moisture conditions and highest cross-sectional seroprevalence), bats were more likely to be seropositive during 2005, the drought year that had near-normal summer temperature and moisture in June and July, than during 2002 and 2004, drought years that remained persistently dry during summer.

Year and day of the season were also implicated in affecting the probabilities of individual bats changing serological status, but in different ways. Adult female big brown bats were more likely to have shown a shift from seropositive to seronegative during 2004 and 2005 in comparison with 2003. Day of the season was implicated as the only factor associated with bats making a transition from a prior seronegative to a seropositive state. This suggests that as the active season progresses in any year, the cumulative likelihood of a bat being exposed to RV and expressing RVNA also increases. This is consistent with models of the dynamics of RV infection in big brown bats from throughout the United States, which show seasonally increasing incidence of rabies cyclically each year [35].

Comparisons with Captive Exposure Experiments

Results from experimental RV infections in captive big brown bats are pertinent to our findings on waxing and waning of RVNAs. Seven wild big brown bats that were seropositive when captured maintained titers during 5 months in captivity and had no RV antigen in brains when euthanized [17]. Three other bats had no RVNA in one of six monthly samplings each, indicating a waxing and waning of antibody production [17]. Two juveniles with positive titers shortly after being brought into captivity during late summer were seronegative at five monthly bleedings thereafter, suggesting waned passive immunity [17]. Substantial antibody titers could be induced in big brown bats through non-lethal exposures to RV or inoculation with RV vaccine, leading to the hypotheses that antibodies were indicative of some degree of immunity, and that bats experience such exposures in the wild [17] (perhaps by receiving small loads of virus transmitted through bites in distal regions, or in transfer of virus in saliva of infected bats in prodromal stages of the disease while scratching and grooming in clusters with non-infected bats). Jackson et al. [18] found that 17 of 20 big brown bats experimentally inoculated with

high doses of RV showed seroconversion. Four bats that survived experimental inoculation maintained titers for >three months, but titers then declined to non-detectable levels at 4.6 months post-exposure. Jackson et al. [18] also suggested that the immune response of big brown bats can prevent a productive RV infection. These findings were followed by experimental inoculation of big brown bats at several dosage levels, then re-inoculation of survivors in secondary and tertiary experiments at 175 and 305 days after the initial exposure [19]. Regardless of dosage, at each experimental stage bats that seroconverted after inoculation had a higher probability of surviving than those that did not, and bats that succumbed but showed RVNA did not seroconvert until 1–6 days before death [19]. Titers in RVNA-positive bats peaked from 31–42 days post-exposure after the first inoculation, but after 175 days 23 of 26 (88%) bats were seronegative; titers remained detectable 6–12 months after secondary inoculation [19]. On the third experimental inoculation (349 days after the first and 174 days after the second), survivors (6 of 10) had antibody titers prior to injection, and showed significantly lower mortality, leading to the conclusion that long-term repeated inoculation of big brown bats with RV may indeed confer significant immunological memory and reduced susceptibility to RV infection [19]. In another experiment, captive big brown bats from our study population were inoculated with either of two big brown bat RV variants circulating locally [22]. Bats that were seropositive when brought into captivity all survived an intramuscular challenge with 10^3 TCID 50 dose of one variant without developing clinical rabies, and subsequently showed increased circulating RVNA; most bats that were seronegative on inoculation with this variant developed fatal rabies [22]. Big brown bats inoculated with the second RV variant all survived without developing rabies, and most seroconverted but with RVNA becoming undetectable after 90 days post-inoculation [22]. Bats that were naturally seropositive at the time of capture did not develop rabies when inoculated with this second variant [22].

Our studies of RVNA in wild big brown bats concur with the above experimental findings in several respects. In the experimental studies captive bats also maintained antibodies for variable periods, with many seropositive for about 5 months, but RVNA were more persistent with repeated exposure [17–19]. We found that some wild individuals consistently showed titers over time whereas others showed variability in RVNA production. Such individual variability may be due to frequency and nature of re-exposures to RV in the wild, as well as to individual variability in general health and susceptibility to environmental stressors. Constantine [71] suggested that some bats may be less susceptible to RV infection and those that develop the disease may be immunocompromised. It is of note that seronegative bats at our study area also had greater infestation by some ectoparasites than seropositive bats, perhaps correlated with greater stress [74]. Experiments demonstrated that wild seropositive bats can have immunity to experimental infection, and that experimental inoculation with a variant of RV found in our study population can lead to circulating RVNA without the disease [22], adding another dimension to the complexity of host-virus dynamics in this system. Bats held in captivity with constant environmental conditions and ready access to food and water in experimental studies are less subject to variable environmental influences, and may show less temporal variability and perhaps longer periods of RVNA production than wild big brown bats. Survival rates and absence of evidence for RV or RV RNA in seropositive wild big brown bats (first demonstrated by Trimarchi and Debbie [3]), and a preponderance of seronegative individuals in most experimentally infected rabid bats (other than cases shortly before death) suggest that in most cases the presence of RVNA in serum should be interpreted primarily as evidence of prior exposure rather than an ongoing RV infection. However, much variability can be expected in expression of RVNA in relation to infection status in big brown bats, as is observed in other mammals [75].

We hypothesized that the phase of reproduction might be an important stress factor influencing seroprevalence. We also anticipated that recent temperature conditions might also influence the presence of RVNA because of implications for daily torpor and its influence on metabolic rates that can in turn influence the immune system of mammals [57–60,76]. Instead we found no support for these predictions, and suggest that variability in RVNA production may be more influenced by year-to-year differences in environmental conditions, especially drought. Prolonged drought may mimic chronic stress, which can impair IgG production and other aspects of immune function in laboratory mice [77,78]. Chronic stress also can elevate baseline cortisol in mammals, and in other species of bats baseline cortisol increases during seasonal food scarcity and periods of poor body condition [79,80]. Although cortisol effects on immunity can be complex, chronic elevation of cortisol can be immunosuppressive [81,82]; restricted energy availability also has negative effects on immune function [83]. We are unaware of any studies of the impact of drought on stress and immune responses of big brown bats. However, some data exist on such responses in other wildlife due to drought, and droughts are associated with negative demographic effects in bats that imply a chronic health effect. Drought had negative impacts on cell-mediated immunity, clutch size, and body mass in passerine birds in New Mexico [84]. Elevated cortisol and impaired innate immunity was documented in wild wallaroos (*Macropus robustus*) during a major drought in Western Australia [85]. Reduced fecundity during a drought year in comparison with a normal year was reported for 3 species of insectivorous bats in southwestern Colorado [40], and female white-striped bats (*Tadarida australis*) ceased reproduction during drought in Australia [86]. Drought years are associated with lower survival of little brown bats in New Hampshire and lower survival and reduced population growth rates in Yuma myotis (*Myotis yumanensis*) in California [39,87]. Drought may exert these demographic effects on bats through decreased insect abundance and resultant energetic and metabolic constraints [39,40]; mounting of immune responses also can be metabolically costly [83,88]. Given this background, we hypothesize that chronic prey shortage and energetic restriction may contribute to suppressed antibody production in big brown bats.

Our hypothesis that drought may be one of the environmental stressors of importance to RVNA production in big brown bats has

negative implications in the context of global climate change. Predictions of global climate change models forecast greater drought conditions throughout much of Western North America [89–91]. Drought has been associated with negative impacts on demographic traits of insectivorous bats as noted above. Our findings suggest the hypothesis that drought has additional negative implications for health of insectivorous bat populations. Alternatively, big brown bats at our study area by coincidence may have been exposed to RV to a greater extent during years when greater seroprevalence was observed. However, passive public health surveillance data and other observations from the study area did not indicate any major changes in the number or proportion of bats with rabies during any year of our study [25].

Acknowledgments

We thank D. Anderson and J. Wimsatt for advice, planning, and oversight during the early stages of the study. C. Andrews and C. DeMattos performed some of the laboratory analyses. L. Ellison conducted the survival analysis. For expertise in field work and sampling bats throughout the project we thank L. Ellison, D. Neubaum, M. Neubaum and R. Pearce. Additional help in sampling bats was provided by S. Almon, T. Barnes, J. Benson, J. Boland, L. Bonewell, M. Carson, K. Castle, S. Cooper, T. Dawes, D. Emptage, L. Galvin, D. Grossblat, M. Hayes, B. Iannone, E. Kennedy, R. Kerscher, J. LaPlante, H. Lookingbill, G. Nance, S. Neils, C. Newby, V. Price, C. Reynolds, L. Taraba, J. Tharp, T. Torcoletti, and M. Vrabely. P. Cryan and two anonymous referees provided reviews of the manuscript. Ideas presented in this paper benefitted from discussions within the Small Mammals Working Group of the Research and Policy for Infectious Disease Dynamics (RAPIDD) program of the Science and Technology Directorate (United States Department of Homeland Security) and the Fogarty International Center (National Institutes of Health). Any use of trade, firm, or product names in this publication is for descriptive purposes only and does not imply endorsement by the U.S. Government.

Author Contributions

Conceived and designed the experiments: CER RAB TJO VS. Performed the experiments: VS CER RAB TJO. Analyzed the data: TRS TJO VS. Contributed reagents/materials/analysis tools: CER RAB TJO VS. Wrote the paper: TJO RAB TRS VS CER.

References

1. Burns KF, Farinacci CG, Murnane TG (1956) Insectivorous bats naturally infected with rabies in southwestern United States. Am J Public Health 46: 1089–1097.
2. Constantine DG, Tierkel ES, Kleckner MD, Hawkine DM (1968) Rabies in New Mexico cavern bats. Public Health Rep 83: 303–316.
3. Trimarchi CV, Debbie JG (1977) Naturally occurring rabies virus and neutralizing antibody in two species of insectivorous bats of New York State. J Wildl Dis 13: 366–369.
4. Salas-Rojas M, Sanchez-Hernandez C, Romero-Almaraz MDL, Schnell GD, Schmid RK, et al. (2004) Prevalence of rabies and LPM paramyxovirus antibody in non-hematophagous bats captured in the Central Pacific coast of Mexico. Trans R Soc Trop Med Hyg 98: 577–584.
5. Steece R, Altenbach JS (1989) Prevalence of rabies antibodies in the Mexican free-tailed bat (Tadarida brasiliensis mexicana) at Lava Cave, New Mexico. J Wildl Dis 25: 490–496.
6. Schatz J, Fooks AR, McElhinney L, Horton D, Echevarria J, et al. (2013) Bat rabies surveillance in Europe. Zoonoses Public Health 60: 22–34.

7. Brookes SM, Aegerter JN, Smith GC, Healy DM, Jolliffe T, et al. (2005) Prevalence of antibodies to European Bat Lyssavirus type-2 in Scottish bats. Emerg Infect Dis 11: 572–578.
8. Harris SL, Aegerter JN, Brookes SM, McElhinney LM, Jones G, et al. (2009) Targeted surveillance for European bat lyssaviruses in English bats (2003–06). J Wildl Dis 45: 1030–1041.
9. Serra-Cobo J, Amengual B, Abellán C, Bourhy H (2002) European bat *Lyssavirus* infection in Spanish bat populations. Emerg Infect Dis 8: 413–420.
10. Serra-Cobo J, López-Roig M, Seguí M, Sánchez LP, Nadal J, et al. (2013) Ecological factors associated with European bat Lyssavirus seroprevalence in Spanish bats. PLoS One 8(5): e64467. doi:10.1371/journal.pone.0064467.
11. Vázquez-Morón S, Juste J, Ibáñez C, Ruiz-Villamor E, Avellón A, et al. (2008) Endemic circulation of European bat lyssavirus type 1 in serotine bats, Spain. Emerg Infect Dis 14: 1263–1266.
12. Perez-Jorda JL, Ibáñez C, Muñoz-Cervera M, Téllez A (1995) Lysaavirus in *Eptesicus serotinus* (Chiroptera: Vespertilionidae). J Wildl Dis 31: 372–377.
13. Amengual B, Bourhy H, López-Roig M, Serra-Cobo J (2007) Temporal dynamics of European bat lyssavirus type 1 and survival of *Myotis myotis* bats in natural colonies. PLoS One 2(6): e566. doi:10.1371/journal.pone.0000566.

14. Pape WJ, Fitzsimmons TD, Hoffman RE (1999) Risk for rabies transmission from encounters with bats, Colorado, 1977–1996. Emerg Infect Dis 5: 433–437.

15. Mondul AM, Krebs JW, Childs JE (2003) Trends in national surveillance for rabies among bats in the United States (1993–2000). J Am Vet Med Assoc 222: 633–639.

16. Patyk K, Turmelle A, Blanton JD, Rupprecht CE (2012) Trends in National Surveillance Data for Bat Rabies in the United States: 2001–2009. Vector Borne Zoonotic Dis 12: 666–673.

17. Shankar V, Bowen RA, Davis AD, Rupprecht CE, O'Shea TJ (2004) Rabies in a captive colony of big brown bats (Eptesicus fuscus). J Wildl Dis 40: 403–413.

18. Jackson FR, Turmelle AS, Farino DM, Franka R, McCracken GF, et al. (2008) Experimental rabies virus infection of big brown bats (Eptesicus fuscus). J Wildl Dis 44: 612–621.

19. Turmelle AS, Jackson FR, Green D, McCracken GF, Rupprecht CE (2010) Host immunity to repeated rabies virus infection in big brown bats. J Gen Virol 91: 2360–2366.

20. Davis AD, Rudd RJ, Bowen RA (2007) Effects of aerosolized rabies virus exposure on bats and mice. J. Infect Dis 195: 1144–1150.

21. Davis A, Gordy P, Rudd R, Jarvis JA, Bowen RA (2012) Naturally acquired rabies virus infections in wild-caught bats. Vector Borne Zoonotic Dis 12: 55–60.

22. Davis AD, Gordy PA, Bowen RA (2013) Unique characteristics of bat rabies viruses in big brown bats (Eptesicus fuscus). Arch Virol 158: 809–820.

23. Neubaum DJ, O'Shea TJ, Wilson KR (2006) Autumn migration and selection of rock crevices as hibernacula by big brown bats in Colorado. J Mammal 87: 470–479.

24. Neubaum DJ, Wilson KR, O'Shea TJ (2007) Urban maternity roost selection by big brown bats in north-central Colorado. J Wildl Manage 71: 728–736.

25. O'Shea TJ, Neubaum DJ, Neubaum MA, Cryan PM, Ellison LE, et al. (2011a) Bat ecology and public health surveillance for rabies in an urbanizing region of Colorado. Urban Ecosystems 14: 665–697.

26. O'Shea TJ, Ellison LE, Neubaum DJ, Neubaum MA, Reynolds CA, et al. (2010) Recruitment in a Colorado population of big brown bats: breeding probabilities, litter size, and first-year survival. J Mammal 91: 418–428.

27. O'Shea TJ, Ellison LE, Stanley TR (2011b) Adult survival and population growth rate in Colorado big brown bats (Eptesicus fuscus). J Mammal 92: 433–443.

28. Neubaum MA, Douglas MR, Douglas ME, O'Shea TJ (2007) Molecular ecology of the big brown bat (Eptesicus fuscus): genetic and natural history variation in a hybrid zone. J Mammal 88: 1230–1238.

29. Turmelle AS, Allen LC, Jackson FR, Kunz TH, Rupprecht CE, et al. (2010) Ecology of rabies virus exposure in colonies of Brazilian free-tailed bats (Tadarida brasiliensis) at natural and man-made roosts in Texas. Vector Borne Zoonotic Dis 10: 165–175.

30. Schmid-Hempel P (2003) Variation in immune defiance as a question of evolutionary ecology. Proc R Soc Lond B Biol Sci 270: 357–366.

31. Hawley DM, Altizer SM (2011) Disease ecology meets ecological immunology: understanding the links between organismal immunity and infection dynamics in natural populations. Funct Ecol 25: 48–60.

32. Ellison LE, O'Shea TJ, Neubaum DJ, Bowen RA (2007) Factors influencing movement probabilities of big brown bats (Eptesicus fuscus) in buildings. Ecol Appl 17: 620–627.

33. Cryan PM, Stricker CA, Wunder MB (2012) Evidence of cryptic individual specialization in an opportunistic insectivorous bat. J Mammal 93: 381–389.

34. Shankar V, Orciari LA, De Mattos C, Kuzmin IV, Pape WJ, et al. (2005) Genetic divergence of rabies viruses from bat species of Colorado, USA. Vector Borne Zoonotic Dis 5: 330–341.

35. George DB, Webb CT, Farnsworth ML, O'Shea TJ, Bowen RA, et al. (2011) Host and viral ecology determine bat rabies seasonality and maintenance. Proc Natl Acad Sci USA 108: 10208–10213.

36. Colorado Climate Center (2013) Drought resources. Available: http://ccc.atmos.colostate.edu/palmerindex.php. Accessed 20 February 2013.

37. National Climatic Data Center (2013) Palmer drought severity index. Available: http://www.ncdc.noaa.gov/paleo/drought/drght_pdsi.html. Accessed 20 February 2013.

38. Snider EA (2009) Post-fire insect communities and roost selection by western long-eared myotis (Myotis evotis) in Mesa Verde National Park, Colorado. Master's thesis, Colorado State University, Fort Collins, CO.

39. Frick WF, Reynolds DS, Kunz TH (2010) Influence of climate and reproductive timing on demography of little brown Myotis lucifugus. J Anim Ecol 79: 128–136.

40. O'Shea TJ, Cryan PM, Snider PE, Valdez EW, Ellison LE, et al. (2011) Bats of Mesa Verde National Park, Colorado: composition, reproduction, and roosting habits. Monogr West N Am Nat 5: 1–19.

41. Colorado Climate Center (2006) Available: http://ccc.atmos.colostate.edu./dataaccess.php. Accessed 22 February 2006.

42. O'Shea TJ, Ellison LE, Stanley TR (2004) Survival estimation in bats: Historical overview, critical appraisal, and suggestions for new approaches. In: Thompson WL, editor. Sampling rare or elusive species: Concepts, designs, and techniques for estimating population parameters. Washington, DC: Island Press. 297–336.

43. Sheffield SR, Shaw JH, Heidt GA, McClenaghan LR (1992) Guidelines for the protection of bat roosts. J Mammal 73: 707–710.

44. Wimsatt J, O'Shea TJ, Ellison LE, Pearce RD, Price VR (2005) Anesthesia and blood sampling of wild big brown bats (Eptesicus fuscus) with an assessment of impacts on survival. J Wildl Dis 41: 87–95.

45. Anthony ELP (1988) Age determination in bats. In: Kunz TH, editor. Ecological and behavioral methods for the study of bats. Washington, DC: Smithsonian Institution. pp. 47–58.

46. Ellison LE, O'Shea TJ, Wimsatt J, Pearce RD, Bowen RA (2007) A comparison of conventional capture versus PIT reader techniques for estimating survival and capture probabilities of big brown bats. Acta Chiropt 9: 149–160.

47. Kunz TH (2003) Censusing bats: challenges, solutions, and sampling biases. In: O'Shea TJ, Bogan MA, editors. Monitoring trends in bat populations of the United States and territories: problems and prospects. US Geol Survey Info Technol Rep ITR-2003–003. 9–19.

48. Willis CKR, Brigham RM (2004) Roost switching, roost sharing and social cohesion: forest-dwelling big brown bats, Eptesicus fuscus, conform to the fission-fusion model. Anim Behav 68: 495–505.

49. Ellison LE, O'Shea TJ, Wimsatt J, Pearce RD, Neubaum DJ, et al. (2006) Sampling blood from big brown bats (Eptesicus fuscus) in the field with and without anesthesia: Impacts on survival. J Wildl Dis 42: 849–852.

50. Bowen RA, O'Shea TJ, Shankar V, Neubaum MA, Neubaum DJ, et al. (2013) Prevalence of neutralizing antibodies to rabies virus in serum of seven species of insectivorous bats from Colorado and New Mexico. J Wildl Dis 49: 367–374.

51. Smith JS, Yager PA, Baer GM (1996) A rapid fluorescent focus inhibition test (RFFIT) for determining rabies virus-neutralizing antibody. In: Meslin XF, Kaplan MM, Koprowski H, editors. Laboratory Techniques in Rabies, 4th ed. Geneva: World Health Organization. 200–208.

52. Dean DJ, Abelseth MK, Atanasiu P (1996) The fluorescent antibody test. In: Meslin FX, Kaplin MM, Koprowski H, editors. Laboratory techniques in rabies, 4th edition. World Geneva: Health Organization. pp. 88–93.

53. Smith JS (1996) New aspects of rabies with emphasis on epidemiology, diagnosis, and prevention of the disease in the United States. Clin Microbiol Rev 9: 166–176.

54. Newcombe RG (1998) Two-sided confidence intervals for the single proportion: Comparison of seven methods. Stat Med 17: 857–872.

55. Marascuilo LA (1966) Large-sample multiple comparison. Psychol Bull 65: 280–290.

56. SAS Institute Inc (2008) SAS/STAT® 9.2 User's Guide. Cary, NC: SAS Institute Inc.

57. Sulkin SE, Allen R, Sims R, Singh KV (1966) Studies of arthropod-borne virus infections in Chiroptera. IV. The immune response of the big brown bat (Eptesicus f. fuscus) maintained at various environmental temperatures to experimental Japanese b encephalitis virus infection. Am J Trop Med Hyg 15: 418–427.

58. Burton RS, Reichman OJ (1999) Does immune challenge affect torpor duration? Funct Ecol 13: 232–237.

59. Prendergast BJ, Freeman DA, Zucker I, Nelson RJ (2002) Periodic arousal from hibernation is necessary for initiation of immune responses in ground squirrels. Am J Physiol Regul Integr Comp Physiol 282: R1054–R1062.

60. Canale CI, Henry PY (2011) Energetic costs of the immune response and torpor use in a primate. Funct Ecol 25: 557–565.

61. Vieira P, Rajewsky K (1988) The half-lives of serum immunoglobulins in adult mice. Eur J Immunol 18: 313–316.

62. Sigounas G, Harindranath N, Donadel G, Notkins AL (1994) Half-life of polyreactive antibodies. J Clin Immunol 14: 134–140.

63. Burnham KP, Anderson DR (2002) Model selection and multimodel inference: A practical information-theoretic approach. New York: Springer. p 514.

64. Arnold TD (2010) Uninformative parameters and model selection using Akaike's information criterion. J Wildl Manage 74: 1175–1178.

65. Burnham KP, Anderson DR, White GC, Brownie C, Pollock KP (1987) Design and analysis of methods for fish survival experiments based on release-recapture. Monogr Am Fish Soc 5: 1–437.

66. Kurta A, Baker RH (1990) Eptesicus fuscus. Mammalian Species 356: 1–10.

67. Bode AV, Sejvar JJ, Pape WJ, Campbell GL, Marfin AA (2006) West Nile virus disease: a descriptive study of 228 patients hospitalized in a 4-county region of Colorado in 2003. Clin Infect Dis 42: 1234–1240.

68. Bolling BG, Moore CG, Anderson SL, Blair CD, Beaty BJ (2007) Entomological studies along the Colorado Front Range during a period of intense West Nile virus activity. J Am Mosq Control Assoc 23: 37–46.

69. Colorado Mosquito Control (2004) 2004 Annual Report City of Fort Collins Mosquito Control Program. Available: http://fcgov.com/westnile/pdf/annualreport04. Accessed 20 February 2013.

70. O'Shea TJ, Johnston JJ (2009) Environmental contaminants and bats: Investigating exposure and effects. In: Kunz TH, Parsons S, editors. Behavioral and ecological methods for the study of bats. Baltimore, MD: Johns Hopkins University Press. pp. 500–528.

71. Constantine DG (1988) Health precautions for bat researchers. In: Kunz TH, editor. Ecological and behavioral methods for the study of bats. Washington, DC: Smithsonian Institution Press. pp. 491–528.

72. Turmelle AS, Allen LC, Schmidt-French BA, Jackson FR, Kunz TH, et al. (2010) Response to vaccination with a commercial inactivated rabies vaccine in a captive colony of Brazilian free-tailed bats (Tadarida brasiliensis). J Zoo Wildl Med 41: 140–143.

73. Messenger SL, Rupprecht CE, Smith JS (2003) Bats, emerging virus infections, and the rabies paradigm In: Kunz TH, Fenton MB editors. Bat ecology. Chicago, IL: University of Chicago Press. pp. 622–679.

74. Pearce RD, O'Shea TJ, Shankar V, Rupprecht CE (2007) Lack of association between ectoparasite intensities and rabies virus neutralizing antibody

seroprevalence in wild big brown bats (*Eptesicus fuscus*), Fort Collins, Colorado. Vector Borne Zoonotic Dis 7: 489–495.

75. Niezgoda M, Hanlon CA, Rupprecht CE (2002) Animal rabies. In: Jackson AC, Wunner WH, editors. Rabies. New York: Academic Press. pp. 163–218.

76. Baker ML, Schountz T, Wang LF (2013) Antiviral immune responses of bats: a review. Zoonoses Public Health 60: 104–116.

77. Silberman DM, Wald MR, Genaro AM (2003) Acute and chronic stress exert opposing effects on antibody responses associated with changes in stress hormone regulation of T-lymphocyte reactivity. J Neuroimmunol 144: 53–60.

78. Murray SE, Lallman HR, Heard AD, Rittenberg MB, Stenzel-Poore MP (2001) A genetic model of stress displays decreased lymphocytes and impaired antibody responses without altered susceptibility to *Streptococcus pneumonia*. J Immunol 167: 691–698.

79. Lewanzik D, Kelm DH, Greiner S, Dehnhard M, Voigt CC (2012) Ecological correlates of cortisol levels in two bat species with contrasting feeding habits. Gen Comp Endocrinol 177: 104–112.

80. Allen LC, Turmelle AS, Widmaier EP, Hristov NI, McCracken GF, et al. (2010) Variation in physiological stress between bridge- and cave-roosting Brazilian free-tailed bats. Conserv Biol 25: 374–381.

81. Sapolsky RM, Romero LM, Munck AU (2000) How do glucocorticoids influence stress responses? Integrating permissive, suppressive, stimulatory, and preparative actions. Endocr Rev 21: 55–89.

82. Baker MR, Gobush KS, Vynne CH (2013) Review of factors influencing stress hormones in fish and wildlife. J Nature Conserv 21: 309–318.

83. Demas GE (2004) The energetics of immunity: a neuroendocrine link between energy balance and immune function. Horm Behav 45: 173–180.

84. Fair JM, Whitaker SJ (2008) Avian cell-mediated immune response to drought. Wilson J Ornithol 120: 813–819.

85. King JM, Bradshaw SG (2010) Stress in an Island kangaroo? The Barrow Island euro, *Macropus robustus isabellinus*. Gen Comp Endocrinol 167: 60–67.

86. Rhodes M (2007) Roost fidelity and fission–fusion dynamics of white-striped free-tailed bats (*Tadarida australis*). J Mammal 88: 1252–1260.

87. Frick WF, Rainey WE, Pierson ED (2007) Potential effects of environmental contamination on Yuma myotis demography and population growth. Ecol Appl 17: 1213–1222.

88. Abad-Gomez JM, Gutiérrez JS, Villegas A, Sánchez-Guzmán JM, Navedo JG, et al. (2013) Time course and metabolic costs of a humoral immune response in the little ringed plover *Charadrius dubius*. Physiol Biochem Zool 86: 354–360.

89. Seager R, Ting M, Held I, Kushnir Y, Lu J, et al. (2007) Model projections of an imminent transition to a more arid climate in southwestern North America. Science 316: 1181–1184.

90. Barnett TP, Pierce DW, Hidalgo HG, Bonfils C, Santer BD, et al. (2008) Human-induced changes in the hydrology of the western United States. Science 319: 1080–1083.

91. McAfee SA, Russell JL (2008) Northern annular mode impact on spring climate in the western United States. Geophys Res Lett 35: 1–5.

23

Effects of Climate Change on Range Forage Production in the San Francisco Bay Area

Rebecca Chaplin-Kramer[1], Melvin R. George[2]*

1 Natural Capital Project, Stanford University, Stanford, California, United States of America, **2** Plant Sciences Department, University of California Davis, Davis, California, United States of America

Abstract

The San Francisco Bay Area in California, USA is a highly heterogeneous region in climate, topography, and habitats, as well as in its political and economic interests. Successful conservation strategies must consider various current and future competing demands for the land, and should pay special attention to livestock grazing, the dominant non-urban land-use. The main objective of this study was to predict changes in rangeland forage production in response to changes in temperature and precipitation projected by downscaled output from global climate models. Daily temperature and precipitation data generated by four climate models were used as input variables for an existing rangeland forage production model (linear regression) for California's annual rangelands and projected on 244 12 km x 12 km grid cells for eight Bay Area counties. Climate model projections suggest that forage production in Bay Area rangelands may be enhanced by future conditions in most years, at least in terms of peak standing crop. However, the timing of production is as important as its peak, and altered precipitation patterns could mean delayed germination, resulting in shorter growing seasons and longer periods of inadequate forage quality. An increase in the frequency of extremely dry years also increases the uncertainty of forage availability. These shifts in forage production will affect the economic viability and conservation strategies for rangelands in the San Francisco Bay Area.

Editor: Ben Bond-Lamberty, DOE Pacific Northwest National Laboratory, United States of America

Funding: The California Energy Commission's Public Interest Energy Research (PIER) Program provided support for this research under the California Climate Change Center. The funders had no role in study design, data collection and analysis, decision to publish, or preparation of the manuscript.

Competing Interests: The authors have declared that no competing interests exist.

* E-mail: mrgeorge@ucdavis.edu

Introduction

California's San Francisco Bay Area is a mosaic of urban and natural lands, and tension exists between the two with scenic beauty attracting development that threatens these prized open spaces [1]. About half of the area of the eight San Francisco Bay counties is classified as rangeland, and these areas account for most of the region's open space. The private ranches on Bay Area rangelands provide a livelihood and a way of life that helps to limit urban sprawl [2]. With a growing population placing pressure on these areas for development and a changing climate posing new threats to rangeland ecosystems, understanding the degree to which their value as working landscapes will be maintained in the future is important to their conservation.

Over the next century California temperatures are projected to rise between 1.7°and 3.0°C for a lower emissions scenario, and 4.4° to 5.8°C for a higher emissions scenario [3]. Downscaled results from global climate models for the San Francisco Bay Area show a lower rise in temperatures, from 1.5° to 3.0°C by 2100 for the lower emissions scenario and 2.5° to 4.4°C for the higher emissions scenario, though considerable variation exists within the region (Fig. 1). Changes in precipitation are more uncertain, with a high degree of variability between different climate models, and even from year to year within the same model, suggesting that the region will remain vulnerable to drought. Overall, the majority of simulations indicate that total annual precipitation will decline,

mostly in the spring months, while winter precipitation will remain relatively stable [4].

The full extent of climate change impacts on rangeland forage production in California and the Bay Area in particular is uncertain. One study [5] modeled the impact of forecasted changes in precipitation patterns on California rangeland production and concluded that areas of the state suitable for cattle grazing would shift, as some areas become wetter and others become drier, depending on the climate model. Statewide, they predicted range forage production would decline between 14 and 58 percent, corresponding to a reduction in annual profits from cattle ranching of between $22 million and $92 million by 2070. Despite this statewide trend, their results for the Bay Area suggested the impacts would be more positive, with production increases projected for Santa Clara, Alameda, Contra Costa, Solano, and Napa Counties. Marin and Sonoma Counties were not included in the model. However, this precipitation-based model did not incorporate rangeland response to warming, and the authors acknowledged that their model may have overestimated the effects of precipitation. While precipitation has been shown to be an important variable for predicting annual rangeland productivity, temperature within the growing season is also important [6]. Precipitation, temperature and forage production data collected since 1935 at the San Joaquin Experimental Range in Madera County have shown that near average production can occur in low rainfall years if precipitation is well distributed and

Change in Average SummerTemperature 1961-1990 -- 2070-2099 (A2 Scenario)

Change in Average SummerTemperature 1961-1990 -- 2070-2099 (B1 Scenario)

Change in Average Winter Temperature 1961-1990 -- 2070-2099 (A2 Scenario)

Change in Average Winter Temperature 1961-1990 -- 2070-2099 (B1 Scenario)

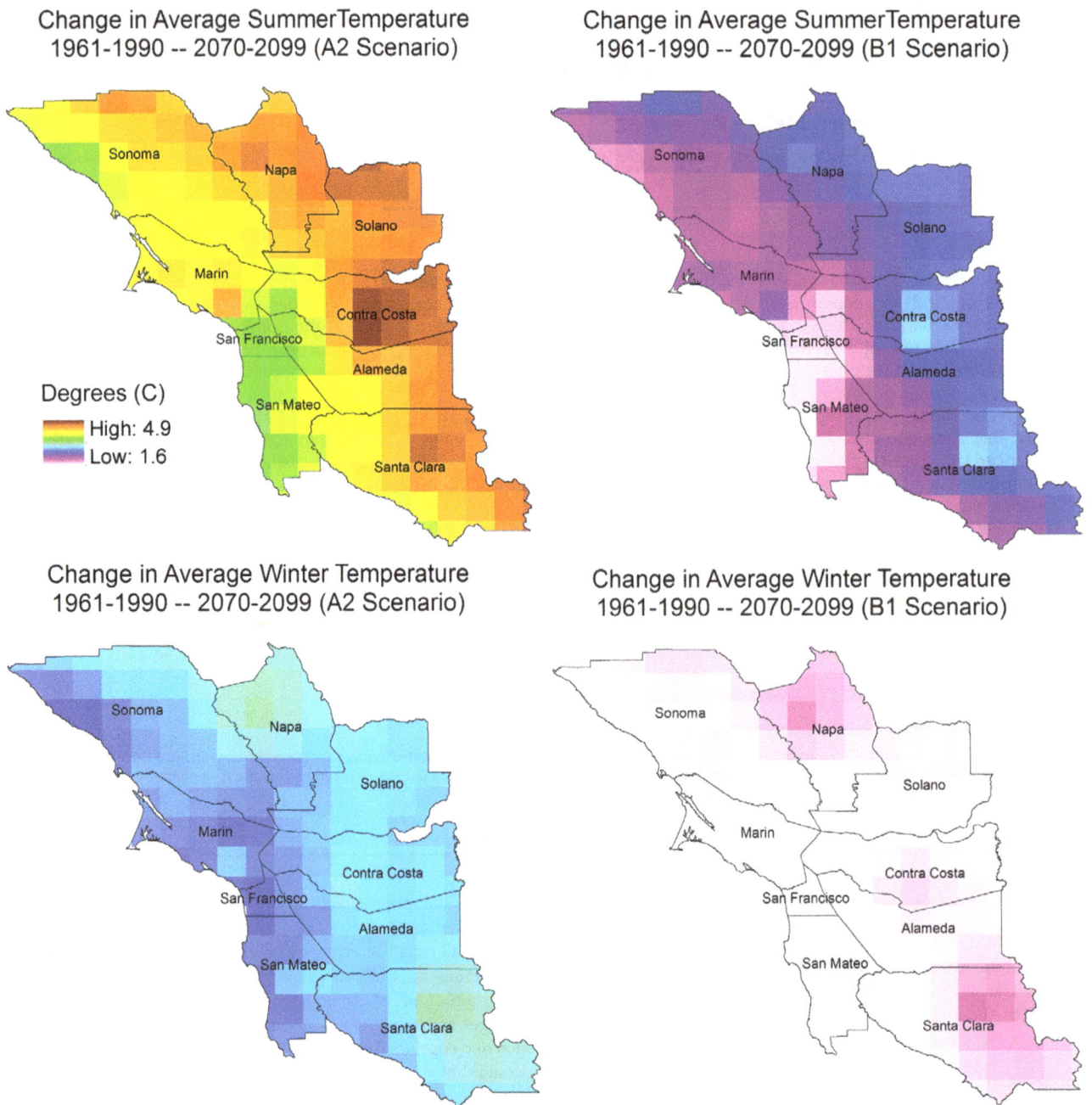

Figure 1. Historical (1961–1990) and projected (2070–2099) average temperatures for summer (June, July, August) and winter (January, February) months in the Bay Area. Temperatures reflect means of four global climate models (downscaled output from CNRM CM3, GFDL CM2.1, NCAR CCSM3.0 and NCAR PCM1).

low annual production can occur in wet years if precipitation is poorly distributed or if temperatures are below normal, as is often associated with wet weather [7]. Thus, integrating changes in both temperature and precipitation could improve forecasts of range-land forage production.

The combined effects of seasonal temperature and precipitation patterns will influence not only productivity, but also growing season length and plant phenology in rangelands [8]. Warming has been shown to increase soil water content by accelerating plant senescence [9], which may interact with changes in precipitation to further affect water availability in rangelands. In fact, grassland

ecophysiology may be less responsive to changes in total quantity of rainfall than to shifts in seasonal patterns of rainfall [10]. Early-season precipitation alone explained 49 percent of the variability in shoot-growth at the University of California Hopland Research and Extension Center (UC HREC), just north of the Bay Area [11], although additional data reduced this explanatory power to 34 percent [6]. Late-season precipitation also has a pronounced impact on Bay Area and North Coast grassland production, shown by increased shoot growth resulting from experimental water additions in the late spring [12,13].

Despite these known links between the effects of warming and precipitation in grasslands, no models incorporating the impact of warming on California rangeland production yet exist. Increases in spring plant production and an extension of the growing season has been predicted for the Great Plains [14]. However, these authors warned that increases in variance may be more important than the mean effect, because uncertainty in predicting plant growth results in suboptimal stocking decisions. They suggested that the increased variance found in their simulations would require carrying capacities to decrease from about 6.5 to 9.0 ha per animal, in order to maintain a 90 percent confidence of not overstocking. Further, more intense management would increase operating costs, and therefore may negate any benefits in forage production. In contrast to the Great Plains, where the growing season begins in the spring months following winter dormancy, the Bay Area rangeland growing season begins with the first fall rains and ends with soil moisture depletion in the spring months. Climate change in the Bay Area may be more comparable to that found in the Mediterranean climate of southern Australia. For this region, lower pasture production has been projected for future climates with lower precipitation and higher temperatures [15].

Incorporating the effects of warming into models of rangeland production in California, and the Bay Area in particular, is an important step in understanding how climate change will affect range livestock production in this region. The economic viability of rangelands is essential to maintaining the natural aesthetic that contributes to the quality of life of the residents of this unique urban-natural interface. This paper reports the projected changes in forage production in response to simulated future temperature and precipitation in the San Francisco Bay Area to better understand how climate change will impact these working landscapes so important to local conservation.

Methods

Study Area

The geographic diversity of the San Francisco Bay Area (hills, mountains, and large water bodies) produces a wide variety of microclimates. Coastal areas are generally characterized by relatively small temperature variations during the year, with cool, foggy summers and mild, rainy winters. Inland areas, especially those separated from the ocean by hills or mountains, have hotter summers and colder overnight temperatures during the winter. The rangelands in the North Bay (with its northwestern most point at 38.9375N, 123.6875W; encompassing Sonoma, Marin, Napa, and Solano Counties) are characterized by higher rainfall, a longer rainy season and cooler temperatures than those in the South Bay (southeastern most point at 36.9375N, 121.1875W; encompassing Santa Clara, Alameda, and Contra Costa Counties). San Jose, at the south end of the Bay averages fewer than 380 mm of rain annually, while Napa, in the North Bay area, can exceed 750 mm. Because range forage production is strongly influenced by temperature and precipitation, there are significant differences in growing season length and productive potential between the North and South Bay areas.

Climate Models

Climate data were acquired from Cayan et al., downscaled from global climate models using a Bias Corrected Constructed Analogues (BCCA) technique to produce two climate scenarios: the lower-emissions B1 scenario and the higher-emissions A2 scenario [4]. Four climate models produce daily temperature and precipitation projections: Centre National Recherche Meteorologique (CNRM) CM3, Geophysical Fluid Dynamics Laboratory

(GFDL) CM2.1, National Center for Atmospheric Research (NCAR) CCSM3.0, and NCAR PCM1. Each climate model was back-cast to simulate historical climate conditions (1961–1990) and represented historical climate data with accuracy [4].

Forage Production Model

Throughout California's 14.5 million acres of annual rangelands, which include the Bay Area grasslands and oak woodlands of this study, temperature is the main constraint to productivity during the growing season. Therefore, precipitation and evapotranspiration drove a simple model to determine growing season length, and temperature and growing season length drove the model for annual forage production.

Daily climate data from the four climate models were input variables for a forage production model reported by Californian researchers [16] who found that growing degree days accounted for 75 to 95 percent of the variation in growing season production (Table 1). This degree-day forage production model was run for all eight model/scenario combinations (described in Climate Models, above), for each of the 244 12 km x 12 km grid cells that comprise the Bay Area region. The mean of the output from the four climate models was taken for the A2 and B1 scenarios, and compared to output for a simulated historical period (1961–1990). All calculations and simulations were produced in the R software package [17].

Modeling the bounds of the growing season was necessary to convert a regression model based on field data into a predictive model to simulate forage production under future climate scenarios. A germinating rain that exceeds 25 mm within one week marks the start of the growing season [16]. There is no similarly well-established climatic phenomenon marking the end of the growing season; in field studies it is determined empirically, by measuring biomass until annual grasses are between the soft and hard dough stage of seed maturity [16]. Therefore, the end of the growing season in this study was simulated using a simple water balance model that was trained using CIMIS weather data for precipitation and evapotranspiration from the University of California Sierra Foothill Research and Extension Center (UC SFREC), 17 miles northeast of Marysville, California. Calculating the point at which cumulative evapotranspiration exceeded cumulative precipitation over a moving window of 60 days best predicted the peak forage date. This simple model generally came within two weeks of actual peak forage date measured at the UC SFREC, rarely extending beyond the end of May. For each year, germination and season end dates were computed according to these methods, and set the seasonal bounds within which forage production was modeled, capturing inter-annual variability in season length.

To simulate forage production, degree-days were first calculated from model-generated minimum and maximum daily temperatures above a base temperature of 5°C using the sine function method [16,18]. Accumulated degree-days (ADD), the sum of all previous degree-days from a given date, were calculated at monthly intervals from germination until the end of the growing season. Monthly standing biomass or total forage production was estimated from ADD using the regression equations from several annual rangeland sites in Table 1 [16]. Absolute forage production varied depending on the chosen equation, but as the relationship is linear, relative measures such as the change in forage production over time were very consistent, differing by only 2 to 3 percent. For this reason, future values for peak forage production are presented in terms of change from historic values.

Growth curves were constructed using the monthly forage production estimates, averaged over the window of historic (1961–

Table 1. Relationship between forage production (y, kg ha^{-1}) and accumulated degree days (x) from 10 sites in four annual rangeland counties [16].

Sample Area	County	Latitude (N)	Longitude (W)	Regression Equation	R^2
1	Yuba	39.330361	121.3476	y = −120+5.2x	0.95
2	Butte	39.384564	121.59456	y = 14+4.4x	0.91
3	Madera	37.088669	119.73461	y = −90+3.8x	0.85
4	Madera	37.089609	119.713978	y = −141+3.1x	0.82
5	Madera	37.089811	119.73698	y = −54+3.9x	0.77
6	Madera	37.095235	119.736424	y = −280+4.9x	0.88
7	Mendocino	38.997164	123.092492	y = 77+2.2x	0.74
8	Mendocino	38.986653	123.08472	y = 138+2.8x	0.76
9	Mendocino	39.006414	123.085347	y = 96+4.1x	0.91
10	Mendocino	39.003462	123.077236	y = 82+2.7x	0.74

1999) and future (2070–2099) time periods for both A2 and B1 emissions scenarios, and then averaged over all rangeland area in a county. These growth curves help to determine which parts of the season have the greatest differences between historical and future conditions, and thus hint at the mechanisms behind the difference. Differences in the first month may indicate that germination date is an important factor. Steeper slopes throughout the middle of the curve would point to the role played by warmer winter and/or spring temperatures. Differences in the slope leading up to the final time step could be at least partially explained by differences in season end date and the length of time for degree days to accumulate in that final period. Months are taken as calendar months, such that if germination occurred on September 23, the month of September would only have one week of forage production. Likewise, if the end of season date is calculated as June 2, the month of June would only have two days of additional forage production added to the overall total.

All model calculations were repeated for every year of simulated climate data from 1961 to 2099, and the data were summarized by taking the means in four windows of time: historic (1961–1999), early century (2005–2034), mid-century (2035–2064), and late century (2070–2099). Changes in peak production and season length from historic to future conditions are presented on maps of rangeland biomes (savannah, grasslands and shrublands) selected from the Existing Vegetation Types layer of the national LANDFIRE dataset [19] and overlaid on the model outputs.

Results

Forage Production

Forage production by the end of the century (2070–2099) increases in each month of the growing season relative to simulated historical forage production (1961–1990), resulting in increased total forage production under future climate conditions throughout the San Francisco Bay Area (Fig. 2). There is little difference between projections for historic vs. early-century (2005–2034) and mid-century (2035–2064) forage production for either scenario. Projected mid-century peak forage production increased only 10 to 13 percent from the historical (1961–1990) conditions for the higher-emissions A2 scenario (Table 2). However, average production for each of the eight Bay Area Counties is projected to increase 24 to 31 percent by the late century (2070–2099) for the A2 scenario, with increases of up to 40 percent in much of Northern Napa and Sonoma Counties (Fig. 3). Projected increases

for these counties in the B1 scenario are slightly higher by mid-century and more modest at the end of the century. However, late-century northeastern Santa Clara County shows an increase above 30 percent for both emission scenarios.

Season Length

The models also predicted changes in growing season length, due to changes in the simulated timing of germination (precipitation exceeding 25 mm in one week) in the fall and soil moisture depletion (evapotranspiration exceeding precipitation over a 60 day period) in the spring. The length of the growing season is projected to markedly decrease under the A2 scenario; two-week shorter seasons can be expected by late-century for much of the Bay Area (Table 2), with parts of Santa Clara showing seasons shrinking by more than three-weeks (Fig. 4). The B1 scenario shows more modest decreases in season length of a few days to a week. With decreasing rainfall, future forage season lengths in eastern Santa Clara and Alameda Counties could drop to as low as 100 days in length, a full 50 days shorter than the shortest season found in Marin or Sonoma Counties.

Inter-annual Variability

The standard errors reported in Table 2 indicate that inter-annual variability in peak forage production and season length is fairly low in the Bay Area, under both historic and future conditions. These standard errors range from 1 to 3 percent of the mean for either of these variables in any county. On the other hand, the maximums and minimums for each 30-year time period range from 38 percent below to 53 percent above the mean forage production and 29 percent above to 46 percent below mean season length (Table 3). More information can be gleaned by considering the variability as the spread of the distribution, the number of data points falling outside a range of 2 standard errors around the mean (Table 4). By this definition, inter-annual variability in peak production declines fairly dramatically throughout most of the Bay Area for both scenarios, due mainly to the number of years that production is above average in the North Bay, while Alameda and Contra Costa shows declines in variability around both sides of the mean for the B1 scenario and in the number of years production is below average in the A2 scenario. Overall inter-annual variability in peak production does not change in Santa Clara, but is skewed more negatively in future climate conditions under both scenarios (more years in which peak

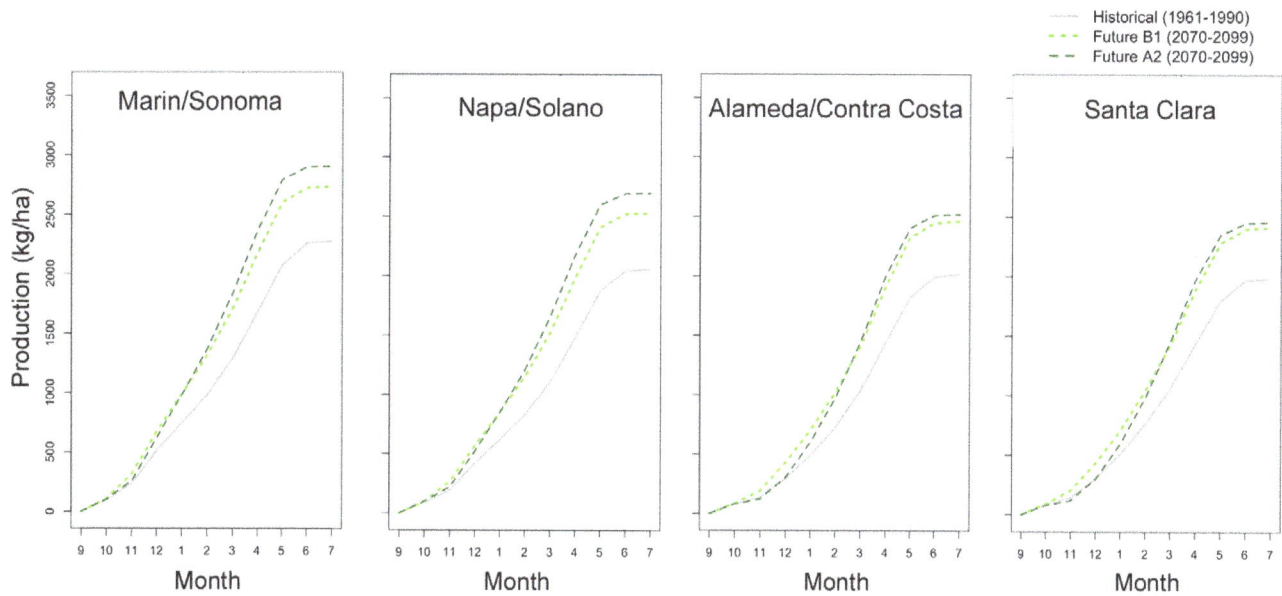

Figure 2. Seasonal growth curves for forage production in different regions under historical (1961–1990) and future (2070–2099, for A2 and B1 emissions scenarios) climate conditions. Growth curves represent the accumulated forage produced on a daily time-step, summarized at monthly intervals as the amount of total forage produced over the season by that date. Each line shows the mean production for all years within each 30-year period and for all cells within each county. Only cells containing rangelands were used (see Figure 3).

production falls below 2 standard errors below the mean occur in the future than historic conditions). Inter-annual variability in season length remains the same or even declines for most counties for the B1 scenario. In the A2 scenario variability increases slightly, with more years falling farther below average season length than historically. The variability is more centered around the mean in all future scenarios than in historical climate

conditions (which were skewed slightly positively; more longer than average seasons than shorter than average seasons).

Drought Years

The climate models predict that some regions in the Bay Area would see some years with no germinating rains and therefore no growing season. Historically this would generally occur in any given location in the South Bay once over a 30 year period. While

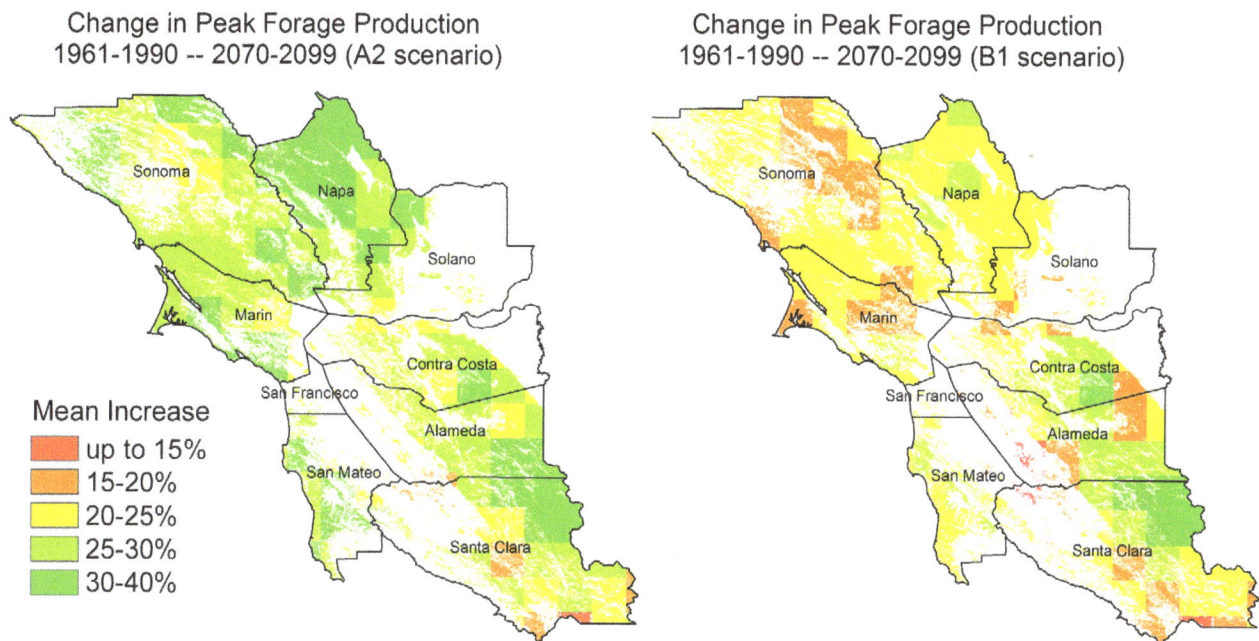

Figure 3. Change (%) in peak forage production by late-century (2070–2099), relative to historical conditions (1961–1990), shown for current rangelands (grassland, savannah, and shrubland) in the Bay Area.

Table 2. Change in peak forage production and season length compared to historic (1961–1990) conditions (mean +/− standard error).

Time Period	Marin-Sonoma		Napa-Solano		Alameda-Contra Costa		Santa Clara	
	Mean	+/−	Mean	+/−	Mean	+/−	Mean	+/−
Change in Forage Production (%): A2 Model								
2005–2034	10.1	2.2	10.4	2.5	13.1	2.8	11.8	2.5
2035–2064	11.9	2.5	13.1	2.9	13.5	3.0	13.0	3.0
2070–2099	27.5	2.8	31.1	3.0	24.9	3.1	23.6	2.9
Change in Forage Production (%): B1 Model								
2005–2034	9.4	2.9	10.1	3.1	9.5	2.9	10.8	3.1
2035–2064	16.2	2.9	19.2	3.2	17.8	3.5	17.1	3.6
2070–2099	20.0	2.6	23.0	2.8	22.1	3.1	21.5	3.1
Reference Historic Season Length (Total Days)								
1961–1990	181.1	2.5	171.8	2.5	152.8	2.78	155.3	2.9
2005–2034	−1.7	2.3	−2.4	2.5	0.3	2.9	−1.2	2.8
2035–2064	−9.7	2.4	−9.2	2.5	−9.6	3.0	−10.9	3.0
2070–2099	−15.5	2.6	−13.7	2.6	−16.3	2.8	−19.4	2.5
Change in Season Length (Days): A2 Model								
2005–2034	−2.9	3.1	−3.1	3.2	−3.4	3.6	−2.5	3.8
2035–2064	−3.3	3.0	−1.9	3.1	−3.6	3.4	−4.6	3.3
2070–2099	−5.1	2.8	−4.1	2.7	−3.3	3.4	−6.2	3.4
Change in Season Length (%): A2 Model								
2005–2034	−0.9	1.3	−1.4	1.5	0.2	1.9	−0.7	1.8
2035–2064	−5.4	1.3	−5.3	1.5	−6.3	2.0	−7.0	1.9
2070–2099	−8.6	1.4	−8.0	1.5	−10.7	1.8	−12.5	1.6
Change in Season Length (Days): B1 Model								
2005–2034	−1.6	1.7	−1.8	1.8	−2.2	2.4	−1.6	2.4
2035–2064	−1.8	1.6	−1.1	1.8	−2.3	2.2	−3.0	2.1
2070–2099	−2.8	1.5	−2.3	1.6	−2.1	2.2	−4.0	2.2

comparisons of model projections revealed that there is a high degree of variation among the four climate models, the mean shows a lower frequency of non-germination years in the southern counties for the B1 scenario compared to historic conditions, and a higher frequency for the A2 scenario. Specific outcomes supported by all models and for both scenarios are that the North Bay is almost entirely unaffected in all time periods and southeastern Santa Clara County experiences more extreme dry years under future conditions.

Discussion

Forage Production

Our model supports the results of earlier efforts incorporating only precipitation into a model for forage production in the Bay Area [5], though the increases in forage production seen here were in spite of, not because of, a shift in precipitation patterns. The agreement between these two models stands in contrast to the projections of a similar Mediterranean climate in Southern Australia [15], where lower forage production was expected under warmer and drier climate. This difference may be due to the fact that precipitation is not as limiting in the Bay Area system as winter temperatures. Projected warming for A2 and B1 scenarios result in higher standing crop throughout the growing season and

at the end of the growing season for the 2070–2099 period compared to the historical period (Fig. 2).

Comparisons of historical monthly production to projected production for the two scenarios reveal that production increases in the A2 scenario reach a maximum (greatest difference from historical conditions) in most parts of the Bay Area by the beginning of March or April (Fig. 5). In the B1 scenario, the maximum change from historical conditions occurs much earlier, by the end of October or November, but the magnitude of the maximum differences between historic and future production are much more variable than in the A2 scenario (25 to 80 percent increases for B1, 30 to 55 percent for A2). This difference between the two scenarios is likely at least partially due to earlier onset of germinating rains in the B1 scenario, as discussed below in Season Length. In the early season, when total forage production is very low, even a small absolute change in production made by a few days or a week of extra production time can make a large difference proportionally.

Season Length

In this model, season length changes with the timing of germination and plant senescence; delayed onset of germinating rains and/or earlier depletion of soil moisture will result in shorter growing seasons. Because forage quality is greatest during the

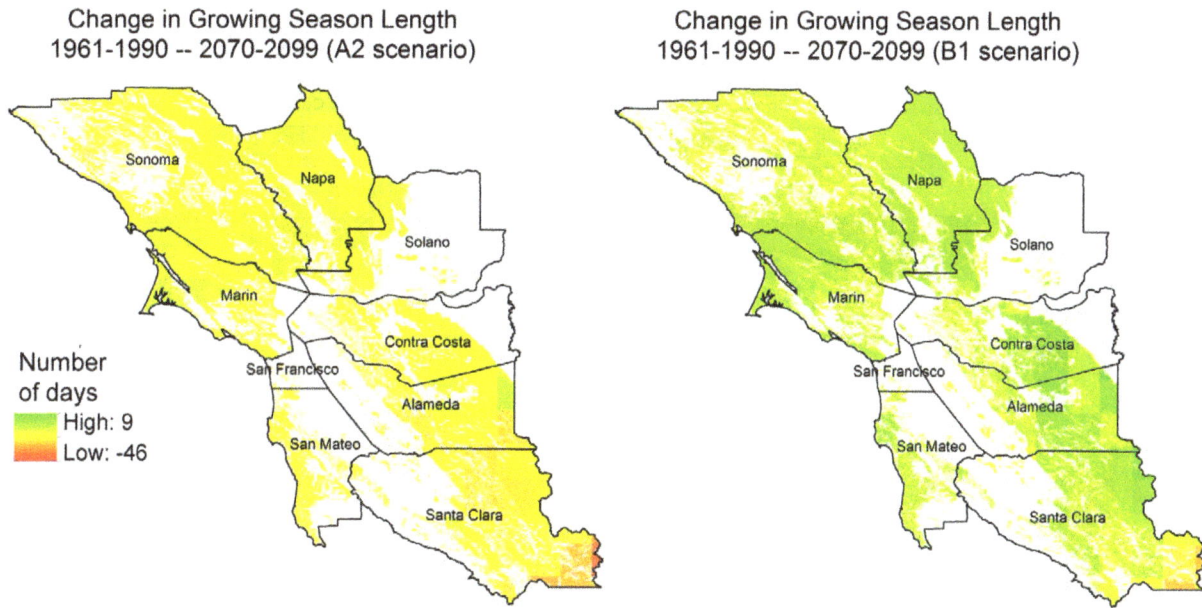

Figure 4. Change in rangeland season length by end-century (2070–2099), relative to historical conditions (1961–1990) for current rangelands in the Bay Area.

Table 3. Maximum and minimum forage production and season length, expressed as percent above and below the mean for the historic period (1961–1990), three projected 30-year periods and all periods.

Time Period	Marin-Sonoma		Napa-Solano		Alameda-Contra Costa		Santa Clara	
	Max	Min	Max	Min	Max	Min	Max	Min
Production A2 Model								
All periods	45	32	50	34	45	28	49	35
1961–1990	23	24	30	25	34	19	33	27
2005–2034	19	24	25	24	26	26	23	17
2035–2064	24	18	30	19	36	23	41	22
2070–2099	28	24	29	26	31	29	34	23
Production B1 Model								
All periods	35	31	39	33	53	38	46	37
1961–1990	23	24	30	25	33	19	33	27
2005–2034	32	22	35	22	30	34	35	37
2035–2064	23	34	27	31	28	30	26	34
2070–2099	25	22	27	23	40	24	34	22
Season Length A2 Model								
All periods	24	23	24	24	26	38	26	32
1961–1990	13	18	17	20	21	17	21	25
2005–2034	15	13	16	11	22	15	20	15
2035–2064	15	18	17	15	25	19	29	17
2070–2099	22	17	24	17	26	32	29	26
Season Length B1 Model								
All periods	22	26	21	25	28	46	24	39
1961–1990	13	18	17	20	20	17	21	25
2005–2034	18	16	20	23	21	36	23	39
2035–2064	15	25	14	21	24	22	21	22
2070–2099	16	15	15	14	28	30	24	29

Table 4. Number of outlier years (>2 standard errors above or below the mean) per time period.

	Marin-Sonoma	Napa-Solano	Alameda-Contra Costa	Santa Clara
Peak Forage Production				
Below historic mean	12	11	13	10
Above historic mean	11	10	10	10
Below mean for 2070–2099 (A2)	10	11	10	11
Above mean for 2070–2099 (A2)	7	8	10	9
Below mean for 2070–2099 (B1)	11	10	10	13
Above mean for 2070–2099 (B1)	8	8	7	8
Season Length				
Below historic mean	9	10	10	7
Above historic mean	12	12	10	11
Below mean for 2070–2099 (A2)	10	12	11	10
Above mean for 2070–2099 (A2)	10	11	10	9
Below mean for 2070–2099 (B1)	11	11	7	7
Above mean for 2070–2099 (B1)	11	10	8	8

growing season [20], periods of adequate forage quality for animal production will be shortened under future climate conditions, despite increases in forage production. The main differences between scenarios and across different regions in this study are due to the timing of germination. In the A2 scenario, the growing seasons in Santa Clara and Alameda are delayed by a week to 12 days compared to historical conditions, whereas the season in the northern Bay Area starts only slightly (2–3 days) later than historically. This intensifies the historical differences between germination dates of the North and South Bay. In contrast, earlier rains mean earlier germination (2–3 days on average, up to a week) throughout much of the Bay Area for the B1 scenario, which almost compensates for the earlier end in the B1 scenario, such that the impact of climate change on season length is more subtle than for the A2 scenario. This earlier start to the growing season also contributes to greater production in the late-century B1 scenario compared to the A2 scenario during the first few months of the season (Fig. 2).

These changes in season length and timing could have major implications for the range livestock industry in the San Francisco Bay Area. For example, delaying the start of the growing season and associated improvement in forage quality could impact the traditional fall calving season, and earlier onset of the dry season could require early weaning of calves in cow calf operations that dominate Bay Area livestock production. If the timing of these key events changes, then breeding and marketing dates may also shift by a few days or weeks. With later germination and earlier end to the growing season, managers will seek to place stock on summer pasture, including public lands, sooner and keep them there longer.

Inter-annual Variability and Drought

Inter-annual variation appears to decline with climate change throughout much of the Bay Area. This suggests that the uncertainty in stocking decisions that were cited as a major concern for climate change in the Great Plains [14] will be less of an issue here, and ranchers will likely not have to change their management response from what they do now to respond to low production years and good production years (see Conclusions). However, this low variability masks outlier years; the minimums of

38% below mean forage production do not include the years in which no forage is produced, due to drought. In fact, the occurrence of these "skipped" forage seasons can be considered an extreme case of inter-annual variability. While there exists a high degree of uncertainty over the Bay Area as a whole, extreme events in future precipitation forecast by the four different climate models are likely to increase in regions already most vulnerable to such events. Droughts so severe that forage is not produced during the growing season can be expected to increase in frequency in parts of the South Bay, which will have serious consequences for stocking decisions and the overall reliability of forage in that area.

Model Limitations and Other Climate Impacts

This simple forage production model was developed from data taken across rangelands that represent much of the Bay Area; UC HREC is ten miles north of interior Sonoma County (east of the Coast Range), and the other two research stations (San Joaquin Experimental Range in Madera County and UC SFREC in Yuba County) are inland sites that are more similar to the eastern portions of the Bay Area counties. However, the coastal regions may not be accurately represented. The curves projected for the historical period in the coastal counties of Marin and Sonoma are steeper than for the inland counties of Alameda and Santa Clara and than those from earlier empirical studies [8], especially during winter months. This difference may be explained by the generally warmer winter temperatures forecast for Marin and Sonoma County rangelands than for Alameda and Santa Clara County rangelands. Therefore, while more coastal data would improve the accuracy of model, most of the coastal effects on productivity are likely temperature effects that should be adequately modeled by accumulated degree days.

One major simplification of the processes involved with forage production in this model is that precipitation is included only to define the bounds of the growing season [16]. Precipitation effects within the growing season are not considered in the model, and it therefore does not respond to midwinter droughts, which may substantially reduce forage production in some years. At most annual rangeland locations in the Bay Area, moisture is seldom limiting during the growing season [7]. The degree day model used in this study was developed using production and weather

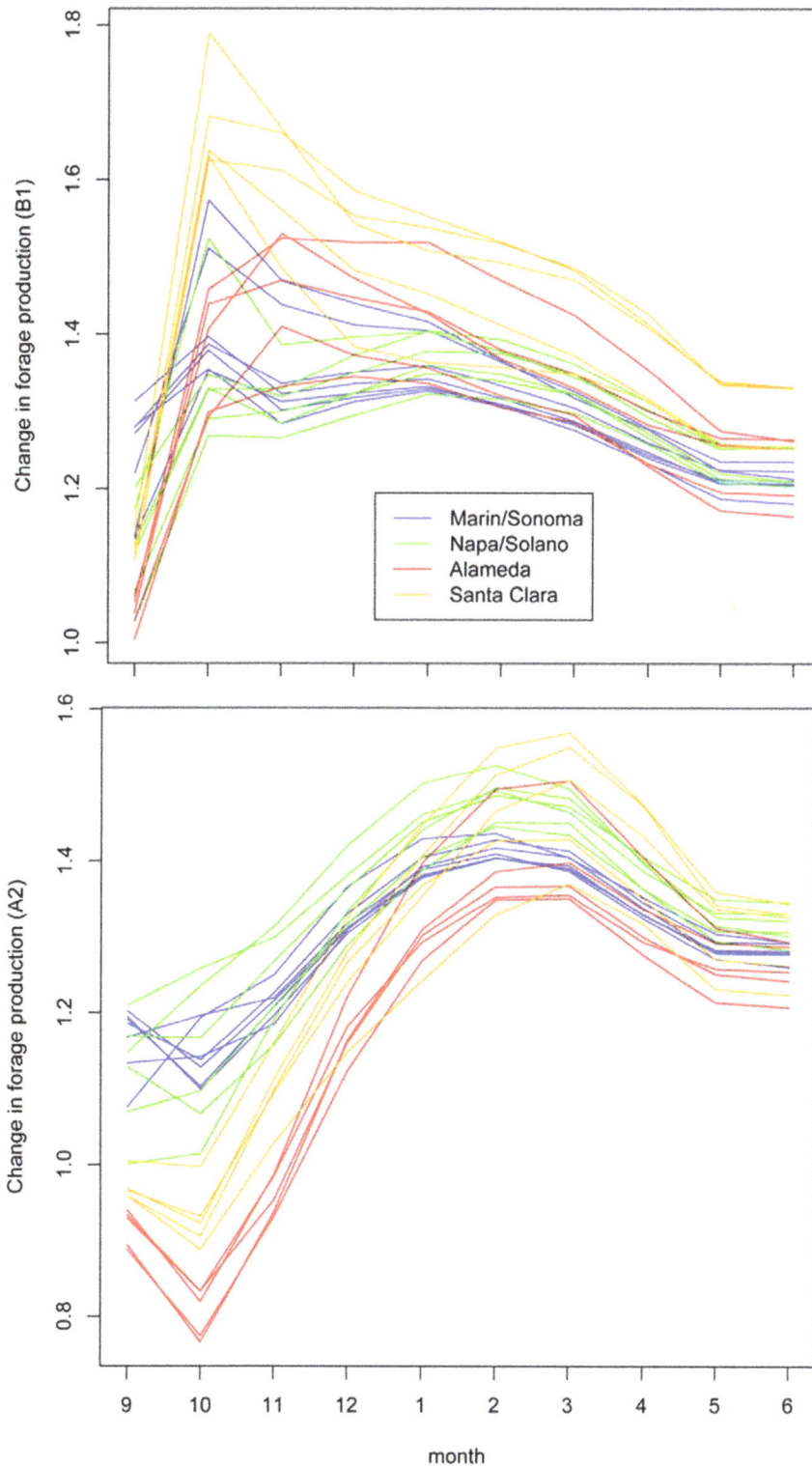

Figure 5. Change (%) in forage production from historical (1961–1990) to future (2070–2099) climate scenarios. Multiple lines of the same color represent different 12 x 12 km grid cells of rangeland in that area.

data from areas that vary quite dramatically in their precipitation regimes, with annual rainfall ranging from 13 to 53 inches [16] Modeled annual precipitation for Marin, Sonoma, Napa, and Solano Counties did not fall outside of this range in any year. Alameda, Contra Costa and Santa Clara Counties did show some

years that fell below 13 inches (330 mm), but these generally accounted for <15% of the total years. The number of years falling below that range did not increase from historical to future conditions in the B1 scenario, which means the model can be applied with confidence to all regions of the Bay Area for this scenario. The number of years precipitation in the A2 scenario fell outside that range increased in Alameda/Contra Costa Counties from 1961–1990 to 2070–2099. Results should therefore be interpreted more cautiously for these counties in the A2 scenario.

Finally, the model leaves out a number of important processes determining forage production that may be significantly altered under future conditions. Elevated atmospheric carbon-dioxide could have fertilization effects that increase the quantity of forage, while simultaneously reducing the quality by diluting the protein content [21]. Future climate could alter evapotranspiration rates, resulting in decreased soil moisture and increased water stress beyond the effects of precipitation [22], further reducing season length and potentially increasing drought frequency. Potential shifts in vegetation states resulting from projected changes in temperature and precipitation can also impact forage production at the landscape level, through shrubland expansion into existing grasslands and long-term conversion of oak woodlands to grasslands [23]. Finally, animal metabolic performance, grazing behavior and availability of stock water can all be expected to change with climate [24,25,26], and while not modeled here, their general decline with temperature will impact the overall viability of the livestock industry in the Bay Area.

Incorporating these additional factors into a model of forage production would result in a more sensitive and nuanced forecast of this important ecosystem service under climate change conditions. Our goal in this research was to apply to future climate scenarios a very simple, empirical model that despite its simplicity explains 75% of the variation in forage production under current conditions in the Bay Area. Future research should compare a simple model such as that presented here with more complex approaches, to determine how well main effects are characterized by the most basic processes. Simple models can be useful for supporting land use decisions in areas where data are limited and/or more advanced processes are poorly understood.

Conclusions

Climate change has the potential to impact the quantity and reliability of forage production, forage quality, thermal stress on livestock, water demands for both animal needs and growing forage, and large-scale rangeland vegetation patterns. This study projects increases in forage production within the growing season counterbalanced by shorter growing seasons. Increased production may result in increased carrying capacity on Bay Area rangelands. However, shorter growing seasons and increased potential for drought will increase risk. One of the primary tools for reducing drought risk is to maintain stocking rate below the carrying capacity of the land, which means the 10 to 25 percent increases in forage production forecast here may not result in substantial increases in stocking rate. Because drought is a regular occurrence

on Bay Area rangelands, especially the south eastern portion that lies in the rain shadow of the Coast Range, ranchers in these areas are already accustomed to coping with periodic drought. For the climate scenarios discussed above, grazing managers will need to strengthen their contingency planning for drought.

Overall, this model has demonstrated that shifting temperature and precipitation patterns must be considered together in order to understand the potential impacts of climate change on rangeland forage production. In a future with higher temperatures and a shorter rainy season, ranchers will need to consider management options for grazing shorter growing seasons and therefore longer dry seasons. Most of the standing biomass remaining during the dry season has senesced and is of poor nutritive quality; an extension of this period means a reduction in the availability of forage that can meet the nutrition requirements of beef cattle. These vulnerabilities to climate change are not as easily translated to economic impacts as total forage production, as each ranch has a unique set of forage sources and operational conditions. Some ranches have the flexibility to transport livestock to forage sources of higher quality during the Bay Area dry season (e.g., irrigated pastures, high elevation meadows, or wetter coastal regions), while others will graze the dry forage remaining in the Bay Area and therefore need to provide supplemental feeds including hay, protein and mineral supplements. Both of these options will increase production costs, reducing already thin profit margins, but how these additional costs will weigh against the projected gains in forage production is not well understood. However, the main message for the effects of climate change on Bay Area ranching is that it will present some opportunities as well as some challenges. The prospect of paying ranchers to graze in order to provide certain ecosystem services such as control of invasive species [27], fire hazard reduction [28], and pollination to nearby farms [29] may be an increasingly important tool to help offset the increased costs of grazing under climate change and to maintain the viability of ranching operations in the Bay Area – and the precious open spaces they support.

Acknowledgments

We thank James Bartolome, Valerie Eviner, Sheila Barry, and Sasha Gennet for their invaluable comments and constructive advice, Brian Galey and the Berkeley Geospatial Innovation Facility for technical assistance, and Mary Tyree and Dan Cayan for their help assembling the climate data.

Author Contributions

Conceived and designed the experiments: RCK. Performed the experiments: RCK. Analyzed the data: RCK. Contributed reagents/materials/analysis tools: RCK MRG. Wrote the paper: RCK MRG.

References

1. Huntsinger L, Hopkinson P (1996) Sustaining rangeland landscapes: a social and ecological process. J Range Manage 49: 167–173.
2. Forero L, Huntsinger L, Clawson WJ (1992) Land use change in three San Francisco Bay area counties: Implications for ranching at the urban fringe. J Soil and Water Conserv 47: 475–480.
3. Cayan D, Maurer EP, Dettinger MD, Tyree M, Hayhoe K, et al. (2006) Climate scenarios for California. Sacramento, CA: California Energy Commission. California Climate Change Center White Paper.
4. Cayan D, Tyree M, Pierce D, Das T (2012) Climate and Sea Level Change Scenarios for California Vulnerability and Adaptation Assessment. Sacramento, CA: California Energy Commission. Publication number: CEC-500-2012-008. Available: http://uc-ciee.org/climate-change/california-vulnerability-and-adaptation-study.
5. Shaw MR, Pendleton L, Cameron DR, Morris B, Bachelet D, et al. (2011) The impact of climate change on California's ecosystem services. Climatic Change 109: 465–484.
6. George MR, Williams WA, McDougald NK, Clawson WJ, Murphy AH (1989) Predicting Peak Standing Crop on Annual Range Using Weather Variables. J Range Manage 42:509–513.
7. George MR, Larsen RE, McDougald NM, Vaughn CE, Flavell DK, et al. (2010) Determining Drought on California's Mediterranean-Type Rangelands: The Noninsured Crop Disaster Assistance Program. Rangelands 32:16–20.
8. George M, Bartolome J, McDougald N, Connor M, Vaughn C, et al. (2001) Annual Range Forage Production. Oakland, CA: University of California. Division of Agriculture and Natural Resources Publ 8018. 9 p.
9. Zavaleta ES, Thomas BD, Chiariello NR, Asner GP, Shaw MR, et al. (2003) Plants reverse warming effect on ecosystem water balance. Proc Nat Acad of Sci 100:9892–9893.
10. Chou WW, Silver WL, Jackson RD, Thompson AW, Allen-Diaz B, et al. (2008) The sensitivity of annual grassland carbon cycling to the quantity and timing of rainfall. Global Change Biology 14: 1382–1394.

11. Murphy AH (1970) Predicted forage yield based on fall precipitation in California annual grasslands. J Range Manage 23: 363–365.

12. Suttle KB, Thomsen MA, Power ME (2007) Species interactions reverse grassland responses to changing climate. Science 315: 640–642.

13. Zavaleta ES, Shaw MR, Chiariello NR, Thomas BD, Cleland EE, et al. (2003) Grassland responses to three years of elevated temperature, CO2, precipitation, and N deposition. Ecol Monog 73:585–604.

14. Hanson J, Baker B, Bourdon R (1993) Comparison of the effects of different climate change scenarios on rangeland livestock production. Agric Systems 41:487–502.

15. Howden SM, Crimp SJ, Stokes CJ (2008) Climate change and Australian livestock systems: impacts, research and policy issues. Austr J of Exper Agric 48:780.

16. George MR, Raguse CA, Clawson WJ, Wilson CB, Willoughby RL, et al. (1988) Correlation of degree-days with annual herbage yields and livestock gains. J Range Manage 41:193–197.

17. R Development Core Team (2011) R: A language and environment for statistical computing. R Foundation for Statistical Computing, Vienna, Austria. ISBN 3-900051-07-0. Available: http://www.R-project.org/.

18. Logan SH, Boyland PB (1983) Calculating heat units via a sine function. J Amer Soc Hort Sci 108:977–980.

19. USGS (U.S. Department of Interior, Geological Survey) (2006) The national map LANDFIRE national existing vegetation type layer. (Last updated September 2006.) Available: http://www.landfire.gov/NationalProductDescriptions21.php

20. George M, Nader G, McDougald N, Connor M, Frost B, (2001). Annual Rangeland Forage Quality. Oakland, CA: University of California. Division of Agriculture and Natural Resources Publ 8022. 13 pgs.

21. Milchunas DG, Mosier AR, Morgan JA, LeCain DR, King JY, et al. (2005) Elevated CO2 and defoliation effects on a shortgrass steppe: Forage quality versus quantity for ruminants. Agric Ecosystems & Environ 111:166–184.

22. Keshta N, Elshorbagy A, Carey S (2012) Impacts of climate change on soil moisture and evapotranspiration in reconstructed watersheds in northern Alberta, Canada. Hydrol Process 26: 1321–1331.

23. Cornwell WK, Stuart SA, Ramirez A, Dolanc CR, Thorne JH, et al. (2012) Climate Change Impacts on California Vegetation: Physiology, Life History, and Ecosystem Change. Sacramento, CA: California Energy Commission. Publication number: CEC-500-2012-023. Available: http://uc-ciee.org/climate-change/california-vulnerability-and-adaptation-study.

24. Hahn GL (1985) Management and housing of farm animals in hot environment. In: Yousef MK. editor. Stress Physiology of Livestock. Ungulates, Vol. 2. Boca Raton, FL: CRC Press. 151–176.

25. Harris NR (2001) Cattle behavior and distribution on the San Joaquin Experimental Range in the foothills of Central California (dissertation). Corvallis, OR: Oregon State University. 199 p.

26. HRC-GWRI (2011) Climate Change Implications for Managing Northern California Water Resources in the Latter 21st Century. Sacramento, CA: California Energy Commission, PIER Energy-Related Environmental Research. CEC-500-2010-051.

27. Huntsinger L, Bartolome JW, D'Antonio CM (2007). Grazing Management on California's Mediterranean Grasslands. In: Stromberg MR, Corbin JD, D'Antonio CM. editors. California Grasslands: Ecology and Management. Berkeley, CA: University of California Press. 233–253.

28. Nader G, Henkin Z, Smith E, Ingram R, Narvaez N (2007) Planned herbivory in the management of wildfire fuels. Rangelands 29:18–24.

29. Chaplin-Kramer R, Tuxen-Bettman K, Kremen C (2011) Value of wildland habitat for supplying pollination services to Californian agriculture. Rangelands 33: 33–41.

24

Conservation Efforts May Increase Malaria Burden in the Brazilian Amazon

Denis Valle[1]*, **James Clark**[2]

1 University Program in Ecology, Duke University, Durham, North Carolina, United States of America, **2** Nicholas School of the Environment, Department of Biology, Department of Statistical Science, Duke University, Durham, North Carolina, United States of America

Abstract

Background: Large-scale forest conservation projects are underway in the Brazilian Amazon but little is known regarding their public health impact. Current literature emphasizes how land clearing increases malaria incidence, leading to the conclusion that forest conservation decreases malaria burden. Yet, there is also evidence that proximity to forest fringes increases malaria incidence, which implies the *opposite* relationship between forest conservation and malaria. We compare the effect of these environmental factors on malaria and explore its implications.

Methods and Findings: Using a large malaria dataset (~1,300,000 positive malaria tests collected over ~4.5 million km^2), satellite imagery, permutation tests, and hierarchical Bayesian regressions, we show that greater forest cover (as a proxy for proximity to forest fringes) tends to be associated with higher malaria incidence, and that forest cover effect was 25 times greater than the land clearing effect, the often cited culprit of malaria in the region. These findings have important implications for land use/land cover (LULC) policies in the region. We find that cities close to protected areas (PA's) tend to have higher malaria incidence than cities far from PA's. Using future LULC scenarios, we show that avoiding 10% of deforestation through better governance might result in an average 2-fold increase in malaria incidence by 2050 in urban health posts.

Conclusions: Our results suggest that cost analysis of reduced carbon emissions from conservation efforts in the region should account for increased malaria morbidity, and that conservation initiatives should consider adopting malaria mitigation strategies. Coordinated actions from disparate science fields, government ministries, and global initiatives (e.g., Reduced Emissions from Deforestation and Degradation; Millenium Development Goals; Roll Back Malaria; and Global Fund to Fight AIDS, Tuberculosis and Malaria), will be required to decrease malaria toll in the region while preserving these important ecosystems.

Editor: Thomas Eisele, Tulane University School of Public Health and Tropical Medicine, United States of America

Funding: Denis Valle was partly supported by a pilot project grant from the Duke Global Health Initiative. No additional external funding was received for this study. The funder had no role in study design, data collection and analysis, decision to publish, or preparation of the manuscript.

Competing Interests: The authors have declared that no competing interests exist.

* E-mail: drv4@duke.edu

Introduction

Deforestation has been a major concern in much of the tropics because of its detrimental effect on biodiversity, atmospheric carbon emissions, regional weather patterns, among other ecosystem services [1]. The Brazilian Amazon in particular has received considerable attention because a large fraction of tropical forest clearing has occurred within this region [2]. This fact has prompted the creation of the world's largest forest-conservation initiative to reduce emissions from deforestation and degradation (REDD+), with an initial pledge of up to $1 billion USD [3], and a commitment by the Brazilian government to reduce Amazon deforestation by 80% [4]. However, few conservation scientists seem to be aware that the Brazilian Amazon also plays an important role in terms of malaria cases and fatalities; almost half of the deaths attributed to this disease in the Americas occurred in Brazil [5,6] and virtually all malaria cases in Brazil originate from the Brazilian Amazon [7,8]. To reduce malaria morbidity and mortality in the region, multi-million dollar initiatives focused on

malaria have also been created (e.g., $5 million USD/year from the Amazon Malaria Initiative [9]; and ~$23 million USD from the Global Fund to Fight Aids, Tuberculosis, and Malaria [10]).

While it is generally agreed that environmental factors play an important role in malaria [11], there are mixed evidence regarding how land cover and deforestation affect malaria in the Amazon region. For instance, proximity to forest fringes [12–16] and land clearing [14,17–24] have both been proposed to explain malaria vector presence, mosquito biting rate and malaria incidence. Yet, the exact role of these factors on malaria incidence has important implications regarding land use land cover (LULC) policies. Based on the evidence of higher malaria risk at recently deforested areas or in areas with active land clearing, it has been suggested that forest conservation can decrease disease burden [25–30]. Based on evidence of higher malaria risk when close to forest fringes, the opposite conclusion has been reached; it has been suggested that the long-term effect of land clearing is to increase the distance of humans to forest edges and thus decrease malaria risk [14,31,32].

These contrasting effects of deforestation have not been studied on a large spatial scale. Here we assess the magnitude of both of these malaria risk factors with a large malaria dataset (totaling ~1.3 million positive malaria tests, gathered over ~4.5 years and over a 4.5 million km^2 region) and evaluate the public health consequences of current and future land use/land cover (LULC) scenarios.

Methods

Malaria Data

The malaria data were collected from January 2004 to August 2008 by the Brazilian malaria surveillance system [33] and are aggregated by month and health facility. A malaria case is defined as an individual that has fever and that has a positive *Plasmodium* spp. detection through microscopy [34]. To the best of our knowledge, this definition has been consistently used throughout the entire 2004–2008 period. Because there are no data on the exact location of each health facility, our approach was to subset the health facilities that are known to be in the urban area and use the spatial coordinates of the corresponding cities as proxies for their location. Determining the approximate location of these health facilities is important to adequately characterize the environmental risk factors to which individuals treated at these health facilities are exposed. We emphasize that despite being classified as urban areas, these are predominantly small cities (i.e., median population size equal to 14,000 people), often surrounded by a considerable area of forest (i.e., 22% of these cities had >50% of their catchment area covered by forests). The surrounding vegetation is critical because it is common for individuals to get infected in the surrounding area (e.g., while participating on selective timber logging, non-timber forest products collection, slash-and-burn agriculture, night fishing, hunting, mining, etc.) but to be diagnosed in the city [35,36]. We further excluded cities that had less than two years of data because it would not be possible to estimate yearly and monthly city-specific random effects for these cities. The final dataset contained approximately half of the original malaria cases (~1,300,000 cases) but covered a similar geographical area (96% of the counties in the original dataset) (a summary description of these data is available in Table S1).

Catchment Area

We adopt a 20-km radius as the "catchment area" around each city and use the precipitation, deforestation rate, and forest cover estimates within this catchment area as our covariates. The size of this catchment area accounts for the malaria vector flight range [13,21], population mobility to and from the surrounding vegetation, and the fact that malaria cases often arise from multiple urban health facilities within a particular city. The same radius has been used elsewhere as the area typically under urban influence in the Brazilian Amazon region [37]. Our results are robust to the use of different radii (i.e., 10, 20, and 30 km) (File S1 and Figure S2).

Covariates

Population size comes from the 2007 Brazilian National Census, aggregated at the census tract level, made available by the Brazilian Institute of Geography and Statistics [38]. Our environmental covariates come from satellite imagery. We used annual forest cover and deforestation rate estimates from the Brazilian Space Agency derived from a semi-automated analysis of Landsat imagery [39]. Estimates of precipitation were derived from the Tropical Rainfall Measuring Mission data ('3B43 Monthly 0.25×0.25 degree merged TRMM and other sources

estimates' product [40]), and average precipitation for a particular month was calculated over all pixels that fell within each catchment area. Based on these precipitation estimates, we also calculated a drought index that has been extensively used to characterize drought in the region [41–43]. We used a one month time lag for precipitation and drought index covariates based on the assumption that water affects the vector mainly through its breeding habitat. Therefore, changes in precipitation or drought should only affect infection risk in the following month since this is the minimum necessary time for the larva to become an adult mosquito, the adult to be infected and finally become infectious. Results did not change substantially when using a two month time lag (data not shown).

Permutation Tests

To compare a particular outcome X (e.g., number of malaria cases per month, onwards simply malaria incidence) for cities with characteristic c_1 versus cities with characteristic c_2 (e.g., high vs. low forest cover), we first calculate the observed difference in means $Dif^{(obs)} = \bar{X}_{c1}^{(obs)} - \bar{X}_{c2}^{(obs)}$. Then, we estimate the probability of an outcome equal or more extreme than the observed outcome under the null hypothesis (i.e., p-value) through a permutation test. To do this, we randomly assign these characteristics to the cities and calculate the simulated difference in means $Dif^{(sim)} = \bar{X}_{c1}^{(sim)} - \bar{X}_{c2}^{(sim)}$. This was done 1,000 times, generating 1,000 values of $Dif^{(sim)}$. We estimate the p-value as $p(Dif^{(sim)} \geq Dif^{(obs)}) \approx \frac{\sum I(Dif^{(sim)} \geq Dif^{(obs)})}{1000}$, where I() is the indicator function that takes on the value of 1 if the condition is satisfied and zero otherwise.

Regression Model Structure

We assessed the effect of forest cover (F_{iy}, percent of catchment area) and annual deforestation rate ($DE_{iy} = F_{iy} - F_{i,y-1}$, percent of catchment area per year) using a Bayesian hierarchical regression approach (i and y stand for city and year). We adjusted for potential confounder effect of climate variables on malaria risk, namely monthly precipitation (P_{iym}) and a drought index (D_{iym}) (m stands for month within year). All covariates were standardized (i.e., centered and divided by their standard deviation).

The number of malaria cases per month (i.e., malaria incidence) is modeled as an over-dispersed Poisson, given by:

$$C_{iym} \sim Poisson(\exp(w_{iym})N_i)$$

where N_i is the population size within the catchment area, and w_{iym} is given by:

$$w_{iym} \sim N(\beta_{0i} + \beta_1 DE_{iy} + \beta_2 F_{iy} + \beta_3 P_{iym} + \beta_4 D_{iym} + e_{im} + e_{iy}, \sigma^2)$$

where $\beta_1, \cdots, \beta_4$ are fixed-effect regression parameters. Additional socio-economic-environmental covariates (e.g., proportion of migrants, age distribution, level of urbanization, gross domestic product, vector ecology, and proximity to large water bodies) tend to be relatively constant within the short time-frame of our malaria incidence dataset (~4.5 years). Therefore, we control for these unspecified city-to-city differences using a city specific random intercept (β_{0i}). We also include a year-by-city (e_{iy}) and a month-by-city(e_{im}) random effect. To complete the model specification, we adopt the usual assumptions regarding the distribution of random effects:

$$\beta_{0i} \sim N(\beta_0, \tau^2)$$

$$e_{iy} \sim N(0, \vartheta^2)$$

$$e_{im} \sim N(0, \gamma^2)$$

and we assume vague hyper-priors for the regression coefficients and variance parameters [44]:

$$\beta_0, \cdots, \beta_4 \sim N(0, 100)$$

$$\vartheta, \sigma, \tau, \gamma \sim Unif(0, 100)$$

We used a Gibbs sampler to iteratively sample from each of the full conditional distributions. Because of conjugacy between likelihood and priors, almost all the parameters could be sampled directly [45]. The only parameters that could not be sampled directly were the w_{iym}, which were sampled with a Metropolis-within-Gibbs step.

We assessed model convergence by running three Markov Chain Monte Carlo (MCMC) chains with over-dispersed initial parameter values for 200,000 iterations. We discarded the first 10,000 iterations as burn-in and retained 500 iterations, systematically sampled from the remaining 190,000 iterations. We visually assessed convergence by overlaying trace plots of these three chains. We also assessed convergence by calculating the convergence statistic R suggested by Gelman and Rubin [46] (Table S2), where R values much greater than 1 indicate lack of convergence. Both, our plots and the convergence statistic R, suggest that convergence has been achieved.

To determine whether our model was over-fitting the data, we performed a validation exercise. In this exercise, we compared the out-of-sample predictive ability of our model versus simpler versions of it, either without the month-by-city random effects e_{im} or without the year-by-city random effects e_{iy}. We fitted these models to 90% of the data and used the estimated parameters to predict the 10% of the data that was left out. The results from the validation exercise (data not shown) and the comparison between the data and the predictive posterior distribution for each city (Figure S1) revealed that our model had an adequate fit. Finally, a preliminary analysis indicated that the assumption of a linear relationship between LULC covariates and malaria incidence was adequate and revealed low temporal and spatial correlation (correlation on Pearson residuals <0.2), suggesting that additional nonlinear terms and parameters to model these correlations do not need to be included in our model. All analyses and figures were created using R [47].

Land Use Land Cover (LULC) future scenarios. To evaluate the long-term effect of conservation strategies in the Amazon basin, Soares-Filho et al. [48] simulated a governance (GOV) scenario and compared it to a business-as-usual (BAU) scenario, revealing that a substantial amount of deforestation (and its deleterious effects) could be avoided. These projections also allow us to evaluate the effect of future LULC trends on malaria. We estimated the ratio of the expected malaria incidence under

the GOV scenario $E(C_{iy}^{GOV})$ and under the BAU scenario $E(C_{iy}^{BAU})$ for each year and city. This ratio $\frac{E(C_{iy}^{GOV})}{E(C_{iy}^{BAU})}$ was calculated using the posterior distribution for the over-dispersed Poisson regression parameters, thus fully accounting for their uncertainty [49].

Results

We find overwhelming evidence that areas with higher forest cover tend to be associated with higher malaria incidence whereas no clear pattern could be found for deforestation rates, when comparing cities with similar population sizes (upper panels in Figure 1). Similar evidence arises when analyzing malaria incidence per person across all cities (lower panels in Figure 1). Using a Hierarchical Bayesian regression, we show that although forest cover and deforestation rate were both positively associated with malaria incidence, forest cover effect was ~25 times greater than that of deforestation rate (Table 1). As a result, the net effect of higher deforestation rates is to *decrease* malaria burden by decreasing forest cover (i.e., increasing the distance to forest fringes). We also find that the number of malaria cases was negatively correlated with precipitation and our drought index, suggesting that drier periods of the year tend to result in higher malaria incidence. These results were robust to alternative definitions of catchment area (File S1 and Figure S2). An alternative model specification, which explores changes in malaria incidence within each city (rather than within and between cities), revealed qualitatively similar results in relation to the LULC variables (File S1 and Table S3).

These findings have important implications regarding LULC policies in the region. For instance, protected areas (PA's) are a cornerstone of current conservation efforts, yet we are unaware of studies that discuss negative health impacts of these PA's on the local population. A simple depiction of our malaria data suggest that cities close to protected areas (PA's) tend to have higher malaria incidence than cities far from these PA's (Figure 2) after controlling for population size, a likely consequence of higher forest cover in these areas. We also evaluated the long-term implications of our findings by comparing a future scenario with reduced deforestation (i.e., governance scenario - GOV) to a future business-as-usual (BAU) scenario. Using our regression parameter estimates, we find that cities with higher malaria incidence in the GOV versus the BAU scenario will initially tend to be concentrated in the south and east portion of the Brazilian Amazon (Figure 3), where roads slated for paving tend to be located. However, by 2050, almost all cities will tend to have higher malaria incidence. A summary of these results indicate that avoiding deforestation through better governance can substantially increase malaria incidence in urban health posts; an average of 10% of prevented deforestation resulted in an average 2-fold increase in the number of malaria cases per month by 2050 (Figure 4). These results raise concern regarding collateral public health effects of conservation policies.

Discussion

We find that drier periods of the year tended to correlate with higher malaria incidence. Similar results have been attributed to decreased survival rate of adult mosquitoes [50] as well as larva being washed away in rivers [51] during the wet season. We refrain from further discussing these seasonal patterns here (we will address them in a separate paper) to focus our discussion on the LULC findings.

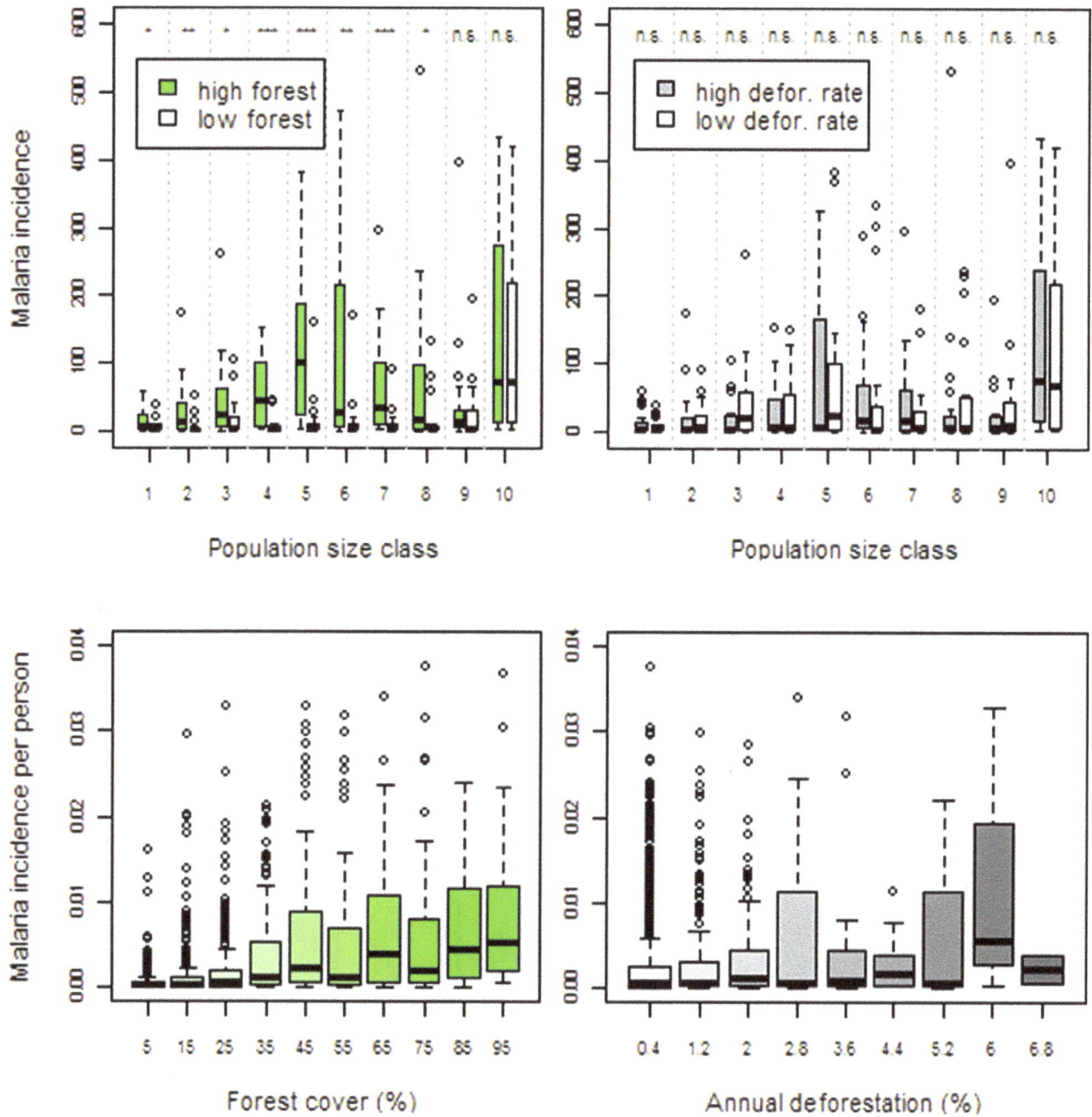

Figure 1. Malaria incidence is higher in areas with more forest cover whereas no clear pattern arises regarding deforestation rates.
Upper panels: Data were stratified into 10 percentile population size classes and average number of malaria cases per month for each year and city was depicted. Within each size class, we compare cities with high (green box-plots) vs. low forest cover (white box-plots) (upper left panel); and cities with high (grey box-plots) vs. low deforestation rate (white box-plots) (upper right panel). Cities with high forest cover (or high deforestation rates) are cities that have forest cover (or deforestation rate) higher than the median for that size class. 'n.s', '*', '**', and '***' are non-significant (p>0.05), significant (0.01<p<0.05), very significant (0.001<p<0.01) and highly significant (p<0.001) difference in means, respectively, based on permutation tests. Lower panels: Mean number of malaria cases per month for each year and city divided by total population as a function of forest cover (lower left panel) and deforestation rate (lower right panel). Note: y-axes were truncated to enable a clearer depiction of the bulk of the data (i.e., less than 0.5% observations were excluded from these plots).

Malaria risk at frontier regions in the Amazon region is often observed to follow a peculiar time trajectory; in the early phases of human settlement, the number of malaria cases soars as naïve settlers arrive and engage in forest related extractive activities, living in precarious conditions. At later stages, as deforestation increases the distance of settlers to forest fringes and economic conditions improve, malaria risk tends to decrease through time

[14]. Our findings regarding the LULC covariates agree with this later stage, suggesting that conservation efforts to decrease deforestation in places where people are already settled might inadvertently increase the number of malaria cases. Some would argue that conservation efforts will also decrease the amount of forest related extractive activities (e.g., fishing, hunting, extraction of non-timber forest products), thus decreasing malaria risk. We

Table 1. Summary of regression parameter estimates.

Parameter	Covariate description	Mean	LCI	UCI
β_1	Annual deforestation rate	0.04	0.00	0.08
β_2	Forest Cover	1.03	0.87	1.20
β_3	Precipitation	−0.07	−0.08	−0.05
β_4	Drought index	−0.06	−0.07	−0.04

LCI and UCI: lower and upper limit of the 95% credible interval.

are skeptical; even if forest conservations efforts succeed in retaining forest cover, hunting and fishing is likely to continue to occur, even within protected areas [52–54].

We note that our finding directly contradicts the growing body of literature that suggests that forest conservation can decrease disease burden [25–30]. This literature often cite the study of Vittor et al. [21,22] conducted in the Peruvian Amazon, as an example of how deforested areas favor the main malaria vector, *Anopheles darlingi*. However, similar entomological studies in the Brazilian Amazon region suggest the opposite pattern for the same

vector species [15,16,55,56], strongly supporting our results. This conflicting evidence might be due to distinct LULC patterns in these regions. In the Peruvian Amazon, swidden-fallow agriculture is the primary driver of deforestation and, as a result, deforested areas are often covered by shrubs and secondary vegetation growth [21,22], whereas in the Brazilian Amazon, forests tend to be substituted by pasture and soy plantations [57].

Malaria occurring in urban areas is often attributed to poor housing and drainage conditions of slums [58,59]. Furthermore, because slums are often located at the periphery of cities and thus closer to forests, this may give rise to a spurious association between forests and malaria incidence. We believe this hypothesis does not explain the malaria patterns we find in the Brazilian Amazon for several reasons. First, slums are rare in Brazilian Amazon cities because these cities are typically very small (i.e., as mentioned earlier, the median population size is 14,000) whereas slums tend to occur in bigger cities where a growing population in limited space gives origin to dense housing, often in hazardous sites. Using the Brazilian government census from 2010, we find that only 12% of the cities in our analysis had slums and that our results in Figure 1 do not change substantially after we exclude the cities with slums (data not shown). Second, the slum effect hypothesis predicts higher malaria incidence in bigger and poorer

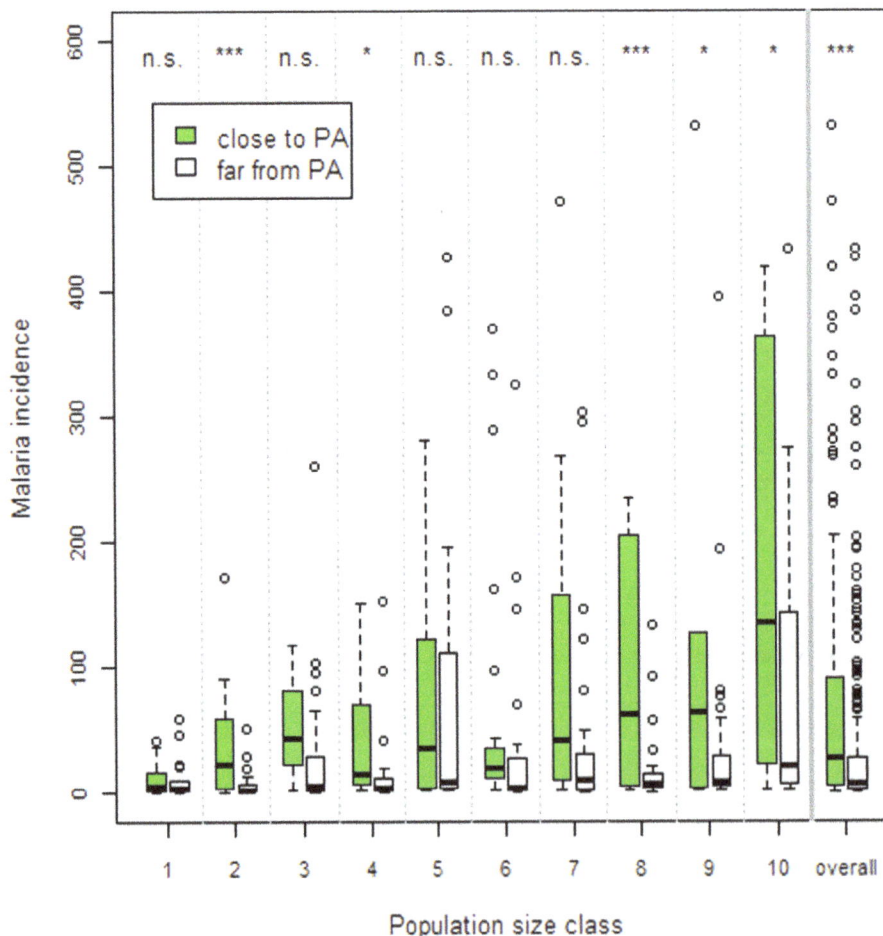

Figure 2. Malaria incidence tends to be higher for cities close to protected areas (PA's). Data were stratified into 10 percentile population size classes and average number of malaria cases per month for each city was depicted. Within each size class, we compare cities close (green box-plots) vs. distant from PA's (white box-plots). Cities close to PA's (i.e., indigenous lands, state and federal parks) are those whose catchment area intersected one or more PA's. 'n.s', '*', '**', and '***' are non-significant (p>0.05), significant (0.01<p<0.05), very significant (0.001<p<0.01) and highly significant (p<0.001) difference in means, respectively, based on permutation tests.

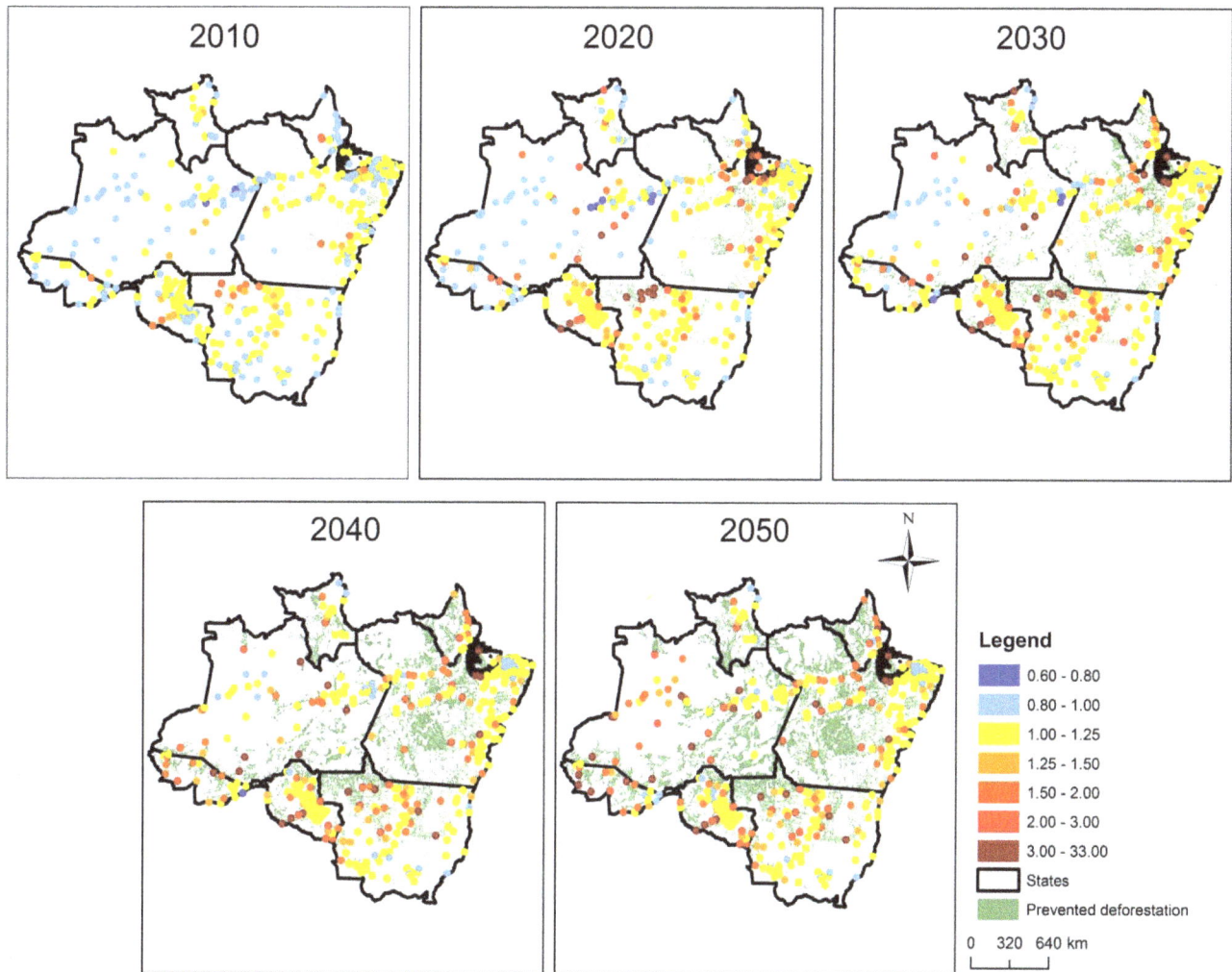

Figure 3. Predicted malaria incidence in urban health posts is higher in the governance scenario than in the business-as-usual scenario. Maps depict the ratio of the expected number of malaria cases per month for each year and city under the governance (GOV) and the business-as-usual (BAU) future LULC scenarios (i.e., $\frac{E(C_{ty}^{GOV})}{E(C_{ty}^{BAU})}$), where values >1 indicate that the GOV scenario results in more malaria cases than the BAU scenario. Areas that were deforested in the BAU scenario but not in the GOV scenario (i.e., prevented deforestation) are depicted in the background for reference. Circles represent the cities in our original malaria dataset.

cities, contrary to the results depicted in Figure 1 and Figure S3. Finally, even after taking into account gender imbalances in the population of each city, we find that the average number of malaria cases per month per person tends to be higher in men than in women, a phenomenon that occurred in 96% of the cities in our dataset. This gender difference in malaria incidence agrees with our hypothesis that forest related activities in the surrounding areas, mostly conducted by men, are the cause of higher malaria rates rather than housing conditions.

Unfortunately, policies that have large effect on LULC in the region (e.g., road opening/paving, creation of rural settlement areas, and the establishment of protected areas) are traditionally perceived to lie in the realm of the Ministries of Environment, Infra-structure, Agriculture, and/or Energy, while the Ministry of Health typically focuses on the delivery of health services [7]. Similarly, global efforts are typically compartmentalized into conservation (e.g., REDD+) and public health (e.g., Roll Back Malaria and GFATM) initiatives. Few studies identify, or discuss

how to address, trade-offs between these global efforts and governmental policies, probably because of the interdisciplinary nature of these trade-offs and the associated ethical issues. For instance, how can one reconcile potential conflicts between the Millenium Development Goals (e.g., goal of combating malaria and the goal of ensuring environmental sustainability)? Although we do not have an answer to this question, acknowledging that these tradeoffs exist is a critical first step towards finding a solution.

Current research and resulting policy recommendations regarding LULC in the Amazon ignore potential public health impacts. For instance, the most frequent policy action to decrease deforestation rates is to create protected areas [60–62]. Several studies suggest, however, that many of these protected areas are established in areas with small deforestation risk [61,63], effectively averting few of the impacts of deforestation. These observations have resulted in recommendations to place these parks in areas more prone to deforestation [4,63,64], which often imply areas with larger human populations, disregarding the

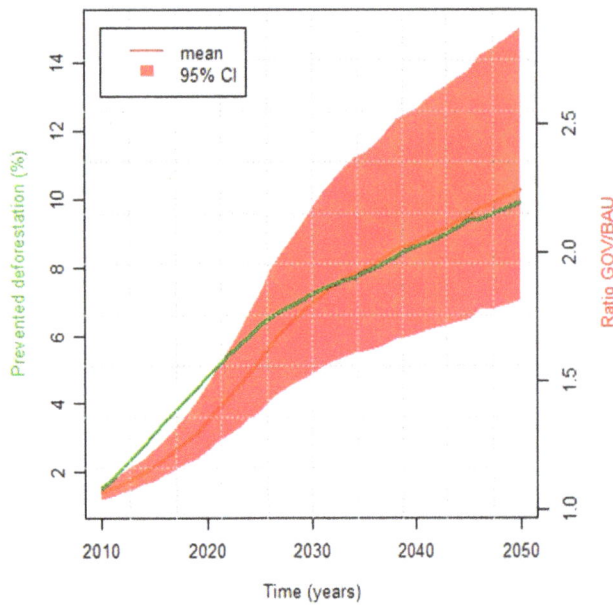

Figure 4. Malaria incidence increase at urban health posts in the governance scenario is predicted to be a direct consequence of prevented deforestation. We depict the relationship between future prevented deforestation under the governance scenario (green line), and the ratio of the expected malaria incidence for each year and city under the governance (GOV) and business-as-usual (BAU) future LULC scenarios (red line) (i.e., $E(C_{iy}^{GOV})/E(C_{iy}^{BAU})$), averaged across all cities. The red polygon represents the 95% credible interval of the average ratio $E(C_{iy}^{GOV})/E(C_{iy}^{BAU})$.

potential for increased malaria morbidity for the local population. Similarly, research acknowledging the negative effects of conservation efforts typically emphasize restrictions on agricultural development rather than the detrimental impact on public health [62,65,66].

One possible interpretation of our findings is that we are promoting deforestation. This is not the case. For instance, large-scale settlement projects in heavily forested areas have resulted in substantial deforestation *and* major malaria outbreaks in the past [14]. Here we argue that deforestation has both negative and positive effects in places where people are already settled, and that the knowledge of these effects is essential for proper LULC and public health planning, particularly in light of the recent ambitious REDD+ targets set by the Brazilian government and four of the Brazilian Amazon states [4]. If conservation efforts (e.g., REDD+) are to avoid this rapid land cover change and its associated adverse effects on several regional-global environmental services (e.g., atmospheric carbon emission, climate and biodiversity), these conservation efforts should, at a minimum, include proper malaria mitigation strategies (e.g., creation of more malaria detection and treatment outposts, distribution of long-lasting insecticidal bed nets, indoor residual spraying) to alleviate their local detrimental effects. Similarly, opportunity costs of reduced carbon emissions through conservation initiatives should take into account their local impact on malaria burden.

Our study has five important limitations. First, we do not take into account potential differences between cities in terms of main malaria vector species, vector ecology and infection efficiency. However, it is well known that collection of entomological data is extremely laborious [67] and therefore logistically impossible to collect over the same geographical scale as our malaria data. Yet,

finding the same overall result over such a vast area by using a separate regression for each city (File S1 and Table S3) gives us confidence that our results are robust to these potential city-to-city differences. Second, in the absence of spatial coordinates of the individual health facilities, we rely on data from urban health facilities aggregated at the city level. Yet, we note that even if individual level data had been available, we would still not have been able to consider many individual-level factors that are known to be important for malaria risk (e.g., mobility, socio-economic status, housing conditions, and occupation) because only a few basic demographic characteristics, such as age and gender, are routinely collected by the malaria surveillance system. Third, in the absence of detailed information for a more accurate modeling of catchment area (e.g., network of unofficial roads [68], origin and mode of transportation of patients, treatment seeking behavior), we relied on relatively arbitrary radii to delimit the catchment area. Fortunately, our results were robust to changes in these radii. We emphasize that these three limitations are typical limitations of studies conducted over large geographical scales (e.g., the area of a single Brazilian Amazon state, Para, is equivalent to the combined area of France and Spain), illustrating the inherent tradeoff between local detail-rich studies, whose results may or may not be generalizable to a wider region, and large-scale detail-poor studies, which reveal broad scale relationships while ignoring many of the local complexities in malaria transmission. Importantly, while site-specific studies have been critical in shaping our knowledge regarding malaria in the region, they may be ill-suited to compare the effect of land clearing to the effect of forest cover because these covariates are often spatially correlated at this scale (i.e., land clearing often occurs in areas with high forest cover). On the other hand, over a large spatial scale, land clearing and forest cover are not highly correlated, allowing us to separately evaluate their effects.

The fourth limitation of our study is that, to avoid spatial extrapolation, our future scenario analysis only considers what would happen to malaria incidence in areas close to where humans are already settled (i.e., the vicinity of urban areas). In these areas, we assume that forests will give place to low intensity cattle ranching and soybean plantations [69,70], thus increasing the distance between people and forest fringes. On the other hand, had we considered new human settlements (e.g., due to human migration to new agricultural frontiers), the BAU scenario might have indicated an initial higher malaria incidence due to an initial decrease in distance to forest fringes. Finally, as with any simulation study, our simulation results critically depend on the implicit assumption that everything else (e.g., age distribution, migratory patterns, patterns of natural resource extraction, climate, etc.) remains constant.

The clear pattern in the data (Figures 1 and 2), the consistency of our findings using alternative model specifications, and the evidence from detailed entomological and epidemiological studies in the region [12–16,55,56], suggest that the association between forest cover and malaria incidence we found is not spurious. Indeed, vegetation management has long been an important strategy to reduce the incidence of malaria [31]. Here we a) show that the effect of forest cover substantially outweighs the effect of deforestation rate (the often cited culprit for malaria in the region) and other climatic variables with a malaria dataset spanning an unprecedented geographical scale; and b) discuss the large-scale multi-sector (i.e., public health, development, and conservation) implications of these findings. Our results suggest caution regarding the widespread assumption that pristine ecosystems will always have beneficial effects for human health [25–30,71–73]. We believe there are undoubtedly numerous ecosystem services from pristine environments; however, ecosystem disservices also exist and need to be acknowledged. Coordinated actions from apparently disparate science fields (e.g., epidemiologists and environmental scientists), government ministries (e.g., Ministry of Health and Ministry of Environment), and the ongoing multi-million dollar conservation and public health efforts in the region, will be required to decrease malaria toll in the region while preserving these important ecosystems.

Supporting Information

Figure S1 Comparison of the data (black line) and the 95% posterior predictive interval (red lines) for 20 randomly chosen cities.

Figure S2 Posterior distribution of the main regression parameters with covariates and population size assessed using three different catchment area radii (10, 20, and 30 km). A line at zero (dashed red line) was added for reference.

Figure S3 Gross domestic product is similar in cities with low and high forest cover. Data were stratified into 10 percentile population size classes and average gross domestic product (GDP) for each year and city was depicted. Within each size class, we compare cities with high (green box-plots) vs. low forest cover (white box-plots). Cities with high forest cover are cities that have forest cover higher than the median for that size class. 'n.s', '*', '**', and '***' are non-significant (p>0.05), significant (0.01<p<0.05), very significant (0.001<p<0.01) and

highly significant (p<0.001) difference in means, respectively, based on permutation tests.

Table S1 Summary description of the malaria dataset.

Table S2 Convergence statistic R [46] for the regression parameters (intercept and slopes for the different covariates) in the main model.

Table S3 Summary statistics of the posterior distribution of the pooled forest cover effect (β_1^{for})and deforestation rate effect (β_1^{def}) for the alternative model.

File S1 Comparison of modeling results using different radii for the catchment area and description of the alternative model specification.

Acknowledgments

We thank Alan Gelfand for statistical advice. We thank the numerous comments provided by the reviewers and Simone Bauch, Meredith Barrett, Bill Pan, Beth Feingold, Matt Kwit, Chantal Reid, Rob Schick, Maria Soledad Benitez, Aaron Berdanier, Kai Zhu, and the Clark lab at Duke University.

Author Contributions

Reviewed and edited the manuscript: JC. Conceived and designed the experiments: DV. Analyzed the data: DV. Contributed reagents/materials/analysis tools: JC. Wrote the paper: DV.

References

1. Myers N, Mittermeier RA, Mittermeier CG, Fonseca GAB, Kent J (2000) Biodiversity hotspots for conservation priorities. Nature 403: 853–858.
2. Hansen MC, Stehman SV, Potapov P, Loveland TR, Townshend JRG, et al. (2008) Humid tropical forest clearing from 2000 to 2005 quantified by using multitemporal and multiresolution remotely sensed data. Proc Natl Acad Sci 105: 9439–9444.
3. Tollefson J (2009) Paying to save the rainforests. Nature 460: 936–937.
4. Ricketts TH, Soares-Filho B, Fonseca GAB, Nepstad D, Pfaff A, et al. (2010) Indigenous lands, protected areas, and slowing climate change. PLoS Biol 8.
5. Roll Back Malaria Partners (2008) The Global Malaria Action Plan for a Malaria-Free World. Geneva: World Health Organization. Available: www.rbm.who.int/gmap/.
6. Murray CJL, Rosenfeld LC, Lim SS, Andrews KG, Foreman KJ, et al. (2012) Global malaria mortality between 1980 and 2010: a systematic analysis. Lancet 379: 413–431.
7. Oliveira-Ferreira J, Lacerda MVG, Brasil P, Ladislau JLB, Tauil PL, et al. (2010) Malaria in Brazil: an overview. Malar J 9: 115.
8. Barreto ML, Teixeira MG, Bastos FI, Ximenes RAA, Barata RB, et al. (2011) Successes and failures in the control of infectious diseases in Brazil: social and environmental context, policies, interventions, and research needs. Lancet 377: 1877–1889.
9. USAID (2012) The Amazon Malaria Initiative: Overview. Available: http://www.usaidami.org/extras/AMIFactsheet1overview.pdf. Accessed 2012 March 19.
10. The Global Fund (2012) The Global Fund: to Fight AIDS, Tuberculosis and Malaria. Available: http://www.theglobalfund.org/. Accessed 2012 March 19.
11. Pruss-Ustun A, Corvalan C (2006) Preventing Disease through Healthy Environments. Towards an Estimate of the Environmental Burden of Disease. Geneva, Switzerland: World Health Organization. Available: http://www.who.int/quantifying_ehimpacts/publications/preventingdisease/en/.
12. Valle D, Clark J, Zhao K (2011) Enhanced understanding of infectious diseases by fusing multiple datasets: a case study on malaria in the Western Brazilian Amazon region. PLOS One 6.
13. Castro CS, Sawyer DO, Singer BH (2007) Spatial patterns of malaria in the Amazon: implications for surveillance and targeted interventions. Health Place 13: 368–380.
14. Castro MC, Monte-Mor RL, Sawyer DO, Singer BH (2006) Malaria risk on the Amazon frontier. Proc Natl Acad Sci 103: 2452–2457.

15. Moutinho PR, Gil LHS, Cruz RB, Ribolla PEM (2011) Population dynamics, structure and behaviour of Anopheles darlingi in a rural settlement in the Amazon rainforest of Acre, Brazil. Malar J 10.
16. Barros FSM, Arruda ME, Gurgel HC, Honorio NA (2011) Spatial clustering and longitudinal variation of Anopheles darlingi (Diptera: Culicidae) larvae in a river of the Amazon: the importance of the forest fringe and of obstructions to flow in frontier malaria. Bull Entomol Res 101: 643–658.
17. Olson SH, Gangnon R, Silveira GA, Patz JA (2010) Deforestation and malaria in Mancio Lima county, Brazil. Emerg Infect Dis 16: 1108–1115.
18. da Silva-Nunes M, Codeco CT, Malafronte RS, Silva NS, Juncansen C, et al. (2008) Malaria on the Amazonian frontier: transmission dynamics, risk factors, spatial distribution, and prospects for control. Am J Trop Med Hyg 79: 624–635.
19. Coimbra CEA (1988) Human factors in the epidemiology of malaria in the Brazilian Amazon. Hum Organ 47.
20. Barbieri AF, Sawyer DO, Soares-Filho BS (2005) Population and land use effects on malaria prevalence in the southern Brazilian Amazon. Hum Ecol 33: 847–874.
21. Vittor AY, Pan W, Gilman RH, Tielsch J, Glass G, et al. (2009) Linking deforestation to malaria in the Amazon: characterization of the breeding habitat of the principal malaria vector, Anopheles darlingi. Am J Trop Med Hyg 81: 5–12.
22. Vittor AY, Gilman RH, Tielsch J, Glass G, Shields T, et al. (2006) The effect of deforestation on the human-biting rate of Anopheles darlingi, the primary vector of Falciparum malaria in the Peruvian Amazon. Am J Trop Med Hyg 74: 3–11.
23. Guerra CA, Snow RW, Hay SI (2006) A global assessment of closed forests, deforestation and malaria risk. Ann Trop Med Parasitol 100: 189–204.
24. Silva-Nunes M, Moreno M, Conn JE, Gamboa D, Abeles S, et al. (2012) Amazonian malaria: asymptomatic human reservoirs, diagnostic challenges, environmentally driven changes in mosquito vector populations, and the mandate for sustainable control strategies. Acta Trop 121: 281–291.
25. Patz JA, Daszak P, Tabor GM, Aguirre AA, Pearl M, et al. (2004) Unhealthy landscapes: policy recommendations on land use change and infectious disease emergence. Environ Health Perspect 112: 1092–1098.
26. Foley JA, DeFries R, Asner GP, Barford C, Bonan G, et al. (2005) Global consequences of land use. Science 309: 570–574.
27. Foley JA, Asner GP, Costa MH, Coe MT, DeFries R, et al. (2007) Amazonia revealed: forest degradation and loss of ecosystem goods and services in the Amazon Basin. Front Ecol Environ 5: 25–32.

28. Keesing F, Blden LK, Daszak P, Dobson A, Harvell CD, et al. (2010) Impacts of biodiversity on the emergence and transmission of infectious diseases. Nature 468: 647–652.

29. Pongsiri MJ, Roman J, Ezenwa VO, Goldberg TL, Koren HS, et al. (2009) Biodiversity loss affects global disease ecology. Bioscience 59: 945–954.

30. Dobson A, Cattadori I, Holt RD, Ostfeld RS, Keesing F, et al. (2006) Sacred cows and sympathetic squirrels: the importance of biological diversity to human health. PLoS Med 3: e231.

31. Keiser J, Singer BH, Utzinger J (2005) Reducing the burden of malaria in different eco-epidemiological settings with environmental management: a systematic review. Lancet Infect Dis 5: 695–708.

32. Singer B, Castro MC (2006) Enhancement and suppression of malaria in the Amazon. Am J Trop Med Hyg 74: 1–2.

33. Ministerio da Saude (2010) SIVEP Malaria. Available: http://dw.saude.gov.br/portal/page/portal/sivep_malaria?Ano_n = 2008. Accessed 2010 March 10.

34. Ministerio de Saude (2005) Guia de vigilancia epidemiologica. Brasilia, Brasil: Ministerio da Saude. 816 p. Available: http://www.prosaude.org/publicacoes/guia/Guia_Vig_Epid_novo2.pdf.

35. Camargo LMA, Ferreira MU, Krieger H, Camargo EP, Silva LP (1994) Unstable hypoendemic malaria in Rondonia (Western Amazon Region, Brazil): epidemic outbreaks and work-associated incidence in an agro-industrial rural settlement. Am J Trop Med Hyg 51.

36. Camargo LMA, Colletto GMD, Ferreira MU, Gurgel SM, Escobar AL, et al. (1996) Hypoendemic malaria in Rondonia (Brazil, western Amazon region): seasonal variation and risk groups in an urban locality. Am J Trop Med Hyg 55: 32–38.

37. Barreto P, Souza Jr C, Nogueron R, Anderson A, Salomao R (2006) Human Pressure on the Brazilian Amazon Forests. In: Mock G, editor. Belem: World Resources Institute. Available: http://www.globalforestwatch.org/common/pdf/Human_Pressure_Final_English.pdf.

38. IBGE (2010) IBGE: Instituto Brasileiro de Geografia e Estatistica. Available: http://www.ibge.gov.br. Accessed 2012 March 19.

39. INPE (2010) Projeto Prodes: monitoramento da Floresta Amazonica Brasileira por Satelite. Sao Jose dos Campos, SP, Brasil. Available: www.obt.inpe.br/prodes/. Accessed 2012 March 19.

40. NASA (2010) 3B43: Monthly 0.25 x 0.25 degree merged TRMM and other sources estimates. Maryland, USA. Available: http://mirador.gsfc.nasa.gov/cgi-bin/mirador/presentNavigation.pl?tree=project&project=TRMM&dataGroup=Gridded&dataset = 3B43:%20Monthly%200.25%20x%200.25%20degree%20merged%20TRMM%20and%20other%20sources%20estimates&version=006.

41. Aragao LEOC, Malhi Y, Roman-Cuesta RM, Saatchi S, Anderson LO, et al. (2007) Spatial patterns and fire response of recent Amazonian droughts. Geophys Res Lett 34.

42. Phillips OL, Aragao LEOC, Lewis SL, Fisher JB, Lloyd J, et al. (2009) Drought sensitivity of the Amazon rainforest. Science 323: 1344–1347.

43. Lewis SL, Brando PM, Phillips OL, van der Heijden GMF, Nepstad D (2011) The 2010 Amazon drought. Science 331: 554.

44. Gelman A (2006) Prior distributions for variance parameters in hierarchical models. Bayesian Anal 1: 515–533.

45. Clark JS (2007) Models for Ecological Data. Princeton: Princeton University Press.

46. Gelman A, Rubin DB (1992) Inference from iterative simulation using multiple sequences. Stat Sci 7: 457–472.

47. R Development Core Team (2010) R: A language and environment for statistical computing. Vienna, Austria: R Foundation for Statistical Computing. Available: http://www.R-project.org.

48. Soares-Filho BS, Nepstad DC, Curran LM, Cerqueira GC, Garcia RA, et al. (2006) Modelling conservation in the Amazon basin. Nature 440: 520–523.

49. Gelman A, Carlin JB, Stern HS, Rubin DB (2003) Bayesian Data Analysis. London: Chapman & Hall.

50. Barros FSM, Honorio NA, Arruda ME (2011) Survivorship of Anopheles darlingi (Diptera: Culicidae) in relation with malaria incidence in the Brazilian Amazon. PLOS One 6.

51. Barros FSM, Honorio NA, Arruda ME (2011) Temporal and spatial distribution of malaria within an agricultural settlement of the Brazilian Amazon. J Vector Ecol 36: 159–169.

52. Peres CA (2000) Effects of subsistence hunting on vertebrate community structure in Amazonian Forests. Conserv Biol 14: 240–253.

53. Peres CA, Dolman PM (2000) Density compensation in neotropical primate communities: evidence from 56 hunted and nonhunted Amazonian forests of varying productivity. Oecologia 122: 175–189.

54. Redford KH (1992) The empty forest. BioScience 42: 412–422.

55. Deane LM (1986) Malaria vectors in Brazil. Mem Inst Oswaldo Cruz 81: 5–14.

56. Povoa MM, Conn JE, Schlichting CD, Amaral JCOF, Segura MNO, et al. (2003) Malaria vectors, epidemiology, and the re-emergence of Anopheles darlingi in Belem, Para, Brazil. J Med Entomol 40: 379–386.

57. Nepstad D, Soares-Filho B, Merry F, Lima A, Moutinho P, et al. (2009) The end of deforestation in the Brazilian Amazon. Science 326: 1350–1351.

58. Tiwari P, Sharma AN (2001) An assessment of socio-demographic correlates associated with cases of maternal malaria. J Hum Ecol 12: 371–373.

59. Unger A, Riley LW (2007) Slum health: from understanding to action. PLoS Med 4.

60. Soares-Filho B, Moutinho P, Nepstad D, Anderson A, Rodrigues H, et al. (2010) Role of Brazilian Amazon protected areas in climate change mitigation. Proc Natl Acad Sci 107: 10821–10826.

61. Joppa LN, Larie SR, Pimm SL (2008) On the protection of "protected areas". Proc Natl Acad Sci 105: 6673–6678.

62. Ferraro PJ, Hanauer MM, Sims KRE (2011) Conditions associated with protected area success in conservation and poverty reduction. Proc Natl Acad Sci 108: 13913–13918.

63. Joppa LN, Pfaff A (2011) Global protected area impacts. Proc R Soc B 278: 1633–1638.

64. Nepstad D, Schwartzman S, Bamberger B, Santilli M, Ray D, et al. (2006) Inhibition of Amazon deforestation and fire by parks and indigenous lands. Conserv Biol 20: 65–73.

65. Ferraro PJ, Hanauer MM (2011) Protecting ecosystems and alleviating poverty with parks and reserves: 'win-win' or tradeoffs? Environ Resource Econ 48: 269–286.

66. Kindermann G, Obersteiner M, Sohngen B, Sathaye J, Andrasko K, et al. (2008) Global cost estimates of reducing carbon emissions through avoided deforestation. Proc Natl Acad Sci 105: 10302–10307.

67. Bousema T, Griffin JT, Sauerwein RW, Smith DL, Churcher TS, et al. (2012) Hitting hotspots: spatial targeting of malaria for control and elimination. PLoS Med 9.

68. Brandao Jr AO, Souza Jr CMS (2006) Mapping unofficial roads with Landsat images: a new tool to improve the monitoring of the Brazilian Amazon rainforest. Int J Remote Sens 27: 177–189.

69. Kaimowitz D, Mertens B, Wunder S, Pacheco P (2004) Hamburger connection fuels Amazon destruction. Bogor, Indonesia: Center for International Forest Research. Available: www.cifor.org/publications/pdf_files/media/Amazon.pdf.

70. Fearnside PM (2005) Deforestation in Brazilian Amazonia: history, rates, and consequences. Conserv Biol 19: 680–688.

71. CBD PAHO/WHO (2012) Background Paper for the Regional Workshop on the Inter-Linkages between Human Health and Biodiversity in the Americas. Convention on Biological Diversity and Pan American Health Organization. Available: http://www.cbd.int/doc/meetings/health/wshb-am-01/other/wshb-am-01-interlink-en.pdf.

72. Barrett MA, Bouley TA, Stoertz AH, Stoertz RW (2011) Integrating a One Health approach in education to address global health and sustainability challenges. Frontiers in Ecology and Environment 9: 239–245.

73. EcoHealth Alliance Consortium for Conservation Medicine. EcoHealth Alliance. Available: http://www.ecohealthalliance.org/programs/26-consortium_for_conservation_medicine. Accessed 2012 Dec 13.

Permissions

All chapters in this book were first published in PLOS ONE, by The Public Library of Science; hereby published with permission under the Creative Commons Attribution License or equivalent. Every chapter published in this book has been scrutinized by our experts. Their significance has been extensively debated. The topics covered herein carry significant findings which will fuel the growth of the discipline. They may even be implemented as practical applications or may be referred to as a beginning point for another development.

The contributors of this book come from diverse backgrounds, making this book a truly international effort. This book will bring forth new frontiers with its revolutionizing research information and detailed analysis of the nascent developments around the world.

We would like to thank all the contributing authors for lending their expertise to make the book truly unique. They have played a crucial role in the development of this book. Without their invaluable contributions this book wouldn't have been possible. They have made vital efforts to compile up to date information on the varied aspects of this subject to make this book a valuable addition to the collection of many professionals and students.

This book was conceptualized with the vision of imparting up-to-date information and advanced data in this field. To ensure the same, a matchless editorial board was set up. Every individual on the board went through rigorous rounds of assessment to prove their worth. After which they invested a large part of their time researching and compiling the most relevant data for our readers.

The editorial board has been involved in producing this book since its inception. They have spent rigorous hours researching and exploring the diverse topics which have resulted in the successful publishing of this book. They have passed on their knowledge of decades through this book. To expedite this challenging task, the publisher supported the team at every step. A small team of assistant editors was also appointed to further simplify the editing procedure and attain best results for the readers.

Apart from the editorial board, the designing team has also invested a significant amount of their time in understanding the subject and creating the most relevant covers. They scrutinized every image to scout for the most suitable representation of the subject and create an appropriate cover for the book.

The publishing team has been an ardent support to the editorial, designing and production team. Their endless efforts to recruit the best for this project, has resulted in the accomplishment of this book. They are a veteran in the field of academics and their pool of knowledge is as vast as their experience in printing. Their expertise and guidance has proved useful at every step. Their uncompromising quality standards have made this book an exceptional effort. Their encouragement from time to time has been an inspiration for everyone.

The publisher and the editorial board hope that this book will prove to be a valuable piece of knowledge for researchers, students, practitioners and scholars across the globe.

List of Contributors

Georgina O'Farrill
Ecology and Evolutionary Biology Department, University of Toronto, Toronto, Ontario, Canada

Kim Gauthier Schampaert
Département de géomatique (KGS), Département de biologie (SC), Université de Sherbrooke, Sherbrooke, Québec, Canada

Bronwyn Rayfield and Andrew Gonzalez
Biology Department, McGill University, Montreal, Quebec, Canada

Örjan Bodin
Stockholm Resilience Centre, Stockholm University, Stockholm, Sweden

Sophie Calmé
Département de géomatique (KGS), Département de biologie (SC), Université de Sherbrooke, Sherbrooke, Québec, Canada
Departamento de conservación de la biodiversidad, El Colegio de la Frontera Sur, Chetumal, Quintana Roo, Mexico

Raja Sengupta
Geography Department, McGill University, Montreal, Quebec, Canada

Jing Tao, Min Chen, Shi-Xiang Zong and You-Qing Luo
Beijing Forestry University, Beijing, People's Republic of China
Silviculture and Conservation of Ministry of Education, Beijing Forestry University, Beijing, China

Clare Duncan, Louise M. McRae and Nathalie Pettorelli
Institute of Zoology, Zoological Society of London, London, United Kingdom

Aliénor L. M. Chauvenet
Institute of Zoology, Zoological Society of London, London, United Kingdom
Division of Biology, Imperial College London, Ascot, United Kingdom

Xin-Ju Xiao
Shenzhen Key Laboratory for Orchid Conservation and Utilization, The National Orchid Conservation Center of China/The Orchid Conservation & Research Center of Shenzhen, Shenzhen, China
Continuing Education College of Beijing Forestry University, Beijing, China

Ke-Wei Liu
Shenzhen Key Laboratory for Orchid Conservation and Utilization, The National Orchid Conservation Center of China/The Orchid Conservation & Research Center of Shenzhen, Shenzhen, China
The Center for Biotechnology and BioMedicine, Graduate School at Shenzhen, Tsinghua University, Shenzhen, China

Yu-Yun Zheng, Yu-Ting Zhang, Guo-Qiang Zhang and Li-Jun Chen
Shenzhen Key Laboratory for Orchid Conservation and Utilization, The National Orchid Conservation Center of China/The Orchid Conservation & Research Center of Shenzhen, Shenzhen, China

Wen-Chieh Tsai
Institute of Tropical Plant Sciences, and Orchid Research Center, National Cheng Kung University, Taiwan, China

Yu-Yun Hsiao
Department of Life Sciences, National Cheng Kung University, Taiwan, China

Zhong-Jian Liu
Shenzhen Key Laboratory for Orchid Conservation and Utilization, The National Orchid Conservation Center of China/The Orchid Conservation & Research Center of Shenzhen, Shenzhen, China
The Center for Biotechnology and BioMedicine, Graduate School at Shenzhen, Tsinghua University, Shenzhen, China
College of Forestry, South China Agricultural University, Guangzhou, China

Abdisalan M. Noor and Robert W. Snow
Malaria Public Health Department, Kenya Medical Research Institute-Wellcome Trust Research Programme, Nairobi, Kenya
Centre for Tropical Medicine, Nuffield Department of Clinical Medicine, University of Oxford, Oxford, United Kingdom

Victor A. Alegana
Malaria Public Health Department, Kenya Medical Research Institute-Wellcome Trust Research Programme, Nairobi, Kenya

Richard N. Kamwi
Office of the Minister, Ministry of Health and Social Services, Windhoek, Namibia

Clifford F. Hansford
National Institute for Tropical Diseases, Tzaneen, South Africa

Benson Ntomwa and Stark Katokele
National Vector-borne Diseases Control Programme, Directorate of Special Programmes, Ministry of Health and Social Services, Windhoek, Namibia

Tian-Zuo Wang, Jin-Li Zhang, Qiu-Ying Tian and Min-Gui Zhao
State Key Laboratory of Vegetation and Environmental Change, Institute of Botany, The Chinese Academy of Sciences, Beijing, P. R. China

Wen-Hao Zhang
State Key Laboratory of Vegetation and Environmental Change, Institute of Botany, The Chinese Academy of Sciences, Beijing, P. R. China
Research Network of Global Change Biology, Beijing Institutes of Life Science, The Chinese Academy of Sciences, Beijing, P. R. China

Fulin Yang
Key Laboratory of Arid Climatic Change and Reducing Disaster of Gansu Province, Key Open Laboratory of Arid Climatic Change and Disaster Reduction of China Meteorological Administration (CMA), Institute of Arid Meteorology, CMA, Lanzhou, China
State Key Laboratory of Vegetation and Environmental Change, Institute of Botany, Chinese Academy of Sciences, 20 Nanxincun, Xiangshan, Haidian District, Beijing, China

Guangsheng Zhou
State Key Laboratory of Vegetation and Environmental Change, Institute of Botany, Chinese Academy of Sciences, 20 Nanxincun, Xiangshan, Haidian District, Beijing, China
Chinese Academy of Meteorological Sciences, Haidian District, Beijing, China

George Wittemyer
Department of Fish, Wildlife and Conservation Biology, Colorado State University, Fort Collins, Colorado, United States of America
Graduate Degree Program in Ecology, Colorado State University, Fort Collins, Colorado, United States of America
Save The Elephants, Nairobi, Kenya

David Daballen
Save The Elephants, Nairobi, Kenya

Iain Douglas-Hamilton
Save The Elephants, Nairobi, Kenya
Department of Zoology, University of Oxford, Oxford, United Kingdom

Andrés J. Cortés
Evolutionary Biology Center, Uppsala University, Uppsala, Sweden
Laboratorio de Botánica y Sistemática, Universidad de los Andes, Bogotá, Colombia

Fredy A. Monserrate
Centro Agropecuario, Servicio de Enseñanza Nacional, Buga, Colombia

Julián Ramírez-Villegas
School of Earth and Environment, University of Leeds, Leeds, United Kingdom
CGIAR Program on Climate Change, Agriculture and Food Security (CCAFS), Cali, Colombia

Santiago Madriñán
Laboratorio de Botánica y Sistemática, Universidad de los Andes, Bogotá, Colombia

Matthew W. Blair
Department of Plant Breeding, Cornell University, Ithaca, NY, United States

Guo-Liang Xu
Key Laboratory of Vegetation Restoration and Management of Degraded Ecosystems, South China Botanical Garden, Chinese Academy of Sciences, Guangzhou, China

Thomas M. Kuster and Madeleine S. Günthardt-Goerg
Forest Dynamics, Swiss Federal Research Institute WSL, Birmensdorf, Switzerland

Matthias Dobbertin
Long-term Forest Ecosystem Research, Swiss Federal Research Institute WSL, Birmensdorf, Switzerland

Mai-He Li
Tree Physioecology, Swiss Federal Research Institute WSL, Birmensdorf, Switzerland

Anthony Manea and Michelle R. Leishman
Department of Biological Sciences, Macquarie University, North Ryde, NSW, Australia

Peter Maes
Medical Department, Water, Hygiene and Sanitation Unit, Médecins Sans Frontiéres, Operational Center Brussels, Brussels, Belgium

Anthony D. Harries
International Union Against Tuberculosis and Lung Disease (The Union), Paris, France
London School of Hygiene and Tropical Medicine, London, United Kingdom

Rafael Van den Bergh, Katherine Tayler-Smith and Rony Zachariah
Medical Department, Operational Research Unit (LuxOR), Operational Center Brussels, Médecins Sans Frontières -Luxembourg, Luxembourg, Luxembourg

Abdisalan Noor and Robert W. Snow
Malaria Public Health Department, KEMRI-University of Oxford-Wellcome Trust Collaborative Programme, Nairobi, Kenya
Centre for Tropical Medicine, University of Oxford, Oxford, United Kingdom

Sven Gudmund Hinderaker
Centre for International Health, University of Bergen, Bergen, Norway

Richard Allan
The Mentor Initiative, Crawley, United Kingdom

Asier Herrero
Department of Ecology, University of Granada, Granada, Andalusia, Spain
Department of Life Sciences, University of Alcala´, Alcala´ de Henares, Madrid, Spain

Regino Zamora
Department of Ecology, University of Granada, Granada, Andalusia, Spain

Marcelo Zeri, Celso von Randow and Gilvan Sampaio
Centro de Ciência do Sistema Terrestre, Instituto Nacional de Pesquisas Espaciais, Cachoeira Paulista, SP, Brazil

Leonardo D. A. Sá
Centro Regional da Amazônia, Instituto Nacional de Pesquisas Espaciais, Belém, PA, Brazil

Antônio O. Manzi
Instituto Nacional de Pesquisas da Amazônia (INPA), Manaus, Amazonas, Brazil

Alessandro C. Araújo
Embrapa Amazônia Oriental, Belém, Pará, Brazil

Renata G. Aguiar
Universidade Federal de Rondônia, Porto Velho, Rondônia, Brazil

Fernando L. Cardoso
Universidade Federal de Rondônia, Ji-Paraná, Rondônia, Brazil

Carlos A. Nobre
Secretaria de Políticas e Programas de Pesquisa e Desenvolvimento, Ministério da Ciência, Tecnologia e Inovação, Brasília, DF, Brazil

Guoyong Ding
Department of Epidemiology and Health Statistics, School of Public Health, Shandong University, Jinan City, Shandong Province, P.R.China
Department of Occupational and Environmental Health, School of Public Health, Taishan Medical College, Taian City, Shandong Province, P.R.China
Shandong University Climate Change and Health Center, Jinan City, Shandong Province, P.R.China

Lu Gao and Baofa Jiang
Department of Epidemiology and Health Statistics, School of Public Health, Shandong University, Jinan City, Shandong Province, P.R.China
Shandong University Climate Change and Health Center, Jinan City, Shandong Province, P.R.China

Xuewen Li
Shandong University Climate Change and Health Center, Jinan City, Shandong Province, P.R.China
Department of Environment and Health, School of Public Health, Shandong University, Jinan City, Shandong Province, P.R.China

Maigeng Zhou
National Center for Chronic and Noncommunicable Disease Control and Prevention, China CDC, Beijing City, P.R.China

Qiyong Liu
Shandong University Climate Change and Health Center, Jinan City, Shandong Province, P.R.China
State Key Laboratory for Infectious Diseases Prevention and Control, National Institute for Communicable Disease Control and Prevention, China CDC, Beijing City, P.R.China

Hongyan Ren
State Key Laboratory of Resources and Environmental Information System, Institute of Geographic Sciences and Natural Resources Research, Chinese Academy of Sciences, Beijing City, P.R.China

Colin E. Studds
School of Biological Sciences, University of Queensland, Brisbane, Queensland, Australia

Smithsonian Conservation Biology Institute, Migratory Bird Center, National Zoological Park, Washington, D. C., United States of America

William V. DeLuca
Department of Environmental Conservation, University of Massachusetts, Amherst, Massachusetts, United States of America

Matthew E. Baker
Department of Geography and Environmental Systems, University of Maryland, Baltimore, Maryland, United States of America

Ryan S. King
Center for Reservoir and Aquatic Systems, Department of Biology, Baylor University, Waco, Texas, United States of America

Peter P. Marra
Smithsonian Conservation Biology Institute, Migratory Bird Center, National Zoological Park, Washington, D. C., United States of America

Sonia Silvente and Miguel Lara
Centro de Ciencias Genómicas, Universidad Nacional Autónoma de México, Cuernavaca, Morelos, México

Anatoly P. Sobolev
Istituto di Metodologie Chimiche, Laboratorio di Risonanza Magnetica "Annalaura Segre", CNR, Monterotondo, Rome, Italy

Ellen L. Fry
Department of Life Sciences, Imperial College London, Ascot, Berkshire, United Kingdom

Grantham Institute for Climate Change, Imperial College London, London, United Kingdom

Pete Manning
Department of Life Sciences, Imperial College London, Ascot, Berkshire, United Kingdom

NERC (Natural Environment Research Council) Centre for Population Biology, Imperial College London, Ascot, United Kingdom

David G. P. Allen and Alex Hurst
Department of Life Sciences, Imperial College London, Ascot, Berkshire, United Kingdom

Georg Everwand
Department of Life Sciences, Imperial College London, Ascot, Berkshire, United Kingdom
Agroecology, University of Göttingen, Göttingen, Germany

Martin Rimmler
Department of Life Sciences, Imperial College London, Ascot, Berkshire, United Kingdom
Department of Ecological Modelling, Bayreuth University, Bayreuth, Germany

Sally A. Power
Department of Life Sciences, Imperial College London, Ascot, Berkshire, United Kingdom
Hawkesbury Institute for the Environment, University of Western Sydney, Penrith, New South Wales, Australia

Index